Sound Improvement Secrets For Audiophiles

GET BETTER SOUND WITHOUT SPENDING BIG

IGOR S. POPOVICH

B.Sc. (El. Eng.)

DISCLAIMER & COPYRIGHT NOTICE

The information contained in this book is to be taken in the context of a general overview, not specific advice. You should not act on the information contained herein without seeking professional advice. Neither the author nor the publisher (or any other person involved in the publication, distribution, or sale of this book) accepts any responsibility for the consequences that may arise from readers acting in accordance with the material given in the book.

Designs marked with a copyright symbol are the intellectual property of their copyright holders and should not be used without their permission. They are discussed here from the educational perspective only. Some circuit and block diagrams of commercial equipment and their test results are used here for review and discussion purposes as "fair dealing," permitted by international copyright laws.

The inclusion of specific DIY projects is simply for illustrative and educational purposes. Since some of their design & construction aspects may be perceived as infringing on one or more existing patents, their inclusion does not constitute encouragement or inducement to the readers.

Most amplifiers involve high and potentially lethal voltages, high temperatures, and other hazards, and so do various test & measurement instruments. By buying this book, you automatically agree to indemnify its author, publisher, and retailer against any claims of any nature and for any reason.

© **Copyright Igor S. Popovich 2021**

All rights are reserved. No part of this publication may be used or reproduced in any form or transmitted by any means, without the prior written permission from the publisher, except in the case of brief quotations in articles and reviews.

Published by Career Professionals Australia

First edition, 2021

Bulk purchases

This book may be purchased in larger quantities for educational, business or promotional use. For special quantity discounts and savings please e-mail us at sales@careerprofessionals.com.au

National Library of Australia Cataloguing-in-Publication Data:

Popovich, Igor S.,
    SOUND IMPROVEMENT SECRETS FOR AUDIOPHILES: GET BETTER SOUND WITHOUT SPENDING BIG
    ISBN: 978-0-6482982-0-5
    1. Electrical engineering 2. Electronics 3. Hi-fi 4. Acoustics
    I Igor S. Popovich II Title

621.3

CONTENTS

1. **WHY YOU SHOULD READ THIS BOOK AND HOW YOU WILL BENEFIT FROM IT** 5

    ● WHY THE NEED FOR A BOOK OF THIS KIND? ● WHO IS THIS BOOK FOR, AND HOW WILL YOU BENEFIT FROM IT? ● THE LANGUAGE AND TERMINOLOGY USED ● HOW TO READ THIS BOOK ● GETTING IN TOUCH

2. **BEFORE YOU BUY AN AUDIO SYSTEM OR COMPONENT - THINGS TO DO & MISTAKES TO AVOID** 19

    ● THE MOST COMMON MISTAKES AUDIOPHILES MAKE AT THE PURCHASING STAGE ● AUDIO REVIEWERS, THEIR REVIEWS AND THE MAGAZINES THEY WRITE FOR ● MARKETING HYPE, MYTHS, AND OUTRIGHT LIES IN THE AUDIO FIELD ● UNDERSTANDING YOURSELF AND YOUR AUDIO PREFERENCES ● BUYING AUDIO GEAR UNHEARD, DISAPPEARING DEALERS AND OTHER PITFALLS ● HOW TO AUDITION AUDIO COMPONENTS AND SYSTEMS ● TIME TO BUY: HOW TO GET THE BEST DEAL ON AUDIO GEAR

3. **WHAT DO WE LISTEN FOR AND WHAT DO WE ACTUALLY HEAR?** 43

    ● HUMAN HEARING 101 ● DISTORTION - THE GOOD, THE BAD AND THE UGLY ● WHY DIFFERENT AUDIO TECHNOLOGIES SOUND DIFFERENT ● WIDE DYNAMIC RANGE: KEY FIDELITY FACTOR #1 ● PRESERVED HARMONIC STRUCTURE: KEY FIDELITY FACTOR #2 ● REPRODUCTION OF FAST TRANSIENTS: KEY FIDELITY FACTOR #3 ● DYNAMIC COMPRESSION: #1 PROBLEM IN SOUND RECORDING AND REPRODUCTION ● THE SONIC SIGNATURE OF MATERIALS ● THE LISTENING ENVIRONMENT AND ITS IMPACT ON THE ENJOYMENT OF MUSIC

4. **CLEANING UP THE POWER SUPPLY TO REDUCE NOISE, HUM, AND INTERFERENCE** 69

    ● MAINS POWER SUPPLY: DIRTY, DISTORTED & DANGEROUS ● ISOLATION TRANSFORMERS AND THE BALANCED POWER SUPPLY PRINCIPLE ● ELIMINATE DC FROM YOUR AC POWER OUTLETS ● AC VOLTAGE REGULATORS AND REGENERATORS ● REDUCING ELECTROMAGNETIC AND RADIO FREQUENCY INTERFERENCE ● UPGRADING AUDIO EQUIPMENT AC & DC POWER SUPPLIES ● BATTERY-POWERED PHONO STAGES, PREAMPLIFIERS AND POWER AMPLIFIERS

5. **CABLES, FUSES, CONTACTS, AND CONNECTIONS** 95

    ● HOW AUDIO SIGNALS PROPAGATE THROUGH CABLES ● INTERCONNECTS ● POWER CABLES ● SPEAKER CABLES ● HEADPHONE CABLES ● COPPER, SILVER, GOLD, NICKEL, RHODIUM, BERYLLIUM AND TELLURIUM: A CRASH COURSE IN AUDIO METALLURGY ● CABLE TERMINATIONS AND CONTACTS ● FUSES

6. **UPGRADING & FINE TUNING THE SOURCES: OPEN REEL RECORDERS, TURNTABLES, PHONO STAGES AND CD PLAYERS** 117

    ● THE ARDUOUS SONIC JOURNEY FROM RECORDING MICROPHONES TO YOUR EARS ● REEL-TO-REEL (OPEN REEL) RECORDERS ● TURNTABLES, TONEARMS AND TRACKING ERRORS ● DRIVE MECHANISMS AND VIBRATION MINIMIZATION ● TURNTABLE SETUP & ADJUSTMENTS ● CARTRIDGE AND STYLI TYPES ● UNDERSTANDING PHONO STAGES ● THE CARTRIDGE-CABLE-PHONO STAGE INTERFACE ● MOVING COIL STEP-UP TRANSFORMERS ● SOUND IMPROVEMENT HACKS FOR TURNTABLE-BASED SYSTEMS ● TUBE BUFFERING YOUR CD PLAYER, DAC, OR MUSIC STREAMER ● ADDING AN OUTPUT TRANSFORMER TO A CD PLAYER, DAC OR MUSIC STREAMER

7. **AUDIO AMPLIFIERS - HOW THEY WORK AND HOW TO IMPROVE THEIR SOUND** 155

    ● UNDERSTANDING AUDIO AMPLIFIERS ● HOW TO EVALUATE AMPLIFIER SPECIFICATIONS ● THE SONIC ANATOMY OF AMPLIFICATION: RULES FOR BETTER SOUND ● PASSIVE PREAMPLIFIERS ● CLASS D AMPLIFIERS ● OTL TUBE AMPLIFIERS ● TUBE ROLLING: CHANGING AND UPGRADING VACUUM TUBES IN AMPS & PREAMPS ● MODIFYING AND UPGRADING AMPLIFIERS AND PREAMPLIFIERS ● BRIDGING AND PARALLELING: TURNING STEREO POWER AMPS INTO MONOBLOCKS

CONTENTS, continued

8. HEADPHONES AND HEADPHONE AMPLIFIERS 189

● TYPES OF HEADPHONE TRANSDUCERS ● HEADPHONE AMPLIFIERS ● DIY PROJECT: SKIPPY, THE MINIMALIST SOLID STATE HEADPHONE AMPLIFIER ● INTEGRATING HEADPHONE AMPLIFIERS INTO AUDIO SYSTEMS ● DIY PROJECT: HEADPHONE INTERFACE BOX

9. LOUDSPEAKER TYPES, TESTS, AND IMPROVEMENTS 201

● DYNAMIC LOUDSPEAKERS ● OTHER SPEAKER AND DRIVER TYPES ● SPEAKER ENCLOSURES AND CROSSOVERS ● CHOOSING A LOUDSPEAKER ● CRUCIAL QUESTIONS AND QUICK SPEAKER TIPS ● HOW TO MEASURE SPEAKERS' SENSITIVITY, IMPEDANCE, AND FREQUENCY RESPONSE CURVE ● SPEAKER MODIFICATIONS AND UPGRADES

10. COMPONENT MATCHING AND AUDIO SYSTEM INTEGRATION ISSUES 229

● GAIN OPTIMIZATION IN THE SIGNAL CHAIN ● MATCHING SPEAKERS AND AMPLIFIERS ● BI-AMPING ● ACTIVE CROSSOVERS ● BI-WIRING ● INTEGRATING SUBWOOFERS INTO AN EXISTING STEREO SYSTEM ● AUDIO SYNERGY ● TUBE PREAMPS DRIVING SOLID-STATE POWER AMPS

11. LOUDSPEAKER POSITIONING 249

● EXPERIMENTAL (EMPIRICAL) SPEAKER PLACEMENT METHODS ● SPEAKER PLACEMENT CHALLENGES - ROOMS OF IRREGULAR SHAPES, FIREPLACES, SLOPING CEILINGS AND OTHER PITFALLS ● FORMULA-BASED SPEAKER PLACEMENT METHODS

12. OPTIMIZING THE ACOUSTIC PERFORMANCE OF YOUR LISTENING ROOM.................. 259

● THE FINAL TRANSDUCER OR THE LAST LINK - THE LISTENING ROOM ● ROOM DIMENSIONS, PROPORTIONS AND RESONANT MODES ● CASE STUDY: OUR LISTENING ROOM ● CASE STUDY: FORMAL LOUNGE AS A LISTENING ROOM

13. ACOUSTIC TREATMENTS 271

● THE PROPERTIES AND BEHAVIOR OF SOUND IN CLOSED SPACES ● SIMPLE ACOUSTIC TREATMENTS FOR LISTENING ROOMS ● POROUS MATERIALS AS SOUND ABSORBERS ● PANEL ABSORBERS ● CAVITY RESONATORS ● BASS TRAPS ● DIFFUSERS ● DIGITAL ROOM EQUALIZATION SYSTEMS

14. MINIMIZING UNWANTED VIBRATIONS & OSCILLATIONS 293

● UNWANTED VIBRATIONS: CAUSES AND REMEDIES ● HI-FI RACKS AND EQUIPMENT PLATFORMS ● CABLE LIFTERS (RAISERS)

15. TROUBLESHOOTING YOUR AUDIO SYSTEM 307

● TYPES OF FAULTS AND THE COMMON CAUSES ● THE BASIC EQUIPMENT - MULTIMETERS AND LCR METERS ● QUICK LOUDSPEAKER CHECKS ● QUICK AMPLIFIER AND PREAMPLIFIER CHECKS ● SIGNAL TRACING METHODS AND PRINCIPLES ● STRANGE AND HARD-TO-PINPOINT AUDIO SYMPTOMS ● TROUBLESHOOTING VALVE (TUBE) AMPLIFIERS

16. THE END MATTER 321

● INDEX ● LITERATURE (FURTHER READING) ● ONLINE AUDIO MAGAZINES ● AUDIOPHILE TUBE AMPLIFIER BOOKS BY IGOR S. POPOVICH ● GUITAR AMPLIFIER BOOKS BY IGOR S. POPOVICH ● OTHER AUDIO-RELATED BOOKS BY IGOR S. POPOVICH

1 WHY YOU SHOULD READ THIS BOOK AND HOW YOU WILL BENEFIT FROM IT

You can't judge a book from its cover, but you can from its introduction.

An astute reader will eventually figure out the author's intentions, preferences and pet peeves, but it is often prudent to outline the basic premise of the book and explain its structure, approach, intended audience and other important aspects beforehand, in its introduction.

An honest author will also mention things and topics some readers may reasonably expect but which are not covered in the book, and explain the reason(s) for such omission.

Apart from introducing the author, the introduction also serves as a preview or an appetizer if you will, to whet the readers' appetite for the main course that follows.

- WHY THE NEED FOR A BOOK OF THIS KIND?
- WHO IS THIS BOOK FOR, AND HOW WILL YOU BENEFIT FROM IT?
- THE LANGUAGE AND TERMINOLOGY USED
- TWO MORE THINGS FOR YOU TO DO
- HOW TO READ THIS BOOK

> "Miss a meal if you have to, but don't miss a book!"
> James E. (Jim) Rohn, self-help philosopher

WHY THE NEED FOR A BOOK OF THIS KIND?

My journey from tube amplification to audio system optimization

Before we start discussing ways of getting closer to that elusive goal of "high fidelity" (whatever that means) and musical reality (or, should we instead be seeking a musical illusion?), an introduction is in order.

I started my journey into electronics as a child, helping my dad with his DIY projects. Dr. Slobodan (Bob) Popovich was a university physics professor by day and an electronics guru by night. His hands-on-training in tube technology started early. To finance his studies, he built and fixed radio receivers, test gear, hi-fi, and guitar amplifiers - all using tube technology, of course.

After completing my university degree in electronics in 1986, I migrated to Australia, where dad joined me in 1995. After two years of initial development, we released our first range of ValveMark tube amplifiers and preamplifiers. In the following years, we designed and developed amplifiers with almost any power tube in use today.

Somehow we managed to jump over numerous design, manufacturing, and marketing hurdles, but then came the system issues. Almost all of our prospective buyers had an audio system already. In some cases, simply replacing their existing amp or preamp (usually of solid-state variety) with one of our single-ended beauties would result in such a significant and immediately noticeable sound improvement that required no further sales efforts.

In other instances, and those are the more memorable ones, there was an obvious problem that even an upgrade to the best tube amplifier in the world would not solve. The most common issues were prospective buyer's existing low sensitivity and thus hard-to-drive loudspeakers, closely followed by poor room acoustics. Short of suggesting replacing their speakers with more efficient and higher impedance alternatives, there was precious little I could do about the first issue. Improving room acoustics, however, was possible in most cases.

My knowledge of and expertise in sound improvement techniques in general and room acoustic, in particular, has been developed through my experiments and listening trials at home, as an audiophile, but also in various listening rooms (including our own) during our efforts to sell amplifiers to local audiophiles.

THE TOP FIVE AUDIOPHILE SYSTEM ISSUES

1. Low sensitivity, low impedance, power hungry and hard-to-drive loudspeakers
2. Incorrect speaker and listening spot positioning
3. Poor room acoustics
4. Suboptimal system integration (component matching)
5. Hum, buzz, hiss and other audible artifacts caused by fluctuating, dirty and distorted mains power supply

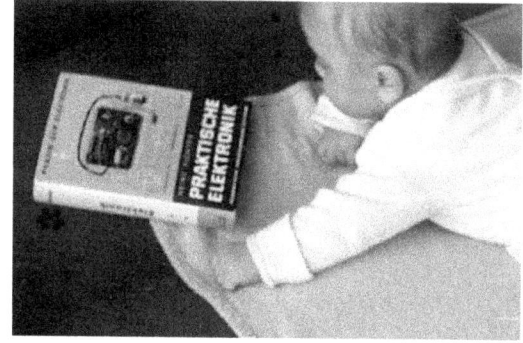

Fig. 1.1. Be careful what kind of a book you give your kids to "read", it may have a profound effect on their life. Here I am in 1963, "reading" my first book called "Practical Electronics", in German, no less.

The ultimate aim of an audiophile system

Ask a dozen audiophiles about the ultimate goal or aim of an audio system, and you are likely to get twelve different definitions. Let's say that we want the system to "make the reproduced music sound its best." The problem word is "best" since it means different things to different people.

An alternative goal could be to expect the system to reproduce the recorded music in such a way as to recreate the original artistic performance. The long and arduous journey the music signal has to undertake between the recording microphone(s) and your loudspeakers makes that all but impossible. Once a musical performance is electronically captured in an analog or digital way, the recording is then usually (apart from some raw or "minimalist" recordings) manipulated to death at the mixing desk and during the mastering process. It has been run through equalizers, expanders, compressors, and a myriad of other electronic devices so that what is left is far removed from the original artistic performance.

The electronic rendition, either digitized as a computer file and streamed, or duplicated as a mass-produced LPs or CD, is then propagated from the source (turntable, CD player, or streamer) through the amplification chain. Finally, it is converted back into sound by the loudspeakers, electroacoustic transducers inside the final audio component, the listening room. All links in the audio chain introduce their sonics into the mix.

Even if captured faithfully during the recording stage, you will never be able to listen to the original acoustic space; you are listening inside a small (compared to the original concert hall or jazz club) and boxy room, a totally different acoustic environment.

So, let's skip many other such definitions of good sound and "fidelity" - there are similar problems with each one of them. Since we will never be able to recreate the actual reality, let's agree that the best we can hope for as audiophiles is a plausible (sounds real) and the enjoyable illusion of that reality.

The seven obstacles on the road to musical Nirvana

I'm an engineer by professional education, and engineers like lists and models. I thought long and hard about this list, and while it doesn't seem like much, it is of paramount importance. I have even decided to structure this book's chapters along the lines of these seven factors or obstacles to good sound. Here they are, in no particular order, certainly not in the order of importance:

1. Recording & streaming quality
2. Audio gear - sources (turntables, CD players, tape decks, music servers), phono stages, line preamps, power amps, loudspeakers, cables, active crossovers
3. The listening room - dimensions, proportions, wall and floor materials, furnishings, acoustic treatments
4. Speaker and listener positioning
5. Power supply quality and equipment that improves it - filters, conditioners, voltage regulators and regenerators
6. Unwanted external noise, RFI (Radio-Frequency Interference), EMI (Electromagnetic Interference), mechanical vibrations, geopathic stress
7. Psychoacoustics - the mood, attention and listening skills of the audiophile

Factor #1 is unfortunately outside of our control, the quality of recordings is a given and there is precious little we as audiophiles can do about it. Thus, it will *not* be covered or discussed in this book.

Audiophiles and even professionals in the audio business pay too much attention to factor #2, the audio gear. After all, that's what 90% of the audio industry produces, so it is understandable they'd concentrate on what brings cash to their coffers.

The remainder of the audio industry focuses on room acoustic software and treatments (#3), the power supply quality improvements (#5), and accessories that help with unwanted vibrations, noise, and interference (#6).

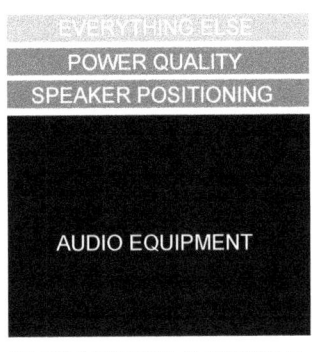
WHAT A TYPICAL AUDIOPHILE PAYS ATTENTION TO

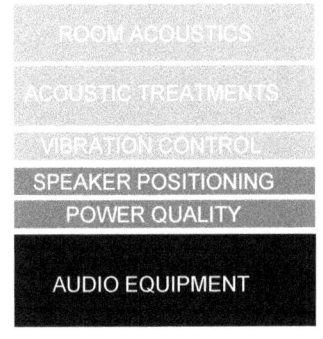
PRIORITIES OF A SOPHISTICATED AUDIOPHILE

Fig. 1.2.
You can tell what type of audiophile someone is by their priorities.

Hiring audio or acoustic consultants to improve the sonic signature of your listening room or an electrician to install a dedicated mains power line for your audio system are clever options, yet, for some strange reason, quite rare. So, the speaker positioning and listening room improvements are left to audiophiles themselves, which is a problem in itself; most have limited knowledge and minimal inclination to deal with such issues.

How is this book structured

The second chapter eases into the subject of buying audio equipment and upgrading your audio system. There are no maths or heavy technical issues here, but it is a crucially important chapter nonetheless - in my view, worth the price of this book just by itself.

Many dangers in buying and selling audio gear are identified, and audiophiles' most common mistakes are discussed. The advice on resisting the relentless marketing hype and understanding reviews in audio magazines follows. We even talk about how to deal with audio dealers (pun intended) and how to audition equipment.

"What do we listen for and what do we actually hear" is a chapter that aims to dispel some audio myths by taking a deeper look into key fidelity factors such as distortion, frequency bandwidth, fast transient reproduction, and dynamic compression. Other issues such as the sonic signatures of various materials and the impact of the listening environment are equally important yet seldom discussed. Finally, we open up a can of worms, namely the controversial issue of why different audio components and the technologies they are based on sound different. Turntables versus CD players, tube versus solid-state amplifiers, and capacitors versus inductors and transformers are just some of the eternal points of disagreement amongst audiophiles.

Every reputable audio designer will tell you that a power supply is the heart of any amplifier. It is impossible to have a top-performing and superior sounding amplifier with a lousy power supply.

Unless battery-powered, all audio components use AC power from wall receptacles to convert it into DC power they need internally (in their power supplies just mentioned). A dirty and distorted power supply cannot result in a clean and convincing sound! Chapter 4 is dedicated to the mains voltage quality, the issue of paramount importance, yet often overlooked by audiophiles who focus primarily on audio gear and seldom on the power that feeds it.

Chapter 5 tries to answer the question "Do cables make a difference?" by answering another: "How much difference (to the sound, not to your financial situation) do audio cables make?" And the answer is - a lot! The related but often forgotten audio signal links such as fuses and contacts are also discussed, apart from power, loudspeaker, and interconnect cables.

Then, in chapters 6 to 9, we follow the signal chain, starting with sources (tape decks, turntables, and CD players), followed by the amplification devices, and finally the transducers, headphones and loudspeakers.

A few pages back we've listed the top five audiophile system issues (from my personal experience). Suboptimal system integration (poor component matching) was #4 on that list. Chapter 10 looks at a few critical aspects of audio synergy, such as the matching of amps and speakers, gain optimization in the signal chain, and various ways of integrating other audio components into it, such as active crossovers and powered subwoofers. Bi-wiring and bi-amping are also of interest to most audiophiles.

Chapters 11, 12, and 13 form a logical cluster and should be read together. We'll cover the basics of speaker positioning first, with particular attention paid to irregularly shaped rooms. Then we'll go deeper into the issue of room acoustics.

Like any complex system, the audiophile setup is only as good as its weakest link, often the listening room. The acoustic behavior and the sonics of small listening rooms are very different from that of large concert halls.

Chapter 12 focuses primarily on the room's bass (low frequency) response and the associated problems such as room modes (resonances) and standing waves. Two practical examples of listening rooms are analyzed together with two different setups, the longitudinal and transversal arrangement.

The next chapter, "Acoustic treatments," deals with bass traps used to tame the unwanted bass artifacts just mentioned and with other types of acoustic panels aimed at the absorption and diffusion of sound. Unwanted vibrations of larger objects in a listening room produce sound waves that bear no harmonic or temporal relationship to the desired sound from the speakers, thus degrading the sonic presentation in various ways. Glass doors and windows, pictures or mirrors on a wall, even sofas, can be detrimental to sound quality.

The vibrations of smaller objects such as turntable tonearm or a power transformer in an amplifier may not be directly sonically detectable but interfere with and degrade their performance, increasing distortion and changing the frequency response and tonal balance of the whole system.

Vibration reduction and minimization measures usually don't involve significant expenses. Still, they result in sonic improvements similar to those achieved by cleaning up the power supply: cleaner, tighter bass, more precise and stable soundstage, the removal or fuzz, grit and haze around notes, and, ultimately, a more relaxed and "natural" sound. All these are elaborated upon in chapter 14, together with hi-fi racks, equipment platforms, and even controversial issues such as cable raisers (lifters).

Audio troubleshooting could be a subject of a whole book, so in Chapter 15, we cover only a few troubleshooting tools and methods every audiophile should have or be familiar with. Quick speaker checks, amp, and preamp checks using a multimeter and LCR meter, and signal tracing principles when there's no sound in one or both channels are discussed.

Two more advanced topics are covered for those who want to go deeper into the subject: troubleshooting tube amplifiers and a few strange and hard-to-pinpoint symptoms.

The question of balance

The main issue in writing this book was deciding how deep to go into the physics and acoustics on one and electronics on the other side, and how much of the mathematics describing such physical behavior to include. Inevitably, every book is a compromise (just as any audio component is). No matter where you draw the line, the material will be too highbrow for some and too pedestrian for others.

Since most readers of this book will not have a scientific or engineering background, the decision was made to keep the maths down to a minimum. Thus, strictly speaking, this book is not scientific or technical, but it is far above the technical level of most popular books.

A dozen or so books on this subject have been published in the last decade, most of which are thin and large font affairs that oversimplify the matter and are more a collection of the author's opinions, prejudices, and platitudes than fact-based practical improvement suggestions.

Their authors' claims and sweeping generalizations leave you perplexed or annoyed since the critical question "Why is that so?" isn't answered very often. For instance, they say things like, "Some audiophiles want to use tube rectifiers. There is nothing that a tube rectifier can do that a 5 cent silicon diode cannot do better." If that was true, why would companies such as Messa Boogie include both types of rectifiers (switch-selectable) in their top models of guitar tube amplifiers? Or, why do the best tube amplifiers (both guitar and audiophile kind) use tube rectifiers rather than solid-state diodes? Such flippant statements are questionable even in a pub, over a pint, let alone in a serious book on an important topic.

Why is this book so technical?

Some of the issues in audiophilia are intuitive, even common sense. Others, however, require the understanding of deeper physical and technical principles which underpin them. That is the reason for the relatively high technical level of this book. The topics and improvements outlined in this book range from small and simple changes anyone should be able to do, without understanding the deeper or technical principles behind them, to relatively complex modifications of audio equipment.

If you are one of those technically proficient audiophiles, someone who has built, repaired and modified numerous audio components, you should be able to perform those more complex improvements yourself.

If you are not, if you lack such knowledge and practical experience, or even a desire to mess around with screwdrivers and soldering irons, don't despair. Such modifications can be easily done by most electronics technicians. As always, look for a qualified and experienced professional who understands what you are trying to achieve and has worked on similar projects before. Money-wise, even paying others to perform such mods should be cheaper than selling your existing gear and buying a new audio component!

I have decided to offer such a wide range of topics of various levels of complexity, primarily for the sake of completeness. Instead of making this book a simplistic popular guide to quick & dirty tweaks (there are a couple of such books around), the aim was to provide you with a comprehensive and in-depth technical guide that will address various issues not just phenomenologically (what works or may or may not work in certain audio systems), but also (if and when possible) in terms of why or why not.

Why neither tuners nor cassette decks are discussed in this book?

Analog tuners and cassette decks had their place in the mid-fi systems of the pre-Internet age. Borrowing and copying tapes and records from your friends or recording your favorite songs from the radio were the only ways to get free or very cheap music (just the cost of a blank cassette). Things are radically different now.

The quality of FM (Frequency Modulation) broadcasts varies significantly, not just because of the limitations of FM technology, but also because the quality of the tapes and LPs used by radio stations varies significantly, as does the quality of the equipment used to play them. Of course, most stations digitized their collections many moons ago, adding another fidelity-reducing step to the broadcast process.

WHO IS THIS BOOK FOR, AND HOW WILL YOU BENEFIT FROM IT?

Audiophiles come in all ages, shapes, sizes, and preferences, not to mention their widely ranging levels of technical interest and proficiency. Some only care about the musical aspects and have no technical knowledge or interest in building or modifying audio gear. Others can perform the basic operations such as soldering, wiring, and simple technical tweaks and modifications.

Finally, there's the ever-growing audiophile group, those heavily into the DIY aspect of audiophilia, some of whom are even more interested in the hardware aspects of the art than in the enjoyment of music.

"Audiophiles don't use their equipment to listen to our music; Audiophiles use our music to listen to their equipment."
Alan Parsons, audio engineer, musician, and record producer, of *The Alan Parsons Project* fame

Amateurs, audiophiles and professionals

While some of you may have a technical background in engineering or acoustics, I suspect most won't. You may be in search of a hobby, and being an audiophile is one of the greatest and most enjoyable hobbies of all. You may be an audiophile who is dissatisfied with constantly upgrading expensive "hi-end" gear that is long on promise but delivers ordinary results, sick and tired of losing large sums of money in that futile process.

No matter your situation, it is reasonable to assume that you are not a professional in the audio field, i.e., an amateur. Before you get offended, the word amateur comes from the Latin "amicus," meaning friend or companion.

French "ami" and Italian "amico" derive from it too, as does the word "amore" or love. Thus, it means a person who does something out of love and not for profit.

Unfortunately, the original meaning got corrupted over the millennia. Now amateur is used in a derogatory manner, instead of "dilettante," a person lacking knowledge, someone who doesn't know what they are doing. This is in contrast to the so-called "professionals," who are supposed to be all-knowing, almost omnipotent. We all know how overrated, ignorant, and even useless many "professionals" are in their professions. So, instead of "amateur," we will use the term "audiophile" in this book.

Dictionaries define an audiophile as a person enthusiastic about high-fidelity sound reproduction. Since high-fidelity is another unfortunate term that is even harder to define, this definition leaves us none wiser. To bring this philosophical pondering to some conclusion, if you feel you would benefit from reading this book and are prepared to invest time, money, and effort into reading it, you are an audiophile!

Open (subjectivist) versus closed (objectivist) minds - which one are you?

Deniers, rationalists, and absolutists claim that if why something is happening isn't known, then what is happening cannot possibly be happening. A medical analogy may help illustrate this dogmatic insanity. We don't fully understand (in many cases, at all) many complex biochemical and neural processes within the human body, yet their effects and benefits are real and undeniable.

Our simplified models of the human body cannot fully explain certain complex happenings. Our test instruments are based on our limited understanding and are thus implicitly limited in their scope and precision. However, just because they cannot register or measure such changes doesn't mean that such complex events are not happening.

For instance, acupuncture and homeopathy were initially rejected because they did not fit our rational, western scientific framework. Some deniers dismiss them as quackery and reject them even today. Yet, they've been helping and improving the medical conditions of millions of people over hundreds and thousands of years.

Likewise, if nine out of ten experienced listeners hear a sonic change when an audio device (amp, preamp, speakers, etc.) is substituted, a tweak is added, or a modification is made, yet that change makes no technical or rational sense, saying "We cannot measure it or explain it; thus it is not happening!" is insane. "It's just our imagination, autosuggestion, or placebo effect playing in our minds!" is the common argument of the absolutist camp. We don't have all the answers, and we never will. How could we, when we don't even have all the right questions?

> THE MEASUREMENT FALLACY
> We measure performance aspects that we know how to measure (in principle), have equipment that can measure them (in practice), and that are fast & easy to measure. We still don't know how to measure things that truly matter!

Science is based on doubt or the lack of blind belief; we only believe in a theory, model, or explanation until a better one comes along or until the old one has been disproved and discredited. "Real" science and scientific disciplines (maths, engineering, and such) are the same in all countries and cultures, and their postulates and verifiable experimental results don't depend on one's worldview, belief system, or frame of mind.

Thus, trying to explain, justify or rationalize all audiophile beliefs (some of which are essentially irrational) under the guise of science is questionable at best and a waste of time and energy at worst.

To gain respectability and acceptance, those whose jobs involve marketing and selling audio products (especially in the "tweaks and accessories" category) don't have to fully explain how and why their products work. They couldn't do that even if they wanted to, even if their products worked, since many don't. A plausible explanation that appears scientific is more than enough. Through autosuggestion, the audiophiles' belief systems will take care of the rest.

From knowing WHY to knowing HOW (or the other way around?)

I still remember some of my fellow first-year electrical engineering students. Even at that tender age, they were already building hi-fi amplifiers, phono stages, and active crossovers, albeit of the solid-state kind (that was in 1982, the dark & backward solid-state era). They would talk about MOSFETs, biasing, intermodulation distortion, and heaps of other things that I had no clue about, as if they were talking about Spanish villages, as the saying goes.

Yet, few of those young "practitioners" passed mathematical analysis, calculus, solid-state physics, and other highly theoretical subjects in the first two academic years; almost none of them graduated four years later, as I did.

Even after graduating in electronics, I still had no clue about the things they talked about and wasn't any closer to designing or building an amplifier. My degree in electronics was a degree in high-level mathematics and lofty mathematical models, with very little practical applicability. Most of the knowledge and know-how I've acquired subsequently through independent study and experimentation.

Almost everything I know about tube electronics I learned from my dad. As a young graduate, I knew a bit about why things were a certain way, but I had no clue how to implement such theoretical models that seemed to me to be of very little value.

The young DIYers knew the rules-of-thumb, or the shortcuts to practicality, but didn't understand why or how those shortcuts or rule-of-thumb were arrived at, by which process, what the limits of such simplifications were, when they applied and when they didn't.

Even after a lifetime devoted to this field, you will not understand *everything*.

For instance, audiophiles and guitar players claim that flat-plate Telefunken ECC83 tubes sound better than their ribbed-plate cousins, made in the same factory and during the same period. Likewise, the mesh-plate 300B tubes are considered superior sounding to their solid-anode brethren.

Fig. 1.3. Apart from the range of practical improvements that you can implement straight away, this book also deals with the underlying principles and fundamentals, so you gain understanding of how things work and why. That should enable you to come up with some improvement ideas of your own!

However, plugged into even the most sophisticated tube tester, both versions of those tubes will measure the same (gain, internal resistance, mutual conductance, etc.). Once working in an audio amplifier, their AC (signal) and DC voltages in significant points will be identical.

Another example is GM70, a bright-glowing Russian high-power output tube pictured on this book's cover. It comes in two variants, with a carbon and copper anode (plate). The former is much cheaper since it is claimed that the copper anode version sounds better.

Assuming such sonic claims are correct (confirmed by thousands of listeners on dozens of different amplifiers), does anybody really know why there is a sonic difference? We may each have a plausible explanation (different materials and geometries sound different), but there can be no conclusive proof of the underlying cause.

Some things simply are, and searching for the whys behind them is an exercise in futility. Despite the claims of the objectivist camp, the agreed-upon parameters of tubes and transistors are not the only parameters that matter. They only describe the overall behavior of such components, while the subtleties have been left unexplored

However, don't let that discourage you. The more you learn, the faster you will realize the limits of your knowledge. You will find an answer to one question but may raise three new questions in the process. You will always have more questions than answers, and that frustrates a particular type of individual, the anal-retentive type.

While conceptualizing and writing this book, I tried to balance two fundamental aspects of the knowledge required by a typical audiophile. Apart from the range of practical improvements that you can implement, a substantial part of this book deals with the underlying principles and fundamentals.

Hence, you wil gain an understanding of how things work and why. That will enable you to come up with some improvement ideas of your own.

Car enthusiasts versus audiophiles: similarities and differences

As you will notice while reading this book, I like using automotive analogies. Audiophile systems share many commonalties with cars. In selling our vacuum tube amplifiers, I found that the easiest way to explain specific issues to audiophiles without much technical knowledge was to draw parallels with the motoring world. Most audiophiles are men, and most men like and understand cars (or at least they think and act as they do).

For instance, the issue of matching amplifiers and speakers is (fundamentally) akin to matching a high RPM (rotations per minute) car engine and the low RPM wheels through the car's transmission.

Yet, when it comes to connoisseurship, I've noticed a striking difference between car enthusiasts, most of whom are deep into fixing and upgrading engines, rebuilding body parts and similar stuff, and audiophiles, who, sadly, lack the understanding of even basic electrical or acoustical concepts and principles.

Thus, this book will require you to deepen your technical knowledge. It's like butterfly or stamp collecting. Anyone can collect them, but without a deeper understanding of what you are collecting, it is a mere surface scratching exercise, devoid of a deeper meaning. To be genuinely knowledgeable about your collection, you must go deeper into biology in the first and history, geography, and similar disciplines in the second instance.

The open- versus closed-architecture audio gear and the decline of technical proficiency amongst audiophiles

Just like many other aspects of our Facebook, Twitter, Instagram, and Snapchat-based lives, the tweaks of this 21st century are superficial. Roll a tube here and a rectifier there. Change the one set of cables to another. Ditch the 50 cents rubber feet that came with your CD player and replace them with Vibracones or a similar audiophile "equipment support devices."

Don't get me wrong - there's nothing wrong with such experiments and upgrades. They almost always result in subtle sonic changes, sometimes even drastic improvements beyond their cost or time and effort required.

Speaking of time, one possible reason for this fast & easy approach could be the fast-paced life and the feeling that we are time-poor, always in a rush and stressed out. So fork out the cash and hope for the best.

Another is the lack of technical knowledge, which has been slowly but steadily diminishing since the 1970s. The increasing complexity of solid-state and digital electronics is another reason audio components are treated as "black boxes," with users having no idea about their internal workings.

Fifty to seventy years ago, the situation was very different. Tubes ruled the audio scene, and a substantial proportion of audio components, especially amps and preamps, speakers, and an occasional turntable, were of DIY variety, either designed and built from scratch or bought as a kit assembled in a couple of afternoons.

An average 1950s and 60s audiophile had a much deeper understanding of the technology and a more intimate connection with their equipment's design and construction. An audio system was tweaked and modified not by buying expensive replacement gear but by changes of the internal components and even by the redesign and rewiring of amps and preamps.

Many businesses were started by improving commercial amps and preamps. For instance, the Audio Research Corporation (ARC) has its roots in its founder, the late William Z. (Bill) Johnson, selling an upgrade to the then ubiquitous Dynaco amplifiers. Their model ST-70 sold more than 350,000 units! The upgrade consisted of cross-coupling the output stage, a type of local negative feedback around the output transformer. Later on, ARC's tube amps featured such topology for many years.

Apart from a more natural and enjoyable sound, this open "architecture," together with a much lower component count and greater simplicity of tube designs, is one of their greatest strengths. It is arguably what saved the tube technology from obsolescence and the main reason behind the considerable resurgence of tube gear in the 1990s, the trend that has lasted to this day. Thus, almost all examples and case studies in this book will revolve around tubes and not solid-state.

By contrast, solid-state (analog) gear, especially digital devices and systems, are of closed nature and thus preclude or at least severely reduce the audiophiles' involvement and tweak/modification fun. Transistors and integrated circuits cannot be changed as easily or quickly as tubes, simply unplugging one and substituting another. Most silicon components are soldered into complex, fragile, and user "unfriendly" printed circuit boards.

Technophiles, technophobes and everyone else in between

It is hard and risky to make predictions, especially about the future, and it's equally tricky to generalize. However, if I had to, I'd say that 20% - 50% - 30% split quite accurately describes the audiophile community.

20% or so of us are technophiles, audiophiles who don't just like the technical aspect of audiophilia but love it. Some would even say they care about the performance of their audio gear more than they care about the music; in other words, they listen to their audio equipment, not to the music.

Although I expect this technical improvement manual to be welcomed and embraced by the technophiles (although most will already possess most of the knowledge within its pages), I didn't write it with that particular group in mind.

Fig. 1.4. The audiophile community is a continuous spectrum of personalities, preferences, and attitudes, with technophiles (the lovers of technology) at one, the technophobes (the scared, technology-averse group) at the other end, and the confused and skeptical majority in between.

My primary aim was to address the fears and knowledge gaps of a typical technophobe, the thirty or so percent who are mortified by the technical aspects of audiophilia, to make them more knowledgeable and thus more comfortable in their dealings with technical specs and operational principles, manufacturers' claims and audio dealer's spiels.

I have also tried to present the often dull and overwhelming (to the technophobes) technicalities interestingly and engagingly, hoping that at least some readers will fall in love (so to speak) with the fascinating world of audio technology in all its aspects - electronics, acoustics, materials science, physics, etc.

Along the way, if the fifty or so percent in the middle embrace the concepts discussed in this book and benefit from them, I would be a very happy chappy indeed.

New? Improved? Superior? How the audio game is played

Capitalism is not based on solidarity and altruism but the rule of capital, perpetual (often mindless) consumption, and narrow specialization (knowing more and more about less and less). It's a game whose aim is to offer more specialized products and services to increasingly discerning, dissatisfied, and narrow-focused customers.

Since this book is about audio, we'll focus on that particular game, but similar games are played in all other aspects of consumerism (cars, houses, white goods, ...) Making audiophiles dissatisfied with what they have and craving "new," "improved," and "hi-end" audio gear is the gist of the marketing aspect of the game.

No audiophile magazine would ever say things like, "Next time, instead of spending $15,000 on a new pair of speakers, consider getting the acoustics of your room right. Their raison d'être is to review a constant stream of newer and generally more expensive audio gear and charge the makers of such equipment large amounts of money to advertise in such magazines. This is not a criticism, just the way things are in capitalism.

The main game of consumerism is the "New & improved is cool, fashionable and superior, old is daggy and inferior." On the one hand, audiophiles are confused and skeptical of the claims made by audio equipment manufacturers and dealers in their relentless pursuit of ever-increasing profits. On the other, they fall in love with superficial cosmetic differences and the desire for the "new."

THE FALLING IN LOVE PRINCIPLE

 We fall in love and choose audio equipment with our hearts, then try to rationalize our decisions and choices by using our brains.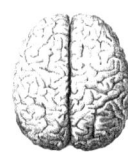

Everybody (including you and me) sells whatever they can (time, knowledge, services, products, and, ultimately, hope) for as much as they can. You cannot blame the sellers for trying to convince you that you need their stuff even if you don't. In a system based on self-interest, it is up to you to educate yourself to critically evaluate their claims and offers and decide which is the case.

Unless you live like a hermit in a wooden hut, there's no escape. We all play the consumerism game at some point in our lives. Ultimately, you have to decide how you want to play it.

After all, there are no pockets on a shroud. When you leave this world, you cannot take anything with you, so you may as well spend your money on meaningful things and experiences. If more audio gear will help you enjoy music more, who am I to dissuade you? Keep "upgrading" your audio system to your heart's content.

Don't be surprised if this book wins in the "Financial Book of the Year" category

 "There is far too much overpriced equipment on the market. Most of it sounds like crap. I just love to fuck the competition."
Roy Hall of Music Hall fame, in the *Stereophile* interview, June 2014

The ultimate goal of this book is to help you get more enjoyment out of your audio system with less money, to avoid spending your fortune on overpriced hi-end products by carefully selecting and matching your audio system, and by making your listening room as good sounding as possible.

I am shocked when the audiophiles I talk to disclose how much they've spent on their (usually very ordinary sounding) audio systems. Typically, ten grand on the power conditioner, six grand on a music streamer, eight grand on a preamplifier, fifteen grand on the power amp, twenty grand on speakers, not to mention cables, turntables, phono stages, equipment racks, ...

This unique technical manual will teach you engineering and acoustic concepts and show you various tips and tricks based on those concepts. It is also a financial book, aiming to save you as much money as possible, enough to buy a new car with such savings after only a few years!

Any fool with deep pockets can walk into a hi-end audio dealership and spend $100,000.- or even more on an audio system. The challenge is to show you how to achieve the same sound quality for $20,000 or even $5,000.- and a few hours or days of your time and effort. Impossible? Certainly not! Keep reading ...

THE LANGUAGE AND TERMINOLOGY USED

Hi-fi, hi-end and audiophile gear, is there a difference?

> "I don't know what you mean by 'glory'," Alice said.
> Humpty Dumpty smiled contemptuously. "Of course you don't - till I tell you. I meant 'there's a nice knockdown argument for you!'
> "But 'glory' doesn't mean 'a nice knockdown argument'," Alice objected.
> "When I use a word," Humpty Dumpty said in rather a scornful tone, "it means just what I choose it to mean - neither more nor less."
> "The question is," said Alice, "whether you can make words mean different things."
> Lewis Carroll, *Through The Looking Glass*

We have already discussed semantic differences on the people side of the equation, namely between amateurs, audiophiles, and "professionals." Now we should clarify things on the equipment or technical side of audiophilia.

The term hi-fi (which stands for "high-fidelity") is relative and therefore debatable. "High fidelity" usually indicates that an audio component measures well. I consider audio measurements to be the least important aspect of audiophilia.

Many amplifiers and other audio components measure well but don't sound good at all. Likewise, some of the best-sounding ones measure poorly and could not be considered high fidelity components at all.

"Hi-end" is equally problematic. Hi-end usually means "at the high end of the price scale," not at the pinnacle of design, construction, or sound quality.

The names and epithets are not the only problematic issue in hi-end audio. No other field suffers from so many myths and misconceptions, so much fluff, and so many furfies! Surprisingly few audio designers, manufacturers, retailers, and reviewers have any engineering background, so ignorance and copycat attitude rule the field.

Secrets? Oh please Igor ...

> "The secret is that there is no secret."
> Sheldon B. Kopp, *If You Meet the Buddha on the Road, Kill Him!*

Now, in the interest of full disclosure, I have to admit that the word "secrets" in the title of this book has been used more as an attention grabber or "hook" than in its true, literal meaning.

In these days of social media and instant communication, there are very few real secrets left. It would be more accurate to call the issues discussed within this book's pages "tips, tricks, methods, discoveries, opinions, and points-of-view, interspersed with the author's raving and ranting," but using that in a title or subtitle would be awkward, and, ultimately, unwise.

Some of these issues, problems, and performance aspects I have never seen discussed or explained anywhere, at least not in print and not covered in any meaningful depth so that they could qualify as secrets. If they had been published or disclosed before, it was done in obscure places, or they've been lost over time.

The prerequisite knowledge and a few alternative (less technical) books on the same topic

I understand that to many audiophiles this book will be too technical. Suppose you are in this category and do not have the time, the capacity, or the motivation to learn the basics of electronics and acoustics as presented here. In that case, there are a few "popular" books that deal with the same subject but at a much more fundamental level.

Good Sound by Laura Dearborn dates back to 1987 but is by no means a dated or irrelevant book. Surprisingly little has changed since! Out of all the titles that follow, it is probably the best-written book.

The Complete Guide to High-End Audio (4th edition) by Robert Harley is a decent and comprehensive book (it better be, it is 578 pages long). His shorter work (240 pages), *Introductory Guide to High-Performance Audio Systems*, is at a much more fundamental level, so I'd only recommend it to audio beginners.

The Audiophile's Guide: The Stereo was written by Paul McGowan of PS Audio. It is the latest of all the titles discussed here (published in 2021), but I cannot for the life of me understand who would learn anything from such a basic yet verbose book (lots of words, but not much substance).

Last but not least, we get to the 2008 book by Jim Smith, titled *Get Better Sound*. The writing style is friendly, but the book isn't as easy to read as it should be. The disorganized and poorly structured chapters and topics result in the lack of flow; instead of a logical or methodological progression, the topics and concepts are all over the place. Although basic, most of the advice is sound, so a neophyte audiophile could find Smith's book of some value.

I wish I could recommend a book that explains the basics of audio electronics' in a friendly and easy-to-understand way, but I haven't come across such a book. For a while, I entertained the idea of adding a few such chapters to this book, but that would require at least 80 additional pages and raise its price by $10 or so.

My books *Audiophile Vacuum Tube Amplifiers, Volume 1*, and *Tube Guitar Amplifiers, Volume 1*, dedicate the first 60-70 pages to the basic electronic circuit theory and the operating principles of electronic components (resistors, capacitors, inductors), vacuum tubes and audio amplifiers. Either of those titles would answer many, if not most, of the technical questions you may have.

TWO MORE THINGS FOR YOU TO DO

Take five to review and recommend this book online

If you've liked this book and benefited from it, the best way to repay a favor is to recommend it to your friends, both in-person and on social media, and to write a glowing online review on Amazon and other online bookshop websites. No, seriously, if you feel that sharing your objective, unbiased opinion with others would be of benefit to them and the audiophile community in general, please take five minutes to do so.

If you disagree with some of my views or find some of my suggestions erroneous, impractical or ineffective in your situation, tell us why and let us know what alternatives have worked better for you. When two people agree on everything, one of them is not needed (or necessary). We learn only through (friendly and respectful) disagreement and dialogue.

Get in touch

If you spot any errors or omissions in the text or illustrations, or should you have any constructive criticism of the book, please get in touch with me.

Also, should you have an interesting story, example, or case you'd like to contribute for inclusion in the subsequent editions of this book, I'd like to hear from you. Let me know what you listen to (equipment-wise), what is your listening room like, what you have done to make it better, what equipment upgrades and room treatments you have tried, what worked, and what didn't.

My e-mail is igorpop@careerprofessionals.com.au

I hope that this book has answered at least some of your questions about getting the best out of your audio system, and I wish you every success on your audiophile journey. See you on the road ...

My e-mail is igorpop@careerprofessionals.com.au

HOW TO READ THIS BOOK

Reading from start to finish or skipping around?

One way to read this book is to go to a section or topic that interests you immediately and then keep jumping back-and-forth to the related issues and chapters. This will, I suspect, be the way the more experienced readers will approach this book.

A more systematic approach is sequential, starting from the beginning and reading in order. This is what I would recommend. Although it seems more time-consuming (since you will read about many issues you may already know a lot about or are not particularly interested in), paradoxically, this approach is often faster. You will not miss anything, and you will not waste time flipping forward and backward trying to clarify an issue that had already been covered but which you've overlooked or perhaps not fully understood.

Whatever you do, don't treat this book as Holy Scripture. Underline or highlight the important parts, write your thoughts and ideas on its margins, sketch things in its blank spaces.

Skipping to conclusions

If you want to understand why some things are the way they are, read carefully and read to understand the technical underpinnings. If, on the other hand, you are only interested in the quick & easy improvements, don't worry; simply skip the technical explanations and focus on the conclusions.

However, I strongly recommend that you read through the whole book and make at least some effort to broaden and deepen your knowledge and understanding of the underlying engineering concepts and the principles behind the suggested improvements.

As with anything in life (taxes, home maintenance, car repairs, financial planning, family health) the more you know, the more you will save. It will also be easier for you to talk to those providing those services for you (doctors, lawyers, accountants, tradespeople, audio dealers) and to get better value from them. The same applies to audio equipment and systems and those who sell them (audio dealers and manufacturers).

The more you know, the easier it will be to ask pertinent questions and make your own conclusions. Instead of being a sitting duck, an easy target that can be manipulated and sold overpriced, unnecessary or simply poor value audio products, you will be an informed scrutinizer, a knowledgeable audiophile who will often know (much) more than dealers themselves.

You'd be surprised how little some audio dealers know. Just like all other trades and professions, they simply go through the motions, never investing the time and effort to broaden their horizons, expand their knowledge and reach the pinnacle of their profession. To find out what kind of dealer you are dealing with, ask them a few poignant questions (this book should give you hundreds of such ideas!) and listen carefully to how they answer.

When you find a dealer worth his salt, who doesn't just want to sell you stuff, develop a closer relationship with him or her, as you'll be seeing them a lot on your audio journey. The best of the bunch have professional pride and would rather lose a sale than sell you things you don't need or something that doesn't match the rest of your system.

The symbols used

Reading a technically in-depth book such as this one isn't easy, so one of the top priorities of an author is to make it user-friendly, in other words, easy to read. My goal was to make this book as short and sweet as possible.

I've done my best to pay attention to important aspects such as the writing style, the progression of the topics, and the book's design and page layout.

I've also tried to break up long blocks of text by interspersing it with photos, diagrams, and frames. Each frame has a symbol in its corner, telling you immediately what type of a frame it is.

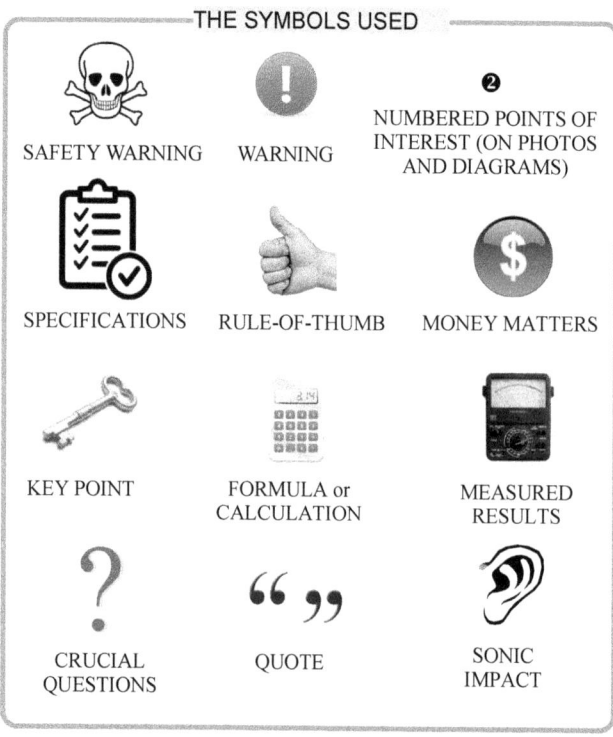

MAGNETIC COMPONENTS

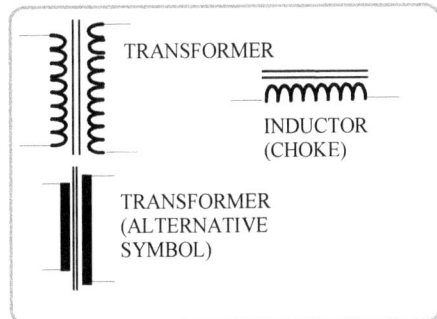

CAPACITORS

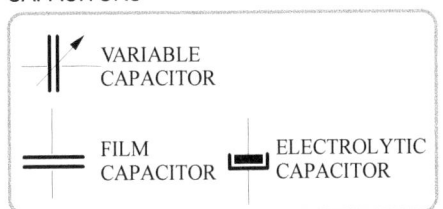

MISCELLANEOUS SYMBOLS

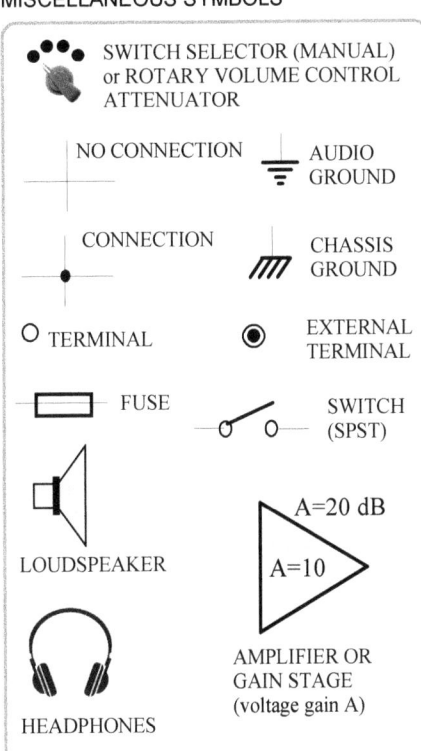

RESISTORS

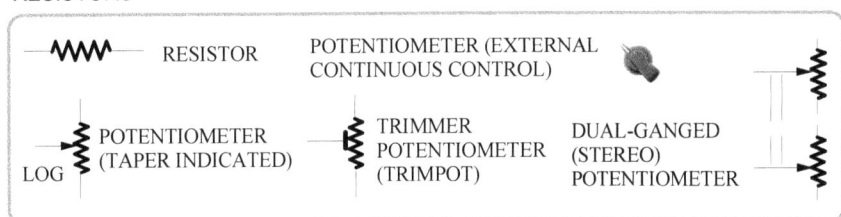

SEMICONDUCTORS & ELECTRON TUBES

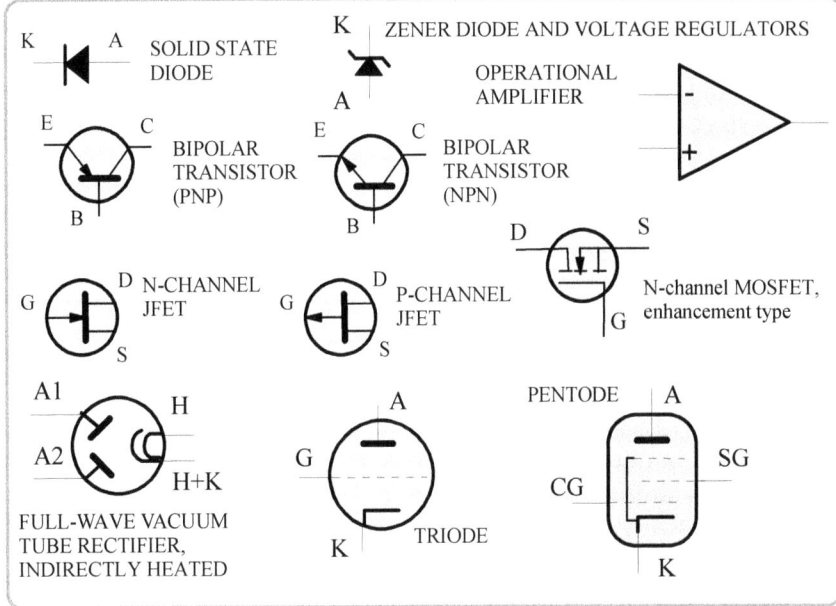

POWER SUPPLIES AND SIGNAL SOURCES

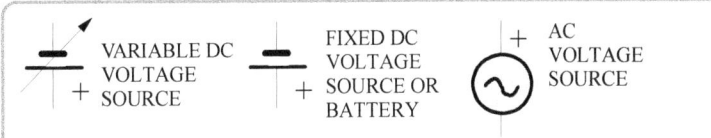

TEST INSTRUMENTS

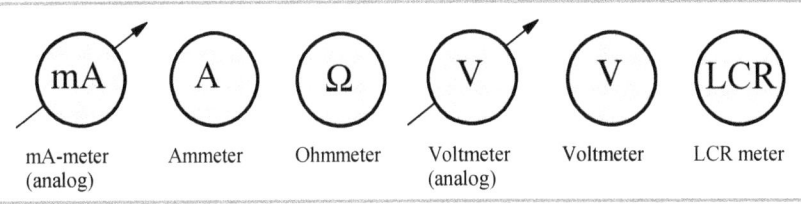

PHYSICAL AND ENGINEERING EXPRESSIONS

V_{AC}	VOLTS ALTERNATING CURRENT
V_{DC}	VOLTS DIRECT CURRENT
V_P	PEAK VALUE OF VOLTAGE
V_{RMS}	RMS OR EFFECTIVE VOLTAGE
V_{PP}	PEAK-TO-PEAK VOLTAGE
P_{MAX}	MAXIMUM POWER LEVEL
P_0	ZERO- OR REFERENT LEVEL

MATHEMATICAL & LOGICAL SYMBOLS

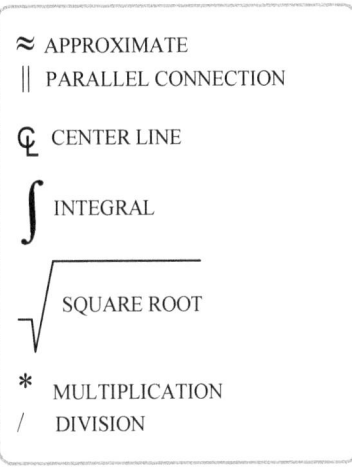

≈ APPROXIMATE
|| PARALLEL CONNECTION
₵ CENTER LINE
∫ INTEGRAL
√ SQUARE ROOT
* MULTIPLICATION
/ DIVISION

PREFIXES - INTERNATIONAL SYSTEM OF UNITS (SI)

FACTOR	NAME	SYMBOL
10^{-12}	pico	p
10^{-9}	nano	n
10^{-6}	micro	μ
10^{-3}	milli	m
10^{-2}	centi	c
10^{-1}	deci	d
10^{1}	deka	da
10^{2}	hecto	h
10^{3}	kilo	k
10^{6}	mega	M
10^{9}	giga	G

The abbreviations used

A

AC   alternating current
AF   audio frequency
AM   amplitude modulation
AMT air motion transformer (tweeter)
ARC Audio Research Corporation
AVR automatic voltage regulator
AWG American Wire Gage

B

BBC British Broadcasting Corporation
BW   bandwidth (frequency range)

C

CCIR The Consultative Committee on International Radio
CCW counterclockwise (turn)
CD   compact disc
CG   control grid (in a vacuum tube)
CLC capacitor-inductor-capacitor filter
COM common terminal
CPS cycles-per-second
CRC capacitor-resistor-capacitor filter
CT   center tap (transformer winding)
CW   clockwise (turn)

D

DAC digital-to-analog converter
dB   decibel
DC   direct current
DF   damping factor
DIN Deutsches Institut für Normung (German Institute for Standardization)
DIL dual in-line package
DIY Do-It-Yourself
DMM digital multimeter

E

EEE experimenter expectancy effect
ELCB earth leakage circuit breaker
ELF extra low frequencies
EM   electromagnetic
EMC electromagnetic compatibility
EPS expanded polystyrene
E/S electrostatic (interference, shield)
EU   European Union

F

FET field effect transistor

G

Gm   mutual conductance of a FET or a vacuum tube
GND ground
GOSS Grain Oriented Silicon Steel

H

HF   high frequencies
HV   high voltage
Hz   Hertz (unit for frequency)

I

IC   integrated circuit
IM   intermodulation (distortion)
IPS inches per second
ITDG initial time delay gap

J

JFET junction FET
JVC Japan Victor Company

L

LCR inductance-capacitance- resistance meter or circuit
LED light emitting diode
LEDE live end dead end
LF   low frequencies
LIN linear scale or taper (potentiometer)
LOG logarithmic scale, output, or taper (potentiometer)
LP   long play (record)

M

MAX maximum
MC   moving coil (phono cartridge)
MDF medium density fiberboard
MIN minimum
MM   moving magnet (phono cartridge)
MOSFET metal oxide FET
MSRP Manufacturer Suggested Retail Price

N

NAB National Association of Broadcasters (USA)
NFB negative feedback
NOS new old stock

O

OEM original equipment manufacrturer
OFC oxygen-free copper
OCC ohno continuous cast (copper)
OTL output transformerless amplifier

P

PCB printed circuit board
PLL phase-locked loop
PP   peak-to-peak (AC signal), push-pull (amplifier)
PRD primitive root diffuser
PTFE polytetrafluoroethylene (Teflon®)
PVC polyvinyl chloride

Q

QRD quadratic residue diffuser

R

RCA unbalanced audio connector
RCD residual current device
RFI radio frequency interference
RIAA Recording Industry Association of America
RMS root-mean-square or "effective" value of an AC signal
RPG reflection phase grating
RPM rotations per minute
RRP recommended retail price

S

SAC sound absorption coefficient
SCR silicon controlled rectifier
SDC sound diffusion coefficient
SET single-ended triode
SG   screen grid (in a tube)
SLO-BLO slow or delay fuse
SMPTE The Society of Motion Picture and Television Engineers (USA)
SNR or S/N signal-to-noise ratio
SP   suppressor grid (in a tube)
SPL sound pressure level
SS   solid state
SUT step-up transformer
SW   short wave (radio communication)

T

THD total harmonic distortion
TVC transformer volume control

U

UHP ultra high purity
UPS uninterruptible power supply

V

VTA vertical tracking angle
VTF vertical tracking force

W

WPC Watts per channel

X

XLR balanced audio connector (3-pins)

2 | BEFORE YOU BUY AN AUDIO SYSTEM OR COMPONENT - THINGS TO DO & MISTAKES TO AVOID

Constant equipment "upgrades" are one of the fundamental aspects of audiophilia. As with cars and other consumer items, audio gear becomes old, unreliable, or surplus to one's requirements, all valid reasons to consider getting something newer and better. However, we often get bored with what we have and fall in love with what we want (notice I did not say "what we need"!)

No matter how you justify spending money on audio goodies, this chapter will take you through some critical aspects of the acquisition game, particularly the pitfalls to avoid if you want to be happy with your purchases and get the best possible value for your money!

- THE MOST COMMON MISTAKES AUDIOPHILES MAKE AT THE PURCHASING STAGE
- AUDIO REVIEWERS, THEIR REVIEWS, AND THE MAGAZINES THEY WRITE FOR
- MARKETING HYPE, MYTHS, AND OUTRIGHT LIES IN THE AUDIO FIELD
- UNDERSTANDING YOURSELF AND YOUR AUDIO PREFERENCES
- BUYING AUDIO GEAR UNHEARD, DISAPPEARING DEALERS, AND OTHER PITFALLS
- HOW TO AUDITION AUDIO COMPONENTS AND SYSTEMS
- TIME TO BUY: HOW TO GET THE BEST DEAL ON AUDIO GEAR

> "Well begun is half done."
> Horace (Quintus Horatius Flaccus), Roman poet

THE MOST COMMON MISTAKES AUDIOPHILES MAKE AT THE PURCHASING STAGE

CASE STUDY: The worst sounding amplifier we've ever made

I still remember one of our prototype power amplifiers with 6L6 tubes in a push-pull configuration. In our test system at the time, it sounded very ordinary. We quickly moved on to single-ended triode amps, which (to me and most other people) sounded better.

The poor push-pull amp had been gathering dust for more than a year, so eventually, I decided to get rid of it, and a local audiophile bought it for $500, below the cost of parts only, a genuine bargain. That was in 1996.

Fast forward to 2009. A local importer of hi-end Chinese-made loudspeakers needed a few good quality tube amps for his showroom, so he approached us. I dutifully loaded the car and brought our best amps over to his shop. And what did I see there? The only amp, proudly displayed in the main listening room, was our orphan, the unwanted child, the above-mentioned amp!

The dealer bought it from its original owner, replaced the tubes we installed 13 years before, added some isolation cones and a premium power cable, and matched it properly with one model from his range of speakers. I could not believe this was the same amp; it sounded so much better, in fact, as good or even better than Cary or Conrad Johnson amps costing upwards of $5,000.-

I felt very proud of my dad, Dr. Bob, the designer and builder of that world-class amp, but I also felt like a loser who sold something so valuable for a pittance. The moral of this story (just in case you've missed it)? It is impossible to correctly judge the quality of a hi-fi component based on its performance in only one system or only one room.

You may be considering selling an audio component that does not go well with the rest of your system and be fooled into thinking that it isn't such a great piece of gear after all and risk selling it way below its actual value.

Or, in a much more common scenario, a shrewd dealer or a private seller has found components that match the audio component he is selling extremely well through lots of experimentation. Now your perception of this component's quality is likely to be much higher than it deserves. Its weaknesses have been cleverly and carefully masked or compensated for by the other links in their audio system.

THE DANGER OF MAKING A DECISION BASED ON AN AUDITION IN A SINGLE AUDIO SYSTEM
It is impossible to properly judge the quality of a hi-fi component based on its performance in only one system and in only one listening room.

This is what smart hi-fi dealers (retailers) do. Unless you are buying the whole system from them, you are unlikely to get such synergy once you bring the component home. Even then (buying the whole, matched, and optimized audio system together), the system may sound much worse in your listening room.

So, never buy straight away; always ask to borrow the audio component and evaluate it with your system in your room.

Most shortcomings in audiophile systems are in the mismatch category - the components simply don't go well together. Even more often, the biggest culprit usually isn't any of these components but the listening room itself. The system just doesn't sound right in that particular room.

THE "YOU HAVE BEEN SET UP" RULE:
Never buy anything based only on how it sounds in a dealer's showroom! Ask to borrow the gear and try it in your room with the rest of your system. If a retailer isn't happy to do so, you can be sure that many of his (more desperate) competitors will be.

Never buy a hi-fi system or a component based only on a review or friend's recommendation!

Apart from cleverly matched setups by shrewd dealers, the biggest pitfall in buying audio components surely must be magazine reviews. As in any profession, there are decent and honest hi-fi reviewers, and there are those with no technical knowledge or any real expertise who, therefore, should not be trusted.

Many are biased towards certain technologies or brands. All have their preferences, and all think they know what things should sound like. The only problem is, it is not what they think that matters; it is what you think.

So, don't blindly follow their advice; read the reviews, but read them critically. Read as many reviews of the equipment in question as possible. Never base your purchasing decisions on only one piece of information, no matter how much you respect or trust its source!

Magazines must make money, and they make most of it from advertising, not from newsstand sales or, heaven forbid, from subscriptions. So, will they publish negative reviews of an XYZ brand amplifier when that same XYZ corporation spends $52,000 per year on advertising with the said magazine? I doubt it!

Many years ago, a hi-fi magazine published a two-page article about one of our tube amplifiers. The only problem is that I wrote the article, and they didn't change a word. Anyone who read it thought that their staff reviewer was full of accolades about our amp, when in reality, I wrote every word of it.

If you think I hold a grudge against hi-fi reviewers, I have nothing against the guys. They have a job to do, and do it based on their values and beliefs. So, I am also warning you against listening to your friends' opinions. Sure, their intentions are most likely good, but you are not your friends. Your idea of a good sound may be different from theirs. Plus, what sounds good in their room, may not sound good in yours.

To illustrate how people make huge mistakes based on their blind trust in reviews, reputable "brand" names and geographic origins of hi-fi gear, let me tell you a story.

CASE STUDY: *The curious tale of the gullible librarian*

A middle-aged librarian came over to listen to our range of amplifiers. He liked what he heard and returned a few days later to listen some more to a pair of monoblocks in chrome finish, strikingly good-looking and even better sounding, producing the sweetest 25 watts per channel most of us have ever heard. You can see their photo below.

Needles to say, I spent a couple of hours with him each time. Then he asked if I could bring a pair of these amps to his place to try them in his listening room. I was happy to do so and spent about three hours setting it all up and listening to various test CDs and music tracks that he knew.

He still couldn't make his mind up and asked if he could keep the amplifiers for a week or so, and again I was happy to oblige. Then he called me one evening saying he had a loudspeaker salesman with him if I wanted to come over (he lived close by) to hear how our amp sounded with the speakers he was considering buying.

I spent a few more hours trying the expensive speakers the other salesman brought over in his station wagon. One model sounded very good indeed, while the other didn't. He didn't buy either the speakers or our amp.

A month later, he asked if I could look at a European tube amplifier he bought but had a problem with. I was just about to tell him where to go and what to do with his new amplifier, but my curiosity was aroused. I had to see and hear the amp he bought instead of ours.

I recognized the amp from ads in electronics magazines. This single-chassis stereo amp was of a push-pull kind, using four EL34 pentodes per channel. Somebody imported it into Australia as a kit from Belgium and assembled it. In the *Australian Hi-fi* magazine review, the reviewer commented on sloppy soldering.

Despite being more expensive, the amp sounded inferior to our monoblocks. The librarian bought it unheard, based on the already mentioned review. He did not know how well the amp would go with his system; it didn't go well at all. To add insult to injury, its toroidal power transformer was buzzing loudly, probably because the amp was designed to operate on 220 V mains in Europe, not on 240-250 V in Australia. Our amps were dead quiet.

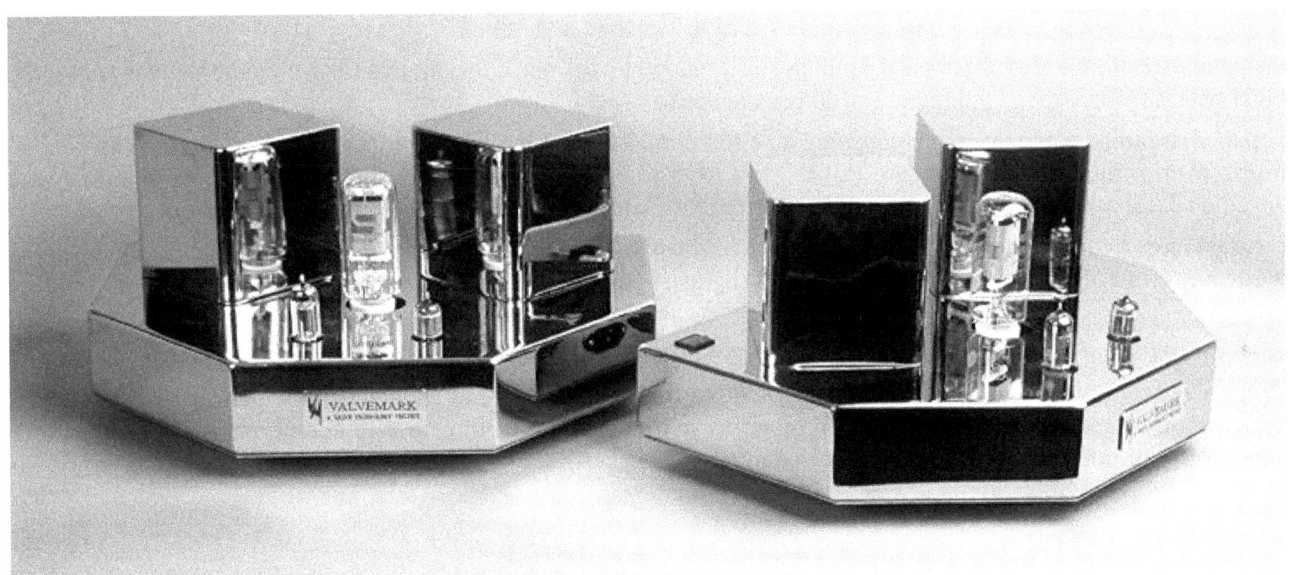

Fig. 2.1. Cumulus monoblocks in full chrome finish with SV572-10 triodes producing 25 glorious Watts of single-ended power. Sleek and smart-looking by day, the magical glow that illuminated the whole room and multiple tube reflections made them irresistible by night.

Finally, wait for this, the guy had the audacity to ask me if we could fix the amplifier for him. I wouldn't do it for a million bucks. The amp was under warranty, anyway, so I suggested he send it back to be fixed.

We were on the other side of Australia, in Perth, so he was reluctant to do so due to the significant cost involved and the risk of damage by often rough and careless Australian "couriers." The warranty was of the RTB (Return-To-Base) kind, so he had to pay for transport costs both ways.

Had he bought our amps, help would be one phone call, and five minutes away, we would have picked the amp up and returned it free of charge, as we did with another client of ours. Yes, I said "client", not clients - in ten or so years that we were making amplifiers, only one ever needed a repair.

I drove home that evening full of some inner satisfaction. I am not a spiteful person or vindictive by nature, I never wish people ill, but I was glad this time-waster had to learn this lesson the hard and expensive way.

Don't judge an audiophile system or component solely by its price or its brand!

Just as the proof of a recipe is in the eating, the proof of all improvement efforts, including component selection and matching, is in the listening. A $15,000 system that sounds terrible (yes, I've heard many of those) or a $50,000 system that sounds mediocre are an utter waste of money. I consider it a crime. That money could feed a whole African village for a few years!

Some shrewd hi-fi dealer sweet-talked the naive audiophile in parting with his hard-earned cash, only to get a frigid, boring sound in return. The brands are reputable, the system looks impressive, but the sound is far from exceptional.

People are suckers for brands, and I'm sorry to say that audiophiles are some of the biggest suckers of all. The librarian in our last story could not believe that a retired university professor, a refugee from the war in Yugoslavia, and his (then) young son could create a world-class audio amplifier in their backyard workshop.

So, he bought a "European" amp, conveniently disregarding the fact that some dealer chasing quick profits imported it into Australia as a kit and put it together in *his* backyard. How's that for a paradox wrapped in an irony?

Just look at the incredibly high prices certain vintage pieces of hi-fi fetch on eBay, for instance, Marantz 7 tube preamp, which often sells for upward of $4,000! There is nothing magical about that preamp; in my opinion, its design is mediocre at best, yet it enjoys a cult status.

A pair of vintage Neumann preamplifier output transformers fetched $1,900 on eBay recently. A pair of Japanese Tango or Tamura output transformers for $3,000? You'll find that extortion on eBay too. And I cannot finish my rant without mentioning $7,000+ for a pair for Western Electric 300B tubes (recent production, not the vintage ones from the 1940s and 50s). Who's mad here?!?

CASE STUDY: Beware of bargains!

Talking about the highly-regarded Tango and Tamura tube amp output transformers, a story comes to mind. A Perth audiophile bought a vintage Japanese push-pull tube amp from eastern Australian states for $2,000, which seemed like a bargain to him. The three steel transformer enclosures had Tango stickers on them. He wasn't happy with its sound and brought the amp over to me and left it to be checked and "fixed."

I tried to listen to the amp for about 10 minutes; it sounded awful, shrill, harsh, irritating. Something was amiss; the tonal balance was wrong, and there was no bass to speak of.

I opened the amp up and reverse-engineered its circuit; it used a standard push-pull topology, so it wasn't difficult to follow even without a circuit diagram. All DC and AC voltages (up to the output stage) checked OK; all tubes were tested on Triplett 3444 tube analyzer and were fine, too. The problem was in the output transformers.

I should have removed the transformer covers to confirm my suspicions and take a few photos, but I didn't want to waste any more of my time on that abomination (I wasn't getting paid for all that work).

The seller promised a 14-day right of return, so I strongly advised the buyer to do so. He was lucky, lost only $100 or so for road courier of the amp both ways, and got his money back.

There are two possibilities. Either those Tango output transformers were of truly shocking quality (which I doubt), or somebody removed the original transformers (to sell them for a profit or use in another amp), and substituted cheap, inferior ones. The second option seems much more plausible.

AVOID IMPULSE PURCHASES (OR WHEN A BARGAIN IS ANYTHING BUT)

A bargain audio component is only a bargain if 1. you like it in the long run, 2. you use it (listen to it) regularly and 3. it fits with the rest of your audio system (synergy).

CASE STUDY: A Cary amplifier that wasn't, or how to avoid buying a lemon

Since you've kept reading up to here and seem to like my stories, here's another one. An audiophile bought a used pair of Cary SLM-100 monoblock amplifiers from a local hi-end dealer. He said they were an absolute bargain and wanted us to give them a thorough checkup, fix anything that needed fixing, and bias the output tubes.

Once the amps were on the bench and opened up, it became evident that almost none of the wiring was original - the amplifiers had been rewired using extremely thin solid core wires used for wire-wrapping. Five of the eight power tubes were USA-made GE 6550, which tested very low, for instance, the mutual conductance was 2.6 - 3.2 mA/V, instead 3.9 (minimum) to 6.0 mA/V (nominal). The other three tubes were new replacements branded Svetlana (Russia) and measured 4.6, 4.7, and 4.8 mA/V.

On one monoblock, the power tubes were pulling 24 mA and 34 mA in one leg of the output transformer and 36mA and 52 mA in the other leg, a DC current imbalance of 30 mA through that output transformer's primary winding. That wasn't enough for that winding to burn out, but it certainly brought the transformer's magnetic core closer to saturation, especially with larger signals (at higher volumes), causing distortion.

Instead of 100 Watts of output power, our measurements showed a maximum of only 55 or so Watts, with a huge (10% +) harmonic distortion. Buying a matched octet of currently produced Russian or Chinese KT88 power tubes would cost AU$600-800.

Luckily we managed to shuffle the eight power tubes around and pair them in such a way to achieve a DC balance in both monoblocks. The sound improved slightly, but the amp still sounded ordinary, even harsh. There was no refinement, no magic, no emotional involvement. We advised the bargain-loving audiophile to sell them ASAP (which he did) and buy something newer and unmolested. His speakers were quite sensitive, and he didn't need 100 Watts per channel of amplification anyway.

> **THE INTERNAL CONDITION OF USED GEAR IS WHAT MATTERS, NOT JUST THE EXTERNAL COSMETICS!**
> Before buying a used audio component, especially an amplifier or preamplifier, ask to see its internals to ascertain if it's in its original condition or if it's been modified.

Identify and remove bottlenecks in your audio system

The concept of diminishing returns is closely related to that of bottleneck, and both feature prominently in audio. For instance, many audiophiles with basic DIY skills buy "hi-end" coupling capacitors at say $200 each (and there may be four of them in an amp!), re-tube the amp at $1,500 a pop, rewire the whole amp using silver wire with Teflon insulation, or "all of the above"!

However, the amp is a $900 Made-in-China concoction, poorly designed and constructed, using small output transformers with low-quality laminations. Its frequency range is thus severely curtailed at both the bass (small size transformers) and at the treble end, where high leakage inductance of the said output transformers and poor sectionalizing of their windings lower the upper-frequency limit of the whole amplifier.

To use a farming expression, this is kind of like putting lipstick on a pig!

The amplifier or the "platform" for those upgrades is a bottleneck, the limiting factor. No matter what we do to it, the sound will improve only marginally, and the point of diminishing returns will very soon be reached.

Beware of the moving target syndrome

The concepts such as diminishing returns and bottleneck don't just apply to individual audio components but your whole sound system as well. The audio components in your system will seldom be of equal quality or perform equally well; there will always be a bottleneck, although such a situation may not be readily noticeable.

Unfortunately, the only way to ascertain which one of the components should be upgraded is by trying out one or more of your components in other people's audio systems and the other way around. If you try two different power amplifiers in your audio system and the sound improves in both cases, your existing power amp could be the bottleneck. Its limitations are holding your system back and preventing it from reaching its full potential.

However, as you start upgrading components, and as soon as you remove one bottleneck, another component will become the next limiting factor. I call it "the moving target syndrome" since the ultimate target is seldom reached. The higher the system's overall performance level, the more pronounced the individual limitations of specific audio components.

This never-ending quest for musical nirvana is an annoyance to some audiophiles but actually a source of joy to others. Let's see how to make this process less error-prone, less expensive, and more enjoyable.

The relentless, often mindless pursuit of the "new"

One of the biggest mistakes you can make as an audiophile is buying new audio components just because they are the latest model, supposedly better than their previous versions (as claimed by the manufacturer or just assumed by the prospective buyers). That is a dangerous assumption to make because, in most cases, it simply isn't true.

Likewise, do not assume that a newer product or a newer release of the same model is necessarily superior across the board. It may be better in certain aspects of its performance, but then the other aspects, equally or even more important, could have been compromised. It seems that in many audio designs, the improvement progression is a zero-sum game: some things (those aspects that designers pay attention to) get better, others get overlooked, and may even get worse with each new model release.

Do audio designers do that on purpose, so they can claim tangible improvements of some parameters and entice us to buy the new, "improved" versions and spend more? Is it simply "limited competence", where they cannot handle all design aspects equally well? Or, could it be "the limited attention quandary," the phenomenon where things we pay attention to get controlled and improved while others are made worse in the process?

This super-simplified timeline is a brief overview of audiophile trends and milestones (analog technology only). The most significant developments and most seminal products happened before 1960. The three decades that followed (1960-1990) were a dark and uneventful age, dominated by awful-sounding early transistor amps.

Although new dawn in the early 1990s started tubes' revival, in that era (that has lasted to this very day), very few truly novel designs or developments took place. Many "designers" are simply rehashing the old circuits from tube manuals or copying commercial designs from the 1950s and 60s. Sadly, the late luminaries such as Tim de Paravicini (Esoteric Audio Research) and Kondo-san of Audio Note fame were notable exceptions.

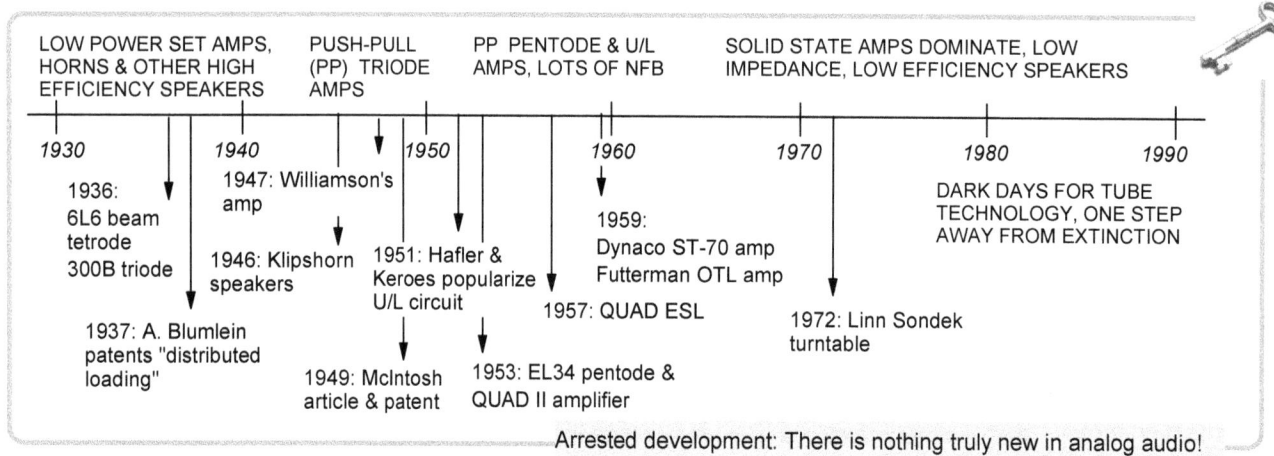

Arrested development: There is nothing truly new in analog audio!

Don't judge a hi-fi system or an audio component solely by its measurements

If you read audio magazines, you may have noticed that some (*Stereophile*, for instance) have a policy of not letting the equipment's reviewer know about the results of its measurements, which another staff member does. In other words, they try to avoid the biasing effect that measurement results may have on the reviewer.

So, a reviewer may be impressed with the sonics of an amplifier (or speakers, or a CD player, or anything else he is reviewing) and give it high marks and lots of accolades. Yet, the measurements and technical aspects of the reviewed piece of gear may be below average, sometimes even very poor.

Single-ended triode amplifiers, especially those without negative feedback, have high harmonic distortion levels and a low damping factor (high output impedance). In many cases, they suffer from a limited (narrow) frequency range or power bandwidth. Yet, matched with the right speakers and for certain types of music, they sound divine.

An audio system or a component should not be judged by its measurements. In fact, it should not be judged (initially) by anything else apart from how it sounds! The key word is "initially"! Of course, sound quality is not the only criterion for buying audiophile equipment, but it is the most important one. If something sounds bad, do you care how cool it looks or if it's covered by a 10-year warranty? I don't.

Only after you are happy with the sound should you consider other factors - the construction quality, the aesthetics (the looks), the upgrade potential, the resale value, reliability, the ease and the cost of servicing, and so on. Warranty terms are of paramount importance - what is covered and what is not, how long the warranty is, is it a local warranty (how close is the service center to you) or do you have to ship it back overseas or across the Australian or American continent, for instance.

How far is far enough, and how far is too far?

Audiophiles have a notorious reputation for constant experimentation, tube rolling, perpetual modifications, component upgrading, and all kinds of minor tweaks and significant component and system-wide changes.

Some of those make no sense at all. For instance, quite a few posts on online audio fora are of the "I bought this amp, can you guys suggest what components (resistors, caps, tubes) I should replace?" kind. This is silly. He has neither listened critically to his new amp in his audio system nor identified its strengths and weaknesses from the sonic, design, construction, or reliability point of view.

By asking such a question, he made it obvious that he has no clue about electronics and is incapable of critical thinking. Yet, he wants to start changing things straight away.

On the other hand, this constant experimentation is part of our quest for that elusive ultimate sound and is thus understandable. The question, however, is how far is far enough, and how far is too far? When is enough enough?

The S-curve is well known in project management. Just like Murphy's law, it applies to most aspects of life. For instance, when you start improving something, often the results aren't immediately noticeable.

You have to persist with your efforts until you overcome the initial period of "inertia" (for the lack of a better word) and reach the linear portion of the curve, where further efforts result in noticeable achievement or improvement.

Common examples in audio would be interstage and output transformers in tube amplifiers. Once installed or replaced, they take some time to be broken in and may sound terrible initially. Likewise, silver audio cables also take quite a few hours to "settle down." Initially, they sound detailed but bright, even harsh on some systems.

Coming back to our S-curve, there usually comes the point where a further investment of time, money, or effort is not resulting in any meaningful improvements. You have reached the "saturation" region, the plateau, the point of diminishing returns. You are investing a lot but getting little or nothing in return.

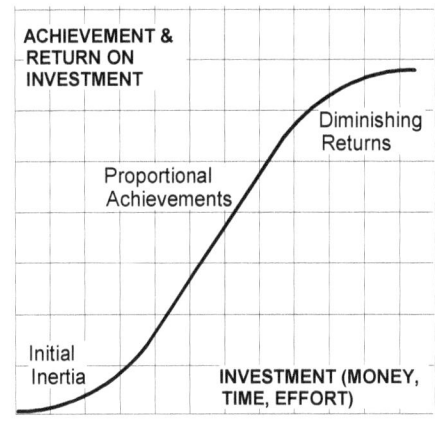

Fig. 2.2. The S-curve applies to most human endeavors, including audiophilia.

The audio know-how and know-why

Almost everything impacts the sound of an audio system, from the obvious significant factors such as the equipment used and the listening room to relatively minor factors such as the accessories and treatments. The problem is that in some cases, even those minor factors could have a significant impact on the sound, and the extent of their impact depends on the rest of the system.

Thus, it is tough to predict if a particular equipment substitution or any other intervention (changing one or more aspects of the system) will improve or degrade the sound and to what extent. The only sure way to be certain is to try things out and listen for any sonic changes.

As in any profession, audio professionals such as engineers, equipment designers and builders, retailers, audio writers/reviewers (and even knowledgeable and experienced amateurs) are better able to "connect the dots." By that, I mean making more reasonable, testable propositions of the "if I change this, that will be the likely or expected sonic result" kind. They aren't psychic and don't possess special powers. They have simply spent more time and effort over the years dealing with such issues, trying things out, experimenting, and drawing conclusions from their experiences. The result is a more developed intuitive feeling for what may or should work and what may not, based on solid know-how.

Knowing what works and how is hard and takes years to learn, but knowing why is a true challenge. I cannot explain many of the phenomena we talk about in this book, at least not thoroughly or concisely. Ultimately, unless you are the anal-retentive type of person and a pathological stickler for pedantry, does it really matter why is something so? I think not. If something works, use it, benefit from it, and enjoy it. Move on. Life's too short to ask why or why not all the time. Only two-year-olds do it constantly, and we all know how annoying they can be.

MAJOR UPGRADES OR MINOR TWEAKS?

Classifying changes into minor tweaks and major upgrades is misleading and counterproductive. Many "minor" tweaks result in major sonic changes (usually improvements), while major (read "expensive") modifications and "upgrades" don't always result in commensurate sonic betterment!

AUDIO REVIEWERS, THEIR REVIEWS, AND THE MAGAZINES THEY WRITE FOR

How audio magazines evaluate and rate components

The difficulty in evaluating, rating, and comparing audio equipment is best illustrated through an example. This is not a criticism of *Fidelity* magazine and their methodology, but rather as an illustration of how problematic and ultimately futile this task is. *Fidelity* uses a 2x2 matrix with four criteria. Since they are paired and on the opposite sides of a simple grid (without any numerical markings), one would assume that the two criteria in each pairing are the opposites of each other.

The first question is, out of many possible factors, why have these particular ones been chosen? Does that mean they are the most important ones? Plus, how can the same factors apply to all types of audio gear? Shouldn't each component type (amplifier, phono stage, loudspeaker, DAC, etc.) have its own quartet of evaluation criteria?

The choice of the terms is also problematic. Dictionaries define analytical as "involving the careful, systematic study of something." Normally, that adjective applies to investigative skills, processes, and rational thinking used by humans. Test equipment is, by its definition, analytical since it's used to analyze physical or chemical properties of substances or devices.

Out of twenty or so synonyms, only a couple can apply here, and these are "precise" and "detailed," presumably as in "capable of reproducing details."

Euphonic is defined as melodic, harmonious, tuneful, or "pleasing to the ear." Vividness is usually associated with powerful visual feelings or strong, clear images in the mind or intensely deep or bright colors. Sonorous means "sound-producing." Here, it is more likely to mean capable of producing "an imposingly deep and full" sound.

Now, going back to the issue of opposites, we have to ask ourselves if this is necessarily so; for instance, why couldn't a component be both euphonic and analytical?

The sonorous-vivid duality is even more problematic.

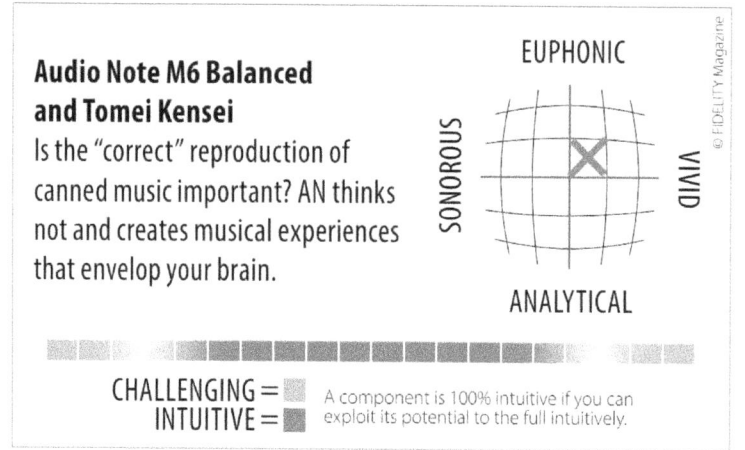

Fig. 2.3. Fidelity magazine uses this graphics to give their readers a quick insight into the general character of audio components.
Copyright: Fidelity International, used with permission

Taken "vivid" to mean bright (but not as in "harsh," rather as in "bright colors," the opposite of gray, drab or boring), and sonorous to indicate "imposing, full-bodied and full range" sound, to me these are hardly the opposites. An amplifier that is lively, dynamic, and "vivid" is usually also "sonorous."

I admit it is possible to have a SET amplifier that is vivid but not sonorous if its dynamics are limited (as they always are, we are talking a few watts of output power here). Its performance at frequency extremes may not be ideal, as in loose and lazy bass, lack of sparkly and sonorous "top end."

To give credit where credit is due, the reviewers of audiophile equipment face a difficult balancing task. They have to reconcile the listening impressions with measurement results, the visual aspect of the reviewed products, their quality of manufacture or workmanship, the reliability, the warranty, and the quality of the after-sales support.

Probably the most critical issue of all, the quality of the design, is an even bigger minefield. Design features such as the basic topology only get mentioned in passing, if at all.

In the short time they spend with a piece of equipment (days and weeks, not months or years), they cannot properly evaluate equipment's reliability, let alone the quality and the inclusiveness of the warranty or the responsiveness during after-sales support.

Reliability can be inferred by a detailed analysis of the design, the currents and voltages, and the component rating in the circuit, but no magazine that I know of pays much attention to that

For instance, if one 300B SET amplifier biases the output tubes at 60 mA and the other at 90 mA, the one with the higher idle current will chew through those tubes much faster, and its long-term cost of ownership will be higher.

Some magazines, such as *Stereophile*, don't use formulas or cumulative scores; others use a basic star system (up to 4 or 5 stars). One German magazine evaluates hi-fi components based on the 70% - 10% - 10% - 10% formula: 70% sound quality, 10% measurements, 10% construction quality, and 10% value for money.

Paradoxically, as soon as you start assigning percentages to different aspects, you raise more questions than you answer. For instance, assigning only 10% for "value-for-money", regardless of how you define this term (another minefield), is questionable. The same can be said for the construction quality, also being only 10% of the score.

A great-sounding component that develops "cold" solder joints, oxidation, and other contact problems will bring its owner considerable grief, diminishing the enjoyment factor, not to mention downtime and expense in getting it fixed. Also, most design and build (construction) issues directly or indirectly impact the sound quality, so things aren't as clear-cut and straightforward as they seem.

The crucial role of audio manufacturers, retailers (dealers) and reviewers - inappropriate or irrelevant comparisons, sneaky words and loaded phrases

When manufacturers stretch the truth, or ignorant dilettantes propagate myths and misconceptions on the web, that is one thing. Still, when professionals, people who are paid to create and disseminate relevant and accurate information and scrutinize products, such as audio reviewers, get their facts wrong, that is something else. An average reader thinks that these guys are in the know and implicitly trusts their opinions.

In *Stereophile* magazine (Jan 2014, page 27), the reviewer (Sam Tellig) wrote this about Unison Research Simply Italy SE amplifier: "Giovanni Sachetti designed the circuit to run the EL34B in pentode mode, but with a sound that closely emulates triode."

That amplifier operates in an ultra-linear mode, not in a pentode mode, and does not "emulate" the sound of triode. That sentence is not just inaccurate but also misleading, for the reader may think that the amp somehow processes the signal to emulate the triode sound - it doesn't.

It gets worse. In the Feb. 2012 issue of *Stereophile*, on page 42, Art Dudley writes: "... Shindo Haut-Brion ($11,000), a stereo amplifier designed around complementary pairs of the famous 6L6 pentode tube." The reviewer managed to make two faux passes in this short partial sentence. Unlike bipolar transistors (NPN & PNP) or FETs (N-channel & P-channel), there is no such thing as "complementary pairs" of tubes. Secondly, 6L6 is not a pentode but a beam power tube. Pentodes have a third or suppressor grid; beam power tubes don't.

In his March 2014 review of Stirling Broadcast BBC LS3/6 loudspeakers, *Stereophile* writer Art Dudley makes a series of faux pas - twice. "All of the LS3/6 drivers are manufactured in Norway by SEAS, presumably to Stirling Broadcast's specifications." Why "presumably"?

If I had to hedge my bets, I would say that a speaker maker would use standard drivers, which are much cheaper than custom-made ones. However, why presume anything? A simple e-mail or phone call to the manufacturer or to SEAS would have answered that question.

Further down the technical description, we read, "The 8.6" woofer has, as you'd expect, a polymer cone ..." Why would one expect that? I don't expect anything, and neither should an open-minded reviewer. By using that phrase, the reviewer implicitly makes it sound as if polymer cones are the most superior of all. They are not.

You may say now, come on, Igor, don't be a language Nazi; being a stickler for linguistic details and nuances is not what this book is or should be about! Perhaps. But audio reviewers are (or at least, should be) professional writers and wordsmiths. Language has to be treated with the utmost respect and used correctly and precisely.

If they start using the language nonchalantly or flippantly, they also contribute to the current slippery slope trend towards its devaluation and debasement. Precise, unambiguous, and well-defined terms should be used by all, especially those who ply their trade with words.

MARKETING HYPE, MYTHS, AND OUTRIGHT LIES IN THE AUDIO FIELD

The three biggest lies in tube amplifier marketing

When reading a book that straddles multiple fields (as this one does, acoustics, electronics, etc.), you can often tell in which areas the author's primary expertise lies. In my case, that is tube electronics. So, let's look at a few problematic claims and pieces of marketing hype from the tube land.

Unfortunately, the same things are present in the marketing of power filtering and conditioning products, acoustic treatments, hi-fi racks, CD players, turntables and all other pieces of audio gear covered in this book. So, once you know what to look for (or listen to), you'll be able to recognize tall tales from miles away.

1. Most amps touted as "all tube" are not all tube.

Even if only silicon diodes are used as rectifiers in the amp's power supply, the amp isn't "all tube." Likewise, if an amp uses solid-state (FET or bipolar transistor) constant current sources or Zener diodes, it cannot be "all-tube" anymore. In most cases, these solid-state components are in the signal path and will impact the sound!

2. The use of printed circuit boards (PCBs) does not make amplifiers more reliable or better sounding.

It just makes them cheaper to manufacture. In fact, the opposite is true. Point-to-point or "hardwired" wired amps (without any PCBs) sound better, are more reliable, and much easier to upgrade or modify later on.

3. Most "Class A" push-pull amps do not work in class A most of the time, only some of the time.

Since many audiophiles do not understand what "Class A" means, the brazen manufacturers can use this marketing strategy with impunity. Class A refers to one of the two possible operating regimes used in audio and *not* amplifiers' quality levels. It means that tubes never go into cutoff - their anode currents never drop to zero.

Single-ended amps *always* operate in Class A; operation in any other class would mean that a portion of the output signal was cut off, which would be unacceptable and make the amp unbearable to listen to.

Class B means that one of the pair (or two of a quad, and so on) of tubes will conduct more and more, but the other will stop conducting its part of the signal and enter a cutoff state.

Class AB is not a class by itself; it's a mix. At low power levels the output stage works in Class A, with both tubes conducting the signal, one more, the other one less. As the output signal increases, the output stage crosses into Class B operation at the point when one tube completely stops conducting, so all the signal is passed by the other tube.

Thus, it is never a complete lie when a push-pull amp is marketed as a Class A amp. All push-pull amps work in Class A at low or very low power levels. The exact crossing point into Class B depends on various design factors.

The seven levels of truth in audio

Here is a simple 7-step model you can use whenever you are evaluating or choosing finished audio products, components, or anything else, for that matter, in any sphere of life. Similar "7 Levels of Truth" apply to real estate, white goods, cars, service providers (housebuilders, consultants, contractors, dentists, accountants, ...) pretty much all goods and services:

THE 7 LEVELS OF TRUTH IN AUDIO

1. TECHNICAL FACTS: Accurate, measurable, and thus universal and indisputable truth.

2. SIMPLIFICATIONS AND MODELS: More or less accurate representations of the reality, but only under certain conditions and within clearly specified and verified limitations.

3. EDUCATED GUESSES: An intuitive feeling of an expert, possibly valid, albeit without any conclusive proof that would verify it.

4. OPINIONS: Of limited value, unless the source is experienced and knowledgeable, as at Level 3, in which case those opinions could serve as a starting point for thinking and further investigation.

5. MYTHS: Started either innocently, out of ignorance, or deliberately, due to one's vested interests, and then propagated by the more-or-less ignorant majority of audiophiles and bystanders ("tube kickers").

6. MISREPRESENTATIONS: Opinions, myths, and half-truths presented as facts and "distinguishing features" by manufacturers, importers, retailers, audio reviewers, or anyone else with vested interests.

7. DELIBERATE LIES: Untruths that are obvious to experienced and educated experts but are often taken as gospel by audiophiles without the knowledge to judge such claims and see them for what they are.

Let's take this book as an example of an audio product. Although you cannot listen to it- just as you cannot listen to equipment racks, acoustic treatments, cables, or power conditioners- since it deals with the topic of audiophilia, it is indeed an audio product. What level of truth does this book sit at?

It certainly deals with facts, so at least some of it should be at Level 1. Building, testing, and auditioning a particular audio circuit, an acoustic panel, or a cable, constitute a verifiable experiment. You or anyone else should be able to reproduce it, and if you do, you should get similar results.

However, to minimize the maths and technical details, simplify things, and make them more user-friendly, there are many models, simplifications, and rules-of-thumb, as in any such book, so Level 2 is also used. This is not an admission of inadequacy; every university course of study uses models and simplifications. Otherwise, it would take 15-20 years to comprehend things to some degree (never fully!) and eventually earn a university degree.

This book isn't of a purely technical nature; it is partly a popular book for non-technically minded audiophiles. I have also often voiced my opinions and hypothesized about the causal or correlational links between issues. Many are my educated guesses and not results of verifiable scientific experiments, so we are down to Levels 3 and 4.

Finally, I believe there are no myths, misrepresentations, or deliberate lies propagated in this book. The key word is *deliberate*.

As in any book, there may be inconsistencies, omissions, and repetitions, there may even be factual errors, but they are by no means deliberate. For instance, as an author, I have no interest in making you buy a particular piece of audio gear instead of a competing product, so why would I aggrandize one and denigrate the other? Which leads us to the underlying issue and the most critical factor, and that is someone's motive.

Did they really say that?

From the Nordost Heimdall 2 power cable product description: "The complex topology of Micro Mono-Filament conductors is especially effective in dissipating mechanical energy present on the AC line, which otherwise enters your delicate electronics rendering dynamic responses sluggish and blurry."

Mechanical energy on the AC line? That can only mean one thing - vibration! Even if it was there, how would such vibrational energy enter the power cord? Through the wall socket? And how would it spread through the power cord into our "delicate electronics"? At what level of truth should we place this marketing spiel?

MARKETING CASE STUDY: Psvane Teflon® capacitors

Psvane film capacitors come in three values. As the values go up, so does the (exorbitant) price. In 2015 the pricing was 0.1 μF/600V (US$99 a pair), 0.22 μF/600V (US$149 a pair), and 0.47 μF/600V (US$199 a pair). Printed on the capacitors' bodies are the words "Reference Cap," a meaningless marketing hype, and "Teflon Copper Foil." These capacitors do use copper foil inside, but the insulating film between the windings of a copper foil is polyester. The only Teflon® is the sleeving around the outside leads, one black, the other red.

Does the printing of "Teflon Copper Foil" constitute misleading advertising, and which of the seven levels of truth would it fall under? Is it a misrepresentation or deceptive advertising (Level 6) or a deliberate lie, Level 7? Had "Teflon Film Copper Foil" been printed, that would be a deliberate lie, no doubt. However, by omitting the word "film," the manufacturer can argue that these are Teflon® caps, since they contain Teflon®, albeit only on the outside, where such insulation of the leads is meaningless and does not make any difference to the sound.

However, unsuspecting DIY constructors or audiophiles "upgrading" their capacitors are likely to automatically assume that Teflon® refers to the most critical aspect of a capacitor (sonically), namely its insulating film. Therefore it could be argued that this constitutes misleading advertising.

MARKETING CASE STUDY: Elemental Watson headphone amplifier

The manufacturer of this budget hybrid headphone amp makes some questionable claims. We will emphasize the problematic phrases *in italics.* In their own words, "By trimming down *unnecessary luxury* and focusing on core competence, we are able to build Elemental Watson in a *dual mono configuration*, a highly touted feature in high-end gear." First, since when is an enclosure, chassis, or case "unnecessary luxury"? Secondly, yes, dual-mono is "highly touted" in the audio fraternity. Still, since this headphone amp uses a common external DC power supply, shared by its two channels, it can not possibly be of dual-mono design.

As for its driving ability, they say, "Followed by class-A power MOSFET output stage, it provides *almost limitless driving current* ..." How can a prospective buyer check that "almost limitless" claim? Well, the output power is specified as 2,000 mW (2 Watts) into 30Ω (although the sticker on the packing box says 1 Watt per channel) and 250 mW into 600Ω. This reduction from two Watts to a quarter-Watt already indicates current limitation at work.

Power is $P = I^2R$, so the unit can supply $I = \sqrt{(P/R)} = \sqrt{(2/30)} = 258$ mA into a 30Ω load, but only $\sqrt{(0.25/600)} = 20.4$ mA into 600Ω, a huge drop of output power. The *almost limitless* current claim is laughable.

Two more common marketing issues are worth pointing out (so many examples in one little product), "pure" class A and "designed in Australia." There's nothing wrong with the latter statement per se, but do audiophiles really care where an audio component was designed? A country of manufacture would be a better predictor of its quality.

There's no such thing as "pure" class A (or "impure" class A, which is implied). An amplification stage or a whole amplifier either operates in class A, or it doesn't. It's like saying "pure" virginity. One is either a virgin or not; there's nothing in between.

UNDERSTANDING YOURSELF AND YOUR AUDIO PREFERENCES

As with all activities, reading and listening to music are both learned experiences. The more we listen, the more we discover and become aware of the subtleties and nuances of both the music itself and its audio reproduction. We develop our preferences, likes, and dislikes. This could be another reason behind audiophiles' quest for better sound, resulting in continuous buying and selling of audio components, endless tweaks, and system upgrades.

Sometimes we know what we are looking for, but more often, the quest reminds us of searching for one's soul mate, whatever that means or whoever that may be. To find your prince, you have to kiss a lot of frogs!

What do you listen for and why?

Every music system is a unidirectional chain of components. The signal propagates from the source through the amplifying chain to the final electroacoustic transducer, the speakers, and the listening room. Every link in the chain affects the sound, some more, others less.

Audiophiles disagree on most things audio. Some argue that the initial components, the turntable, and the phono stage, or the CD player and the preamp (in digital systems), are more important than the components down the line; others argue the exact opposite. However, the audio chain does not end with the listening room; it has to include the subject or the listener. Before we form the impression of a particular system, how it sounds, the sound has to pass through another transducer, and that is the human ear.

The sensitivity of the human hearing apparatus varies with the sound frequency, and the ear itself distorts, creating intermodulation components that were not present in the reproduced sound.

Hearing is subjective; that is why the science that studies related phenomena is called psychoacoustics. Our psyche, including our moods, affects the way we hear things. Some people hear more, some less. Some are sensitive to the bass frequencies; others pay more attention to the high-frequency end of the audio range. Depending on their mood, even the same person will judge the same amplifier and system differently.

When we build an amp and take it to the listening room for evaluations, I don't form my opinion just during the night time listening, when the volumes are generally lower. The amp will sound different during the daytime, or when the output levels are higher, or when I'm tired, when the mains voltage is high or when it's low. These are just some of the factors in play. Listening evaluations are like wine tasting, which means highly subjective.

I wish I had a dollar for each time a prospective buyer asked me, "Igor, which of your amplifiers will best match my XYZ speakers?" How a particular amplifier will sound with a specific loudspeaker cannot be predicted with any degree of certainty, just as the following link in the acoustical chain between the loudspeaker and the listening room cannot be established beforehand without the actual listening test.

As you listen to different sources, amps, and speakers, in other rooms and at different times, your preferences will gradually form if they haven't done so by now. You will have a pretty good idea of what kind of sound you like, what matters to you more, and what matters less.

An audiophile system is a reflection of its owner's personality, his taste in musical reproduction. It illustrates his priorities and idiosyncrasies, what he values and pays attention to. It is part of his identity, saying to those who see it and listen to it, "This is how I believe music should sound and how we should experience it!"

Different audiophiles use different criteria to judge audio components and systems

At the start of our amplifier-making business, neither Bob nor I had any experience in selling, and for us, that proved to be the most difficult task of all. Designing and building highly complex amplifiers was relatively easy by comparison. We bravely decided not to go through retail outlets (hi-fi shops) but to sell them ourselves, through magazine advertising and direct mail campaigns.

This true story illustrates an important point. A well-educated, upper-middle-class couple come to our demonstration room one evening to audition a tube amplifier. That was in 1996, the first year we were "in business." The amp was a single-ended triode design with PL519 power tubes.

It was a prototype unit, so its finish wasn't great. It was used for demonstration purposes, meaning it had fingerprints all over it, minor scratches, and dust. They didn't seem to notice, or, if they did, they said nothing.

They listened politely to a couple of demonstration CD tracks I played for them and then shyly asked if they could play a particular CD the lady pulled out of her handbag. It was a good quality amateur recording of their daughter's school choir. They only wanted to hear one particular song. In fact, not even the whole song - they only listened to about thirty seconds of it, just one specific passage.

The husband stood up and said, "Igor, this is the first amplifier which enabled us to hear our daughter's voice clearly and distinctly from the other voices in the choir. On all other amps we auditioned, we could not hear her at all; it was all smudged and messed up!"

It was a nice compliment. I had no idea what other amps they auditioned, I was too excited, so I forgot to ask. I was prepared for a long, tedious negotiation session. Even worse, when people praise your product, that usually means they don't even think about buying it; otherwise, they would start finding faults with it to bring the price down.

What happened next was a true miracle (I wish it happened more often). The husband and wife looked at each other, and while he was taking their CD out of the player, she opened her handbag, flashed a wad of cash, and bought the amplifier on the spot. No negotiation, no hassle, and no hustle. To quote the great Chuck Berry, "it goes to show you never can tell!"

The Yin and Yang of audiophilia

Choosing audio components or systems has a lot in common with choosing a marriage partner. Far too often, we fall in love and make hasty decisions only to get disappointed sometime later. We got attracted by certain aspects of an amplifier or loudspeaker but overlook a myriad of other, equally crucial questions or aspects of its performance. We look at things from a short-term perspective and ignore the long-term implications.

The ancient Chinese Yin-Yang dichotomy is another way of looking at this issue. The Yin outlook or aspect is soft, feminine, emotional, adaptable, and based on feelings. The Yang is male, hard, forceful, controlling, and logical.

The design of audio components such as turntables, amplifiers, and loudspeakers is both an art and science. Each designer and equipment maker must balance the two ends of the scales, just as every audiophile must discover where on that spectrum his preferences lie.

There is nothing wrong with a male person having feminine or Yin qualities; in fact, it is normal and may even be desirable in certain situations and aspects of life. And don't worry, paying attention to and getting in touch with your feelings does not make you less of a man, quite to the contrary.

It all boils down to getting to know your expectations and preferences. What aspects of musical reproduction are important to you? For instance, prominent, forceful, internal-organs-slamming bass is a male or Yang expectation, just as the search for liquid, enchanting, "flowing like honey" female vocal reproduction is a Yin trait.

Of course, many of us expect all aspects of a component's or system's performance to be equally present or well-executed. Still, the reality is that certain classes of audio gear and even individual components amongst them are usually clearly identifiable as either Yin or Yang.

A high-end turntable, machined of solid pieces of metal and put together like a Swiss clock, is an exquisite precision mechanism and often even an art object. It invokes nostalgic feelings and mesmerizes us with its rotation, reflections, and analog or Yin nature. A CD player or a solid-state amplifier does not invoke such emotional feelings just by looking at their utilitarian enclosures, but many tube amps do.

Ideally, a particular design should balance both aspects. For instance, a tube amplifier should be of sound design (no pun intended), properly built and reliable, all left-brain (Yang) attributes. It should also be pleasing to the eye and ear, look good, and be pleasurable to use, all right-brain or Yin qualities.

Technical or left-brain aspects of design are easier to judge and evaluate. Technical performance, build quality, and reliability are still open to a certain degree of interpretation, but not nearly as much as the right-brain aspects are. The beauty, after all, is in the eyes of a beholder.

Thus, what seems to be a stunning and unusual piece of audio engineering to one audiophile may seem a butt-ugly or over-the-top piece of kitsch to another.

BUYING AUDIO GEAR UNHEARD, DISAPPEARING DEALERS, AND OTHER PITFALLS

CASE STUDY: A cautionary tale about buying online

A local audiophile visited our demo room on a few occasions, so I spent quite a few hours taking him through our range and setting various amps up for him to audition. He was a friendly Greek guy, a tradesman. He eventually narrowed his choice down and asked me to bring one amp to his place to try with the rest of his system.

He had a few computers in his audio room, which was, believe it or not, his study and bedroom! All were connected to the same power outlets, resulting in the loudest buzz I have ever heard. The number of wall warts and other switch-mode power supplies was mind-boggling, and there was nothing I could do to fix the problem.

In desperation, I asked him to bring a long extension cord and plug it into one of the power points on the other side of the house. Miraculously, that instantly eliminated the buzz.

He was happy with the sound and that I fixed his system for free and pulled out $1,500 in cash, although the agreed price was $1,700.- The original price of that amp was $3,000.-, it was the last one (discontinued model), but it sounded incredible, one of the best we have ever made. Mildly offended, I refused to budge, so I packed the amp and drove back home.

Considering the number of hours I had spent with him, possibly ten or so in total, and the hourly rate I should have charged (but I couldn't), he was practically getting the amp for free! Obviously, he didn't see it that way and did not appreciate my knowledge, advice, and all the effort I had invested in that failed sale.

A month or so later, he called me, telling me that he bought a $1,400 made-in-China tube amp but that it didn't sound nearly as good as ours. He was on his own, with no one to blame but himself. I could have helped him improve his setup, possibly even modify and upgrade his new amp to sound half decent, but why would I? For the sake of $300 that he "saved," he was stuck with an amplifier that he didn't like or enjoy.

The Disappearing Dealer Syndrome

The rapid disappearance of high-end audio dealers, a direct consequence of rising commercial rent levels and lower online prices of the same audio gear, is one of the crucial audio developments of the first two decades of the 21st century. The inability to talk to an audio retailer and benefit from their expertise and advice makes it so much harder to audition audio components and to make an A-B-C comparison of those on your short list.

If you live in a remote place, your troubles are even bigger. To hear anything, either brand new or secondhand, you have to travel hundreds of kilometers. Having lived in a few remote mining towns in Western Australia, the hi-end systems I've seen and heard in those places consisted of individually superior components. Mining jobs have always been very well paid, so their owners spared no expense in getting the latest and, if not the best, at least the most fashionable gear at the time. You could find Martin Logan and Apogee speakers, Cary and Audio Research amplification. However, almost all of those systems sounded very ordinary, even sterile and anemic to me.

The problem was in component matching and system integration. Their owners bought them at different times from different sources and relied on the blind hope that they will go well together. After only a few minutes of even casual listening, it was evident to me that they didn't.

For those of us in large cities, there's usually a steady stream of used gear offered for sale, but the three or four models on your short list will never be available at the same time. Unless you have an extraordinary long-term memory and can remember the sound of a component you had auditioned two, three, or more months ago and can compare it with the alternative component you are auditioning now, the meaningful comparison is impossible.

Plus, even if you decide that the gear you had heard all that time ago is superior to the component you are currently auditioning, chances are it had already been sold.

The expensive but low-risk way to audition multiple audio components concurrently

So how to minimize the risk of buying a component that will not fit well into your current audio system? The only way I can think of is a multistep approach.

First, perform due diligence. Do your research, read the reviews, audition as many components in your price range at audio shops and private sales and listen to your friends' systems. Then, narrow your list down to three or four components that you liked, both visually and sonically, and that fit your budget.

The problem is some of those will be current models available from an audio dealer; others may be older models offered for sale privately and only available sporadically. Getting them together at the same time for A-B-C auditioning will be on the hard side of the impossible.

Assuming you have a sufficient stash of cash, buying all of those components and auditioning them together in your listening room is the way out of this predicament! Yes, buy all four pairs of speakers on your list, all three tube amps that you liked, or all four phono stages you are considering, whatever you are looking for. It may take you a few months to get them all (depending on how rare such models are and where you live), but keep chasing them down until all are in your possession. Then, audition them concurrently with the rest of your audio system.

Once you decide on the one(s) you want to keep, sell the rest of them over time. Don't put all of the unwanted ones on the market at once. Financially that is never a good idea. List them online one by one.

For local buyers who will come to your home to see and hear them, hide your chosen component (the one you are keeping) and set the unwanted ones one-by-one in your system, making them sound the best you can. Don't tell prospective buyers that these came up short in the A-B-C-D shoot-out or that you bought something else. Just say that your musical tastes have changed, and you are gradually changing your whole system over.

Obviously, this approach would be impractical with highly-priced gear; buying three pairs of speakers at $20-30,000 a pop would be enough to break up even the strongest of marriages! And that is assuming you have $60,000 in spare cash just lying around.

If you've negotiated hard and obtained that gear at rock-bottom prices (new or used, it doesn't matter), you should be able to recoup your initial costs. In some cases, especially with the in-demand models continuously appreciating (gaining in value) on the used market, you may even profit upon their resale a year or two later.

YOU MAKE A PROFIT WHEN YOU BUY, NOT WHEN YOU SELL

A financial parallel between audio gear and cars is striking. Most audio components' value will depreciate over time (but not as rapidly as cars do). However, some notable exceptions (a Linn Sondek turntable, anyone?) will make you a tidy profit when you sell them. While some such models are overhyped and don't deserve their stellar reputations, many do - buy them!

Why aren't there (more) audio consultants or system integrators?

Although some audio equipment manufacturers occasionally provide free system integration advice, as does the ever-shrinking pool of hi-end retailers (dealers), judging by poorly matched audio components in most systems and the almost total disregard for room acoustics problems, this kind of service is sorely missed. Audiophiles prefer to experiment and discover things by themselves (by the most primitive method, trial & error) rather than paying somebody else (a system integrator or audio consultant) to do it for them.

Now, I, of all people, should value the DIY approach, but such an option assumes that the person doing it himself has the knowledge and know-how required. Sadly, most audiophiles don't (although most think they do)!

It never ceases to amaze me how audiophiles would pay tens of thousands of dollars for a component or just a marginal upgrade (a pair of power tubes for $7,000 ?!?). Yet, they wouldn't pay $500 to $1,000 for an expert to come over for a few hours to optimize and voice their system or fix a few glaring problems.

Even when such advice and service is provided for free, most prospects don't appreciate it. That could also be a part of the problem. In general, people don't value free advice or free anything.

In a sense, this book is my attempt to "right such wrongs," so to speak. I hope you will use your newly gained knowledge and confidence to tackle the thorny issues of system integration, optimization, and room acoustics improvements. Just like professional system integrators, you should know how vocals, instruments, and various types of music sound. Upon listening to an audio setup, you should be able to identify its problems and shortcomings and overcome them (or at least minimize them) by applying a few simple yet effective measures and tweaks.

Anyone can spend $22,000 on a new pair of speakers (or anything else in the signal chain) in a delusional hope that such an upgrade will somehow solve all their sonic problems. It won't.

Consider the resale value of the gear you are buying

Real estate gurus like to say that you don't make money when you sell; you make money when you buy. That means that if you buy cheap, below the property's true market value, you will most likely make a decent profit when you sell it. I would go one step further and say that you make money both when you buy low and sell high.

The same principle applies to buying and selling anything, including hi-fi equipment. A true audiophile never buys anything to make money; they buy to enjoy the sound. Still, nevertheless, considering how many times audio aficionados buy and sell gear, the concept of resale value should be first and foremost on their minds.

This is not a book about the financial aspects of audiophile systems. Still, this primacy principle applies to sound improvement as well, and it goes like this: It is usually much easier and cheaper to avoid issues and problems with your hi-fi system before you buy it than to fix them after the purchase. If you prefer medical analogies, I have another fundamental principle for you:

> THE "PREVENTION IS BETTER THAN CURE" PRINCIPLE
> Proper equipment selection and matching before purchase leads to longer satisfaction and fewer problems after the purchase.

Synergy is not just some dirty word from boring management seminars

The ultimate goal of every audiophile or audio system integrator is an integrated or coherent whole. Some call it system synergy and express it as "1+1=3". We all know that mathematically that makes no sense, but life is not mathematics. What that intriguing equation is saying, we need to select and combine the components so that the whole system sounds better than we would believe possible when listening to or evaluating each component in isolation or as part of some other, less optimized system.

Proper component matching is the first step towards system synergy. Again, what "properly" means is open to interpretation. One possible definition would mean that components are combined so that the result (the sound of that particular system in that particular room) is pleasing to the audiophile.

Synergy is one of those concepts that are hard to describe and even more difficult to define, but you will recognize its presence when you hear it in an audio system. More often than not, you will be immediately struck by its absence. More on synergy later on in the book.

I don't have enough money to buy the whole system at once, how should I prioritize?

This is one of the most common yet most difficult problems to solve and probably the main reason audiophiles buy and sell audio gear so often. We all have to start somewhere, and one or more audio components in your setup will certainly be inferior to others.

The critical task of every audiophile is to impassionately keep analyzing their system, identifying the current bottleneck (the weakest link), and upgrading equipment in that priority order.

This is akin to a moving target shooting practice; as soon as you upgrade one component (say an amplifier), another link in the signal chain will show its limitations, for instance, a mediocre turntable or speakers. That is why audio system optimization is a never-ending process. If only the exasperated wives (spouses, to be politically correct) could understand such irrefutable reality.

You can go in the downstream direction, buying the best CD player or turntable you can currently afford, then a preamp, power amp, and finally the speakers. Or you can go upstream, buying the speakers first, the power amp second, and so on.

I would take the latter path and buy the speakers and the amp first. Even the best turntable, CD player, or music server in the world will not sound very good if hooked up to a $500 amp and $800 pair of speakers.

However, from personal experience, a decent "budget" CD player or music streamer will sound fine with a great amp and an excellent pair of speakers. That will give you time to save enough to buy the source component that will be a worthy pairing with your first-class amplifier and loudspeakers.

HOW TO AUDITION AUDIO COMPONENTS AND SYSTEMS

Critical listening versus listening for pleasure

There are two types of listening, listening for pleasure and critical or purposeful listening. Critical listening is predominately a left-brain, logical and deliberate activity, an analytical process during which you pay close attention to and analyze various aspects of the musical performance.

That can be from a musical standpoint, analyzing the artist's performance, or from the equipment point-of-view, judging various aspects of the audio component or system you are evaluating. In this book, we are only interested in the latter kind.

Listening for pleasure is a more emotional, right-brain activity, where we tend to (or at least try to) relax and enjoy the pathos, the emotional aspect of music as a whole. If any aspect of the system's performance is substandard or draws attention to itself and thus stands out, the illusion is spoilt and the overall enjoyment diminished.

For instance, if the bass is muddy or boomy (one-note mess) or the treble is harsh and irritating, it is almost impossible to "roll with the flow" and let yourself be carried away by the musical envelopment.

Critical listening is invaluable when auditioning a piece of equipment, for instance, an amplifier, loudspeakers, or turntable you are considering buying. We then deliberately try to "take apart" or compartmentalize various aspects of its sonic performance and analyze them individually.

That could be a sense of pace and rhythm, bass, midrange and treble balance, macro and microdynamics, imaging and soundstage 3-dimensionality and stability, or a myriad of other vital factors.

When listening for pleasure, we take in only the overall musical illusion that the component or system creates, how it makes us feel and how much enjoyment it provides. It is like falling in love (or lust) with a person; normal people don't deliberately look for that person's flaws and perform endless psychoanalysis on them.

The danger in critical or purposeful listening

In critical or purposeful listening, we pick apart the reproduced sound into its different aspects, paying attention to one at the time. Human attention span is limited. The attention cannot be in more than one place at a time.

The problem with purposeful listening is that we hear what we pay attention to. The more we concentrate on one particular aspect of the audio system's or component's performance (soundstage, transparency, microdynamics, bass definition, midrange smoothness, etc.), the more change we notice.

It is almost impossible to distinguish if, say, component A sounds better (in that regard) than component B, or are we hearing a change (for better or worse) just because we are listening differently, more attentively.

Listening for pleasure is more of a relaxed right-brain activity, emotional and holistic. In that frame of mind, details or aspects of the sonic reproduction become noticeable only when they lack in some sense (weak bass, veiled soundstage, poor microdynamics), or when the audio system overemphasizes some of them, making them too prominent, so they stand out like the proverbial sore thumb.

That could be harsh treble, boomy, muddy, one-note bass, too sweet or syrupy midrange, lack of dynamics and punch (lackluster dynamics), and a myriad of other audio ailments.

When auditioning audio components, we need both types of listening. I usually listen holistically first to determine the overall character of the audio system or a component and then focus more and more on individual aspects of its performance (purposeful listening for specific issues).

The "Experimenter Expectancy Effect"

Confirmation or expectation bias is a tendency of experimenters (researchers and audiophiles) to look for or emphasize information that confirms their hypothesis (expectation) while overlooking the information that contradicts it, resulting in incorrect interpretation of the results and invalid conclusions. One way of increasing research validity is using double-blind experiments, known in audiophilia as "double-blind listening tests."

The experimenter expectancy effect is a similar social research phenomenon, where people (participants) are involved. The researcher's or experimenter's expectations can influence participants' behavior, which increases the likelihood that they will respond in a way that confirms the experimenters' hypotheses or expectations. In other words, subtle (and often not so subtle) clues from the person running the experiment are picked up by test subjects, bias their choices, and compromise the research results.

This phenomenon is also alive and well in audiophilia, where the prevalent attitude is not the skeptical "When I hear it, I'll believe it!" approach, but a rather subjective "When I believe it, I will hear it!" However, the personal preferences of an audiophile do not have to be statistically valid, as research results must be.

If you believe that a cable swap, tube substitution, or any other change in the audio chain makes a positive difference (and thus constitutes an improvement), and if you are prepared to pay for it, who am I to argue?!?

AUDIOPHILIA IS NOT A SCIENTIFIC DISCIPLINE

For most of us (except audio reviewers, equipment manufacturers, dealers and others in the audio business), audiophilia is a hobby, an enjoyment or a pastime, not a scientific or a rigorous research discipline, so personal preferences of audiophiles do not have to be statistically valid, as research results do. Ultimately, there is no point arguing about personal preferences.

The adaptation problem

In our amplifier-making days, when trying new designs, as soon as my dad and I would finish a prototype, we'd perform the initial listening test. To me, that was the most important 5-10 minutes of the whole process, for the amplifier's qualities and weaknesses would usually be immediately apparent (well, most of them).

For some reason, the longer I listened, the more I'd get used to its sonics, and while its outstanding qualities would remain noticeable, somehow its shortcomings would become less and less so. It was as if my brain would accept such "negatives" and then disregard them while focusing on the outstanding aspects, the positives. Only after switching to another amplifier and not listening to the first one for a while would the difference in their sonics become apparent again.

The same happened with room acoustics. When visiting prospects to try one or more of our amps with their systems, the listening room's sonic character (mostly deficiencies and problems) would be revealed immediately. However, the more time I spent there, the more I'd get used to how the room sounded.

It seems the owners of those audio systems suffered from the same syndrome - most of those audiophiles were blissfully unaware of how ordinary even bad their rooms and audio setups sounded. That would be obvious only to someone with a "fresh pair of ears," listening to such audio systems for the first time.

Some psycho-acoustic adaptation mechanism is definitely in play here, and I mention it here mainly because this syndrome impacts your auditioning.

Both short-term and long-term listening evaluations are necessary

The common audiophile wisdom is that only long-term evaluations are credible, while initial impressions are fickle and misleading and should be disregarded. Sure, longer-term evaluations are necessary because some shortcomings only manifest themselves after prolonged listening.

However, due to the adaptation problem just mentioned, much more attention should be paid to the initial, immediate impressions when hearing a system or audio component for the first time. If the sound strikes you as being "wrong," or if it doesn't immediately appeal to you, you should ask yourself why that is the case and try to remember that initial feeling for later analysis.

When I evaluate tube amps and preamps (my specialty) or when I advise audiophiles on their room acoustics, I only listen to a few of my test pieces of music while my ears are still sensitive to the new environment, before I get used to the room's acoustic shortcomings and peculiarities. I suggest you do the same in your auditioning, for after a while, your mind will acclimatize, and you'll become impervious to various acoustic anomalies of the room you are listening in or of the equipment you are auditioning.

> **LONG-TERM EVALUATIONS OR THE INITIAL IMPRESSIONS?**
> The common wisdom is that only long-term evaluations are credible. However, while they are necessary, much more attention should be paid to the initial, immediate impressions when hearing a system or audio component for the first time due to the adaptation problem.

The danger of quick evaluations and A-B comparisons

Some audiophiles rely on initial impressions and immediately noticeable differences between the components under evaluation. However, short-term comparisons can be misleading because the sonic signature of one audio component can and does bias the listener's perception of the other. For instance, in an A-B comparison, if one preamplifier is very much on the bright side, another may sound warm and mellow in comparison, while, in effect, it isn't. Once brought in and connected to your system, that preamp that sounded "soft and subtle" at the dealer's or private seller's setup may still sound too harsh or edgy for your liking.

Another problem with relying only on quick comparisons and immediate impressions is best illustrated with an example from interpersonal relationships. Have you ever met a person who you immediately liked, but the more time you spent with them, and the more you got to know them, the less you liked them? That happens to audiophile components and systems, too.

It also works in reverse. You may not like the sound of a particular sound that much initially, but the more you listen to it, the more it "grows on you." To continue with the analogy of the relationship, I've met a few women that initially I wasn't attracted to; they didn't seem that beautiful or interesting, but the more I got to know them, the more I liked them, and the more attractive I'd found them.

As with people, the sonic qualities you were attracted to initially often irritate you in the long run. For instance, the more detailed, transparent, or revealing a component or system sounds initially, the more likely it is that you will experience the listener's fatigue later on and perceive it as harsh, bright, and irritating in the long term.

What should we listen for?

Models are helpful visual and logical tools that help us organize factors, aspects, and issues. However, they are highly subjective, seldom comprehensive, and either too complex or too simplistic. In most cases, they are not scientifically validated but simply a personal view of their creator.

On the next page, you have a suggestion for a "Listening Evaluation & Auditioning Sheet" you could adopt and use when auditioning equipment. It is based on the 5-factor model. The factors I chose are tonality, transparency & resolution, imaging, dynamics and musicality.

Alternatively, feel free to devise your own, using criteria you deem to be the most important. What evaluation factors would you choose and why?

Don't worry if your model does not make complete sense or if there are overlaps, as mentioned. Not even a Ph.D. thesis in acoustical engineering would be sufficient to cover all these issues in a meaningful way. A much broader and thicker treatise is needed, one that I have no intention of writing. Audio should be experienced through listening, not pontificated upon or studied in minute detail!

Change (evaluate) only one component at a time

When I said that audiophilia is not a scientific discipline, that didn't mean we should forget about sound (no pun intended) engineering and logical principles and methods and go about equipment evaluation and selection in a haphazard, willy-nilly way.

One of these universal principles forms a bedrock of all scientific evaluation and research, which is the principle of a singular cause. The only way to show that a certain factor, cause, or independent variable "X" has a statistically valid impact on the result or dependent variable "Y" is to 1) hold all other possible factors that could impact on Y constant, and 2) change only variable X and observe (measure) what happens to Y.

The same kind of scrutiny and methodological consistency must be present in audio evaluations and comparisons so that only one factor is changed at any time. For instance, don't change an amplifier and its power cord or speakers and speaker cables together. Remember, cables are components of an audio system, and they do affect its sound (sometimes significantly), so you will never know how much each of the two changes contributed individually.

Interconnects and power cables could be an exception to this rule. Due to their cumulative effect, it is easier to notice sonic changes if all interconnects and all power cables are changed simultaneously. Of course, all interconnects, and power cables should be identical (of the same type).

LISTENING EVALUATION & AUDITIONING SHEET

TONALITY

SPECTRAL BALANCE	1	2	3	4	5	6	7
		bass dominant		balanced			top heavy
FREQUENCY RANGE	1	2	3	4	5	6	7
		narrow		average			wide
COHERENCY	1	2	3	4	5	6	7
		disjointed		fully integrated and coherent			
NOISE-HUM-HISS	1	2	3	4	5	6	7
		noticeable		quiet			not audible

TRANSPARENCY & RESOLUTION

DETAIL RETRIEVAL	1	2	3	4	5	6	7
		low		average			high
CLARITY	1	2	3	4	5	6	7
		hazy & fuzzy		average			expansive
RESOLUTION	1	2	3	4	5	6	7
		low		average			high
SIBILANTS	1	2	3	4	5	6	7
		harsh & irritating		noticeable			smooth

IMAGING

SOUNDSTAGE	1	2	3	4	5	6	7
		narrow & shallow		average			expansive
LOCALIZATION (FOCUS)	1	2	3	4	5	6	7
		poor		average		pinpoint accuracy	
LISTENER ENVELOPMENT	1	2	3	4	5	6	7
		poor (narrow front)		front & sides			all around

DYNAMICS

DYNAMIC RANGE	1	2	3	4	5	6	7
		too compressed		unremarkable			wide
TRANSIENTS	1	2	3	4	5	6	7
		slow & sluggish		average			fast
BASS CONTROL	1	2	3	4	5	6	7
		loose & boomy		average			fast & tight

MUSICALITY

EMOTIONAL INVOLVEMENT	1	2	3	4	5	6	7
		low - lacks passion		so-so		high - enjoyable	
NATURALNESS	1	2	3	4	5	6	7
		artificial sound		hi-fi sound			sounds "live"

NOTES & COMMENTS:

_____

The piano test

We all have our preferred types of music, and I genuinely believe that specific audio components reproduce certain types of music better than others. For instance, simple jazz or acoustic music sounds best on low-powered triode amplifiers. Their renditions of human voice no other amplifier technology can match.

However, play large-scale orchestral works at loud levels, as they should be played, and many such amps start falling apart (not literally, of course). It is hard to recommend certain types of music to use as test pieces for those two reasons. You will have to devise your own playlist of test songs. The music streaming services make that super easy; you no longer have to burn a test CD and carry it with you.

Personally, no matter how much I like the sound of an acoustic guitar and other acoustic string instruments such as mandolin, these aren't as useful as test pieces as some other instruments, mainly piano and some brass and woodwind instruments.

A guitar sounds nice and tonally rich on most audio systems. Still, a percussion instrument such as acoustic piano will separate the champions from "also runs."

While bass notes on a piano are sounded by its hammers striking a single, thick string, the treble notes are produced by three strings to add energy and power that a single string could not generate. Since the three strings are never identical and the hammer cannot strike them simultaneously, this adds a specific tone and harmonic character to each piano. An audio system that can reproduce such nuances is a sure winner.

A human voice, both male and female (different tonal registers and thus testing different frequency bands of an audio system), are also indispensable when testing the midrange reproduction.

Solo instrumental pieces or longer passages within orchestral tracks (solo flute, trumpet, saxophone) are the most revealing. Some pros in the audio and room tuning business also like to use solo violin and cello recordings.

The shape of music - ADSR parameters

Each musical instrument or a family of similar instruments has a distinctive sound "envelope" or "pattern." The envelope generally consists of the attack, decay, sustain, and release sections.

The attack is a sharp transient between the moment a piano key is pressed (the percussive effect) until the sound reaches its maximum intensity (peak). The sound then decays (reduces in volume or loudness) until it reaches a certain sustain level, during which it stays more or less constant (or drops gently).

The release is the tail of the sound wave that fades to zero (silence) once the key is released.

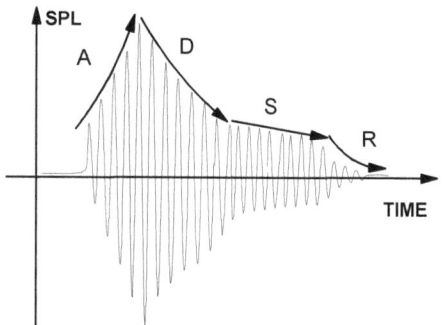

Fig. 2.4. The generic sound envelope illustrating its four sections - attack, decay, sustain and release.

The attack transient is often short & sharp; it's the best predictor of an audio component's fidelity. The length of the release is also useful; on superior systems, it should not end abruptly but fade away gently. Learn to recognize both and listen for them in equipment evaluations.

Test recordings - CDs and LPs

There are dozens, perhaps even hundreds of test CDs and LPs in circulation. Apart from record labels such as *Chesky, Opus, Groove Note*, and *Sheffield Lab*, numerous hi-fi manufacturers have also released their own demo/test CDs over the years. *Marantz High-End Audiophile Test Demo CD, Denon HiFi Check CD, McIntosh Demonstration Reference Disc*, and S*onus Faber Test and Demo CD* are just a few in a long line.

Audio magazines such as *Stereophile* and *Hi-Fi News and Record Review* are additional sources of test records and CDs. As a rule, while most test CDs contain a few selected music pieces, some also include technical tracks, such as pink noise bands, 1/3 octave bands, sweep tones, sine, and square wave spot frequencies, equipment burn-in signals, solo instrument, and vocal tracks (piano, guitar, flute, saxophone, drums, bongos, violins, cellos, and violas, etc.) and various other tests (bass resonance, dynamics, imaging, resolution, etc.) and even sound effects.

On the other hand, test LPs mostly contain test signals to be used for various turntable and cartridge adjustments. For instance, the *Hi-Fi News and Record Review* test record has the following tracks: Channel Identification, Phasing (Voice Alternately In Phase And Out Of Phase), Pink Noise -20dB Left only, Pink Noise -20dB Right only, Bias Setting (300Hz tone L+R @ +12, +14, +16 and +18dB), three tracking ability tracks, Cartridge/Arm Lateral Resonance Test (sweep from 25 down to 5Hz), Cartridge/Arm Vertical Resonance Test (sweep from 20 down to 6Hz), Cartridge Alignment (Azimuth) Test, and Residual System Noise (unmodulated grooves section).

The pairing problem

One of the main issues in auditioning and reviewing audio components is pairing; in other words, the rest of the system used to evaluate a certain audio component. Should we use other components in the same price range as the auditioned device or more (often much more!) expensive systems? Is auditioning a $2,500 amplifier with $10,000 speakers and a $12,000 CD player justified?

Some say such "hi-end" or at least high-priced paired components are more likely to showcase the true capabilities of the amplifier and also to expose its flaws and limitations. Others would disagree, claiming that an audiophile considering such a purchase is more likely to have or plan to acquire $3,000 speakers and $1,200 CD player, so reviewing the same amplifier with components in that lower price range is more honest and realistic.

I think you should audition audio components priced slightly (up to 50% more expensive) and even significantly above your budget, those priced at twice or three times your budget.

That way, you will gain a feeling for and understanding of what is sonically possible. As you work your way down towards your target budget, you will do some soul-searching and decide on what aspects of the audio performance you are willing to compromise and which are non-negotiable must-have qualities.

We are, of course, assuming that a $13,000 amp you are using as a benchmark is significantly better sounding than the $4,000 amp, which is within your budget and which you are auditioning. Surprisingly, that isn't always the case. In that instance, you will be ecstatic that your $4,000 choice sounded so close and even better in some aspects than the $13,000 benchmark.

The listening order

The same thinking can be extended to auditioning when buying a whole audio system at once. Let's imagine that you are one of the lucky ones to live in a city that still has a few audio dealers left in business. You visit all three. For some reason, the first one you dislike immediately and walk out. The other has a strange product range, nothing that appeals to you, but the last one has three auditioning rooms, beautifully set up with brands and models that make you drool and wish you were rich, single, or both.

That dealer has purposefully anchored each room in a specific price range, let's call them The Low, The Mid, and The Top Room. Don't get stuck on the figures; they are just examples. Just as paying $6,000 for a budget system in the Low Room seems expensive and is out of the question for some, $46,000 for the no holds barred system in The Top Room is laughably cheap to those whose audio budget could extend to a few hundred thousand dollars.

We have already decided that you should visit and audition all three systems, so the only question now is in which order you should do so? There are six possible sequences, 1-2-3, 1-3-2, 2-1-3, 2-3-1, 3-1-2, and 3-2-1, but since this isn't a course on probability and statistics, and I'm not some sadist to torture you with all such permutations and thus waste your precious time, lets narrow our choice down to the two most likely ones.

Would you start with The Low Room and finish with The Top Room (1-2-3) or the other way around (3-2-1)? What are the pros and cons of each option? Does such listening order matter at all?

I know I'm about to disappoint you (please don't return this book and ask for a refund, I have children to feed), but my star sign is Libra, and the most challenging thing for us Librans is to make decisions. Since we have that gift (or a curse?) of seeing all sides of an issue, we often agonize for hours and then decide that we want to keep sitting on the fence and that the best and safest decision is not to decide at all!

After hearing The Top Room, some audiophiles (I had a friend like that) would never be satisfied with The Low and even The Mid Room. They'd go home without buying anything. Others would be able to put things into perspective, thus making the order less relevant. They'd most likely start with The Low and progress onto the Mid Room, and, then, time and curiosity permitting, perhaps even spend twenty or so minutes in The Top Room.

If I had to choose between the two, I'd choose the 1-2-3 sequence. That way, knowing that you can afford the system in The Low Room, you can justify spending a considerable time there to get a good feel for the "baseline," and then, once in the Mid and Top Rooms, only listen to certain sonic aspects where they are superior, or where the system in The Low Room did not meet your expectations.

"But what if my budget is in The Middle Room ballpark or higher?" you may ask. We have already answered the first possible question. You should spend some time in The Top Room anyway, to see what you'd be missing on. How about The Low Room, should you "waste" any time in it? Again, it depends on your knowledge and experience as an audiophile. If you've listened to many such basic systems, skip it, you know what to expect. Otherwise, spend some time there to put your mind at ease, as a reassurance why you are spending 2-3 times as much.

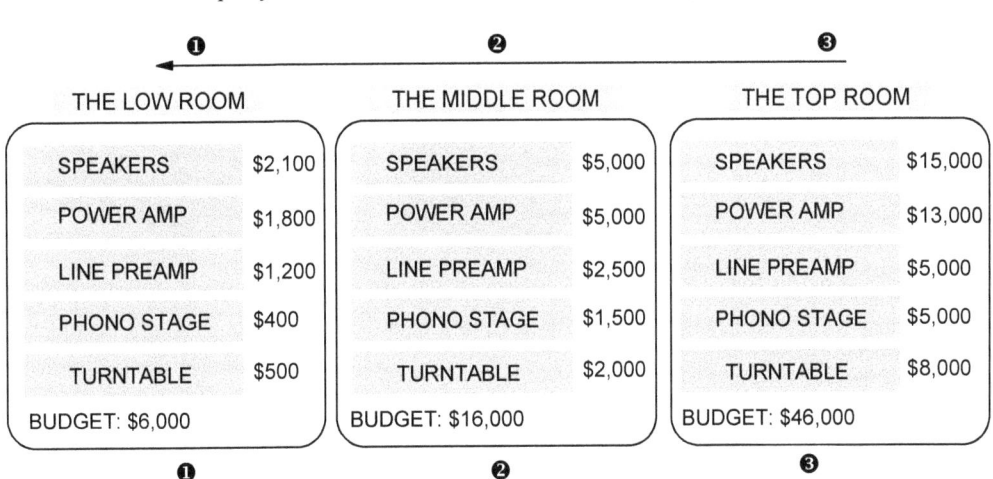

Fig. 2.5. When buying a whole audio system, should you listen to all three setups, and if you should, in which order?

How to ensure consistent loudness levels through A-B or A-B-C evaluations

You may have heard or read about A-B listening tests and the oft-repeated rule that no matter what component is being evaluated (amplifier, preamplifier, CD player, phono stage, etc.), the listening tests must be done at the same level of loudness. So, in serious A-B or A-B-C tests, it isn't enough to swap amplifiers and roughly adjust the volume. A tone generator, a test record, or a test CD must be used (with calibrated tones), together with an SPL (Sound Pressure Level) meter to ascertain that is the case.

Assuming we are comparing amplifiers, the conventional wisdom says that the louder amplifier will sound better. Is that really the case? It will probably sound more "confident" (How do you measure confidence in a piece of equipment?) and even more dynamic, but what about distortion?

About 250 years ago, Italian composer and violinist Giuseppe Tartini discovered an important psychoacoustic phenomenon. When two notes of different frequencies are played simultaneously, a listener can "hear' additional tones whose frequencies are the sum and difference of the two frequencies. We now call this type of distortion intermodulation distortion, which seems to be caused by the nonlinearity of the inner ear. In his book "On the Sensations of Tone as a Physiological Basis for the Theory of Music," published in 1862, Hermann Helmholtz showed by experimental and mathematical analysis the existence and nature of these combination tones and that they increase steeply with the rising intensity of the fundamental tones.

So, we could hypothesize that increasing volume levels would also sharply increase the IM distortion in our ears and make us perceive the louder amplifiers as harsher or more irritating, at least on a subliminal level. This could be compared to listening to some solid-state amplifiers, which sound nice initially, but soon listener's fatigue sets in, and the longer you listen, the more "uneasy" you feel. Instead of being drawn deeper and deeper into the musical pathos, as it happens with the best tube amplifiers, you are compelled to stop listening altogether.

To ensure identical volume levels during auditioning, mount an SPL meter on a standard camera tripod at the listening location and at tweeter height (most tripods have an adjustable height feature). Using your favorite songs, adjust the volume of the existing system to an average listening volume you usually listen to. Play a pink noise track from a CD or music streamer and measure the SPL.

After swapping component A (preamp, amp, speakers, cables) with component B, play the same pink noise track again and adjust the volume control on an amp or preamp until you measure the same SPL as before the change.

SPL (Sound Pressure Level) meters

SPL meters aren't expensive, and if you do lots of equipment auditioning and buying and selling of audio gear, you should get one.

An electret condenser microphone (1) converts the sound pressure into voltage, which is then amplified, filtered, and displayed on an analog or digital display (2). The range switch (3) indicates the dB figure that must be added to the meter's reading (4), the analog meter on the right indicates around -1dB on the 60 dB range, meaning its reading is 59 dB.

The newer models display the overall figure directly (5), although the range switch or button still must be adjusted up or down. The accuracy's in the order of ±2 dB, which is more than adequate for our purposes as audiophiles.

At least two weighting options are offered (6), usually "A" and "C" and "Fast" and "Slow" display response (7).

For the new kids on the block, some apps do the same job, for instance, NIOSH Sound Level Meter app for iPhones, which can be used with both internal (not approved for "legal" measurements) and external (calibrated and approved) microphones.

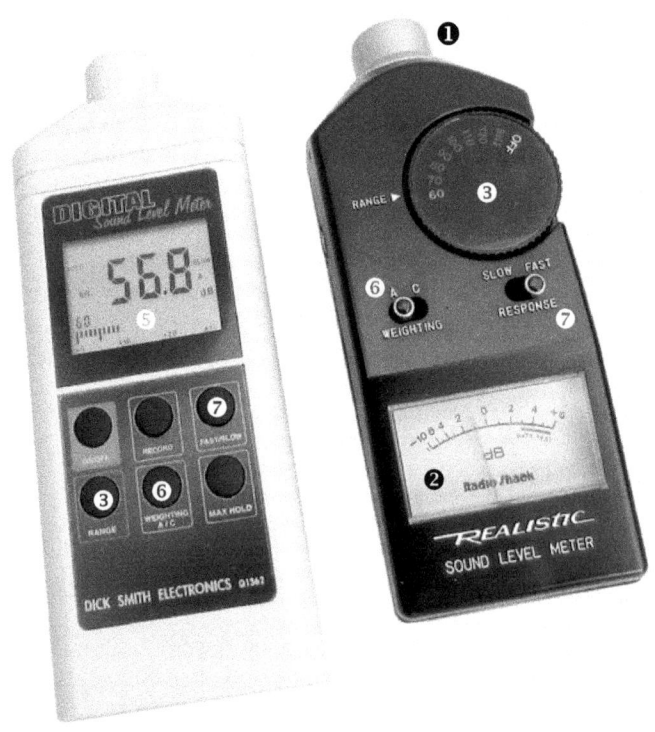

Fig. 2.6. With smartphone apps that can perform such measurements, both analog and digital SPL meters are now distinctly "old school."

A-weighting has amplitude-frequency characteristics that attenuates low and high frequencies and limits the meter's response to midrange frequencies where the human ear is the most sensitive. Thus, it is used mainly for noise checks in health & safety applications. There is also a signal output of 1 V_{PP} (open circuit, full scale at 1 kHz) with less than 2% distortion (4).

To give you a feel for the figures you will encounter in these tests, keep in mind that the sound intensity varies as the pressure squared. Ten times higher sound pressure level (second row from the top in the table below) results in a 100 times higher intensity ratio, expressed as a 20 dB.

Sound pressure ratio	Sound intensity ratio	dB
1:1	1:1	0
10:1	100:1	20
100:1	10^4:1	40
1,000:1	10^6:1	60
10^4:1	10^8:1	80
10^5:1	10^{10}:1	100

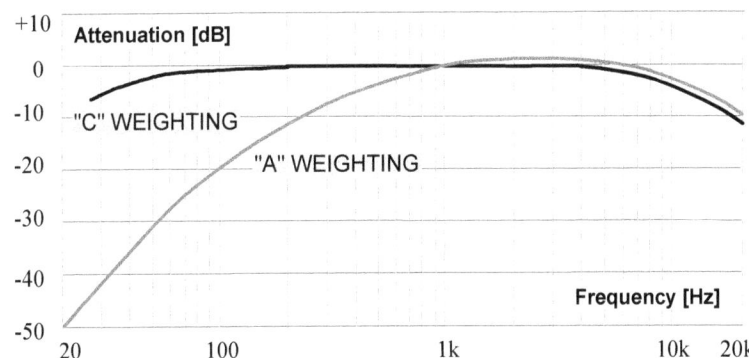

Fig. 2.7. Sound pressure and sound intensity ratios and their decibel equivalents.

Fig. 2.8. The A-weighting attenuation curve is mostly used for health & safety noise checks, the C-curve in audio tests

TIME TO BUY: HOW TO GET THE BEST DEAL ON AUDIO GEAR

To be a good negotiator, you must understand the money side of the business or individual you are dealing with. Asking a retailer whose profit margin is 20% for a 20% discount means he will lose money on that deal and is likely to perceive your offer as insultingly low. However, if a retailer has a wide profit margin, as audio dealers generally have (up to 50% of the retail price), asking for a 20% discount could be a smart move on your part.

Let's have a brief look at the ballpark figures of the audio industry.

How are audio components priced

As in any business, there are many retail models and pricing structures in the audio industry, so we can't cover them all here. Some manufacturers sell directly to customers, so they are the exporters, distributors, and retailers. Some retailers are also the distributors of specific brands, so those two frames should be merged in the diagram below.

Different product categories often have other pricing structures and profit margins. So, for instance, a dealer may be making much more money on the sales of cables and accessories than amps and speakers. In short, please understand that the following is a gross simplification, a ballpark only applicable to a typical 3-party model.

The standard rule-of-thumb pricing model (1) says that the recommended retail price (RRP) is double the wholesale price (WSP), or what a distributor will charge the retailer, meaning the retailer doubles the wholesale price.

Another model (2) says that the retail price is roughly five times the manufacturing cost. In our case below, it costs the manufacturer $2,000 to produce a certain component (amplifier, speaker, turntable). To that, he adds his profit, arbitrarily chosen here to be $1,000 of 50% of the manufacturing cost (50% markup), arriving at the manufacturer's price (MPR).

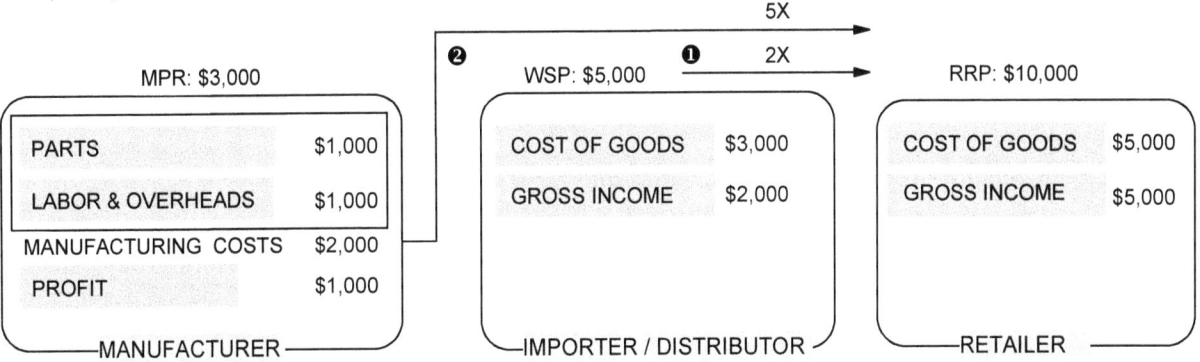

Fig. 2.9. A simplified example of the audio retail chain, illustrating the two standard pricing methods as applied to a $10,000 audio component

In this case, the importer/distributor adds $5,000/$3,000 = 1.67 or 67% on top of his cost (what the manufacturer is charging him). You may be thinking now, "Well, the dealer has $5,000 to play with; I can push him for a $3,000 or even $4,000 discount!". However, if he sells it to you at $10k, the $5k is his gross income. From that, he has to pay all fixed and variable costs such as wages, insurance, rent, utilities, cleaning and maintenance, bank interest, and all other expenses.

Now, notice that I've said "if" he sells it to you at that price. Most dealers accept trade-ins because prospective buyers demand them. So, he may give you a $3,000 trade-in on these $10,000 loudspeakers and is now stuck with your old speakers. It may take months to sell them, and if he miscalculated (misjudged) their resale value, he might even have to sell them at a loss, say for $2,000 instead of $3,000 he "paid" you.

How to negotiate with audio dealers (retailers)

> "There is no such thing as a fixed price!"
> Igor S. Popovich in his book "Loser No More!", subtitled "Negotiate better and win more often - at home, one the job and in business"

In Australia, many products have the so-called Recommended Retail Price (RRP), which is usually set by the manufacturer or importer. Of course, retailers can sell below or even above that price; notice the word "recommended." It pays to view all prices that way as something flexible and open to negotiation.

Start low, you can always keep offering more, but you can never go in the other direction! Don't be embarrassed if you are perceived as a tough bargainer or even stingy by the dealer, so what? Gently and politely explore options and boundaries to see what you can get away with. Ask lots of "what if ..." type of questions.

Be patient and persistent; take your time. Even if you spend half an hour negotiating (usually such negotiations take less than that) and manage to reduce the price by $500, you have just been paid a thousand bucks an hour! For most of us mere mortals, negotiation is the highest-paid activity we will ever undertake.

Don't fall in love with a piece of gear; if you do, don't show it. If a dealer sees that you love what he's selling, forget about discounts and incentives and be prepared to pay the full price. Keep repeating to yourself, "This isn't the only deal around; there are plenty of other (amplifiers, turntables, speakers, ... whatever you are buying). If I don't get the price down to what I want to pay, I am prepared to walk away."

As for the attitude you need to project, if you look timid, confused, gullible, or too easygoing, you will be treated accordingly - as a sitting duck, an easy prey. To be treated with respect and professionalism, be informed (or at least seem so), look determined and demand the best service and the best price.

Why would audio retailers (dealers) reduce their price?

- To get you as a new client, expecting future purchases from you
- To clear stock, slow-moving, obsolete, discontinued, superseded (by a new model), demo unit
- To improve their cash flow. They may have urgent bills to pay, or the business is very slow, so better to sell it to you even at no profit at all (at cost) than to lose the sale.
- As a quantity discount. If you and two other friends place an order together, demand a higher discount.
- To achieve their sales quote. Some manufacturers give dealers stepped quantity discounts, so they have to sell a minimum number of units or a minimum total value per month or quarter to get the same or higher discount next month or quarter (the "accounting period"). They may be one or two units short, so again, better to sell it to you even at cost than to lose the sale and with it the substantial future discount from the manufacturer.
- To win the price war with a competitor and increase their market share. This is less likely in the audio business but is a compelling dealer motivator with other goods and services.

Understand the big picture - value dealer's time, effort and expertise

If you think I am advocating that you squeeze audio dealers on price no matter what, you are wrong. Short of moving your lawn or marrying your ugly sister, some will do anything to make you and keep you happy. They will spend hours demonstrating things to you, advising you and guiding you through the auditioning and selection process. Many will even be glad to load their wares into their cars or vans and bring them to you, to loan you components for extensive auditioning, and to pick them up afterward. You have to value such efforts and, if you will, "price" it in your calculations.

3 | WHAT DO WE LISTEN FOR AND WHAT DO WE ACTUALLY HEAR?

Each material used in audio equipment has different mechanical, electrical, and acoustic properties, and so do electronic components - tubes, transistors, resistors, capacitors, transformers. This simple and indisputable fact may explain why different audio technologies sound different (tubes vs. transistors, turntables vs. open reel recorders vs. CD players, electrodynamic vs. electrostatic speakers, etc.)

Different materials and audio technologies distort and compress audio signals in different ways, which are two of the most significant factors that determine how an audio component or a system using such materials and technologies deals with various audio frequencies and fast-changing signals, and, ultimately, how it sounds.

A wide dynamic range, the absence of dynamic compression, preserved harmonic structure, and the faithful reproduction of fast transients are identified and discussed in this chapter as the key factors determining audio reproduction fidelity and how close we get to the illusion of live music.

- HUMAN HEARING 101
- DISTORTION - THE GOOD, THE BAD AND THE UGLY
- WHY DIFFERENT AUDIO TECHNOLOGIES SOUND DIFFERENT
- WIDE DYNAMIC RANGE: KEY FIDELITY FACTOR #1
- PRESERVED HARMONIC STRUCTURE: KEY FIDELITY FACTOR #2
- REPRODUCTION OF FAST TRANSIENTS: KEY FIDELITY FACTOR #3
- DYNAMIC COMPRESSION: #1 PROBLEM IN SOUND RECORDING AND REPRODUCTION
- THE SONIC SIGNATURE OF MATERIALS
- THE LISTENING ENVIRONMENT AND ITS IMPACT ON THE ENJOYMENT OF MUSIC

> "Observations have gone far beyond our ability to explain. We haven't found any measurements to describe what we are hearing."
> J. Gordon Holt, founder of *Stereophile* magazine

HUMAN HEARING 101

The ear as an amplifier and A/D converter

The human ear and associated audio "circuitry" is a complex and sensitive instrument. While its frequency range spans the full four decades (from 20 to 20,000 Hz), its dynamic range is even wider; it can detect and process sounds from 0 dB to 120 dB, a full six decades!

Our auditory system is comprised of three main parts or subsystems. The external or outer ear starts with the earlobe or pinna, the visible part of the ear. It collects sounds and passes them through the ear canal to the tympanic membrane, a.k.a. eardrum.

The ear canal is akin to a resonant tube. About 25 mm (1") long, it resonates at a frequency of around 3-3.5 kHz, making the ear most sensitive in the midrange.

The eardrum is a vibrating membrane whose task is to convert sound pressure variations into mechanical motion (1-2 mm excursions).

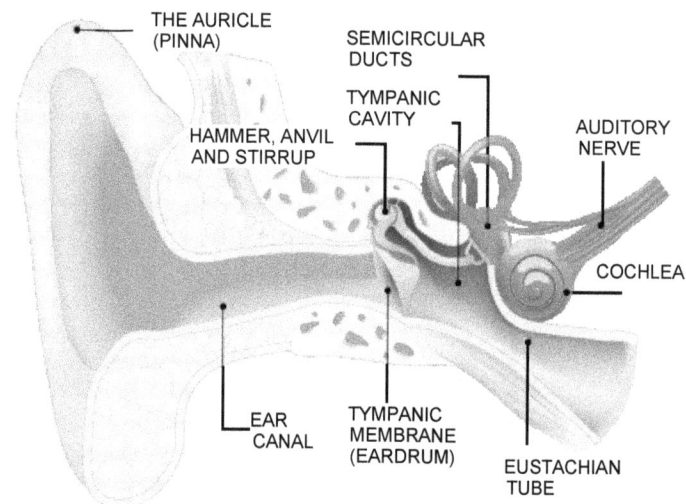

Fig. 3.1. The simplified anatomy of the human ear, with the main components marked and named

The middle ear consists of the bones called the ossicles (the three are popularly known as the hammer, the anvil, and the stirrup) located in an air-filled tympanic cavity. In contact with the eardrum, these bones act as a mechanical lever system that further amplifies the force from the eardrum about 30 times.

The inner ear is made up of the cochlea and the auditory nerve. The cochlea is a complex coil-shaped organ (2 and 3/4 turns) filled with fluid. When subjected to pressure changes from the middle ear, it produces an electrical signal, which is then transmitted to the brain through the auditory nerve. That nerve is actually a bunch of 4,000-odd nerve fibers, and the signal the cochlea produces is not a continuous or analog signal but a series of short and sharp pulses. The greater the SPL (sound pressure level), the higher the number of pulses sent to the brain.

The cochlea could be thus considered a sophisticated analog-to-digital (A/D) converter, converting analog pressure changes into digital electric pulses sent to the brain.

Why do audio systems sound different at lower volume levels?

The "equal loudness" or Fletcher-Munson curves, illustrated here, show how human hearing varies with signal frequency and sound pressure levels (SPL). In audiometry, the unit of loudness is a phon, but in acoustics, a decibel (dB) is used. The curves for various phon levels have a similar shape. Two things are apparent.

Our hearing is frequency-dependent. Human ears are most sensitive in the frequency range between 500 Hz and 5,000 Hz, where most of the human speech content falls. This is the critical "midrange."

The highest sensitivity is around 3.5 kHz (1). Thus, audio designers should strive to minimize the distortion in the midrange. Their magical midrange could explain the fact that single-ended triodes amps have a relatively large following.

The "magic" and "warmth" of triodes are mostly (but not exclusively as the objectivist brigade would like you to believe) caused by pleasant-sounding 2nd harmonic distortion.

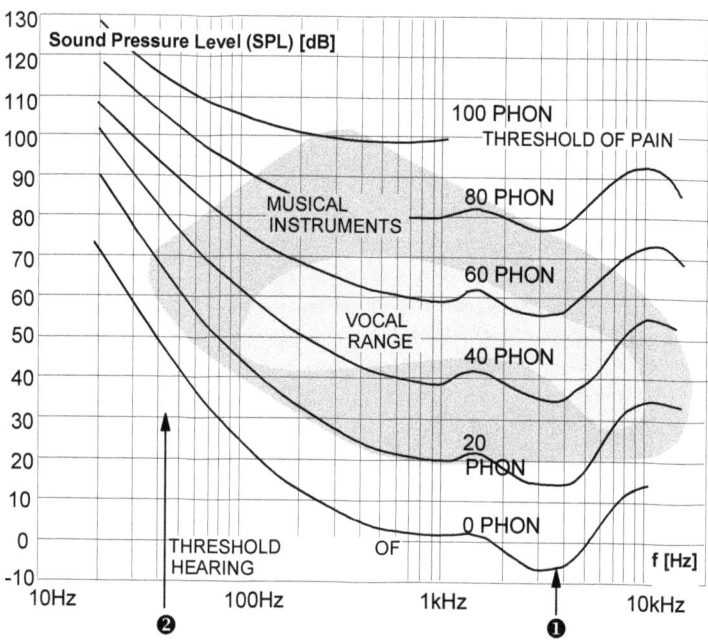

Fig. 3.2. The equal-loudness curves, ISO226:2003 revision of Fletcher-Munson curves

Secondly, as the frequency of sound drops towards the bass region, higher and higher SPL levels are needed to achieve the same perception of "loudness" by human test subjects (2). These curves were derived by asking people to adjust a volume to achieve the same subjective level of "loudness." Bass usually needs to be boosted at lower volumes to restore the original balance across the frequency range.

The same conclusion applies to higher frequencies, although to a lesser extent. As a result, bass frequencies need to be amplified more than the midrange signals to achieve the same perception of loudness, especially at lower loudness levels. This becomes a factor of great annoyance to audiophiles, whose hi-end amplifiers and preamplifiers usually have no tone controls, so it is impossible to compensate for this nonlinearity. Thus, even the best audio system will sound "thin" and lacking in the bass at low levels.

The importance of auditioning at your usual listening levels

I've just returned from a visit to a local hi-end audio shop. Two setups were proudly displayed. Very expensive Sonus Faber speakers were driven by Audio Research tube amps (monoblocks) in one while a McIntosh hybrid amp (tubes in the preamp stages, transistors in the output stage) was the pièce de résistance in the other. Both pairings sounded "clean," but ordinary, even anemic. Both amps were of push-pull persuasion with plenty of power, so the lack of dynamics or emotional engagement was surprising. Perhaps it was the music played or my mental state at the time, which reminds me of another issue.

Many speakers need a certain amount of power to "come alive"; they don't sound their best when played at low volumes. So, if you listen primarily at low volumes, you must find a speaker that will sound its best at such low levels. The drawback of some of those speakers is that they suffer from distortion and compression at higher volumes. You have to decide early on what is more important to you and search for speakers of that type.

THE "JACK OF ALL TRADES" MYTH
No audio component or system sounds equally good at all volume levels and with all types of music. A "jack of all trades" in audio is a myth; the reality is closer to the second part of that saying, "but the master of none!"

Daytime and nighttime sound? A proposition not as silly as it seems

Many audiophiles have noticed that their systems sound better in the evening and at night. "Better" means smoother, more detailed, with "darker" silent passages, and, ultimately, more musical and enjoyable.

There are three possible explanations, mechanical, electromagnetic, and psychoacoustic. The mechanical or acoustic factor is the background sound level (or rather noise level), which is usually much lower during the night. Neighbors, home appliances, outside traffic, and myriad other noise sources all quiet down in the evening and deeper into the night, making it easier to hear lower-level details and nuances from your speakers.

Electromagnetically, there is usually less EMI (Electromagnetic Interference) on the mains line, meaning a cleaner power supply to your audio components. The levels of RFI (Radio Frequency Interference) could also be lower.

Finally, we shouldn't ignore the psychoacoustic aspect of night listening. The two previous technical factors notwithstanding, it is possible that nocturnal music sounds better simply because we are more relaxed at night. Everything will sound better and more enjoyable when the listener is in the right frame of mind, and, likewise, even the best audio system will not sound as it should when we are stressed out, absent-minded or tense.

Ultimately, whatever the reason(s) for this variation in sound quality, the lesson here is to do with critical listening or component evaluation. For the reasons just mentioned comparisons between components or whole systems where one was listened to during the day and the other at night should be avoided.

DAYTIME AND NIGHTTIME LISTENING EVALUATIONS
Audition equipment at different times of day and when in different moods. Such situational factors are extremely important to form an accurate picture or impression of an audio component or system.

The visual distraction problem

Our hearing is affected by our vision, perhaps because our attention is divided or we may be distracted and influenced by visual clues. What we hear (and, perhaps more importantly, what we don't hear) is affected by what we are seeing. One possible explanation points to the common factor: our brain, which simultaneously processes visual and aural information (apart from many other tasks).

So, for critical listening and serious evaluations, always listen with your eyes closed and your mind relaxed.

DISTORTION - THE GOOD, THE BAD AND THE UGLY

Mechanical, electronic and acoustic distortion

An audio system is a serial chain of components. It starts with an analog (turntable or tape deck) or digital source (CD player, DAC, music streamer, etc.), collectively called source components since they produce the audio signal, which is then amplified through one or more amps and preamps before being fed to an electroacoustic transducer known as a loudspeaker. The speaker produces mechanical waves (changes in air pressure), so the direct sound from the speakers and the reflected sounds, contributed by the room, are finally combined in our ears, the final transducers. They convert mechanical energy (sound) back into electrical signals that get processed by the brain.

Each of these links distorts and adds its own "sonic signature" to the mix. These distortions can be *mechanical*, as with phono cartridges and unwanted vibrations reaching them and other audio components, *electronic* (inside amps and preamps), *acoustic* (loudspeakers and listening rooms are the highest distorting components of all) and even *aural* or *psychoacoustic*. Our ears, being imperfect acoustic detectors, and our complex hearing mechanism (the drum, cochlea, and all those inner workings) add their own distortion to which we are generally oblivious.

While there is a pleasantly sounding distortion that humans don't object to, and often even like, one wish of each audiophile, and indeed, one of the goals of this book is to reduce the sonically more malign types of distortion.

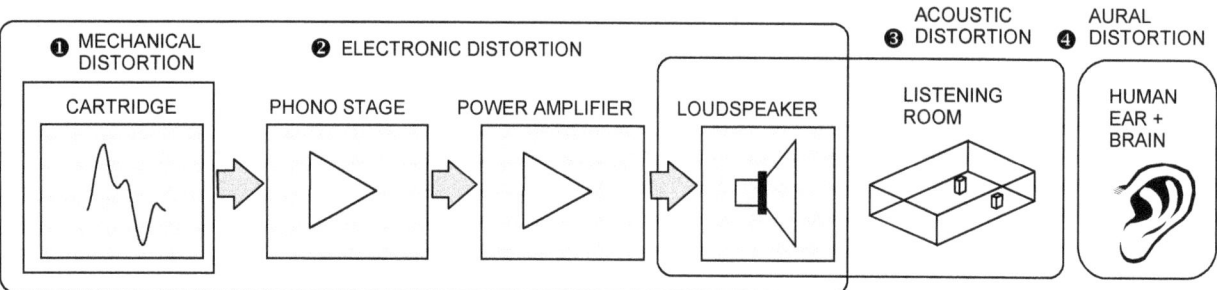

Fig. 3.3. Various types of distortion in an audio system.

Harmonic distortion (THD)

Assuming a simple sine wave signal at the input of a real amplifier (in contrast to an "ideal amplifier" that does not distort at all), the output will also be a sine wave but slightly distorted. Such a distortion cannot usually be detected visually, by observing the waveform on an oscilloscope, for instance, except in the cases of severe distortion. Nevertheless, should such a signal be brought to the input of a distortion or spectrum analyzer, the presence of new harmonics will be detected.

Real amplifiers generate harmonics due to their nonlinear input-output or "transfer" characteristics. For the transistor fanatics in our midst, bipolar transistors are even less linear than vacuum tubes. The most linear of all amplifying devices is a triode.

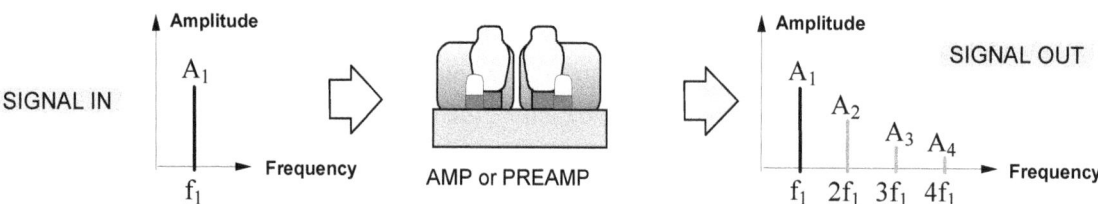

Fig. 3.4. Harmonic distortion is the generation of higher harmonics of the original signal.

Usually, the second harmonic (of twice the original signal's frequency) has the highest amplitude, followed by the third, the fourth, and so on. The relative amplitudes of various harmonics depend on the amplifying device (triodes distort differently from pentodes and beam tubes), the power level at which the measurement is taken, the design of the circuit (single-ended or push-pull), and many other factors.

Intermodulation (IM) distortion

To explain this type of distortion, let's look at the way it's measured. Two pure sine wave signals are mixed and fed into an amplifier. One is of a lower frequency f_1 (usually the mains frequency, 50 or 60 Hz), the other of fifty times higher frequency f_2 (for instance $f_1 = 50$ Hz and $f_2 = 2{,}500$ Hz). The amplitude of the lower frequency signal is adjusted to be four times higher than the amplitude of the higher frequency signal ($A_1 = 4A_2$).

A real (imperfect) amplifier will also generate two unwanted signals (or "sidebands") of the higher frequency signal. One will be of f_2-f_1 frequency (or 2,500-50 = 2,450 Hz in our case), the other will have a frequency f_2+f_1 (or 2,500+50=2,550 Hz). In severe cases of distortion, second sidebands will be also be generated, f_2-2f_1 and f_2+2f_1.

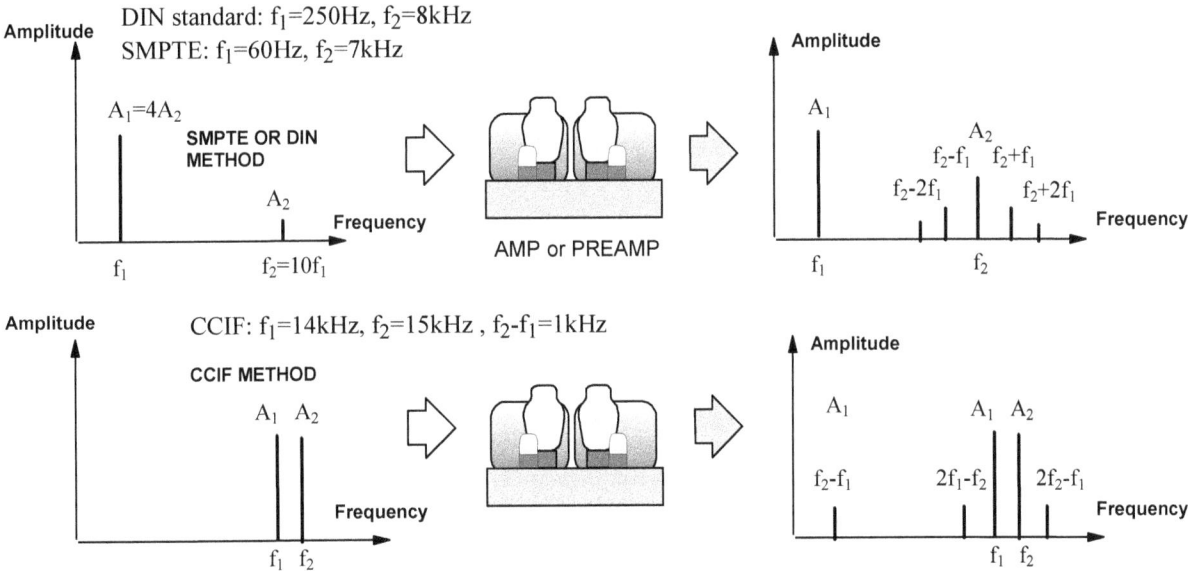

Fig. 3.5. Two different methods of IM measurements, SMPTE (and DIN) versus CCIF

The SMPTE method uses a moderately high HF signal, usually 6 or 7 kHz, while the LF signal is simply the mains voltage (50 or 60 Hz) and the four times higher amplitude than the HF signal), usually taken from a secondary winding of the mains transformer (to save money on the second oscillator that would otherwise be required). German DIN standard uses slightly different frequencies, 250 Hz and 8 kHz, for which two oscillators are obviously needed.

The CCIF standard uses a mix of two signals of the same amplitude and close frequencies, 14 and 15 kHz. The lower sideband $f_2-f_1=1$ kHz then falls in the center of the amplifier's midrange, where the human ear is the most sensitive! The second sideband $2f_1-f_2 = 13$ kHz is in its treble range, which would also be noticeable, primarily as a harsh and "brittle" treble sonic signature. IM distortion is much more unpleasant to the human ear than harmonic distortion.

Delay distortion

The phase angle between the input and output signal of a typical real amplifier stays constant through the midrange frequencies but changes significantly at frequency extremes. The illustration shows a complex input audio signal comprising of the fundamental tone and its 3rd harmonic. If the third harmonic lies in the frequency region where the phase angle changes from its midrange value, it will be shifted or "delayed" in phase by the angle θ.

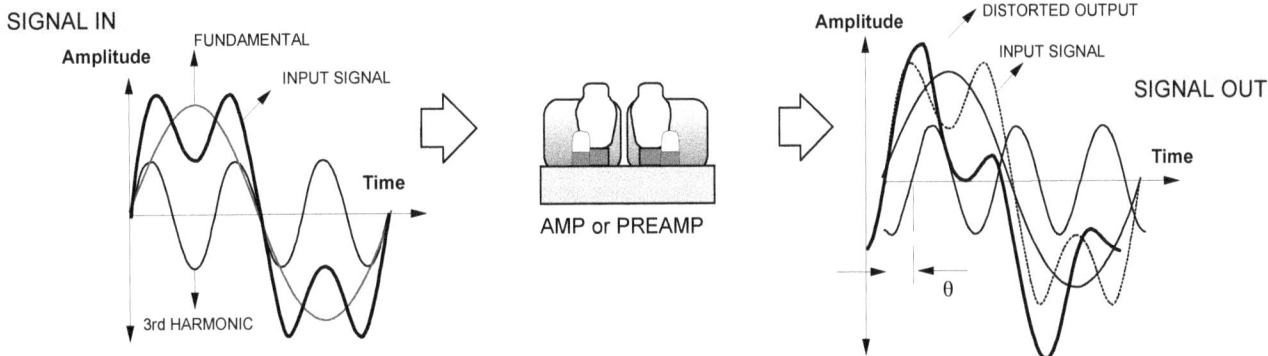

Fig. 3.6. Delay distortion results in a severely distorted signal in the time domain. Luckily, it is benign to the human ear.

Their sum, the output signal, will differ from the waveform of the input signal. This kind of distortion is called a phase or delay distortion. Looking at how different the shape of the resultant output voltage is from the input waveform, you'd think that this kind of distortion would be the most serious (malign) of all. Paradoxically, our ears do not object to this kind of distortion; in fact, they don't even detect it.

Amplifier distortion in time- and frequency domains

Now that we understand the concept of harmonics in a two-dimensional or time domain, we can study it in a 3D space and introduce the frequency domain. The amplitude-time graph shows a distorted periodic waveform which can be broken down into two components, the fundamental harmonic and the second harmonic, of twice the frequency of the fundamental ($2f_1$). There is also a phase shift α (alpha) between the fundamental and the second harmonic. This is the "time domain" in which we see the amplitude and phase relationships of these three signals. The test instrument that displays signal waveforms in the time domain is called an oscilloscope.

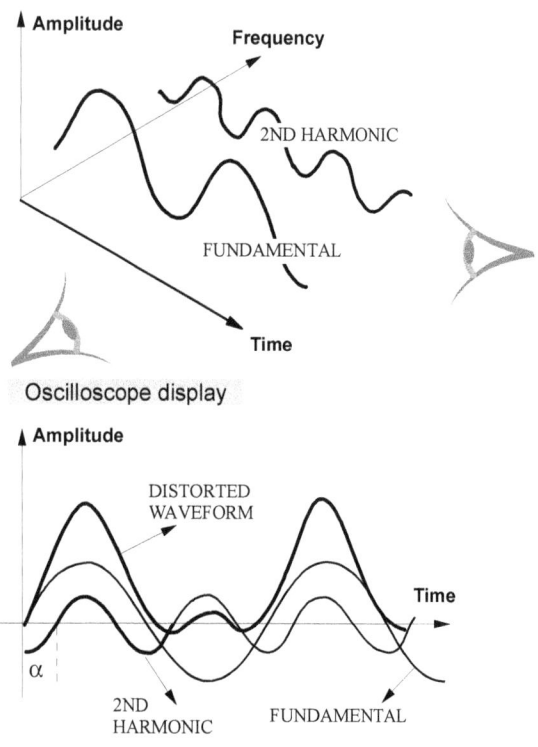

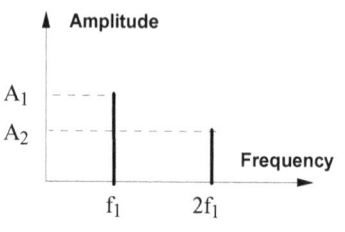

Fig. 3.7. The relationship between the time and frequency domains in a 3D space

Spectrum analyzer display

If we depict these harmonics in a three-dimensional space and add frequency as the 3rd dimension, it would look like this. The projection on the A-f plane will result in the $f_1 + 2f_1$ spectrum, as illustrated. This is the frequency domain. We don't see the waveforms anymore (we know they are sinusoidal signals, anyway), but we see the frequencies, the absolute and relative amplitudes of all harmonics, something we don't see in the time domain. Test instrument that displays amplitudes of signal harmonics in the frequency domain is called a spectrum analyzer.

Therefore, the two depictions illustrate different aspects of the same signal. In this case, we have only the fundamental (1st harmonic) and the second harmonic, but there'll generally be higher-order harmonics present.

WHY DIFFERENT AUDIO TECHNOLOGIES SOUND DIFFERENT

Triode versus pentode

The spectral diagrams below answer one of the most common questions in tube audio: Why do triode amplifiers sound better than those using pentodes? Let's look at two single-ended tube amplifiers of similar output power capabilities, one with 2A3 triode, the other with 6F6 pentode, and their distortion signatures.

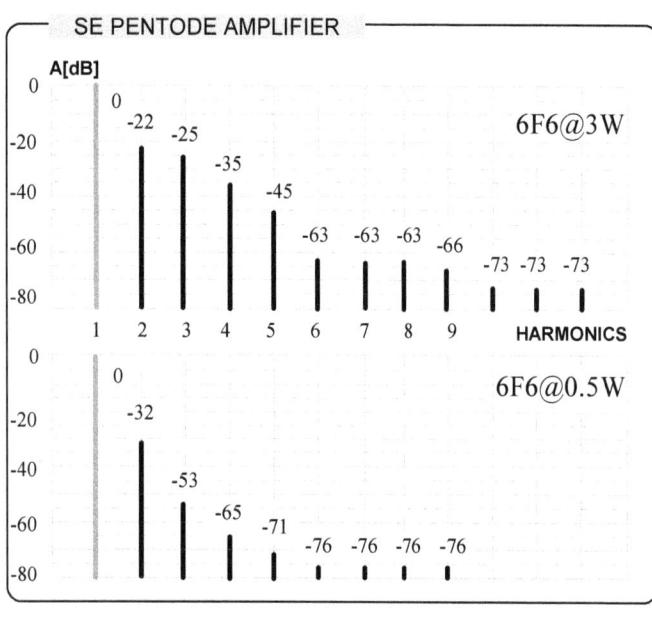
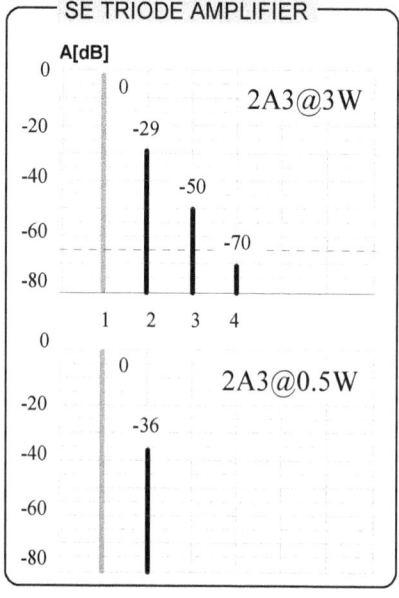

Fig. 3.8. Harmonic distortion spectra of two low powered SE amplifiers, 6F6 pentode and 2A3 triode, for two output power levels.

The spectra of 2A3 single-ended amplifier at half-a-watt and three-watt power outputs show a dominant second harmonic. At lower power levels, that is the only harmonic measurable. The third and fourth only appear when the amp approaches its maximum power of 3.5 watts. The harsher-sounding 3rd harmonic never exceeds the harmonically pleasing second harmonic and is "masked" by it.

The pentode's sonic signature is very different. Even at low power levels, the 3rd and the 5th harmonics are present. At higher power levels, odd harmonics increase rapidly (from -53 dB to -25 dB, in the case of the 3rd harmonic), and that is what makes pentode amps sound shrill and harsh compared to triodes.

Analog versus digital

Most, if not all of the various disputes in audiophilia would be avoided if only the warring parties acknowledge the fact that systems and components may not sound better or worse in *all* aspects (only in some). Still, they certainly sound *different*. The following comparison is a highly simplified yet illustrative and pretty accurate summary of the differences between analog and digital recordings and equipment.

The strength of the analog technology is at low to moderate gain or volume levels. Its resolution and level of detail drop at higher dynamic levels due to its narrower dynamic range and saturation/distortion phenomena, and at very low signal levels, as the noise floor is approached, where noise (tape hiss and such) start to mask the audio signal.

Digital technology assigns the maximum resolution to the highest dynamic levels. As the resolution drops due to fixed quantization steps, it loses focus and definition at the lower signal levels, where analog technology is superior.

The human hearing apparatus is not fully analog. It starts that way but then acts as an A/D converter, with the brain receiving packets of "digital" pulses through the auditory nerve. Could it be that digital audio systems sound fatiguing and unnatural due to the complex interaction between our internal A/D conversion (designed to process natural sounds, which are all analog) and that of the digital audio system?

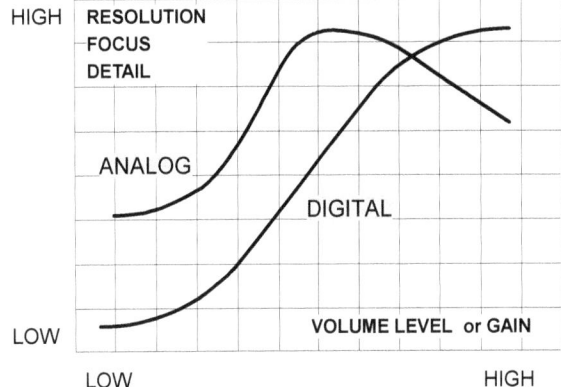

Fig. 3.9. One aspect of the analog-digital divide is *qualitatively* illustrated, resolution vs. volume level. Please don't measure it with a ruler and send me emails asking me to justify the relative levels of the curves!

Tube versus solid-state amplification

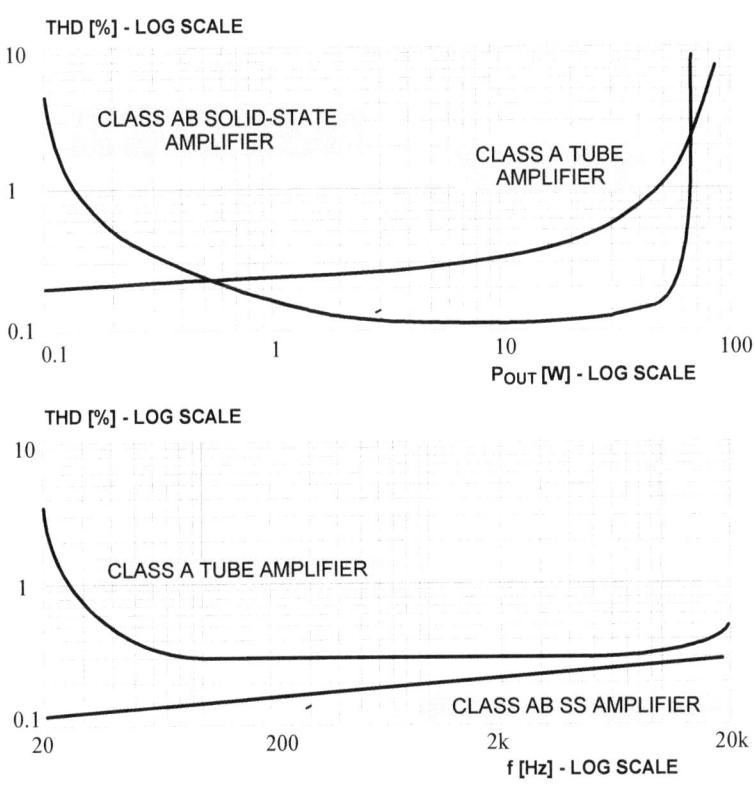

Most audiophiles would agree that tube and solid-state (transistor) amplifiers sound different. Is one category inherently better sounding than the other is another matter of endless debates, divisions, and denigration (of the other camp). Making wide-sweeping statements is risky and often counterproductive since it borders on generalization. That is why I said "different" instead of "better." Even that statement will be challenged by many readers, who claim that all well-designed amps sound the same. Hint: They don't!

I still remember an interview with the late doyen audio designer Tim de Paravicini, in which he claimed that he could design and build a tube amplifier to sound like a great solid-state amp and vice versa, and I entirely agree with him. Two examples come to mind.

Fig. 3.10. THD versus output power characteristics of a typical Class AB solid state and Class A tube amplifier and THD versus frequency characteristics of a typical Class AB solid state and Class A tube amplifier.

In my teenage years back in Europe, I owned a few different mid-fi systems by Grundig and Siemens, which were fully solid-state but sounded very musical and tube-like. Then, early this century, we designed and built a 300B amplifier with active tube HV regulation. It sounded great but unlike most typical 300B amps. It had many qualities of an excellent solid-state Class A amplifier, but it still retained some of that 300B magic, without that over-the-top syrupy coloration some 300B designs suffer from.

Although they have numerical scales, don't get stuck on the figures, the graphs on the previous page are meant to qualitatively illustrate two of many reasons why tube and solid-state amps sound different.

First, notice that solid-state amplifiers' distortion at low power levels is relatively high, and then it comes down with an increase in output levels, only to jump suddenly once the clipping levels are reached.

The distortion of a typical tube amplifier is low at low power levels and slowly creeps up with increasing output levels. Clipping is gradual and soft, not sudden unless high levels of feedback are used, in which case it resembles solid-state clipping.

Perhaps that explains why tube amps sound better at low power levels and why their distortion at high power levels often goes unnoticed due to its gradual nature. When solid-state amps reach clipping levels, one would need to be totally deaf not to notice such unpleasant harshness.

The second graph illustrates that THD (Total Harmonic Distortion) in solid-state amps increases steadily with increasing frequencies. Tube amps distort more at very low frequencies, especially at high power levels, due to the limitations of their output transformers. After that, the harmonic distortion is more or less constant through the mid-band, increasing only in the upper-frequency range above 10 kHz.

CASE STUDY: Tube sound for $299? Keep dreaming ...

In his book *The Audio Expert*, Ethan Winer says, "... distortion similar to that of vacuum tubes can be created using a few resistors and a diode, or a simple software algorithm." Just in case someone had missed the point, on page 65, he repeats it: "If I decide I want the sound of tubes, I'll add that as an effect later."

Semiconductor ("solid-state") diodes and transistors sound differently from tubes. However, tube sound is much more than simple distortion. You can add distortion, but you cannot add the other sonic aspects of vacuum tubes. Such simplistic statements always make me seriously question someone's depth of understanding.

A half-page ad in the *Stereophile* magazine shows a $299 iFi tube buffer using a single 5679 vacuum tube. This is obviously a typing or typesetting error; 5679 is a duo-diode, the 5670 tube is used in the current model. The tagline is "ADD THE RICHNESS OF TUBES TO ANY SYSTEM". The capital letters are theirs.

I agree that a tube buffer will add "richness", which is simply a pleasing harmonic distortion if Ethan Winer is right. Even if there are other sonic aspects added as well, the presumably solid-state system (why would you add a tube buffer to a tube-based audio system?!?) will never sound like a tube-based system. You may add "richness," but you cannot remove the harshness, the sterile character, or the fatiguing aspects of solid-state and digital audio systems.

If adding a three hundred bucks buffer was the solution, audio reviewers and their magazines would be out of business, as would tube equipment manufacturers and even yours truly - you wouldn't need this or any other book on audio. You'd buy a cheap Class D digital amplifier, hook it up to iFi tube buffer and enjoy your music. We all know how far removed from reality such a delusion is.

I'm sure iTube2 (the current product) is a well-designed and engineered device and that it may improve the sound of some lesser audio systems. As such, it represents an incredible value-for-money. However, it will not make a cold, uninvolving and fatiguing system sound like a magical, directly heated triode-based system, for instance.

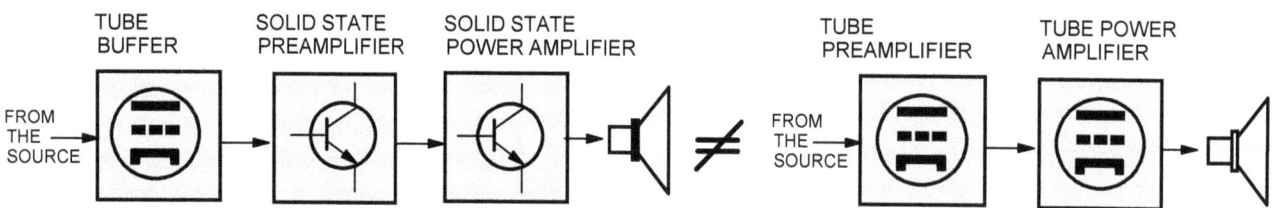

Fig. 3.11. Despite claims to the contrary, such two audio systems will never sound the same!

THE "SONIC IMPACT MYSTERY" AXIOM

There are so many factors that impact the sonics of audio components that we don't even know what many of them are, let alone how and why they affect the sound the way they do.

Audio components that work in different ways sound different!

So far, when considering the possible reasons for the sonic difference between amplifiers with tubes and transistors, and bipolar transistors, in particular, we focussed on their distortion and frequency behavior. However, although nobody knows for sure what other phenomena may be in play, let me raise a few possible causes.

First of all, the mechanism of bipolar transistor operation is highly complex. Electronics students spend a year or two studying it and are none wiser. In vacuum tubes, there is simply a flow of electrons in one direction, controlled by the negative bias voltage at its grid, which acts as a "brake." Onto that DC bias voltage, a variable (AC) signal is superimposed. We can conceptually explain a tube's operation in a couple of sentences. You can break the glass of a tube and take it apart; parts are clearly visible and identifiable.

In contrast, transistors are tiny blobs encased in plastic or metal housing, nothing to be seen there. Inside, electrons move one way while the spaces they've just left or evacuated ("the holes") flow or move in the opposite direction. There are majority and minority carriers, storage delay time, reverse recovery time, transit time, and similar limitations related to the accumulation of carriers in different P-N junction regions (a transistor has two).

THE "DIFFERENT PHYSICS, DIFFERENT SOUND" AXIOM
Electronic parts (components) on the micro- and audio components at the macro-level sound different because they work on different physical principles.

Could constant temperature changes in transistor parameters be the cause of their inferior sound?

Transistors are very sensitive to changes in temperature. All of their static and dynamic parameters change with temperature, the most critical being current gain, collector leakage current, and power dissipation, and that causes constant shifts of the operating points of amplification stages. Such instability increases distortion, affects the bandwidth, and lowers the maximum output power. All these are bound to have a negative impact on the way audio signals are reproduced! Tubes are not sensitive to temperature changes and don't suffer from thermal runaway.

Half of the transistors and passive components (resistors, capacitors, diodes) in a typical transistor amplifier don't perform any signal handling or amplifying function. Some are necessary for temperature compensation, others for overvoltage, and other types of protection.

Except mechanically (cannot be as easily broken as tubes) transistors are much more sensitive and easier to destroy electrically. Some can even be destroyed by static electricity on your hands while handling them.

Why are transistor circuits more complex than vacuum tube ones, and why do they need to use much higher levels of negative feedback?

A bipolar transistor is a bilateral device, which means that its input and output circuits are not independent, as they are with tubes, which are unilateral devices. Bipolar transistors don't have a high input impedance like tubes, meaning any changes at the output causes a change at the input and vice versa. In short, unlike tubes, the input resistance is a function of the load resistance, and the output resistance depends on the source impedance.

The very name transistor comes from "transfer resistor" since resistances in the input and output circuits are transferred across. This interactive nature of the whole circuit makes transistor audio design more complex than designing with tubes, which are electrically much simpler. Again, due to those reasons, transistor circuits are much more complex than their tube equivalents, and simpler circuits sound better!

By their behavior, transistors are much closer to being a current source, while tubes (especially triodes) are voltage sources. Thus, voltage amplification circuits with tubes (such as line-level preamps) should be simpler and better sounding than those using transistors.

The only two applications where current amplification is needed is in moving coil phono preamps, where the input signal is current from an MC cartridge, and in the output stages of power amps, where current gain and low output impedance is needed and where high internal impedance tubes cannot be used without the problematic output (impedance) transformers.

Since vacuum tubes are much more linear amplifiers than transistors, to linearize their designs (read: "to minimize distortion"), transistor circuit designers must use high levels of negative feedback ("degeneration"), typically 20-30 dB. While it achieves some improvements, namely lower distortion, reduced output impedance, and wider frequency range, negative feedback is detrimental to the overall sonics

The whole premise of NFB, to bring a percentage of the distorted output signal back to the input of an amplifier or an amplification stage and mix it there with the "clean" or undistorted signal, is problematic. The distortion has already happened, so the reactive NFB mechanism is acting after the fact, and as such is always lagging behind.

The test instruments we use tell us that NFB succeeds in achieving its aims, but music is not static, and our ears are fundamentally different from the simple test instruments on our workbenches. Negative feedback works well in industrial servo control systems (such as your car's cruise control). Still, our hearing apparatus can detect its sonically detrimental effects, something that cannot be measured but can easily be heard.

A few commercial tube amplifiers have an adjustable NFB, and magazine reviewers and most owners of such amps prefer listening with zero or very low NFB settings. Any amp or preamp that uses negative feedback can be modified, and the negative feedback can be made user-adjustable.

Replace the fixed resistor R_1 with a potentiometer of a higher resistance (as illustrated). For example, if R_1 had a value of 22 kΩ, replace it with a 100 kΩ pot. When its wiper is in the far right position, the whole R_1 is bypassed, and the feedback is at its maximum ($R_1=0$), and the feedback voltage is the whole output voltage $V_F = V_{OUT}$. By turning the pot, the slider moves to the left so more and more of R_1 is in the circuit, lowering the feedback voltage V_F.

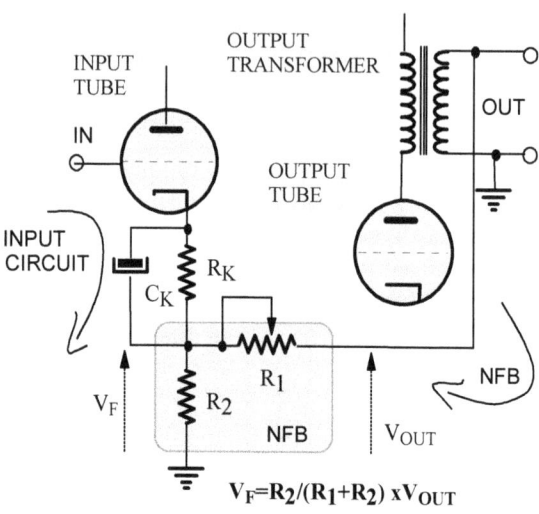

Fig. 3.12. One way of applying global negative voltage feedback in a tube amplifier, in series with the input voltage (across the un-bypassed cathode resistor R_2).

How about FETs?

FETs (Field Effect Transistors) have electrical characteristics similar to vacuum tubes (meaning they are voltage-controlled voltage sources). Their operating principle is very different from bipolar transistors, so they don't suffer from many drawbacks discussed above. However, they are still semiconductor devices whose sonics may still suffer from the effects of signals propagating through silicon rather than through vacuum.

> **THE AUDIO SIGNAL PROPAGATION PROBLEM**
> Just as audio signals propagating through copper or silver cables with different types of insulation sound different, why wouldn't they sound different propagating through vacuum and solids (semiconductors)?

Push-pull versus single-ended amplification stages

Most tube and solid-state amplifier output stages use a push-pull topology, where one device (tube, FET, bipolar transistor) amplifies the positive part while the other handles the negative part of the signal. The signal is thus "chopped" into two and then assembled. This is done to achieve much higher output power levels than what is possible with single-ended output stages. However, two fundamental problems plague such amplifiers.

No matter how precisely matched the output devices are, they are never matched 100%. They may be matched in one operating point but never across the whole operating range. There is always some imbalance and distortion at the crossover point when one device stops conducting the signal and the other takes over. Paradoxically, this distortion is often higher at low power levels (where it is more noticeable and where we listen at most of the time) than at higher output power levels.

For their proper operation, push-pull output stages depend on the quality and symmetry of the preceding phase splitter stage, where one signal is split into two signals of an identical amplitude but inverted or opposite phase (180 degrees phase shift between them). Just as the output stages are never completely symmetrical or ideal, neither are the phase inverters. Despite the variety of phase inverter designs and topologies available to the designer to choose from, they all have inherent flaws or weaknesses.

Even more insidious is the fact that (if perfectly balanced) push-pull stages cancel all their even harmonic distortion artifacts (2nd, 4th, ...) and leave only odd harmonics (3rd, 5th, 7th,...). Since the even harmonic distortion is pleasant-sounding and the odd ones are harsh and unpleasant sounding, this exposes the odd harmonics (removes the even harmonics that "mask" them), making their harmonic structure unnatural and sonically undesirable.

> **THE FLAWS AND FOIBLES OF PUSH-PULL OPERATION**
> Sonically, push-pull is an inherently flawed topology, with an unnatural and undesirable harmonic structure consisting of unpleasant sounding odd harmonics, "unmasked" by the cancellation of pleasant-sounding even harmonics. Furthermore, its proper operation and performance depend on both static and dynamic balance, plus a myriad of other factors, all of which are hard to achieve and maintain over time.

Why amplifiers that measure well don't necessarily sound better?

Oh boy, I know I'm opening a can of worms and that such an important issue cannot possibly be covered in half a page. I also understand that the absolutist camp will attack me with all they've got, but I don't care; I'm an intrepid fellow, so here's my $2 worth of wisdom.

The first issue is the lack of standard load, which would emulate a typical loudspeaker, most likely because there is no such thing as a typical loudspeaker. We test amplifiers using a purely resistive dummy load, while loudspeakers are complex loads with highly frequency-dependent impedance.

They also generate a counter electromotive force (EMF), which is fed back into the output stage of an amplifier and interacts with it in the most unpredictable ways.

Many tests (frequency range, input impedance, output impedance and damping factor, harmonic distortion, maximum power) use only a single test frequency of fixed amplitude, while music signals are complex waveforms comprised of dozens of rapidly changing harmonics.

Total harmonic distortion measurements do not make a distinction between pleasant-sounding even harmonics and harsh and irritating odd harmonics. The spectral analysis gives us the relative amplitudes of all harmonics. Hence, such a test is more meaningful than simple THD measurements. However, despite knowing the number of dB levels of each harmonic, we still cannot predict how the overall spectral signature would sound to our ears.

Finally and ultimately, we measure things that are easy and convenient to measure, not necessarily those that have the greatest impact on sound fidelity! The most important aspects such as rhythm and pace, the sense of "presence," "musicality," "microdynamics," or "transparency" cannot be correlated with measurements at all.

> THE WINE QUALITY ANALOGY
> Just like chemical analysis results do not and cannot define a superior tasting wine, measurements cannot define or identify a superior-sounding audio device.

Measurements versus musicality

Measurements have their place in audiophilia as a design and troubleshooting tool but are of minimal use in predicting the ultimate sonic performance of audio components.

In our tube amplifier designing and manufacturing days, creating a new model would often start with a new preamp or power tube (the one we had not used before). Voltage gain, internal resistance, and mutual conductance for preamp tubes, and anode (power) dissipation, voltage gain, and internal resistance for power (output) tubes were of most interest.

Then we'd mentally (based on experience with similar tubes) explore possible circuit topologies, those that would bring the best out of the tube(s) in question. Often two or more circuits would be built, and their performance measured on the bench (no listening evaluation yet) and compared.

Once we were happy with the selected circuit's measured performance (wide frequency range, low distortion, fidelity in the reproduction of a square test wave, etc.), we'd build it as a prototype. This meant using a chassis or enclosure (no more breadboard or temporary hookups on the bench), with proper topology and optimal positioning of tubes, passive components (resistors, capacitors, and chokes), and power and output transformers.

The first listening evaluations would then take place. Most other amplifier and loudspeaker designers follow a similar approach, which involves an iterative process called voicing, the coarse (initially), and then fine-tuning of the design until the desired sonics are achieved. Such changes may include trying out various output transformers (with different core sizes, winding arrangements, and primary impedances) or other critical components' values (resistance, capacitance, inductance).

Fine-tuning is usually a tryout (auditioning) of various types (technologies) of resistors and capacitors in critical positions, for instance, paper-in-oil versus film & foil coupling capacitors or carbon film versus non-inductive wirewound resistors.

> WHAT DO MEASUREMENTS TELL US ABOUT THE SOUND OF AN AUDIO COMPONENT?
> Not much. While the measured results may point to certain flaws in their design and possible limitations in their pairing with other components (due to insufficient gain or output power, low input or high output impedance, etc.), the measured figures are poor predictors of an audio component's sonic character. In other words, it is impossible to tell how something will sound based on its measured results!

WIDE DYNAMIC RANGE: THE KEY FIDELITY FACTOR #1

Most of the contemporary popular recordings (rock, pop, etc.) have been severely compressed - the race to produce the loudest-playing music has led to an abysmal quality of modern digital recordings, especially when it comes to the ever-shrinking dynamic range - the difference in loudness between the loudest and quietest passages.

One usually becomes aware of this sad state of affairs when listening to superior recordings of classical music. Its dynamic range is startling, especially that of full symphonic works, with unexpected crescendos scaring the pants off the unwary casual listener such as myself (I'm not really into classical music).

This led me to try to identify the critical fidelity factors, those that are necessary for the recorded and reproduced music to sound as close as possible to the live performance. I've identified four: a wide dynamic range, the absence of dynamic compression, preserved harmonic structure (only possible with a wide frequency range of audio components), and the faithful reproduction of fast transients.

The first two are interrelated (a wide dynamic range is impossible to achieve with amplifiers and speakers suffering from dynamic compression), as are the last two. An amplifier with a wide dynamic range will usually have a fast rise time and reproduce fast transients well.

Understanding headroom and clipping

Music signals generally have very high peak-to-average ratios, 25:1 or even more, depending on the type of music and the quality of the recording (the compression introduced by the manipulation during recording and mastering). Sadly, many recording and mixing/mastering engineers compress the music so much that it becomes unlistenable!

The "average" music level is also called the "nominal" program level.

Both "headroom" and "margin" are expressed in dB since they are merely ratios of two SPL levels, not absolute levels such as the average, peak, and clipping levels. The signal-to-noise ratio is the ratio of the average program level and the noise floor.

Although audiophiles usually talk about "headroom," it's worth emphasizing another parameter of audio fidelity, the one that separates truly superb amplifiers from the minnows and "also runs," and that is the "margin" or "cushion." Such reserve between the top of the musical peaks and the clipping level is crucial.

Generally, an amplifier with higher headroom and "margin" will sound better. Again, as mentioned elsewhere in this book, the very gradual (hard to notice) and asymmetrical distortion curves of tube amplifiers produce the pleasant-sounding 2nd harmonic distortion, in stark contrast to the rapid clipping of solid-state amplifiers.

Thanks to their gradual saturation characteristics, magnetic tapes also produce primarily the 2nd and 3rd harmonics, one of the main reasons for the popularity of open-reel recorders and their reputation for natural and engaging sound.

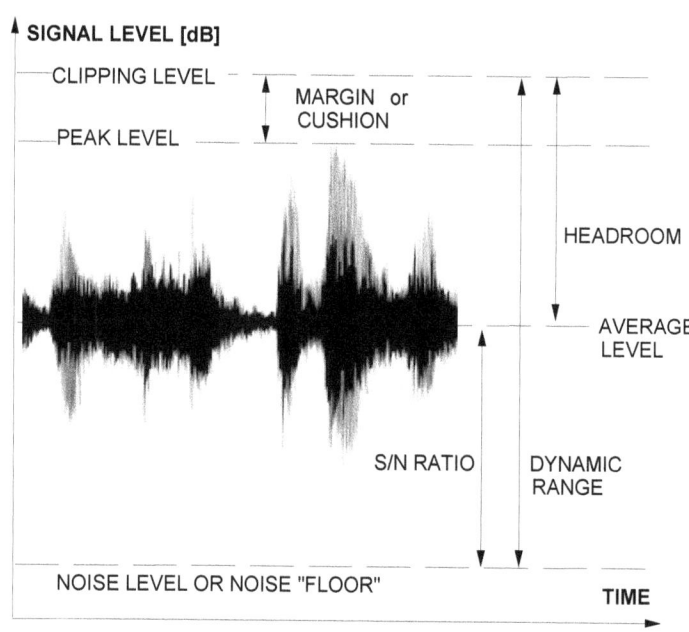

Fig. 3.13. The average and peak music levels in relation to the noise floor and clipping levels, with the definitions of the S/N ratio, the dynamic range, the headroom and the "margin" or "cushion".

How to calculate the headroom of an amplifier

Most audiophiles don't have an intuitive feel for audio power levels, sound pressure levels, and their interdependence. To get a better idea of the figures involved, let's look at two amplifiers and their headroom.

A SET 300B amplifier with 10 Watts maximum output power is driving speakers of a highish 93 dB/W efficiency. At a typical listening level, the amplifier's power output has been measured as 0.2 Watt. Thus the headroom is $HR = 10\log(P_{MAX}/P_0) = 10\log(10/0.2) = 10\log 50 = 17$ dB!

Notice that "the typical listening level" will be different for each one of us, depending on how loud we like our music to be, but also on the sensitivity of the speakers used and the acoustic parameters of the listening rooms. Again, very few things in audio can be considered or studied in isolation; almost everything is context- and system-dependent.

A second case is a 300 Watts per channel Musical Fidelity solid-state amplifier, driving low-efficiency speakers, say 83 dB/W (otherwise, why else would anyone even think of buying a 300 WPC amplifier?). Let's say that at a typical listening level, the amplifier's power output was measured as 25 Watts. Thus the headroom is HR = 10log (P_{MAX}/P_0) = 10log(300/25) = 10log12 = 10.8 dB!

We see a perhaps paradoxical situation where an amplifier that is 300/10 or 30 times more powerful operates with a much lower headroom than what solid states fanatics derogatory call a "flea power" amp, 10.8 versus 17 dB! Of course, the 10 dB difference in speakers' sensitivity made all that difference, not the two amplifiers by themselves. With the 300 Watt amp driving the sensitive speakers and cruising along at 0.2 Watts, its headroom would be HR = 10log (P_{MAX}/P_0) = 10log(300/0.2) = 10log1,500 = 31.8 dB!

In these examples, we assume that both amplifiers will reach the clipping point at their nominal rated power levels; this may or may not be true. Some amplifiers, such as almost all tube amplifiers, clip at higher power levels than those proclaimed by their manufacturers as nominal output power levels.

For instance, most of our 300B amps, which we rated at 10 Watts (to avoid controversy and deriding online comments and "please explain" requests by some smart ass tire kickers), were capable of producing 15 or even 18 Watts at the point of clipping, further increasing their headroom.

THE "QUANTITY OVER QUALITY" MADNESS
If the first watt sounds like shit, why would you want 99 more of them?!? It would be like a diner complaining that a restaurant meal wasn't just foul-tasting, but the portion was tiny, too.

Tone-burst dynamic headroom (dynamic power) test

Static tests using sine and square waves are useful in the basic design and troubleshooting of amplifiers and other audio components, but they don't tell us the whole story. We use these test signals not because they resemble the music signals (in fact, they are as much unlike music signals as possible!) but because these signal sources are relatively simple and affordable and because these tests are easy to perform and interpret.

A sinusoidal signal has a small peak-to-average power ratio; the difference between the peak and average power is only 3 dB, while music signals have peak-to-average short-term power ratios of 14-18 dB (25-63 times) or more. Thus, during a quiet musical passage, the power demand on an amplifier may be 1 Watt, but during loud crescendos, the peak power demand shoots up to 50-100 Watts or even higher.

Two amplifiers of similar power output but using different power tubes and different circuit designs may both sound very good at low to moderate volumes, yet when pushed to their limits, one sounds a little harsher but still OK, while the other distorts terribly and sounds awful. The explanation is mostly in their different overload abilities.

A tone-burst signal is controllable and repeatable, but, most importantly, it resembles music signals much better than simple waveforms. The (a) waveform shows the test signal, a sinewave tone burst.

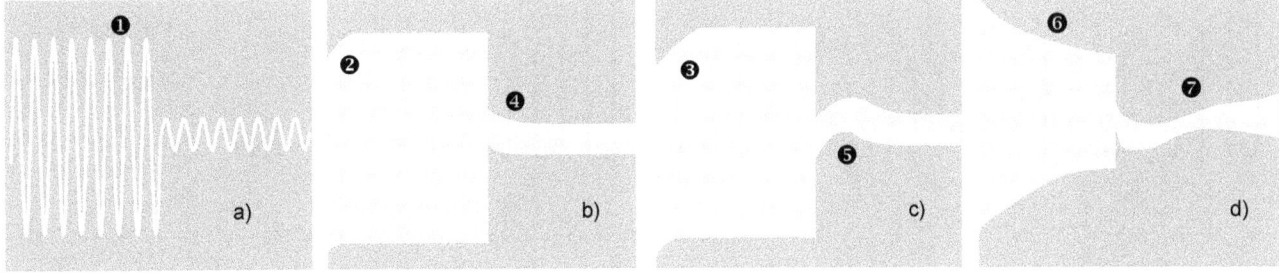

Fig. 3.14. The waveform of the sinewave tone burst test signal (a) and the responses of three different amplifiers, (b) being the best, followed by (c) and (d) the worst of the three

The larger amplitude (1) is adjusted until it takes the tested amplifier into an overload of 2-5 dB. Responses of three different amplifiers are illustrated.

In b), the amplifier uses coupling capacitors driving the output tubes' grids, as does the amplifier in c), so there is a noticeable rise time (2) and even slower response in (3). Both amps cope with the high amplitude signal, meaning their power supplies had a sufficient capacity to keep up with the overload situation. The amp b) recovers faster from overload (4), amp c) takes longer to get back to a straight line (5).

The amplifier in d) is inferior; it cannot cope with the overload signal, as indicated by the sagging top (6). Its power supply is running out of steam (cannot provide the current demanded by the load), and the recovery (getting back to the steady-state lower amplitude signal) is long and laborious (7).

PRESERVED HARMONIC STRUCTURE: KEY FIDELITY FACTOR #2

Harmonics and spectra

To understand sound reproduction, we need to understand the spectra of signals. A square wave, for instance, is comprised of an infinite number of sine waves, but only odd harmonics (1st, 3rd, 5th, 7th, etc.) are present; there are no even harmonics. The 1st harmonic (called "the fundamental") is a sine wave of the same frequency as the square wave, the frequency of the 3rd harmonic is three times the frequency of the first, and so on.

Notice how the amplitude of the harmonics drops off in an exponential fashion. Although the illustration only shows spectral components up to the 11th harmonic, the harmonics continue indefinitely. The amplitudes of harmonics are $A_N = A/N$, so the amplitude of the third harmonic is 1/3 of the fundamental's amplitude, the fifth harmonic has a 1/5 of the 1st harmonic amplitude, and so on.

This other illustration shows the contribution of only the first three odd harmonics, the 1st, the 3rd, and the 5th. Notice how their sum (thicker line) already resembles a square wave. As the successive higher harmonics are added, the waveform will more and more approach the shape of a square wave.

Although amplifiers are modeled, analyzed, and tested using sine waves of a fixed frequency, there are no such pure tones in nature. Sounds have complex waveforms, comprised of sine waves of various amplitudes and frequencies.

A spectrum of a signal is depicted in an X-Y manner with frequency on the horizontal or X-scale and the amplitude of various harmonics on the vertical scale, usually in dB. Two examples of musical instrument spectra are depicted here, violin's and flute's.

Notice that the amplitudes of some of the harmonics are higher than the fundamental tone; that is perfectly normal. Also, not all harmonics are depicted in the violin's case, only those up to 1,500 Hz.

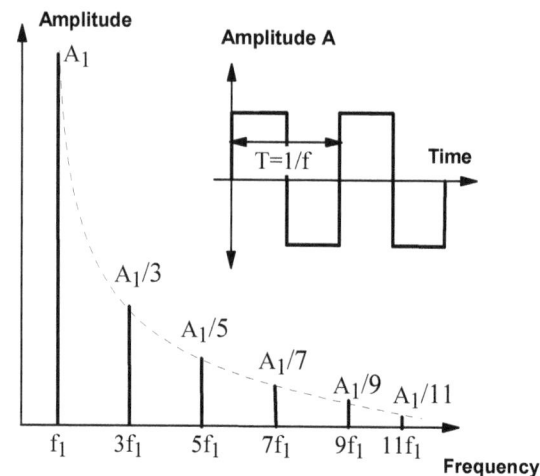

Fig. 3.15. The spectrum of a square wave

Fig. 3.16. The fundamental A_1 and first two harmonics (A_3 and A_5) of a square wave added together.

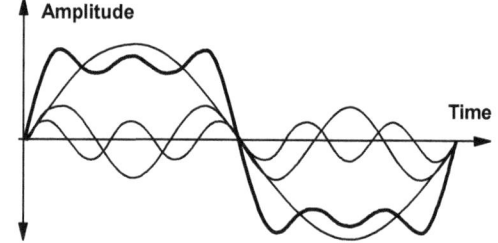

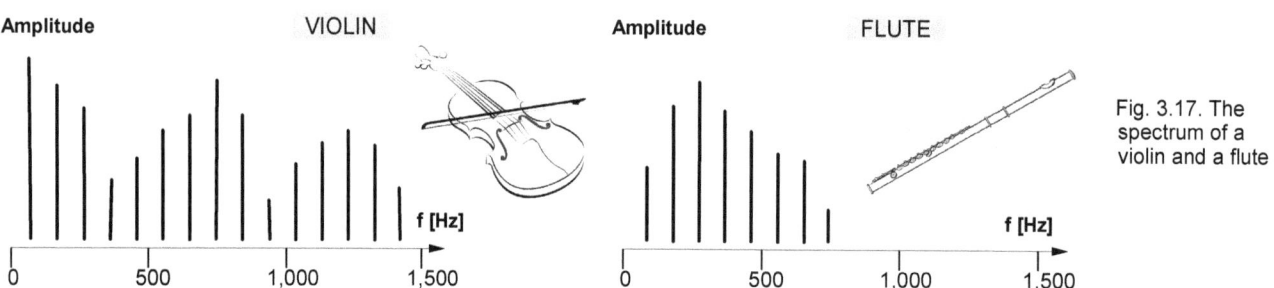

Fig. 3.17. The spectrum of a violin and a flute

The amplitude and phase characteristics of real amplifiers

An ideal amplifier would have an infinitely wide frequency range and amplify all frequencies, from DC (f=0) to infinity. Real amplifiers and preamplifiers behave like bandpass filters (frequency curve on the next page). While they amplify the midrange band of frequencies well, their performance at frequency extremes suffers.

At a certain low frequency f_L, usually in the region of 10-30 Hz, the amplification factor (gain) drops to 0.707 or 71% of its midrange value A_0 (specified at 1 kHz). As the frequency is increased, at a certain frequency f_U the gain again drops to 71% of its midrange value. These are -3dB points, also called "half-power" or corner frequencies.

How does output power drop by half if the voltage drops to 71% or $1/\sqrt{2}$ of its midrange value? It's all a matter of mathematics: The power on the load R is $P = V^2/R$, so the midrange power is $P_0 = V_0^2/R$ and its half-power is $(V_0 * 0.707)^2/R = (V_0/\sqrt{2})^2/R = V_0^2/2R = P_0/2$!

The frequency characteristic of a typical amplifying stage is illustrated here using a linear vertical scale, so it is not in dB but in absolute numbers.

The midrange gain is $A_0=30$ and the phase shift is $180°$, meaning the output signal lags behind the input signal by $180°$. This amplifying stage inverts the phase of the signal. In popular speak, the input and output signals are "out of phase".

The phase characteristic of an ideal amplifier would be a straight line, there would be no phase shift between the input and output signals, or such a phase shift would be constant, i.e., the same for all frequencies. As with the signal's amplitude, the phase relationships also change at frequency extremes.

At low frequencies, the phase lag is increased, and at high frequencies, the lag is decreased. Notice that at our half-power or -3 dB points, frequencies f_L and f_U, the phase changes +/-45° from the midrange phase. At f_L the phase is $180°+45° = 225°$, and at f_U the phase is $180°+45° = 225°$.

The "typical" gain vs. frequency curve shown above is anything but typical. Typical is the word textbook writers and university professors use to illustrate a point or present a simplified argument. Many commercial amplifiers have much more malign A-f characteristics, and I am not talking just about sub-$1,000 budget models.

Look at the A-f curves of the Cary CAD-805 amplifier, published in the "Audio" magazine's review (right).

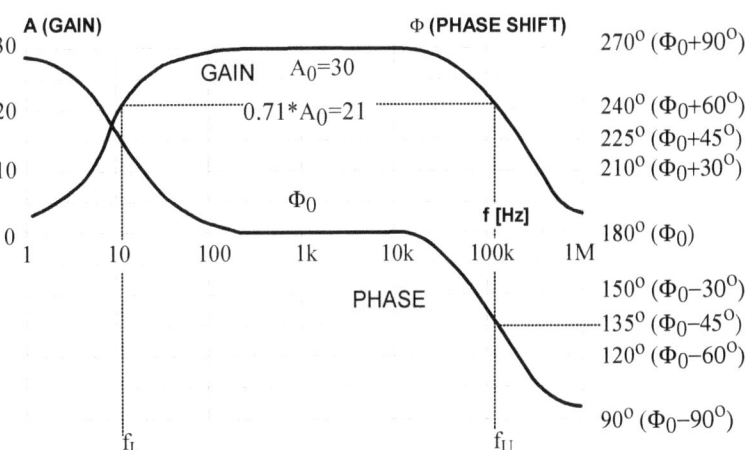

Fig. 3.18. Typical amplitude versus frequency and phase versus frequency characteristics of a real amplifier and the meaning of -3dB points f_U and f_L

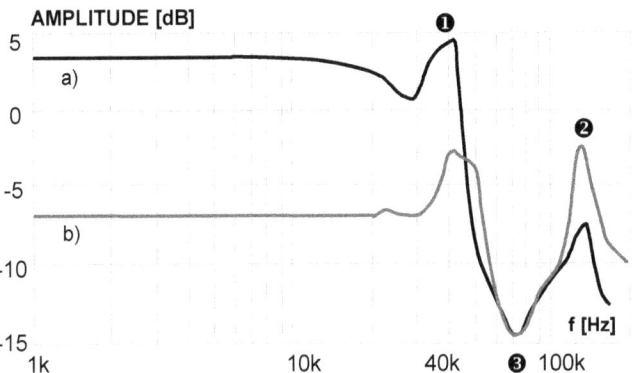

Fig. 3.19. Gain vs. frequency of Cary CAD-805 amplifier for two NFB settings
a) feedback control fully anti-clockwise - minimum feedback
b) feedback control fully clockwise - maximum feedback
Source: *Audio*, July 1995 review

Negative feedback is adjustable by a potentiometer, but no matter how much NFB is used, the amplifier exhibits two amplitude peaks and a severe dip at around 70 kHz (3). The peaks are due to the output transformer resonances at about 40 kHz (1) and 130 kHz (2), where leakage inductances react with distributed parasitic capacitances.

The "wideband versus limited high frequency extension" debate

But Igor, you may be protesting now, such peaks and dips are above the upper audible limit of 20 kHz; we cannot hear such ultrasonic frequencies anyway! Well, we can't hear them directly, but indirectly we hear their "artifacts," their interactions with the audible frequencies, so we should care about the ultrasonic behavior of amplifiers!

Although 20 Hz - 20 kHz is the "official" audio range, audiophile components sound much better if their frequency response extends at least 2X higher for power amps and 5X higher for the preamps. A similar (and even more important) requirement applies to the low-frequency range. A superior power amp should have its lower half-power or -3 dB frequency fL as low as possible; 10 Hz would be great, although that is very difficult to achieve due to the limitation imposed by the output transformers. Preamplifiers can be designed to go down to 4 or 5 Hz since most don't use output transformers and the power levels are low because they only provide voltage amplification.

This may seem strange for at least four reasons. First, at the source, if you look at the frequency range of musical instruments, none of their fundamental harmonics exceed 4 kHz (next page).

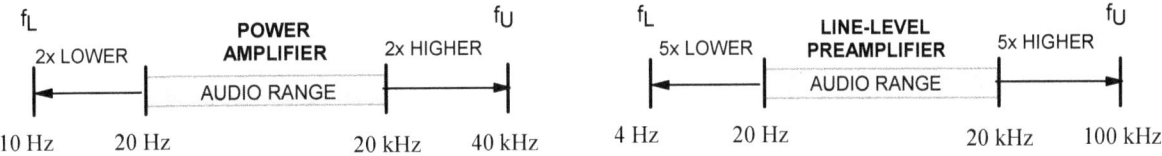

Fig. 3.20. Rules-of-thumb for the target frequency range of a quality power amplifier (LEFT) and a line-level preamplifier (RIGHT).

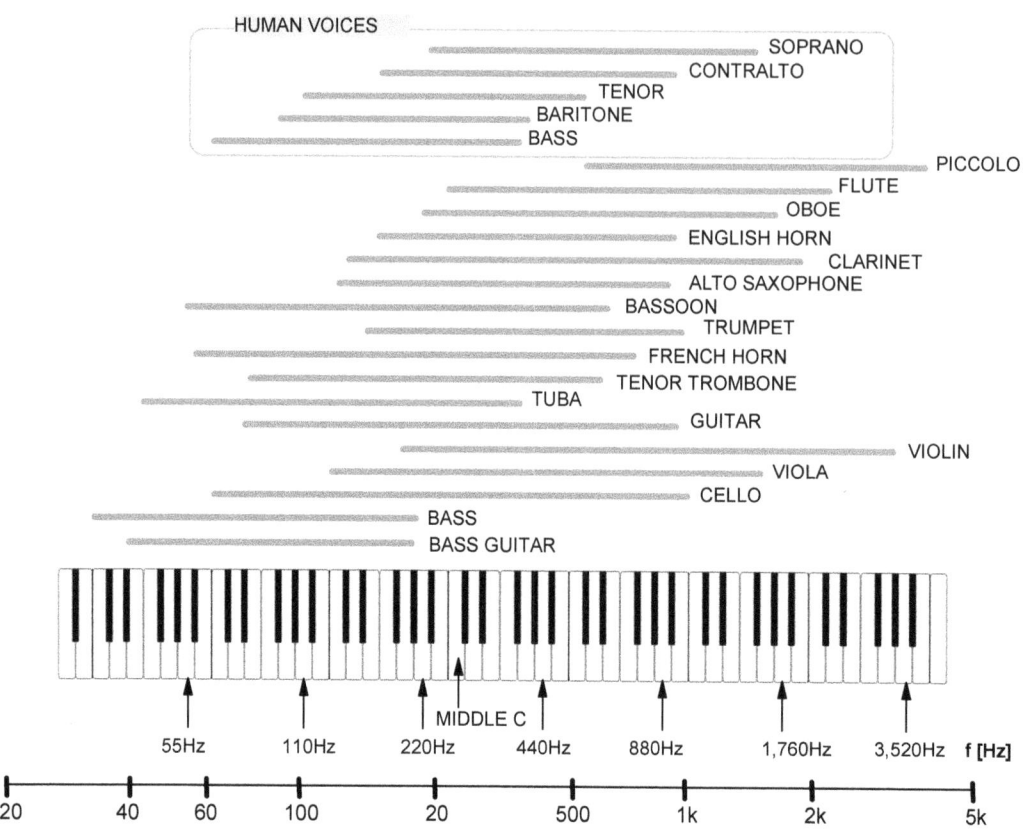

Fig. 3.21. The frequency range of various instruments and human voices compared to the piano scale

Secondly, at the end of the musical chain and the bottom of the frequency spectrum, no loudspeaker can reproduce 20 Hz, let alone 5 Hz. The higher frequencies can be easily reproduced. Tweeters can and do go much higher than 20 kHz; just ask your dog or cat how the music you listen to sounds to them.

Thirdly, the recorded material, be it an LP, a CD, or a digital file, is limited in its frequency range; most cannot reproduce anything over 20 kHz due to sampling limitations in CD players and similar confines for other media. Finally, even if those sources and transducers could reproduce a wider audio spectrum, no human can hear frequencies below 20 Hz or above 20 kHz, so why would we impose such challenging design goals on ourselves?

The first argument is misleading; it mentions only the fundamental tone. While the highest notes played on a violin or piano are only around 3 kHz and 4 kHz respectively, their harmonics extend well into the ultrasonic range (above 20 kHz). Although you cannot directly hear those high-frequency overtones, you will hear a different tonal presentation with them present and absent.

It is the harmonics and their relative amplitudes and interactions that define specific instrument sounds. Ultimately, the short answer is that "wideband" amps and preamps generally sound better than those whose frequency ranges are narrower.

Over the years, we have noticed a positive correlation between a wide frequency range (at the upper-frequency end) and the transparency (or "the air around the instruments") of our tube amplifiers. All went beyond the "prescribed" 20 kHz minimum, the upper limit of human hearing. Some had a bandwidth that extended up to 30, 50 or even 70 kHz. Assuming that the preamp stages were designed and constructed correctly so that their frequency extension was way above 100 kHz, the upper cutoff frequency f_U was determined primarily by the output transformers' limitations. Transistor amps have no such limitations so their bandwidth can extend to 200 kHz or even higher.

The sense of separateness or "air" around the instruments was also dependent on other factors - the power tubes used, the topology chosen, and the output transformer's design. Generally, the wider the frequency bandwidth of an amp, the higher its transparency level. Amplifiers by other brands also conformed to that rule. A few commercial tube amplifiers of poor design, incompetent construction, and using substandard components had upper-frequency limits in the order of 12 kHz, which made the sound muffled, lifeless, and lacking the sparkle.

In my view, widening the frequency range beyond the rational limits into the ultrasonic region (which some experts and amp designers claim to be totally unnecessary) can be heard, not directly, of course, but rather "detected" or "sensed" as a sense of ease, openness, and transparency, sometimes even as an improvement in the imaging and sound staging.

However, the wideband school of audio design is not universally accepted. Some audio designers limit the upper-frequency extension of their designs, claiming that such amplifiers sound softer and more musical. There is some truth in their claim that significantly extending the bandwidth only gives prominence to higher harmonics. The odd ones (5th, 7th, etc.) are particularly objectionable (harsh sounding and irritating).

It is not just the higher harmonics by themselves, but also the intermodulation distortion products that lie in the 20-60 kHz band. The limited bandwidth amplifiers would significantly cut off or attenuate such artifacts, but wider bandwidth amplifiers would reproduce them at full levels!

HF extension needs to be limited if high levels of negative feedback are used, which can become positive at such high frequencies and make amplifiers unstable. So, the designs that produce low THD and IM distortion levels and use low levels of negative feedback will probably benefit from a wide bandwidth. In contrast, poor designs (distorting and oscillating) will suffer from it.

The expectation versus reality, or "The booming bass fallacy"

The proliferation of bass-reflex loudspeakers and high-powered solid-state amplifiers has led many audiophiles to get used to and then expect a heavy, punchy bass. Likewise, the advances in tweeter technologies and near-field listening in (smallish) listening rooms have resulted in a prominent top-end (high frequencies).

Since live concerts are rare in many parts of the world, audiophiles have developed an unrealistic construct, a false expectation of high fidelity sound. An exaggerated, poorly defined bass with recessed midrange and overly prominent treble has become a derogatory definition of the "hi-fi" sound.

Those who prefer such elevated frequency extremes are disappointed when they hear a live concert, where tonal balance can, unfortunately, vary with the listener's seating location. The bass may not be "punchy" or boomy, while the highs are never overbearing or harsh (unless one sits in the first row close to the orchestra).

The same shock happens when they hear superb horn speakers driven by single-ended triode amps. "The bass lacks slam," they say, or "there's no bass at all," but the bass is there, and the longer they listen, the more they realize they've been listening to artificially pumped-up bass at home!

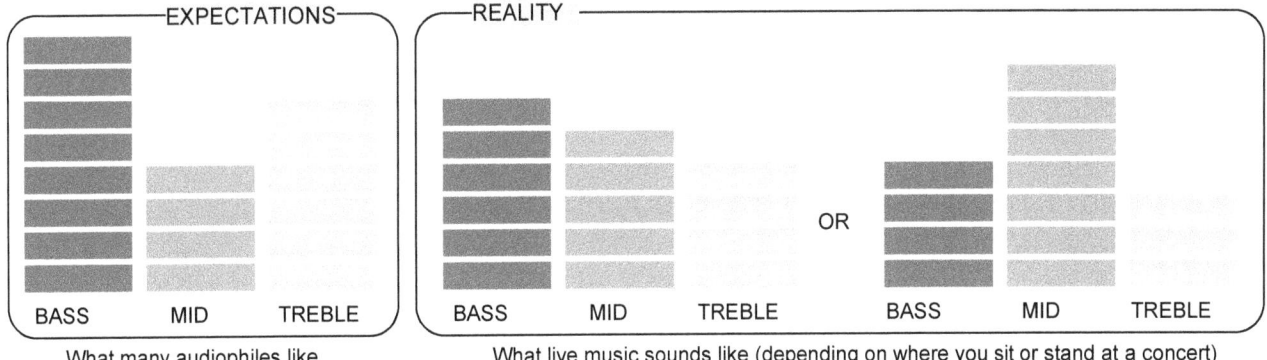

Fig. 3.22. Most audiophiles do not listen to the "natural" sound but a tonally exaggerated "expectation."

REPRODUCTION OF FAST TRANSIENTS: KEY FIDELITY FACTOR #3

If you ask audiophiles to identify the most critical factors that affect the sound quality of an audio system, you'd hear issues such as the frequency range, the residual hum level, or harmonic and intermodulation distortion. While these are all important for the fidelity of sonic reproduction, to human ear-brain audio decoding apparatus, the most important clue that we are listening to live or reproduced music is the transient response.

The music signal is a complex waveform whose amplitude varies enormously, primarily due to the presence of short transients or abrupt changes in the signal's amplitude. The presence of such transients in live music is hard to reproduce in a listening room environment, and that is what separates a great audio system from an ordinary one. Inferior quality components (amps, preamps, speakers) cut off or compress the high peak transients, plus their slew rate is limited, meaning they take far too long to reproduce steep rising vertical edges of transient signals.

Square-wave reproduction as an indicator of a component's or system's transient handling fidelity

There are two ways to evaluate the transient-handling ability of audio devices such as amplifiers and preamplifiers. The easiest to perform is the square-wave test. A square wave signal from a function generator is fed into the device, and an oscilloscope is used to observe the shape of the voltage on its output.

Let's look at oscillograms (oscilloscope displays) of the SET EL34 "Look" amplifier (below). The 100 Hz response is slanted (1). The slope indicates how low in frequency the amp goes; the flatter the top (the less slanted the response), the better the amp's or preamp's reproduction of bass frequencies.

The 1 kHz response is poor, significant "ringing" (2) due to the low resonant frequency of the output transformer. Also, the rise time of the leading edges is very long (3), the amp is slew-rate limited, or simply "slow," it cannot cope with fast transients (sudden changes in signal levels).

Its 10 kHz response better shows its very long rise time (4) and an overshot plus ringing, followed by an equally long fall time (5). At 20 kHz, the output is triangular instead of a square; the whole amp behaves like a giant capacitor, "integrating" the signal. The upper -3 dB frequency f_U of this amp is very low (we measured $f_U = 21$ kHz)!

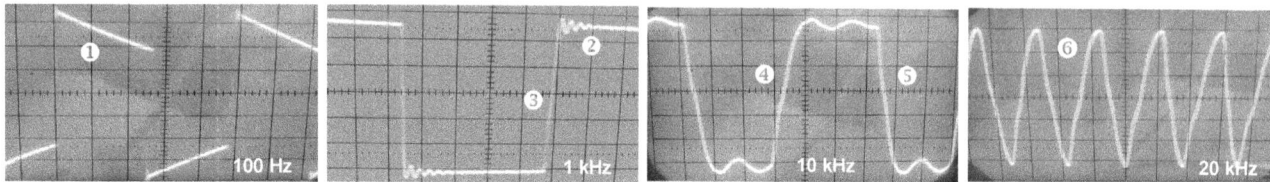

Fig. 3.23. Square wave response of a poor-quality audio amplifier at few typical test frequencies.

Before you purchase an amp or preamp, study its square wave test results. Typically, at test frequencies around ten times lower than the amp's upper cutoff frequency, there will be a noticeable rise time (3) and rounding of the leading and falling edges of the reproduced signal (4). They tell you about its fidelity at high frequencies.

With test frequencies around ten times higher than the amp's lower cutoff frequency, there will be a drop in the ideally horizontal tops of the square wave (1); the slope tells you about the amp's fidelity at low (bass) frequencies. The higher the drop, the worse its bass will be.

Below are oscillograms of four single-ended 300B amplifiers reproducing a 10 kHz square wave test signal. The top two amps have similar square-wave responses, Allnic with a higher overshoot, Cayin with a slightly smoother but much slower response. Sophia Electric's amp has the narrowest bandwidth of all four, so even at a relatively low frequency of 10 kHz, the square wave response has degenerated into a wobbly mess.

Keep in mind that we are not talking about the sonic merits of these amplifiers, only about one aspect of their fidelity. It is possible that the worst measuring amplifier (in this case Sophia Electric) in some aspects could sound better than the one with the best test results (our own design).

I still remember our SET power amplifier using RS1003 output tubes. As soon as the prototype was finished, before any benchtop tests and design refinements, I had a quick listen and loved the sound.

It had presence, it conveyed emotion, it was dynamic, punchy, and above all, "natural." It sounded like performers were right in front of you in a live concert.

When I returned it to the test bench, I got a rude awakening - it was the worst measuring amplifier of all we've made over the years. The square wave was warped, the distortion was high. It measured very well after some fine-tuning and sounded great, but the magic and that "organic" sound were gone.

Does that mean that there is no correlation between the measured results and sonics? Or, heaven forbid (the objectivists will now start getting heart palpitations), that the two can even work in "reverse", that higher distortion levels result in a more natural or "organic" sound?

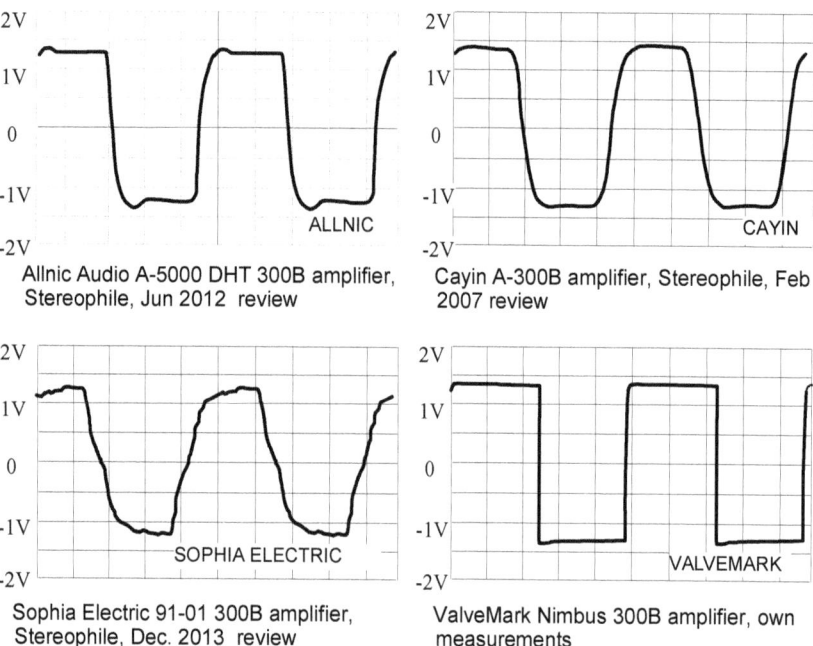

Allnic Audio A-5000 DHT 300B amplifier, Stereophile, Jun 2012 review

Cayin A-300B amplifier, Stereophile, Feb 2007 review

Sophia Electric 91-01 300B amplifier, Stereophile, Dec. 2013 review

ValveMark Nimbus 300B amplifier, own measurements

Fig. 3.24. 10 kHz square wave reproduction by four 300B SET amplifiers. Partially redrawn with permission from *Stereophile* (www.stereophile.com)

DYNAMIC COMPRESSION: #1 PROBLEM IN SOUND RECORDING AND REPRODUCTION

In 1996 I asked one of the hi-end audio dealers in Perth, Australia, to evaluate the prototype of our first single-ended tube amplifier (10 Watts per channel from triode-connected PL519 power tubes), with a view of them selling it once in full production. The guy was no fan of low-powered triode amps; he hooked it up to the first pair of (inefficient) speakers he could find and played a passage from "1812 Overture" by Tchaikovsky.

Needles to say, the famous cannon fire sounded like somebody popping open a can of soda, the amp flopped, and the sound fizzled. His evaluation lasted a full minute! That was the most challenging test of any amplifier and a true "mission impossible" for a low-powered triode amp. It simply could not deliver the high current pulse required by the woofers to reproduce such a demanding music material faithfully.

To understand why such dynamic compression happens in amplifiers, let's start by looking at a typical vacuum tube voltage-amplification stage.

Compression due to tubes' or transistors' nonlinear nature

An ideal dynamic transfer characteristic of an amplification stage would be a straight line. The input signal at the control grid would be "reflected" into the anode circuit (the output) without any distortion.

Compared to the ideal (linear) transfer characteristic, triodes' parabolic curve results in an asymmetric waveform of the anode current and the ultimate output signal, the anode voltage. The negative (bottom) half of the current sine wave is widened and flattened (1), with the slight narrowing of the positive half as well (2).

This type of asymmetrical distortion results in the addition of even harmonics to the original signal; of those, the second harmonic is the most prominent. Since the anode voltage is of the opposite polarity ("phase") of anode current, its positive leaks are smaller (5) than the negative ones (6).

Notice how the choice of bias (negative DC voltage on triode's control grid) is crucial for maximum signal swing and minimal distortion. In this case, the quiescent point Q (with idle DC current I0 flowing through the tube) is too low; the bias should be reduced (made more positive), as indicated by the gray arrow (4), so the operating point is raised to Q1 and I_0 is increased, as shown.

Further compression (together with a sudden jump in distortion) happens when the output voltage reaches the upper or lower limits (points X & Y on the transfer curve). At point X, the anode current drops to zero, while at point Y it reaches its maximum (zero bias).

If the input grid signal keeps rising ("overdrive"), the output signal (anode voltage) cannot follow it in proportion to the amplification stage's gain. The output signal is compressed, with its peaks flattened or "clipped," the reason this condition is called "clipping." Guitar players love distortion, but it's clearly undesirable in audiophilia.

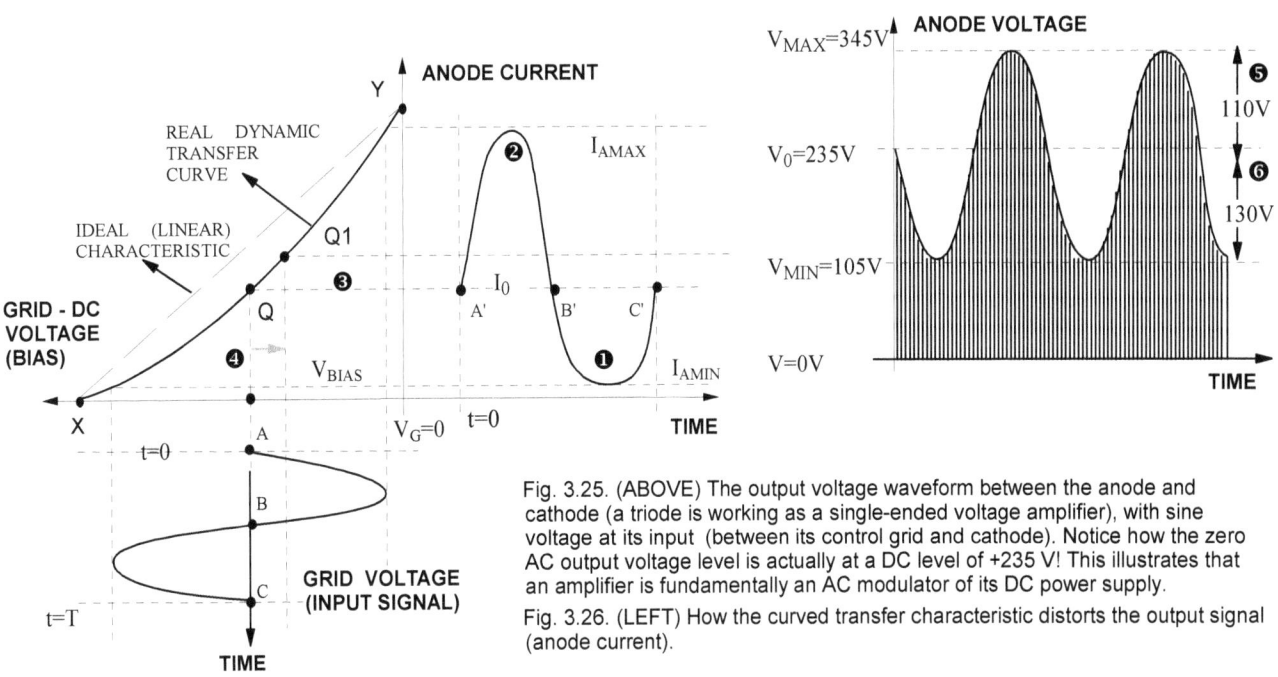

Fig. 3.25. (ABOVE) The output voltage waveform between the anode and cathode (a triode is working as a single-ended voltage amplifier), with sine voltage at its input (between its control grid and cathode). Notice how the zero AC output voltage level is actually at a DC level of +235 V! This illustrates that an amplifier is fundamentally an AC modulator of its DC power supply.

Fig. 3.26. (LEFT) How the curved transfer characteristic distorts the output signal (anode current).

Dynamic compression due to power supply limitations

While the preamp stages only provide voltage gain, the output stages of an amplifier must also have enough current capacity to drive current-hungry voice coils of loudspeakers, especially woofers, where most of the power is dissipated. Compression also happens when the output stage reaches the maximum current the amp's power supply can provide and cannot supply any more.

The best DC power supply is a battery bank of infinite current supplying capacity. With solid-state gear (especially phono stages and preamplifiers), assembling such a bank is quite feasible since both the required voltages and currents are relatively low. Automotive batteries can supply 300 Amperes of current for short periods. No amplifier current draw would come close to those figures, so a battery bank could be considered an infinite current reservoir. Plus, their internal resistance or impedance is very low, almost negligible.

In comparison, all other power supplies, such as the electronic power supplies inside preamps and amps, are limited in terms of the current they can deliver, both continuously and instantaneously. To make things worse, such power supplies have a high internal impedance. As the load (the audio section of the amplifier) starts drawing high current, more and more voltage is lost on the internal impedance of the power supply, and less and less is available to the power amplification stage. This causes distortion and dynamic compression of louder music passages.

Dynamic compression caused by AC power filters, conditioners and regenerators

There are various ways to improve the quality of the AC power entering audio components. High frequency can be filtered out; ferromagnetic resonators can be used to stabilize a fluctuating voltage. In extreme cases, the mains voltage can be rectified, filtered into a DC voltage, then regulated. Then, a DC-AC converter would generate a completely new sinewave voltage of low distortion and stable frequency - the operating principle behind power regenerators. We will talk about AC power filters, conditioners, and regenerators later on in the book.

A stiff and stable power network is theoretically an infinite source of current, although practically limited by the current carrying capacity of its distribution network transformers and cables. Mains filters, conditioners, and regenerators are inherently of limited current capacity. Both the levels and the rate of increase of the current they could deliver into demanding loads such as power amplifiers are limited. This AC behavior is analogous to the limits of the internal DC power supplies inside audio components, which also causes distortion and dynamic compression of louder music passages.

Dynamic compression in loudspeakers

The two main types of loudspeaker compression are the voice-coil thermal compression and the displacement or speaker cone excursion compression.

The displacement compression happens when the driving signal pushes the speaker cone close to and even beyond its maximum excursion limits. Dynamic compression causes a change in tonal balance. If a woofer is overdriven, its contribution relative to the midrange driver (if used) and the tweeter is reduced. The speaker will sound brighter or even shrill, with more prominent midrange and treble frequencies.

When the driving signal current flows through a speaker's voice coil, it heats it up. The resistance of a voice coil may be very low (a few ohms). Still, since the currents flowing through it can be pretty high, the power dissipated into heat on such a physically small coil of wound wire becomes significant. As a result, the temperature of the coil increases substantially. That's why many tweeters use ferrofluid cooling. They are much smaller than woofers (the built-up heat cannot quickly dissipate) and can easily burn out due to overheating.

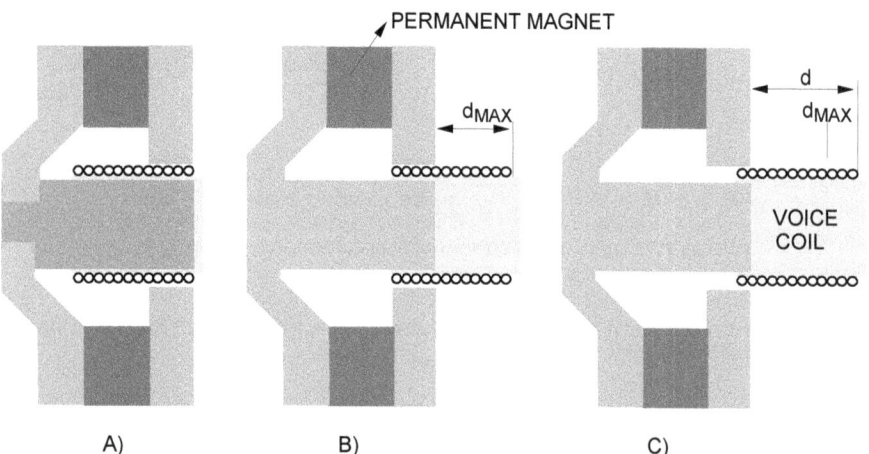

Fig. 3.27. How loudspeaker's voice coil travels during compression and overload

A) No audio signal - speaker's voice coil (VC) at rest (fully retracted)

B) The largest allowable excursion d_{MAX} of the voice coil and the attached cone (not shown)

C) Audio signal too large, the VC travels beyond its allowable maximum limit (d larger than d_{MAX})

NOTE: The spider (voice coil and cone support), the cone and the dust cap are not shown to preserve clarity

The resistance of a hotter copper voice coil at a temperature T will rise to the value $R_T=R_0[1+\alpha(T-T_0)]$, where R_0 is the resistance at a reference temperature T_0 (usually given in tables for 20 °C) and α (alpha) is temperature coefficients of resistance for the particular metal. For copper $\alpha=4.041*10^{-3}$ [1/°C], so if the DC resistance of the voice coil is 5.6 Ω at 20 °C, at 70 °C will rise to $R_{70} = 5.6[1+0.0004041(70-20)] = 6.73$ Ω, an increase of 20.2%!

Dynamic compression and distortion caused by the use of negative feedback

When a global negative feedback signal is brought back from the amplifier's output to its input, the linear region (2) of input voltages becomes wider than without it (1), so if you are driving that power amp with an active preamp, for instance, it will be able to accept larger input signals without clipping. The transfer characteristic of the input stage is "straightened," and the linear region is widened by the use of negative feedback.

However, once clipping occurs (the flattening or compression of signal's peaks), as illustrated here, it is more abrupt. The output signal (anode current of a tube amp) now resembles a square wave and introduces harsh sounding odd harmonics.

You have been told that NFB reduces distortion; it does, but only within the linear operating range of the amplifier. Once clipping occurs, negative feedback significantly increases distortion!

The problem isn't limited to tube amps and preamps; it happens in solid-state stages as well, even more so. NFB widens the linear region of operation but makes clipping harsher (more abrupt) and much more unpleasant sounding.

For this reason most single-ended triode amplifiers don't use negative feedback. When overdriven, their clipping is gradual, and such natural dynamic compression and distortion often pass by unnoticed.

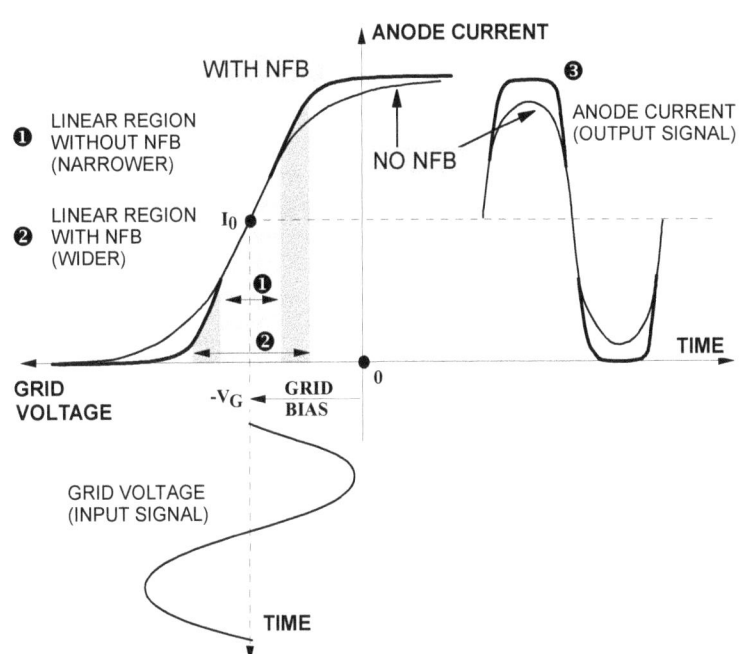

Fig. 3.28. Negative feedback widens the linear region of operation but makes clipping harsher (more abrupt) and much more unpleasant sounding. Outside the normal operating range, once clipping occurs, negative feedback significantly increases distortion!

> **HOW WRONG CAN ONE BE?**
> "Actually, negative feedback is a very good thing. The more negative feedback we can realistically inject into an audio power amplifier design without encountering stability problems, the better the power amplifier will perform in almost every aspect."
> G. Randy Slone in his book *High-Power Audio Amplifier Construction Manual*

Dynamic compression in turntables and reel-to-reel tape decks

Compression also happens with analog sources such as turntables and reel-to-reel tape machines. The limited trackability issue plagues cheaper styli and cartridges (but is present in all, even the most expensive ones), causes the signal peaks to be flatted, thus compromising the critical reproduction of transients.

In tape decks, the frequency response varies with the recording level, compressing high-frequency signals. This is due to the magnetic tape's limited coercivity and retentivity at such high signal levels.

The highly nonlinear hysteresis curve of all magnetic materials is another culprit. An increase in signal amplitude in the saturation region does not result in a proportional increase in the recorded level on a tape, a textbook example of dynamic compression.

THE SONIC SIGNATURE OF MATERIALS

Speakers, equipment racks, and listening rooms sound different because they are made of different materials

The art of audio is not just about electronics and acoustics; it is also about materials. Each material has its sonic or acoustic properties. There is a good reason why violins and other string instruments are made of wood and not out of metal or why brass instruments are made of, well, brass!

Speaker drivers with their cones made of different materials (paper, aluminium, polypropylene, flax fibers, fiberglass, or Kevlar®) sound different, just as speaker cabinets made of MDF (medium-density fiberboard) sound different from those made of solid mahogany, for instance.

Equipment racks will impart a sonic character to the audio components that sit on its shelves, depending on what those shelves are made of. A preamp or turntable sitting on a glass shelf will sound different from the same component sitting on a solid marble or thick timber shelf. A shelf is an integral part of an audio system!

Likewise, you and your audiophile friend may have identical audio systems and listening rooms of the same dimensions, even down to similar acoustic treatments on your walls, but your systems could sound very different. The answer is in the different materials used to construct your two listening rooms' walls, floors, and ceilings.

Your house is of double brick construction (external walls) with single brick internal walls. Your floor is a concrete slab over which a carpet was laid. Your friend lives in a timber house with hollow stud walls covered by gyprock sheets. His floor is also timber, raised off the ground on studs. While the walls and the floor of your music room are solid and rigid, his are effectively large resonant panels that will vibrate when the music is played. They will absorb some of the sound energy, resonate at specific frequencies, and re-radiate such energy back into the room.

In essence, his room will behave almost like the resonant body of an acoustic guitar, while yours would be akin to the solid body of an electric guitar. The two will sound very different. We will talk more about that issue in the "Optimizing the Acoustic Performance of Your Listening Room" chapter (pages 259-270).

ALL THINGS WERE NOT CREATED EQUAL a.k.a. THE WOODEN LAW
You cannot make a violin out of a tree that grew to be used as rake handle. The same applies to people.

The dielectric theory of sound

Audiophiles tend to hate capacitors with passion (especially the electrolytic kind!), claiming they are detrimental to sound quality. This disdain is probably the main reason behind the resurgence of directly coupled and transformer-coupled designs. Transformers sound different from capacitors, just as pentodes sound different from triodes or tube amps from transistor amps. Perhaps their sonic signature (read "distortion") is more pleasing to the human ear. However, as we will see soon, replacing coupling capacitors with direct coupling or interstage transformers solves one problem but creates two or more new ones.

Capacitors are inevitable; they are everywhere. As a matter of fact, you are sitting inside a huge capacitor right now. The wiring in the roof of your house is one conductive plate, the earth or ground is another, and you and everything around you (including the air) is the dielectric between the two plates. Even if you remove all the coupling capacitors in an amplifier, there will still be dozens of capacitances left. These are parasitic capacitances between the cables, components (resistors, capacitors, inductors, tube sockets), and chassis. Another issue is tubes' internal capacitances, and there is sweet nothing you can do about them.

All other components, from potentiometers to filtering chokes, also have parasitic capacitances, and all those capacitances will impact the sound of an amplifier. You cannot see them or touch them, they are not discrete capacitors but parasitic (unwanted) capacitances distributed across cables and components.

The sound of a capacitor, or any other component for that matter, is determined by two main factors. One is its conductive parts, the cathode, the grid and the anode of the tube, the conductors in a cable, the metal plates and leads of a capacitor, the winding wire of a transformer.

The other factor is the dielectric between those conductors. The same cable, say silver-plated OFC (oxygen-free copper) interconnects, will sound different with a Teflon® (PTFE) jacket and a PVC jacket. The same output transformer will sound different if a paper is used for insulation between its winding sections instead of plastic film such a Mylar® or Kapton®.

The sound of magnetic components such as audio transformers will also depend on the properties of their magnetic core (laminations). A transformer with GOSS (Grain-Oriented Silicon Steel) core will sound different from an otherwise identical transformer wound on an ordinary 3% silicon steel core or a 49% nickel core.

Different resistor and capacitor technologies, construction methods and materials sound different

Every component imparts its sonic signature onto the signal. However, since this claim cannot be verified by measurements, many audio "experts" claim that all resistors and all capacitors sound the same. Any audio designer and amplifier builder would disagree.

For instance, carbon composition resistors may be noisy and grainy, but they sound the warmest of all.

You wouldn't use them in preamplifiers or phono stages, where the signals are minuscule, and signal-to-noise ratios are critical, but they sound soft and musical in power amplifiers. On the debit side, they don't age well, their values drifting significantly over time, and their tolerances are the poorest of all resistor technologies.

Carbon film resistors are superior in terms of noise yet sound softer and warmer than metal film resistors. To me, metal-oxide film resistors don't seem suitable for audio, imparting a cold character and harsh edge to the sound.

As for film capacitors, film & foil caps are sonically superior to the metallized-film type. The plastic film, which acts as a dielectric between metal electrodes, is usually polypropylene, polyester, or polystyrene. Some capacitors use paper as a dielectric (traditionally impregnated in oil), hence their name PIO (Paper-In-Oil) capacitors.

PIO is a vintage technology that precedes the use of plastics. Oil loses its insulating properties when subjected to high temperatures, and paper is inferior to plastics in every way, so these capacitors don't last nearly as long as the plastic film types. However, many audiophiles prefer the sound of PIO caps, making them highly sought-after!

Why do interstage transformers sound better than coupling capacitors ?

> "Why would you buy a step-up transformer instead of just sticking with a high-gain phono preamp? Because, with exceptions, transformers offer better sound - providing more drama, more color, and, especially, greater touch and impact."
> Art Dudley in his "Listening" column, *Stereophile,* March 2017

Most preamplifiers and power amplifiers (both solid-state and tube) involve more than one amplification stage. In some cases, the stages can be directly coupled (connected), but usually, a film capacitor is used as a coupling element to separate DC (direct) currents and voltages in those stages. Different DC voltages bias each stage differently so they can operate properly, and since capacitors don't pass DC currents, there is thus no interaction between the stages in DC terms; only an AC signal is passed through.

The same aim can be achieved by using interstage transformers. Transformers don't pass DC voltages or currents between their primary and secondary windings and are thus perfect for galvanic isolation between amplification stages. However, interstage transformers are physically larger, heavier, and much more expensive than coupling capacitors. "Much more" in this context could mean up to 500 or even 1,000 times more! If a good quality film capacitor costs $2.-, good interstage transformers cost $1,000+ each.

As we have seen on page 29 (Psvane film capacitors), there are "audiophile" capacitors brazenly priced at $200 or even more, narrowing the pricing gap between them and interstage transformers.

Due to their high cost, interstage transformers are used only in the most expensive tube amplifiers and preamplifiers, where their sonic superiority to coupling capacitors can be fully appreciated. Nobody knows why inductive coupling through a transformer sounds better. One possible reason is a higher distortion of interstage transformers (compared to coupling capacitors) and the even harmonics they generate that sound euphonic to us humans.

With transformers, there is either no phase shift between primary and secondary signals (they are "in phase") or the phase of the signal is inverted (180° phase shift).

On the other hand, capacitors introduce a 90° phase shift between the AC current flowing through them and the AC voltage drop across them. Although I haven't seen any research that would corroborate this view (not even a serious discussion in a book or audio forum), a quarter-wavelength phase shift (90°) may be more sonically objectionable than the half-wavelength (180°) shift.

Ultimately, it may not be just the distortion or the phase shift behind the sonic differences, but also the physical mechanism of the signal transmission. Audio signals propagate through capacitors by electrostatic polarization of their dielectric. Transformers operate on the electromagnetic induction principle; a signal flowing through one winding creates a magnetic flux in the magnetic core that couples the two windings, and that changing flux induces a voltage in the secondary winding.

The nature of the transmission and the interaction between the input and output signals of these two components are fundamentally different.

Breaking in equipment - myth or reality?

Most audiophiles believe that all audio equipment and even individual electronic components (on the micro-level) change their sonic signature after a few weeks or months of use. There are also some absolutists, such as Ethan Winer, the author of *The Audio Expert*, who still dismiss such possibility.

Again, a car analogy is in order. Most brand new cars take a few months to "settle down" and generally drive better, smoother, and quieter after 3-5,000 km.

The primary issue of engine break-in is the settling of piston rings against the cylinders' walls. Still, the brakes, the hydraulic systems, air-con compressors, and many other mechanical and hydraulic devices also perform better after the initial settling period.

By extension, mechanical transducers with moving parts, for instance, microphones, phono cartridges, headphones and loudspeakers, even turntables, should also be subject to the settling-in syndrome and thus benefit from a burn-in period. The real debate is about electronic components and systems, those seemingly without any moving parts.

Do electronic components (tubes, transistors, capacitors, inductors, transformers, etc.) sound different after fifty or a hundred hours of usage? I believe they do. Such changes in sonics can be detected even when changing resistors or capacitors but are very noticeable with complex audio equipment such as amplifiers, speakers, and CD players.

There are many possible reasons for the change in sonics over time. Continuous cycles of heating & cooling (thermal expansion and contraction) are just one possible explanation in the case of both transistors and (especially) vacuum tubes. Subtle changes in the structure of their materials are another.

Transformers, chokes, and capacitors exhibit "memory" effects (hysteresis, for instance), where the state of the magnetic and dielectric circuits depends on their previous operating conditions. If that isn't a temporal change, I don't know what is.

The most heated break-in debate seems to center on cables, with skeptics proclaiming, "It's just a piece of wire for goodness sake, it cannot possibly be directional, and there is no way its sound could change over time!' Perhaps paradoxically, speaking from personal experience, the most noticeable break-in changes were in the sonic signature of cables, especially interconnects and speaker cables.

Over the years, I've used silver-plated Teflon®-insulated wire on many occasions to make both interconnects and speaker cables. Initially, I disliked their sound as being too bright and unnaturally sparkly. Still, after fifty or so hours, they would mellow down and become much more enjoyable to listen through, superior to "ordinary" copper cables with PVC or polyethylene dielectric.

Another class of components that often don't sound that great initially are audio transformers such as MC step-up transformers and tube amplifier output transformers. Still, the most significant break-in effect was noticeable in the case of interstage transformers.

> THE SONIC VARIANCE BETWEEN SEEMINGLY IDENTICAL COMPONENTS
>
> Complex components of a mechanical nature (with many interrelated parts), especially if built by human hand (phono cartridges, vacuum tubes, loudspeakers, and headphones), can significantly vary in sound quality, even when made in the same factory. Auditioning one such component (well broken in) and buying another, still wrapped and "virginal," has resulted in the disappointment of many an audiophile.

THE LISTENING ENVIRONMENT AND ITS IMPACT ON THE ENJOYMENT OF MUSIC

Geopathic conditions and their impact on the sound in your listening room

Geopathic stress is the umbrella concept dealing with a wide range of earth energies from natural (geological) and man-made features. The flow of underground water streams and currents, the magnetic and other impacts of the moon cycles, natural radioactive radiation from subterranean mineral deposits, ionospheric activity, solar flares, and magnetic grid lines across the surface of the Earth are just some of the natural phenomena that impact humans and other living organisms on Earth.

There's also a strong indication that the remaining (lingering) energies from previous land and house occupants and events (especially the violent ones such as battles and murders) can be felt by some more sensitive people. Even the skeptics who dismiss such possibility feel uneasy and restless in such places, unable to relax or sleep.

Electromagnetic and technopathic stress from various industrial and telecommunication sites and systems, including your own household emissions (TVs, microwave ovens, mobile phones, etc.), are much less controversial. Still, their impact can be even more severe, including long-term health problems. Fatigue, stress, persistent ill-health, and even cancer have all been linked to the so-called "sick house syndrome."

Even real estate agents would be able to tell you a tale or two about properties that are hard to sell (people immediately feel uneasy there), and when they are sold, the owners don't stay there for long. Pay more attention, and you will see 5-10% of properties back on the real estate market every few years, changing owners perhaps five times in ten years, while other properties stay under the same ownership for 20, 30, 50 years, or longer!

Many animals instinctively avoid such places, dogs being the best example. They seek positive or "yang" energy. Interestingly, cats prefer such spots and places of "negative" or "yin" energy. So, if your cat stays away from your listening room, but your dog loves it, it's a good sign.

The research in this fascinating field started relatively late, at the end of the 1920s, but since then, many studies have shown clear adverse effects and ill-health linked to such geopathic stressed places and phenomena, including significantly higher rates of cancer in people who lived or slept there for prolonged periods of time.

Schumann resonances

Extra-low frequency (ELF) electromagnetic waves naturally occur in the space between the surface of the Earth and the ionosphere. The 7.83 Hz is considered the fundamental so-called "Schumann frequency," but other frequency peaks have been detected and measured, around 14, 20, 26, 33, 39, and 45 Hertz.

Researchers have discovered an uncanny similarity between spectral power density profiles and patterns of the ionospheric resonances and the electrical activity of the human brain. They've even identified periods of synchronicity when the human brain frequencies would synchronize with or "tune into" the Schumman's resonant frequencies, just as two oscillatory systems can do in physics experiments.

It is not clear what impact such ELF waves have on humans. Still, it is believed that Schumann resonances influence the biological oscillators within the human brain so much so that their absence or alteration results in mental and physical health problems.

Recent research has focused on the impact of man-made electromagnetic "smog" (electromagnetic radiation by power lines, lighting, radio communications, and appliances such as computers, TVs, cell phones, radios, digital devices, air conditioners, etc.) on the natural geomagnetic activity. Significant changes in the intensity and stability of the Schumann waves were detected. Even the resonant frequencies have been altered to those that humans have not been used to, which could lead to changes in human brain rhythms and the disruption of our own bio frequencies, increasing the frequency and severity of various acute and chronic diseases.

Schumann generators

Experiments described and documented in scientific journals have studied, among other aspects, the influence of artificially generated low amplitude 7.83 Hz sine waves on fundamental cellular processes and identified their protective effect during oxidative stress. The natural Schumann waves are a superposition of many individual time-delayed signals and are thus much more complex than the pure sine wave produced by a simple generator.

Such generators have been marketed and sold as "relaxation" devices to "soothe nerves and improve sleep," and even for offices, to improve "inspiration and work efficiency." Such relaxation and sleep improvement devices are typically priced under US$50.- but those aimed at audiophiles are much more expensive.

Acoustic Revive was the first vendor to release their model RR-77, followed by the more powerful RR-777 and the "ultimate" RR-888, with "an improved accuracy of the waveform and a low noise circuit," priced at US$650-700.

Acoustic Revive claims that the unit "neutralizes electromagnetic waves generated from an audio equipment and external harmful radio waves. So interference of electromagnetic waves and radio waves that exist among equipments is disappeared (sic)". They even claim that "it neutralizes harmful standing waves. And it has the feature of improving the viscosity of the air in the listening room."

Poor English aside, I doubt that a single frequency signal of very low intensity can neutralize anything. It is such ambitious yet unsubstantiated claims that lead the absolutist camp to completely dismiss many important issues and inflict irreparable damage to the advancement of our collective audio cause.

It isn't exactly known how such devices improve the quality of the sound, but alpha brain wave pattern stimulation is the most likely mechanism. The human brain generates alpha waves during a calm, meditative and relaxed state. Ultimately, audiophiles report a subtle but noticeable improvement while using such devices.

Intrigued, we constructed our own Schumman generator, one producing a square waveform and another one with a sine wave output. The square wave generator made listeners unable to relax and enjoy the music; everything sounded harsh and strident. The sine wave generator seemed to have a mild but positive impact on most listeners.

The effect of geopathic stress on audio equipment

It is possible, indeed very likely, that such natural and "paranormal" phenomena affect our mental and physical state as listeners and thus impact our enjoyment of reproduced music. There's, however, another possibility that just as humans (biological systems) are affected, audio equipment and other complex technical systems could be affected themselves.

In their research paper "Effect of Geopathic Stress and its correction on human body and machinery breakdown", published in Sept. 2014 in *Landmark Research Journals Medicine and Medical Sciences*, A. Poddar and S. Rana found that geopathic stress zones not only negatively affect the health of people who inhabit them but also result in a significantly increased breakdown rate of machinery and electronic equipment in such zones.

Geopathic stress zones in their investigations were detected by the Lecher antenna, a mechanical instrument based on the resonance principle that works as a more elaborate type of dowsing rod, capable of detecting and even measuring a wide range of radiation fields.

How to dowse your music room

There are many good books on energy dowsing (or water divining as it's also known), so we will only make a brief summary here. Take two L-shaped copper rods (other materials can also be used), but such copper rods can be cheaply bought online, just as we did.

Study the principles of dowsing from materials online or one of the many books just mentioned. Holding the shorter end of the L-rods, one in each hand and parallel to each other (see the photo on the next page), relax your mind and your muscles. Don't grab the rods tightly or clench your fists; hold them very gently so they can rotate freely.

Walk around your listening room until the rods either cross over each other (turn inwards) or open up, away from each other. Mark these spots on a large printed plan of your room. Join the dots (literally!), and you will obtain a grid of regular, parallel, and perpendicular lines, as illustrated here on the plan of our listening room.

The spacing of the regular Hartmann lines that you'll divine in your room may differ from ours. These are not of particular concern anyway, although it would be best if your listening position (the "sweet spot) isn't directly on one of the lines, and especially not at a point where two lines cross.

In some cases, if there is significant underground water flow or other geological faults (irregularities) in one part of the room, you may get an irregular line crossing the room at a different angle (illustration on the previous page).

These fault lines are of much more concern since they result in negative or malign energy, detrimental to human health. Again, if your listening position falls on such a fault line, consider changing your setup and moving your listening position and equipment as far away from the fault line as possible.

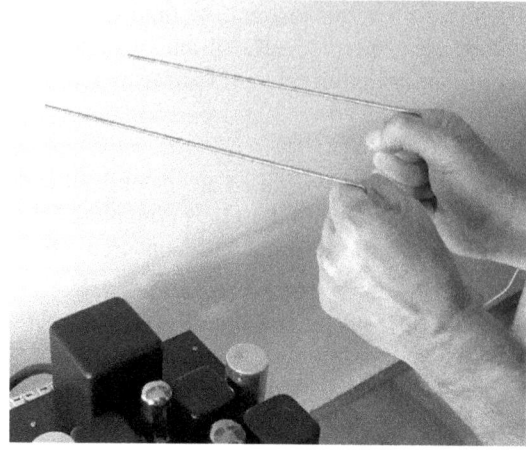

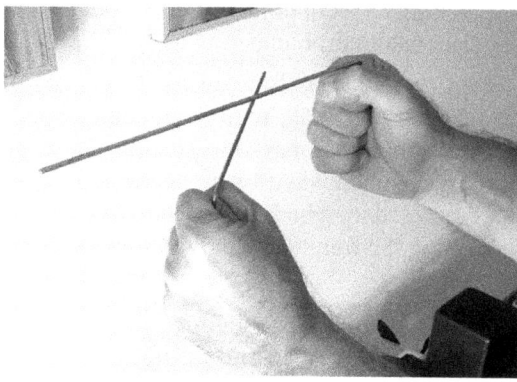

Fig. 3.29. Divining our music room with two copper L-rods in parallel position, just before a grid line, and at the grid line's outer edge, behind the equipment rack, marked (1) on the plan. Hartmann grid lines, fault lines, underground water channels, and power centers are indicated by the rods crossing or splitting apart. Which one of these is present is a matter of further, detailed investigation.

Fig. 3.30. Dowsed energy lines in our listening room showing regular Hartmann lines and one irregular fault line.

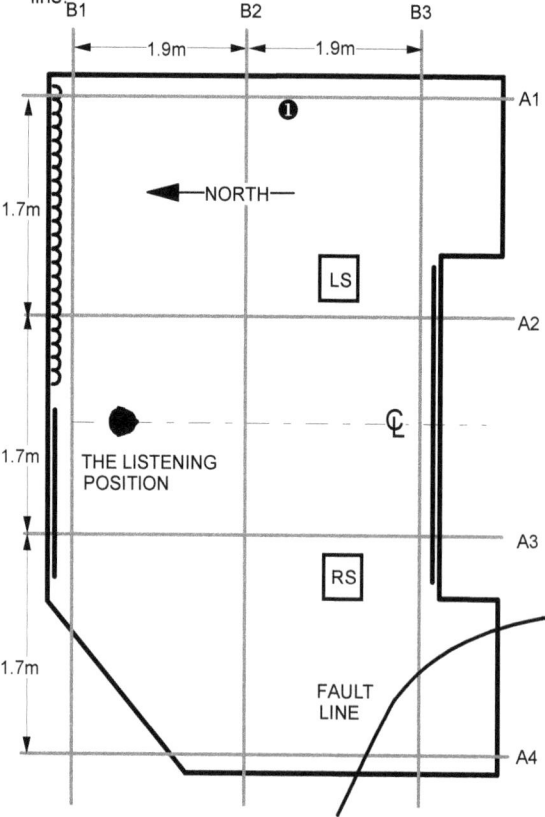

4 CLEANING UP THE POWER SUPPLY TO REDUCE NOISE, HUM, AND INTERFERENCE

Every reputable audio designer and builder worth his salt will tell you that a power supply is the heart of any amplifier; it is impossible to have a top performing and superior sounding amplifier with a lousy power supply.

Likewise, with so many audio components powered up by the mains voltage, its quality is of paramount importance. A dirty and distorted power supply cannot result in a clean and convincing sound!

Since phono stages, preamplifiers, power amplifiers, active crossovers and a myriad other audio components work by "modulating" a DC voltage derived from the AC mains supply, the cleaner (less noise, interference and "grunge") and more stable the mains supply, the better sounding (cleaner, more dynamic, more "relaxed" and musical) your audio system will be!

- MAINS POWER SUPPLY: DIRTY, DISTORTED & DANGEROUS
- ISOLATION TRANSFORMERS AND THE BALANCED POWER SUPPLY PRINCIPLE
- ELIMINATE DC FROM YOUR AC POWER OUTLETS
- AC VOLTAGE REGULATORS AND REGENERATORS
- REDUCING ELECTROMAGNETIC AND RADIO FREQUENCY INTERFERENCE
- UPGRADING AUDIO EQUIPMENT AC & DC POWER SUPPLIES
- BATTERY-POWERED PHONO STAGES, PREAMPLIFIERS AND POWER AMPLIFIERS

> "A dirty and distorted power supply cannot result in a clean and convincing sound!"
> Igor S. Popovich, audio philosopher and the person you should blame for all errors, omissions and stuff-ups in this book

MAINS POWER SUPPLY: DIRTY, DISTORTED & DANGEROUS

Ideal versus real mains voltage

The ideal AC supply is an undistorted sine wave (without any harmonics) of a fixed and stable frequency, with no DC component or DC "offset" either. Your mains supply could be close to this ideal or very far from it, depending on where you live. How can you tell if you live in a place that belongs to a "name and shame" power quality list? I devised a straightforward yet highly accurate test: if your local department store or hardware shop sells various models of power conditioners, voltage regulators, and "power stabilizers", you most definitely do.

Even if you live in a more "civilized" part of the world, the quality of your mains will range from acceptable to shockingly bad. It will depend on the size of the power grid (the larger the grid, the more stable or "stiff" it usually is), the distance from your place to local industrial areas (full of nasty power polluters), and the overall quality of the power authority's infrastructure - power generators, transformers, transmission and distribution lines.

The voltage waveform illustrated below (redrawn from the photo of our oscilloscope screen) clearly shows a significant distortion. The tops of the sine wave are chopped off, but the positive peaks are distorted differently from the negative ones. That asymmetry will cause a significant DC component, which can create all kinds of problems in your audio system.

We cannot see frequency fluctuations from a static waveform such as this one (one snapshot frozen in time), but they are always present; the only question is how serious that fluctuation is.

In some 50 Hz (the mains frequency) cities, the frequency varies between, say, 47 and 52 Hz, which isn't that bad, compared to other places, where it may fluctuate between 45 and 53 Hz (usually the frequency dips are more pronounced than the peaks).

We can do nothing about frequency fluctuations except using power regenerators, which totally recreate the mains voltage of a stable and constantly controlled frequency. Voltage regenerators are also a solution for the problem of significant DC voltages present on AC power lines. In fact, they are a solution for all mains power problems. However, they are expensive, complex devices can and do fail (just as all inverters do, as in inverter-type air conditioners), and their power capacity is always limited. More on them shortly.

Luckily, there are much simpler and cheaper solutions when it comes to the DC on AC lines problem, as we'll see soon.

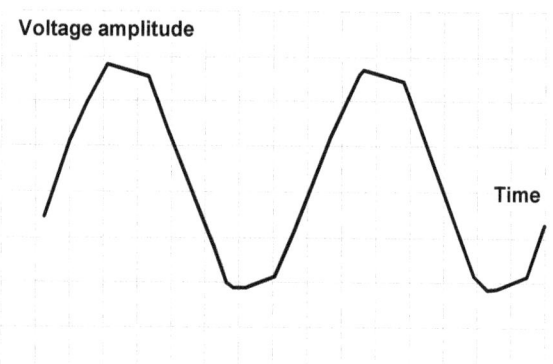

Fig. 4.1. The oscillogram of severely distorted mains voltage, typical in some areas of Australian cities, close to industrial estates and heavy industries

Evaluating the quality of your mains voltage

This line type of EMI meter does not measure the electromagnetic fields in the surrounding air, like traditional EMI meters, only the electromagnetic interference on the power (mains) circuit. More specifically, on a single-phase AC power line, such as the one that is supplying the power sockets in your listening room.

Once plugged in, it measures total line noise in milliVolts peak-to-peak. The maximum that can be displayed is 1,999 mV (almost 2 Volts) in the frequency range of 10 kHz to 10 MHz. Some similar models only operate up to 700-800 kHz, which is too low, with the most harmful noise frequencies in the 1 - 10 MHz range.

The internal speaker (1) acts as an audio indicator, reproducing the amplitude-demodulated waveform of the interference. This can help identify the source of the EMI signal, such as AM or short-wave radio stations, motors, or electric arcs from welders.

In 2020 this model sold on eBay for around US$90 (including postage from China).

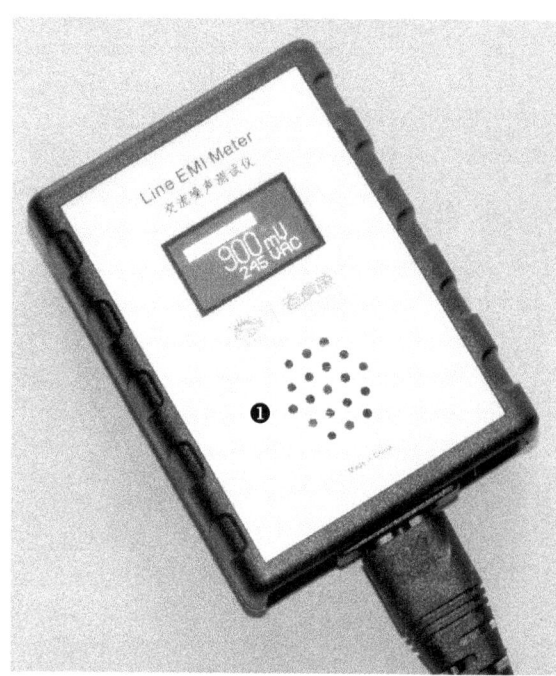

Fig. 4.2. 0.9 Volts (900 milliVolts) of superimposed electromagnetic interference in the 10 kHz to 10 MHz frequency band on our mains power line (245 V) is a pretty bad result.

CLEANING UP THE POWER SUPPLY TO REDUCE NOISE, HUM, AND INTERFERENCE

Install a dedicated mains power supply circuit to your listening room

The mains supply to your audio components is one of the major impediments to achieving the ultimate sound. In old houses, built in the 1930s, '40s, and '50s, when copper was scarce due to wartime use as a strategic material (it is still classified as such in Australia and subject to a very high government tax), the mains wiring is most likely on a thin side.

Also, the purity of these old copper cables may not be as good as it is today. So, it pays to rewire your whole house, or, at least, to run a dedicated modern power cable from your house's switchboard to your listening room. Builders try to save every penny, and even modern installations are likely to be substandard. When we had our house built, the drawings showed all 18 power points on the same circuit!

I paid additional fees for a three-phase supply to the house to be installed and asked the builder to split the power points into two separate circuits, each on a different phase. The power points used for dishwasher, microwave oven, laundry (washer and dryer), kitchen appliances (mixer, toaster, etc.), and study (computers, printers) are on the "dirty" circuit, while the bedrooms and the music room are on the another, "clean" circuit.

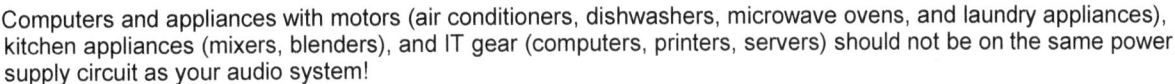

> SEPARATE DIRTY LOADS FROM YOUR HI-FI CIRCUIT
> Computers and appliances with motors (air conditioners, dishwashers, microwave ovens, and laundry appliances), kitchen appliances (mixers, blenders), and IT gear (computers, printers, servers) should not be on the same power supply circuit as your audio system!

The earthing issues

There have been cases where all efforts to clean up the mains power supply to the listening room have failed, and the cause was found to be noise and interference propagation through the earth connection.

Most houses have their own earth stakes driven relatively deeply into the ground. Such a stake (1) is connected via the earth wire (3) to the network of water pipes, which are usually made of copper, an excellent conductor of electricity. That effectively grounds the water supply and the main switchboard earthing bus, where all earth (ground) wires from power points and light fixtures terminate.

Unfortunately, the crimped connections, such as the two pictured, (2) and (4), deteriorate through oxidation, which increases their contact resistance. In some cases, such a connection may even be completely lost. In such a situation, find an electrician with advanced knowledge of earthing issues and principles. Get them to measure the earthing resistance and fix the problem for you if it's found excessive. A new earth wire and a deeper stake should be installed, and all deteriorated crimped connections replaced.

Fig. 4.3. The earth stake (1) is connected to the garden tap copper pipe (2) and to the switchboard (a meter or so away) via the multi-stranded earthing conductor (3).

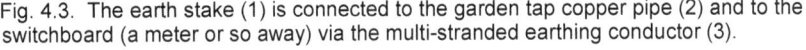

SOLUTION: *Ensure the power cables have a correct AC polarity*

Australia, New Zealand, China, and Argentina use a 10A-rated plug with two flat 1.6 mm thick blades, set at a 30° angle to the vertical. These are live and neutral pins and are insulated with plastic covering (white in the photo but usually black). The earth pin is longer, making contact first while plugging it in and breaking the contact last when unplugging a device, a safety feature.

As with most other major world plugs, there is also a 2-pin version without the earth (ground) pin, but that type can only be used on double-insulated devices and can only be plugged in one way.

If unterminated with the IEC plug, as in the photo, most cables come with the earth lug already attached to the earth wire (5).

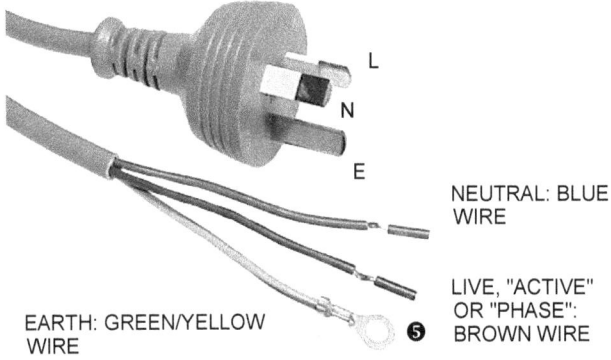

Fig. 4.4. The Australia & New Zealand mains plug and wiring color-code

Fig. 4.5. Top-to-bottom: European Schuko plug, United Kingdom plug, and USA 3-pin plug.

Most European Union countries use the so-called "Schuko plug," the name coming from the German "Schutzkontakt" or "protection contact." There is no earth pin; the earth clips of different shapes and sizes are positioned on both sides of the plug. The two uninsulated round pins in the horizontal plane are L (live) and N (neutral). The pins aren't insulated since the EU socket is recessed, a feature that prevents finger contact.

The problem with such plugs and sockets lies in their symmetry. Since they can be mated in two different ways, the only way to ensure optimal sonic results is to try both plug orientations and *listen*!

Used mainly in the UK, Ireland, Cyprus, Malta, Malaysia, Singapore, Hong Kong, and some Middle Eastern countries, this plug has rectangular pins and an integral fuse. As with the AU plug, the L & N pins are insulated, and the uninsulated earth pin is longer. The cable color-coding is identical to AU and EU cables.

The USA or, more accurately, "North American" power cable (let's not forget Canada and Mexico) is usually rated at 15A and $125V_{AC}$. The ground (earth) pin is longer than the line and neutral flat blades, so the device is grounded before the power is connected.

While older audio components used permanently wired power cords, modern designs mostly use detachable cables. At one end is the plug of the country the amps are sold in (such as those discussed above), and at the amplifier end, the cable is terminated by the IEC plug.

Notice the standard wiring, more precisely, the orientation of the "live" and "neutral" contacts. The audiophile power cable in the photo below has its L (live) and N (neutral) wires reversed.

The orientation of the plug on the photo is the same as the plug on the right, so the voltage between the contact on the right (neutral) and the middle contact (earth or ground) should be zero volts or close to it (not more than a couple of volts). However, we measured a full 244 V (our mains voltage in Perth, Australia).

Having a reversed AC polarity power cable in your audio system could have audible effects, depending on the wiring of the audio component the power cable is used with. To check if your power cables are wired correctly, get a multimeter (on the "AC Volts" range) and measure the AC voltage between the center pin of the IEC plug and one of the end pins. One pin should be at around 100 V (Japan), 120 V (USA), 230 V (Europe), or 240 V (Australia), the other at zero Volts or thereabouts.

Assuming your audio components are wired correctly in the factory (or during your DIY construction), all your power cables should be wired the same, correct way.

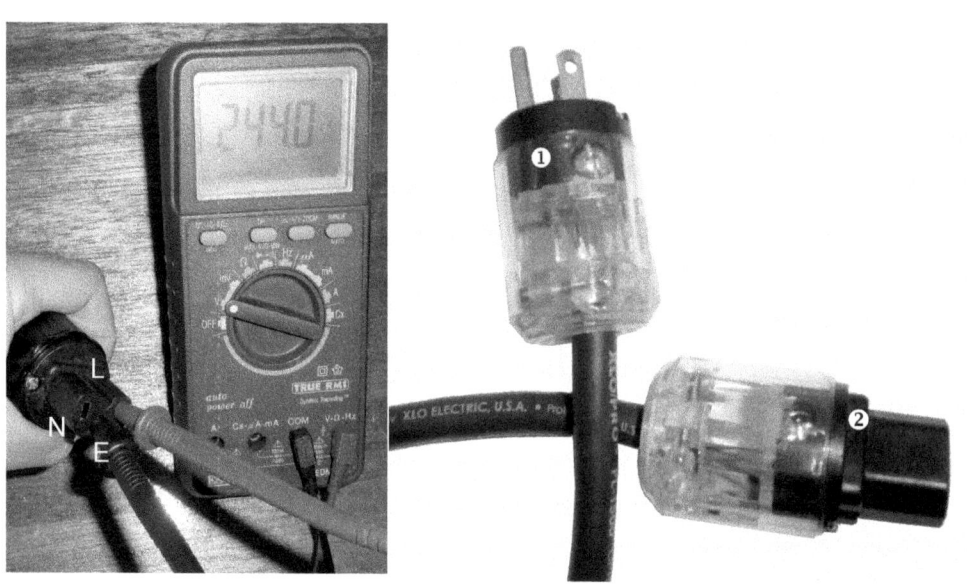

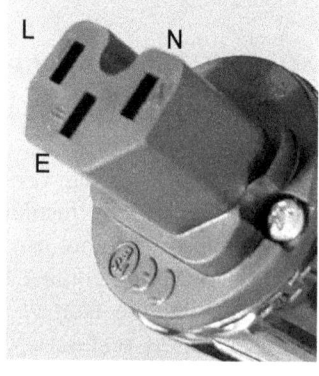

Fig. 4.6. How to check the power cable wiring for AC polarity. If incorrect, open one end, either the male plug (1) or the female plug (2), and reverse the live and neutral wires.

If they are not, open the improperly wired ones at one end, either the male plug side (1), the one that plugs into the wall socket, or the female plug side (2), that plugs into an audio component, and reverse the live and neutral wires. As we have seen on the previous page, these will be either black and white (in the USA, for instance) or blue and brown (in most other countries) wires.

Safety precautions and audio equipment handling rules

Audio components are fragile and expensive electronic devices, so careful handling is in order. Also, since they always involve lethal voltages and often even high temperatures, a few basic safety rules should be understood and consistently adhered to! Here they are, in no particular order:

- Audio amplifiers, especially tube amps, involve voltages that may be lethal, so precautions must be taken to mitigate the risk of an electric shock if you are testing them, modifying, and working on "live" circuits. Get an electrician to install in the switchboard of your house or workshop a device that goes under many names: ELCB (Earth Leakage Circuit Breaker), GFI (Ground Fault Interrupter), or RCD (Residual Current Device). In Australia, these are now mandatory by law, but electrical laws in many other countries are much laxer.
- Consider using an isolation transformer to power up your audio system. That will all but eliminate the probability of an electric shock and may even improve the sound of your audio system by reducing noise and interference and cleaning up the power supply to some extent - more on that on the next page.
- Vacuum tubes get hot very quickly. Never touch or handle hot tubes with bare hands. Use a thermally insulating kitchen glove.
- Tubes and high-powered solid-state amps (especially those working in Class A) get very hot, making them serious fire hazards. Remove all flammable material from their immediate surroundings and make sure there's nothing placed on top of them. They should always be placed on the top shelf of your audio rack.
- Never leave your audio system on when you are not around. Unplug all components when away on holidays.
- Treat tubes gently; one moment of lapsed concentration may cost you hundreds of dollars. When pulling tubes out, do no jerk them left and right; pull them straight out vertically.
- Amplifiers are heavy; bigger ones can weigh 30, 50, or even 70 kg! Exercise caution when lifting them up to your hi-fi rack and whenever you move them around. Bend your legs (your knees), not your back.
- Keep children and pets away from your hi-fi rack, and, if possible, ban them totally from your listening room.
- Never use water or chemical cleaners to clean your audio components. Dust them off and vacuum regularly to minimize the ingress of dust into their chassis (enclosures).

A critical safety upgrade for vintage amplifiers and other audio components

Most amplifiers, preamplifiers, and other pieces of audio or test equipment built before the 1970s in the USA used a two-pin mains plug. There was no earth pin, so these units were not earthed (or grounded) at all (1). Furthermore, these 2-prong plugs are reversible, so instead of the neutral being connected to the metal chassis, you can end up with the live 115-120 V_{AC} on it.

Over the years, the capacitors connected between the phase and neutral (2) and the neutral and the chassis (3) become leaky or completely short out, which allows the mains voltage to end up on the metal chassis. These capacitors, affectionately called "widow makers," are a serious hazard and must be removed. Our example, the Scott 299D tube amplifier, had two of those, C211 and C212.

These days only X2-rated and safety approved capacitors can be used in such critical positions (the short circuits cannot develop).

With such vintage audio pieces, you must perform this safety conversion or pay an audio technician to do it for you.

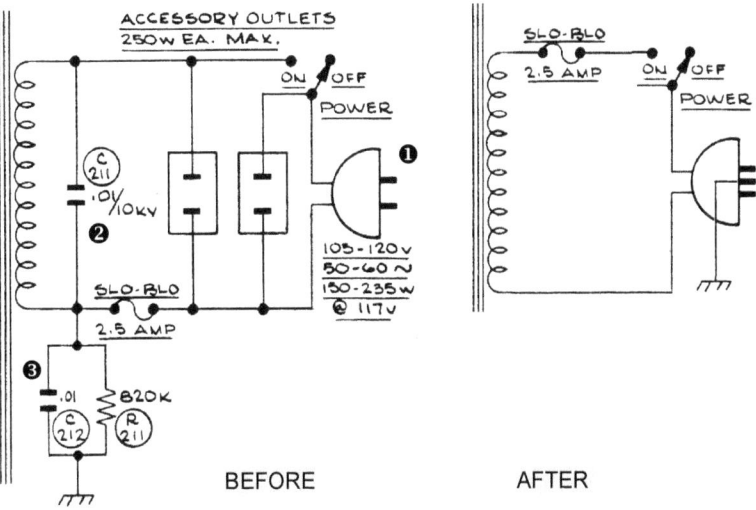

Fig. 4.7. The mains circuit of Scott 299D tube amplifier in its original state and after safety modifications.

Bin the two-core mains cable, the auxiliary 2-pin power outlets, and the two-pin plug, and replace them with an approved 3-core mains cable and plug. Remove any capacitors and resistors connected between phase and neutral and ground, such as C211, C212, and R211 on the Scott 299D diagram. Connect the earth lug via an insulated green or green-yellow wire to a dedicated bolt with good galvanic contact with the metal chassis.

This upgrade does not apply to double-insulated or "floating" devices, such as many CD players and DACs. Earthing (grounding) such instruments may render them inoperative or even cause unwanted short circuits.

The most dangerous piece of advice of all times

Cary 805AE amplifier's user manual says this in its "Troubleshooting Guide" section: "Hum or 'buzzing' in speakers Cause: ground loop Remedy: Install 2-pin adapter in AC cord to float the ground."

You should never "float the ground" by using such "cheater plugs" or by disconnecting the earth (ground) pin from the power cord or the amplifier. This would disconnect the metal chassis from the ground conductor in your electrical installation and make the amplifier not just illegal but also deadly dangerous. If a fault develops and the amplifier starts a fire, a subsequent investigation may result in your insurance claim being rejected.

Should a short circuit develop, the chassis may end up at a high DC voltage of 400-1,200 V (depending on the amplifier type and design) or an AC mains voltage of 100-240 V (depending on your country's mains voltage level). Even if a "ground fault interrupter" or "earth leakage circuit breaker" is installed in your switchboard, such protection would be ineffective since there is no connection to the earth conductor. Touching the chassis would result in a possibly lethal electric shock.

> **SAFETY WARNING**
> Never, ever use "cheater" plugs or disconnect the earth pin from a 3-pin power cord or an audio component!

ISOLATION TRANSFORMERS AND THE BALANCED POWER SUPPLY PRINCIPLE

How isolation transformers work

Isolation transformers galvanically isolate the primary and secondary windings. Touching the secondary L_S (Live) or N_S (Neutral) terminal would not result in electric shock since these are floating, not "referenced" to the ground, so the path through the human body cannot be closed, and circulating currents cannot flow.

As indicated by the arrows, one fault in the secondary (load) side, such as a short circuit to earth (or the metal case, which is earthed), is tolerated. It will not endanger the user since it will simply "earth" one of the secondary terminals. However, the isolation properties of the transformer would be lost since that would reference the secondary winding to the ground.

To have two faults simultaneously is highly unlikely. Even if it does happen, it will short-circuit the transformer's secondary winding, and the secondary fuse (4) would blow or the primary circuit breaker (3) would trip. Notice that only the transformer is "floating" (its windings); its metal case (enclosure) is always grounded or earthed (5)!

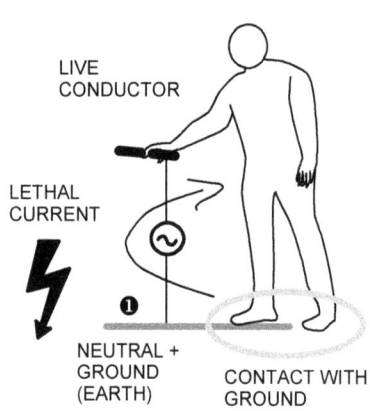

Fig. 4.8.1 Ordinary mains installation with neutral and earth bonded together at the switchboard (1). Once the body closes the circuit by touching a live conductor, lethal currents flow!

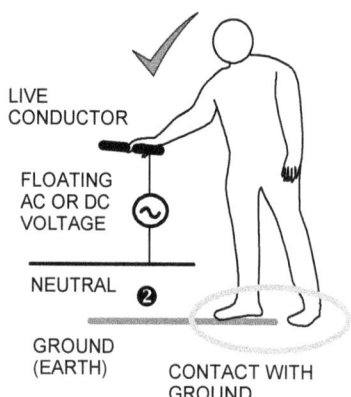

Fig. 4.8.2 The "floating" system when isolation safety transformers are used. The phase & neutral are not referenced to ground (2), so touching a live conductor is harmless.

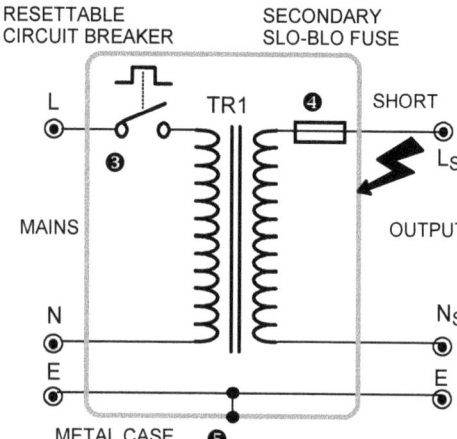

Fig. 4.8.3 A short circuit (between Ls or Ns and the metal enclosure) on the secondary side would not endanger the user but the isolation properties of the transformer would be lost.

Ordinary (unbalanced) versus balanced power supply

While beneficial from the safety perspective, most noise, interference, and voltage spikes propagate right through ordinary isolation transformers. In this case, the turns and the voltage ratio is 1:1 (the standard connection of the isolation transformer), so the secondary spike is of the same amplitude and phase as the original one on the primary. The two black dots indicate the winding ends at which the voltages are "in phase."

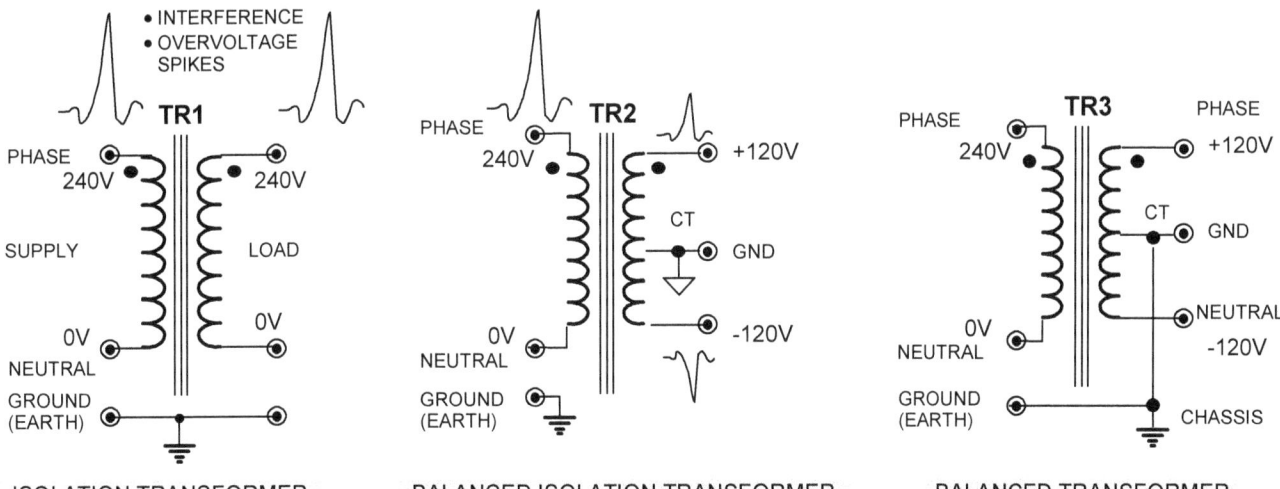

Fig. 4.9. Balanced isolation transformer = balanced transformer + isolation transformer

Transformers have limited bandwidths, meaning they only transmit AC signals of relatively low frequencies between the primary and the secondary. This is good news since interference signals of high (or "radio") frequencies will be strongly attenuated. Thus, the unbalanced or "single-ended" isolation transformer reduces noise and interference as well. The good news doesn't stop there; there is an even better option.

A balanced power supply is inherent in a mains transformer with a secondary CT (Center Tap). We still have 240 V_{AC} in Australia & New Zealand, 230 V in Europe, or 120 V in the USA at the secondary, but this time the center tap is earthed or grounded, not one of the ends of the secondary winding. The CT is more positive than the -120 V terminal, and it's more negative than the +120 V terminal.

The voltage spike is now transformed into two smaller (approx. half the amplitude) inverted spikes out-of-phase with respect to the common ground (0V) at the center tap CT. Therefore, being of equal but opposite polarity, they completely cancel each other. The same happens to radio frequency interference (RFI) and other interference signals. The result is a much cleaner power supply, quieter (or "darker") sonic backgrounds, and better sound! For 120V_{AC} mains (USA, Canada, etc.), the secondary voltages will be +/-60 V_{AC} or +/-50 V_{AC} in Japan, where the mains voltage is 100 V_{AC}. In other words, always +/- half of the required mains voltage.

Be they of unbalanced or balanced variety, transformers pass only AC (alternating current) power or signals between their primary and secondary windings. DC (direct current) voltage or current present on the primary side will not be present on the output, the secondary side. This means that transformers automatically act as DC blockers, which is another piece of good news; no add-on DC blockers are required.

Commercial balanced isolation transformers

"Audiophile" balanced power supplies are generally priced above $2,000.-, yet most only have a toroidal transformer inside a nice flashy machined aluminium box. Luckily, there are cheaper alternatives. This 500 VA balanced "isolation" transformer (photo on the next page) sells for around AU$200 (US$150 at the time of writing), including the international courier from China, we were intrigued and decided to get one.

Sadly, however, as with many made-in-China products, this device was not as advertised. It is a balanced transformer (balanced secondary winding), but it is not an isolation transformer! Why? Because it is connected as per the diagram in the top right corner of this page.

Referring to the photo of its insides (next page), the ground on the input side (5) is connected to the chassis via black wire (6) and metal brackets (7) and to the ground (center-tap) on the load side, so the CT (Center Tap) of the secondary winding is connected to primary "earth," and all the isolation properties are lost!

Notice also a pair of ordinary film capacitors connected between the "phase" and "neutral" on the load side, (8) and (9). These are not approved for such critical use (from the safety and fire rating regulations point of view) and are illegal in such an application.

As already mentioned, only specially-approved X2-rated capacitors can be used in such positions, although here they aren't necessary at all and should thus be removed!

Again, a typical compromised design coming from behind the bamboo curtain. Sadly, there are many such examples around, not just power filters and products but also amplifiers and preamplifiers.

Over the past decades, we've had many such "half-baked" products, mostly tube amplifiers, preamplifiers, and phono stages, on our test bench.

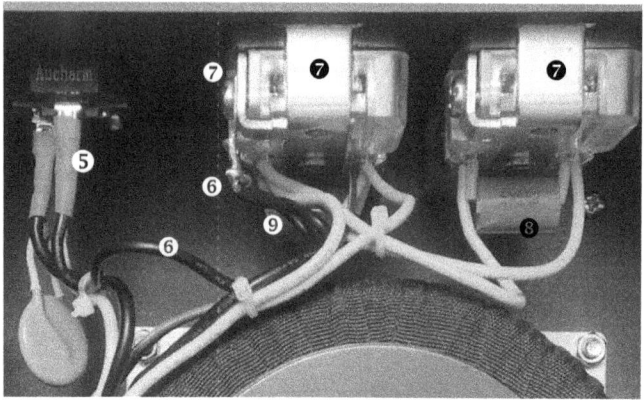

Fig. 4.10. 500VA unbranded, made-in-China balanced "isolation" transformer
1) Front-mounted On-Off switch
2) Rear IEC power inlet with a fuse
3) Toroidal transformer
4) Four USA-style power sockets (outputs)

THE SONIC BENEFITS OF A BALANCED POWER SUPPLY

Implementing a balanced power supply by adding a balanced transformer was one of the best investments described in this book. The sound improved significantly, in almost every way, from the bass speed and definition to sound stage and microdynamics.

Quite a few are featured as case studies (on how not to do it!) in my books *Audiophile Vacuum Tube Amplifiers, Vol. 1-3*. Some well-designed, professionally built, and excellent sounding made-in-China audio products represent an incredible value for money. However, there are also many unsafe (illegal), unreliable, and awful sounding ones!

Why should the mains AC voltage supply be optimized?

You may be thinking now, "Sure, this balanced supply idea sounds good (literally, no pun intended!), but my local AC voltage is way too high (or too low). Is that an issue?" Well, it sure is.

Most audio components were designed for a specific input voltage, except those that use switch-mode power supplies, which almost always operate on a wide range of input voltages. You'll recognize those by simply inspecting the audio component's nameplate. It should say something like "Input voltage:100 - 240 V_AC". In that case, you don't have to worry about your mains voltage being too high, too low, or constantly fluctuating; the power supply will automatically and continuously adjust its operation no matter what AC voltage is present at its input!

For audio components with analog power supplies designed to operate at a single AC voltage and frequency, ideally, your local mains AC voltage should match their nominal voltage. Unfortunately, that is rarely the case! For instance, to save money, instead of using three different internal power transformers (for 220 V, 230 V, and 240 V mains) in their audio gear, some manufacturers use transformers designed for 220 V but declare their products suitable for 220-240 V, a practice that could be considered fraudulent!

First of all, many of those transformers will buzz and overheat at 240-250 V Australian mains, resulting in a short and stressed life of such audio components. Secondly, with the mains AC voltage too high, the transformer's secondary AC voltage will be much higher. Consequentially, all DC voltages inside an amp with such a transformer will also be too high. The operating point of various amplifying stages will be changed from what its designer envisaged, thus changing the sonics of the amp, usually for the worse.

Tube heating voltages are another vital issue. There may be only one, say if all tubes use 6.3V heating, or several (for instance, 6.3V for the preamp tubes and 5.0 V for 300B power tubes). They will be proportionally higher, too, and that will significantly shorten the life expectancy of all the tubes in the amp.

Underheating tubes (say with 5.9 V instead of 6.3 V) is not such a serious issue, although the output power may be reduced and tubes' life somewhat shortened as well.

> **THE MANUFACTURERS' OBFUSCATION**
> Many audio equipment manufacturers specify a range of mains voltages their amps, preamps, and other gear will operate on, the specs saying something like "Mains: 220-240 V_{AC}". Sure, the audio component may work at any voltage within that range, but its output voltage, maximum output power, distortion, and sonic qualities will vary significantly. So, which mains voltage within that range will result in the optimal operation and best sound?

How to use step-up/step-down transformers

Most step-up & step-down transformers have only one input and one output (a fixed voltage ratio). Some, however, have an "input voltage selector" (IVS) at the back, which should give you a bit more flexibility in adjusting the output voltage to the level required by your audio components. How much difference does that feature make?

Let's say you live in Europe, where the voltage has been standardized to 230V. With the IVS plugged into a 220 V socket, you will get the same voltage as your input voltage at the "Output 220 V"(2).

With input voltage (the mains voltage in your house at that particular time) at, say, 227 V, that is what you will get at (2). So, we are talking voltage ratios here, not the absolute values.

Are you going to get 110 V at "Output 110 V" (3)?

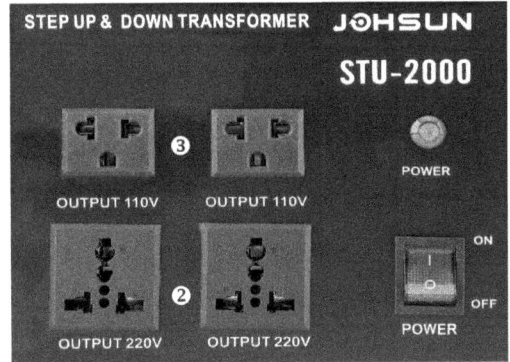

Fig. 4.11. Some step-up/step-down transformers have multiple taps at the primary winding, selected by plugging in the input voltage selector jumper into one of the four receptacles

No, you will get half (220/110) of your input voltage or 227/2 = 113.5 V. So, the principle of the usage of such transformers is to select the IVS closest to your measured (actual) mains voltage to get the same voltage at (2) and half of that voltage at (3).

There are other possibilities. For instance, with IVS plugged in the "240 V" position but with 227 V actual input voltage, you will get less than 220 V at (2). You will get 220/240*227 = 0.917*227= 208 V. Likewise, with IVS plug in the 200 V position but with 227 V actual input voltage, you will get more than 220 V at the output! Since 220/200 =1.1, you will get 10% higher voltage, or 1.1*227 = 249.7 V. This would be useful for equipment designed for 240 V, such as some Australian-made audio components, for instance, for which 227 V would be a bit too low.

If you are in the USA, there is only one IVS position of any use to you, "110 V," so you won't get any flexibility out of this transformer. Whatever your mains voltage is, that's what you will get at (3) and double that at (2). So, 117 V in, 117 V out at (3), and 234 V out at (2).

How to use variable autotransformers

For precise adjustment of the AC voltage, you'll need a variable autotransformer, usually referred to as "Variac" (a trade name of one particular brand).

This made-in-China "variable voltage regulator" is rated at 2 kVA and designed for 220 V_{AC} voltage input. Just like a choke, an autotransformer has only one winding and works on the auto-induction principle. The "variable" part in its name comes from the fact that the output tap is not a fixed point but taken from a rotating metal arm that scrapes along the windings (horizontally), and, since it is attached to an insulated knob (4), it can be adjusted by the user. The arm can be driven past the point on the winding to which the input voltage is connected.

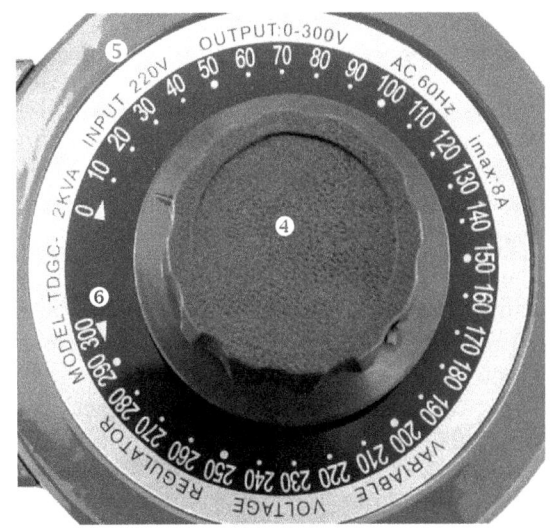

Fig. 4.12. All variable autotransformers have a coarse scale on top. Some also have a digital display of the output voltage, a very handy feature.

Since there are no primary and secondary windings as such, the output voltage can be higher than the input. In this model, with 220 V_{AC} at the input (5), the output can go as high as 300 V(6)! Thus, a step-up "boost" of voltage can be obtained. Since its scale was marked based on the assumption of 220 V_{AC} voltage at its input terminals, the indication of this variable transformer will be wrong at any other input voltage. Luckily some models, like the one pictured below (1), have a digital display of the output voltage, a convenient feature.

How to hookup a basic AC supply/filtering chain

Now you have a variable autotransformer to manually adjust the AC supply voltage for your audio system and a balanced transformer. Since such transformers usually only have a single output, while you need half a dozen or more, you'll also need to add a power distribution board. Forget cheap & nasty ones from your supermarket or hardware store. They have poor quality contacts inside their sockets, so arcing and even fires result.

Fig. 4.13. A variable autotransformer (1), followed by a balanced isolation transformer (2) and a filter/power distribution board (3)

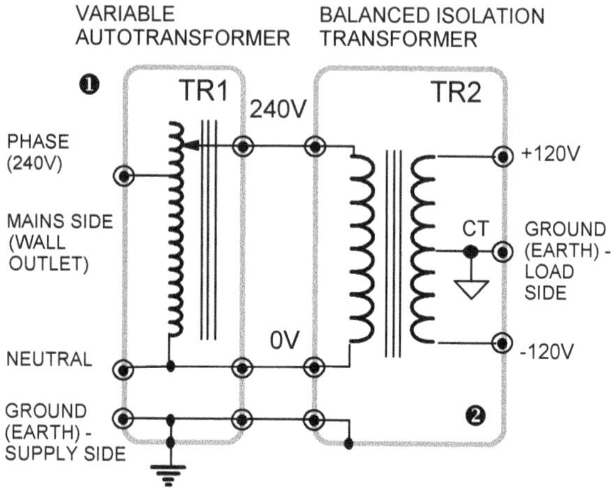

Fig. 4.14. Circuit diagram of the setup (power distribution board not shown)

Invest in a better quality power distribution board with an integral multiple filter and overvoltage protection inside. We'll have a look at one example in a minute.

There is only one correct order of stringing the three required devices. The variable autotransformer should be the first in the power supply chain, and the balanced transformer should plug into it, never the other way around. The power distribution board goes last, followed by all the audio components (loads).

FINDING THE ELECTRICAL "SWEET SPOT"

Using a variable autotransformer to vary the supply voltage from, say, 220 to 250 V (in a 240 V country such as Australia), or from 105 V to 130 V in the USA, you should notice significant sonic changes. A few famous guitar players discovered that trick and used variable autotransformers on stage to find the right tone and distortion character of their guitar amps. WARNING: This method will increase all voltages inside tube amps, including the heater voltage, and shorten the tubes' life.

The anatomy of a commercial filtered power board

Weiduka AC8.8 is a 15A-rated power filter board. While the basic model displays only the mains voltage, this more expensive version (around AU160.- with shipping) also shows the power draw in amperes and Watts and the power consumption in kWh.

There are two banks of five power sockets; the three USA-standard ones are soldered onto the large printed circuit board (8) while the two universal ones (at the rear) are hard-wired. These accept various plugs, USA, Australian and European. The two banks (with four outlets each) are individually switchable (4) at the back of the unit.

Fig. 4.15. Just like many made-in-China amplifiers and other audio products, the Weiduka AC8.8 power board is impressive looking on the outside but very basic on the inside.

The IEC socket for an input power cord (1) and a resettable circuit breaker (2) are at the back, followed by a varistor overvoltage protection on a small attached PCB (3). Each single-stage filter consists of one resistor, four capacitors, and a common-mode choke (5). The two output leads are taken through a ferrite ring, and that is all folks.

As expected, the marketing blurb is full of wild claims and inaccuracies, such as "A unique patented non-inductive filtering ..." but two large inductors are in plain sight (5)! A bit later, the description mentions "... together with its two-inductor-current filter ...". There's nothing unique or patented here; it is an ordinary inductive-capacitive filter in a fancy-looking enclosure.

Since there is so much space inside, we wanted to add a DC blocker. Still, due to the awkward construction and the arrangement of two large printed circuit boards, that would be impossible without totally rewiring the four outlets on PCB (7) and the removal of the filter on the other PCB (5).

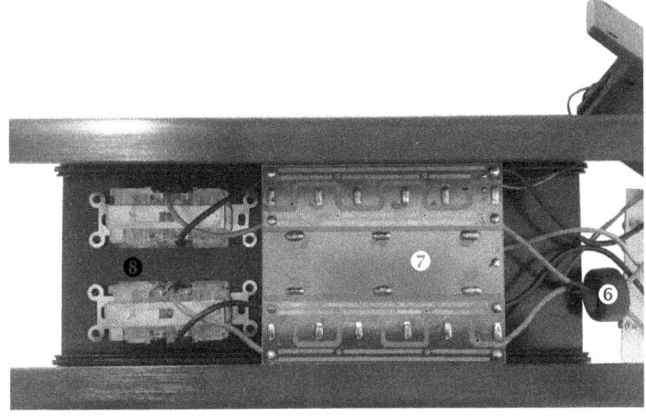
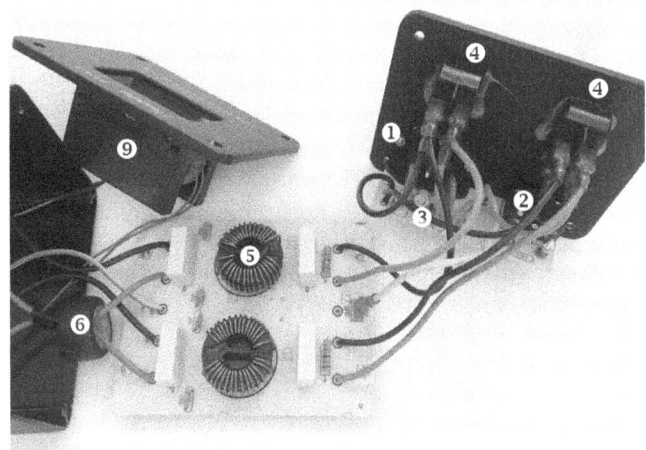

Fig. 4.16. The internal view of Weiduka AC8.8 power board

ELIMINATE DC FROM YOUR AC POWER OUTLETS

Why are a few DC volts on an AC line such a problem?

Adding a DC component to the AC mains voltage by the appliances just mentioned causes significant DC currents flowing in the primary windings of power transformers inside CD players, amplifiers, and most other audio components. This alters their operating regime and can even saturate the magnetic cores of power transformers, seriously impeding and degrading their proper operation.

Let's say there are 2 volts of DC voltage on your mains line. Two volts doesn't seem much compared to 117 volts, let alone 230 or 240 volts, so why is a small DC voltage a problem if present on AC mains outlets?

If you take a multimeter, set it to measure resistance, and connect it between the phase and neutral pins on your amplifier's mains IEC socket, you will measure the DC resistance of its mains transformer. Of course, the on-off switch has to be in the "ON" position, and the power cord has to be unplugged (amplifier de-energized). You will get a very low value, 5-20 ohms (lower values in the USA and other 117 V mains countries, higher value in Europe, Australia, and other 230-240 V countries).

On this particular tube amplifier (right), we measured 10.5 ohms. How much DC current will flow in the primary winding of this amplifier's mains transformer?

The Ohm's law says $I = V/R = 2/10.5 = 0.19$ Ampere. The lower the primary resistance, the higher the DC component of the current. 190 mA of DC current flowing through the primary winding of a small mains transformer (the size found in hi-fi components) doesn't seem much. Still, it is more than enough to move its operating point from the linear section of the magnetizing curve to the flat part of the curve, a nonlinear region known as the magnetic core saturation (see page 82).

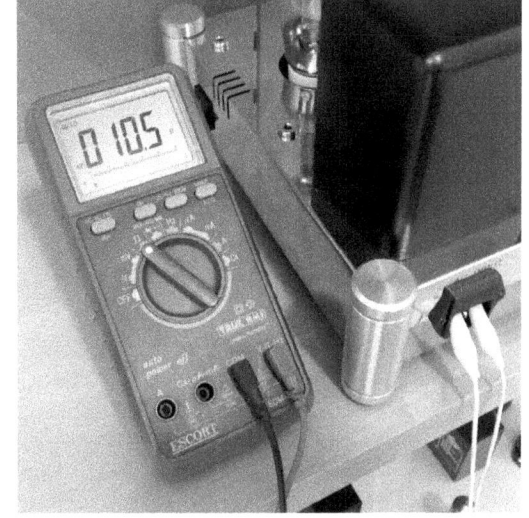

Fig. 4.17. Measuring the DC resistance of tube amp's power transformer's primary winding (together with the fuse, the on-off switch and associated internal mains wiring, but their resistance is much lower, enough to be considered negligible) using a multimeter.

This is an undesirable condition where the magnetic properties of the transformer's laminated core are compromised and where its performance is seriously degraded. The output (secondary) voltages will be severely distorted and reduced in amplitude.

A large toroidal power transformer in one of our stereo amplifiers would occasionally start buzzing loudly for a few seconds, sometimes even a few minutes, and then it would stop. Others, smaller in size, buzzed all the time. The buzz correlated with varying DC voltage levels in the AC outlet, depending on the time of day and the quality of the AC power, which varies when large industrial loads and many air conditioners come online.

Due to the nature of their construction (no air gap of any kind), toroidal transformers are much more susceptible to DC saturation than EI or C-core transformers! So, if your amp or preamp uses them, you are more likely to have a similar problem and are a prime candidate for adding a DC blocker to your amp or the whole audio system.

A simple electronic DC blocker

This simple DC-blocker will improve the performance of your preamplifier, amplifier, or even a whole audio system. It can be built into an amp or preamp or installed into a power distribution board that powers all of your audio components.

The four diodes and four electrolytic capacitors are connected in two paralleled branches. Then the whole blocker (with two terminals, X and Y) is inserted into the "hot," "phase," or "live" mains supply line. Notice the required polarity of the electrolytic capacitors!

The silicon diodes' current capacity should be adequate for the load current (since the total load current flows through them). All capacitors are 4,700 μF, 16V working voltage or higher. There will be a slight AC voltage reduction of around 1.2 V, a sum of two diodes' conducting voltage drops, so that the outgoing AC voltage will be 1.2 Volts lower than the mains voltage, but that is a negligible difference.

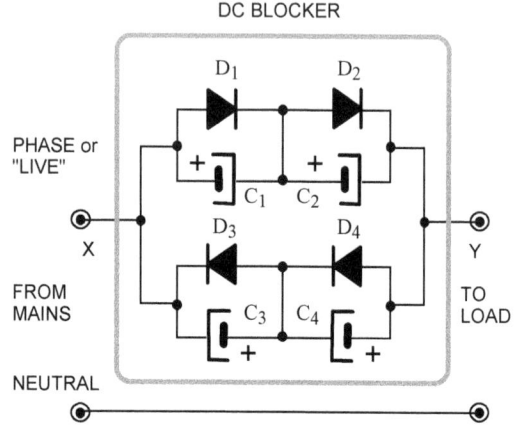

Fig. 4.18. One version of a DC-blocking circuit

Fig. 4.19. 1560G diodes have a continuous current capacity of 15 A, enough for most audio systems, and can pass up to 150A for short periods of time.

Decoding diode's basic parameters is easy. For instance, 1560G (illustrated on the right) has the following parameters: V_{RRM} = 600 V (peak repetitive reverse voltage or DC-blocking voltage), V_F = 0.85 V (forward voltage drop), $I_F(AV)$ = 15A (Average Rectified Forward Current) and I_{FSM} = 150 A (Non-repetitive Peak Surge Current).

For proper heat dissipation, diodes should be mounted on heatsinks. Since one of the diode's ends is electrically connected to its metal case, each diode must have its own heatsink; they should not be connected to the same heatsink unless insulating substrates are used.

DIY PROJECT: Power distribution board with over-voltage protection, RFI filtering and DC blocking

Using the highest quality, hospital-grade US-power outlets, we housed our mains distribution board in a standard 1U (unit) high 19" rack chassis. Since a standard IEC power socket with an integral switch and fuse (1) was used, the supply cable can have any type of plug on the other end. In our case, it has an Australian plug so that it can be plugged into an Australian wall socket. Why do we use US-style plugs and sockets on 240V mains voltage?

US-style plugs and sockets are made in many different designs and quality levels, including hospital grade. Australian ones are generally not available in the "audiophile-grade" quality levels, and the standard ones sold in shops (shonky plastic) don't take large diameter mains cables.

Fig. 4.20. Audiophile power distribution board built into a standard 1U (unit) high 19" rack chassis

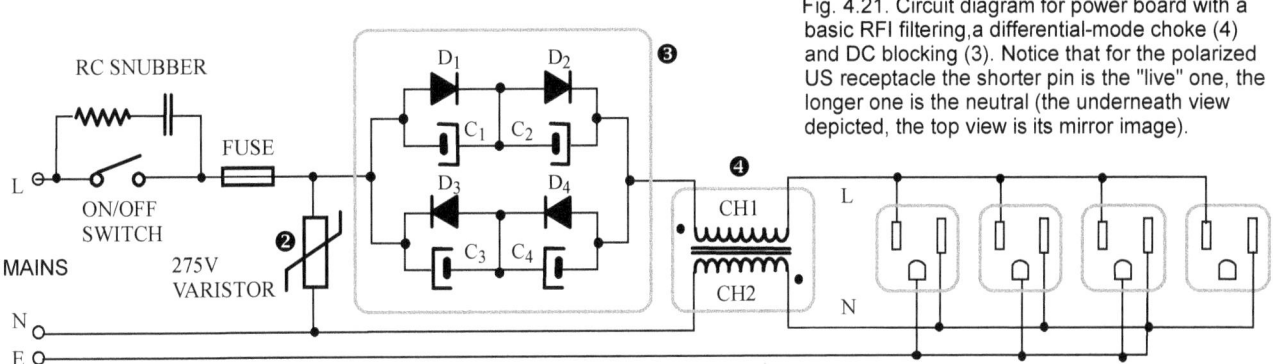

Fig. 4.21. Circuit diagram for power board with a basic RFI filtering, a differential-mode choke (4) and DC blocking (3). Notice that for the polarized US receptacle the shorter pin is the "live" one, the longer one is the neutral (the underneath view depicted, the top view is its mirror image).

Plus, many audiophile-grade mains cables are sold on the internet, most with the US- or EU-style plugs, very few with Australian plugs. Finally, this prevents others in your house from plugging in appliances that should not be plugged in here, such as hairdryers and iPad or mobile phone chargers. It is your board, for hi-fi only.

You can include as many "modules" or options as you like: overvoltage protection in the form of MOV or Metal-Oxide Varistor (2), switch arcing protection (RC snubber across the on-off switch's contact), dual RF choke filtering, and RF beads. The one module you should always include is a DC eliminator. Filtering DC voltages out will make one of the most significant improvements to the sound of your system of all.

If you use an isolation transformer (with separate primary and secondary windings), be it of the unbalanced or balanced type, you do not need this electronic type of DC blocker. By their very nature, isolation transformers do not pass DC voltages from their primary to secondary windings. However, autotransformers (Variacs) do not provide DC blocking, so any DC current present will flow through their sole winding, to which the load is also connected!

AC VOLTAGE REGULATORS AND REGENERATORS

In some cases, the AC power supply is so poor (fluctuating voltage, unstable frequency, distorted waveform, over-voltage spikes, lots of DC and high-frequency interference on the supply lines) that none of the measures outlined so far will completely clean it up. Although this will improve the sonics of your audio system, for best results, more drastic (which also means more expensive!) measures must be used.

There are various ways to improve further the quality of the AC power entering audio components. High-frequency interference can be filtered out (more on how filters work in the next section), and AVRs (Automatic Voltage Regulators) can be used to stabilize a fluctuating AC voltage. In extreme cases, when everything else fails, AC regenerators can be the last resort. They provide a fixed frequency and a stable and undistorted AC voltage supply, but at a price, higher complexity, and high cost. Let's have a look at their operating principles and pros & cons of various types.

Servo type automatic voltage regulators (AVRs)

A servo voltage stabilizer is a variable autotransformer (VAT) whose rotating arm is driven by a servo motor. The motor movement is controlled by a PID (proportional- integrative-derivative) feedback controller (CNTR), which though the voltage sensing circuit (VS) compares the load (output) voltage with a reference voltage level. It then produces an error signal that determines in which direction the motor will turn (the output voltage needs to be lowered or raised) and how much. The DC motor is driven by a power amplifier.

The response time of this electromechanical system is slow, so it cannot smooth out fast changes (fluctuations) in voltage. The sliding contact eventually wears out and needs replacement, as do DC motor brushes (commutators). Dust, dirt, and corrosion are issues of concern. This slow and outdated technology isn't recommended for audio systems.

Tap-changing AVRs

Servo tap-switching AVRs use a motor to move a sliding contact from one secondary tap to another. Tap changing alters the transformer's voltage ratio and its output voltage. This is a crude and slow method since tap changing takes time and produces finite voltage changes, not smooth, continuous ones.

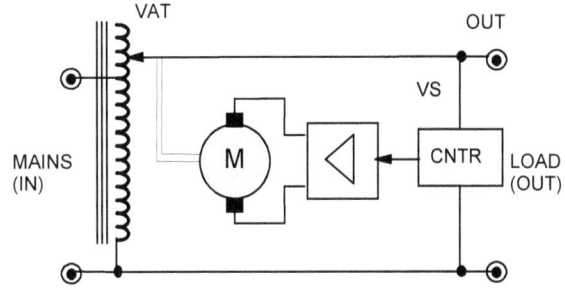

Fig. 4.22. Servo-type voltage regulator with a variable autotransformer (VAT). Notice that the supply and load side are galvanically connected (no isolation) so any DC offset voltage is passed straight through.

Electronic tap-changing AVRs dispense with the motor and use silicon controlled rectifiers such as thyristors and triacs as switches to turn one tap on and all other taps off. Tap changing is almost instantaneous, much faster than with the servo type.

In the "full power " type AVR the whole load current flows through such electronic switches, which results in increased losses and reduced reliability. SCRs are easily damaged or destroyed by overload and inrush currents.

The series transformer type electronic tap-changing AVR solves that problem by removing the SCR switches from the load current path and interposing a series transformer into the load circuit.

Ferroresonant voltage regulators

In contrast to power transformers in audio components, the transformer (1) in a ferroresonant voltage regulator does not operate in the linear region. Its magnetic core is deliberately taken into the saturation region by its "tank" circuit (2).

In the linear region, the magnetic flux (the vertical axis of the magnetization curve) increases in proportion to the increase in magnetizing current flowing through the transformer's primary winding. The secondary (output) voltage is proportional to the magnetic flux, and the primary current is proportional to the input voltage.

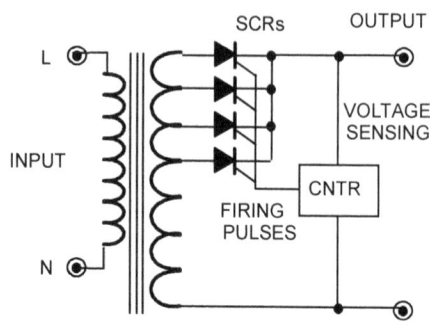

Fig. 4.23. Block diagram of a full power electronic tap-changing AVR

Fig. 4.24. Block diagram of a series transformer type electronic tap-changing AVR

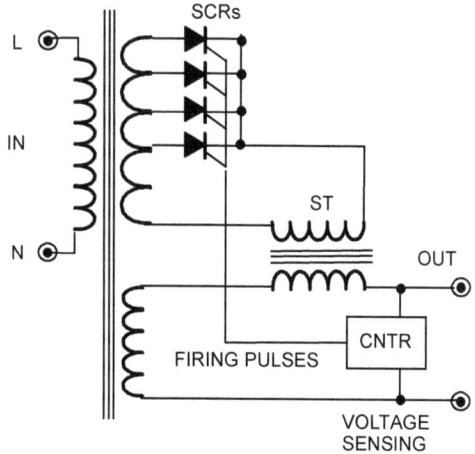

Fig. 4.25: A simplified (in-principle) diagram of a ferroresonant AVR. In contrast to power transformers in most other electronic devices, including audio components, the transformer in a ferroresonant voltage regulator does not operate in the linear region. Its magnetic core is deliberately taken into the saturation region.

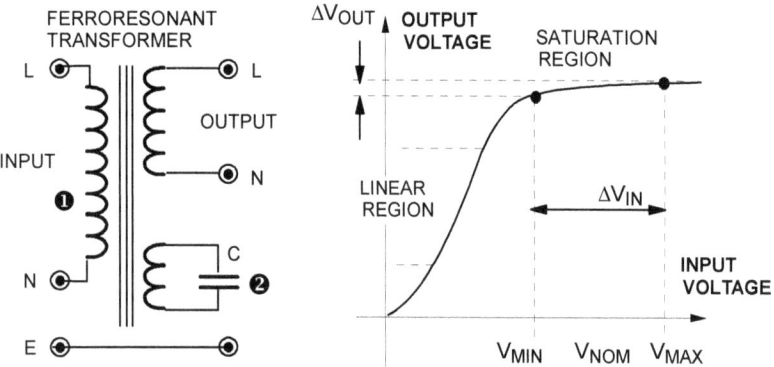

In the saturation region, a significant change in input current, caused by large fluctuations of the input voltage from V_{MIN} to V_{MAX} results in a minimal change in the output voltage ΔV_{OUT}, as indicated on the diagram below. Thus, a significant degree of voltage regulation and stabilization is achieved.

Ferroresonant transformers provide decent isolation of the load from the noise, surges, and interference on the mains line. The drawback of this mode of transformer operation is a dismal operational efficiency. To make matters worse, the lower the load as a percentage of the AVR's maximum power capacity, the lower the efficiency.

For instance, at 80% load, its efficiency could be around 70%, meaning that 30% of the power drawn from the mains is wasted into heat and 70% delivered to the audio system. However, if the AVR is oversized, say the audio load is only 30% of its capacity, the ferroresonant transformer's efficiency will plummet to around 40%!

While voltage regulators achieve their main aim, they create a significant amount of noise, primarily through the harmonics-rich switching transients. As is often the case in medicine and audiophilia, the cure is worse than the disease. Except for a few rare cases, mainly in countries with poor quality power grid, most audiophiles would be wise to avoid this class of power conditioners altogether.

Double-conversion voltage regulators or "regenerators"

AC regenerators rectify the AC mains voltage, filter (smooth-out) the rectified pulses into a DC voltage, and regulate it, so it doesn't vary with time and the load. A DC/AC converter, consisting of a stable oscillator and a microprocessor controller, produces a sine wave voltage of low distortion and stable frequency. This voltage is then amplified by a power amplifier that can provide the rated load current.

And this is where the good news end. A stiff and stable power network is theoretically an infinite source of current, although practically limited by the current carrying capacity of its distribution network transformers and cables. Mains filters, conditioners, and especially regenerators are inherently of limited current capacity. If undersized for the particular audio application, which means they are unable to meet the current demands of the audio system, most notably the power amplifier(s), a severe dynamic compression will result.

This AC behavior is analogous to the limits of the internal DC power supplies inside audio components, which also causes distortion and dynamic compression of louder music passages.

Due to their complexity, power regenerators are more expensive than filters and power conditioners. However, there are also some brazenly expensive "audiophile-grade" (whatever that means) power conditioners.

Some regenerators give the user a choice of operating frequency. USA manufacturer PS Audio erroneously calls this useful feature "power factor," which is obviously wrong, power factor being something completely different. The oscillator's frequency can be varied from 50 to 120 Hz.

Can audio components designed for 50 or 60 Hz mains frequency work on 70, 90, or even 120 Hz? Of course they can. The only internal components affected are the power transformer and the rectifier/filter that follows it.

The product of frequency f and magnetic flux density B must remain constant, so if we set the regenerator to produce a 100 Hz sinewave instead of 50 Hz, the doubled frequency means the flux density inside the power transformer will be halved; the transformer will run cooler and quieter. Many audiophiles report improved sonics when using their regenerators at frequencies of 80-90Hz.

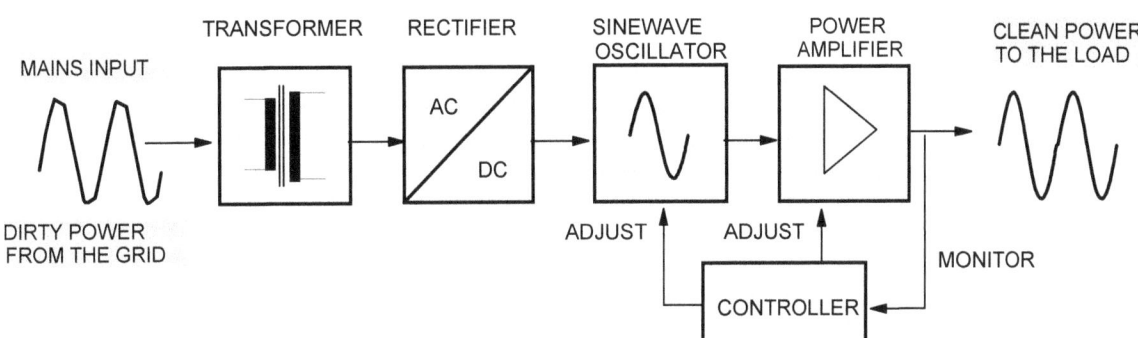

Fig. 4.26: The block diagram of a typical AC voltage regenerator

Do I really need a filter, voltage regulator or regenerator?

From my personal experience in Australia, in 80% or more cases, there is no need for elaborate power conditioners or regenerators at all if a few simple measures (all covered in this chapter) are implemented. The situation in Europe and North America is similar.

The 20 or so percent of problem cases are audiophiles living close to large industrial estates in cities or remote mining areas, with heavy or fluctuating loads, especially those with locally generated power (small diesel or gas turbine power stations), not connected to a larger power network.

Even in the cities, the recent proliferation of solar panels and mains-synchronized inverters seem to have a negative impact on power quality. The local power utility generates less baseload, and the constantly increasing percentage of local solar generation depends on the atmospheric conditions and is highly fluctuating.

Add to that the fact that almost all currently sold reverse-cycle air conditioners are of the inverter-type, and you have a recipe for a worrisome DC offset and significant injected levels of high-frequency artifacts (harmonics) by such inverters. Both DC offset and HF harmonics are highly detrimental to the sonics of audio systems.

The increased series impedance as an impediment to current delivery

All filters, regulators, and regenerators discussed so far increase the impedance of the AC supply. This issue affects high-current loads such as power amplifiers, primarily those operating in classes AB and B, whose current draw varies with the output power. Class A amplifiers are not affected since their DC power draw is constant, and so is their current draw from the mains.

This drawback has led to the appearance of a distinct class of power conditioners that have no AC impedance raising inductive or capacitive components in their power supply path.

Antigone by DR Acoustics is a simple, passive unit, with each of the three conductors (Live, Neutral, and Earth) passing through a quartz-filled carbon fiber tube, or "Quartz Crystal Filter."

In their products such as PowerCell SX and PowerCell 12UEF (a "non-current-limiting line conditioner"), Synergistic Research also uses carbon fiber tubes containing "folded electromagnetic cells." I would call it "rolled-up," but that's semantics. These line conditioners adopt an active approach, also including a "ULF (Ultra Low Frequency) Field Generator," which to me seems to be a Schumman generator (see page 67).

For more info, see patents US6545213 and US8658892, available online.

Why a single power filter, conditioner or regenerator may not be enough

A single power filter, conditioner, or regenerator may solve the dirty mains problem, but audio components themselves also pollute the mains side. Digital components (Class D power amplifiers, D/A converters, CD players, and streamers) may feed digital noise and interference back into the rest of the audio components if they are all connected to the same output of an AC power filter, conditioners, or regenerator (1).

Power amplifiers that use CLC filtering in their power supplies draw high amplitude current pulses from the mains supply (2). The spectrum of such spikes contains significant high order harmonics, which are then fed back (3) into all other audio components connected to the same regenerated supply.

In contrast, the current drawn by amplifiers that use LC filtering is continuous and smooth, generating very little detrimental harmonic content. This is the most likely reason amplifiers with CLC filtering sound inferior to those with LC filtering, which don't have such pulses to "smudge" and defocus their sonic presentation.

The solution is to use at least two regenerators, one for digital components, the other for analog ones. An even better option is using three separate regenerators, one for digital devices, another one for low power analog components (phono stages, active crossovers, line-level preamps), and the largest capacity one for power or integrated amplifiers.

Some regenerators and conditioners have three or more separate outputs, so in that case, there can be no interaction between audio components powered up in such a fashion. For instance, Audio Magic's Oracle power conditioner has three banks of four outlets each for digital components, low-level analog phono- and line-stages, and higher current loads such as amplifiers.

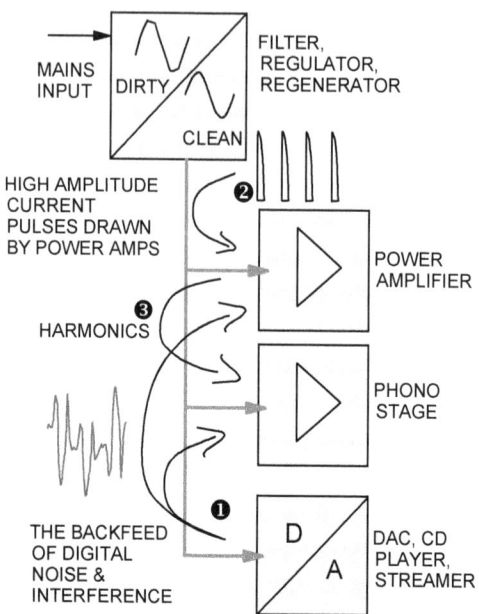

Fig. 4.27: Digital components may feed digital noise back into the rest of audio components connected to a single power line

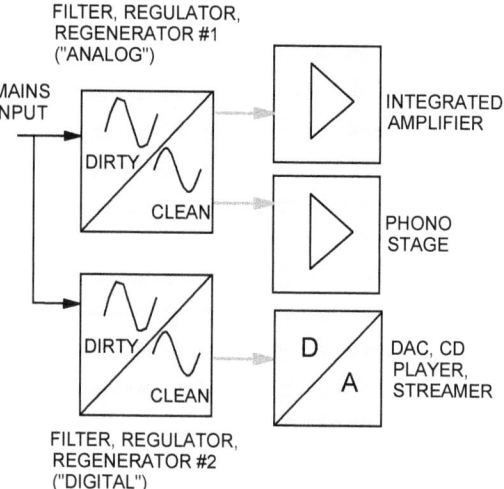

Fig. 4.28: One of many possible solutions is to use two smaller AC power filters, conditioners, or regenerators, one for digital components, the other for the analog ones

Fig. 4.29: Some AC power regenerators have three or more separate clean outputs, so there can be no interaction between loads

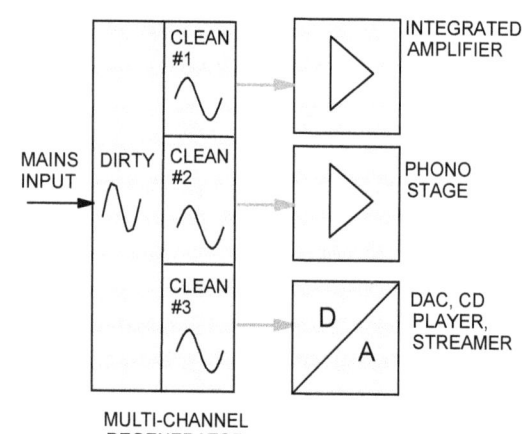

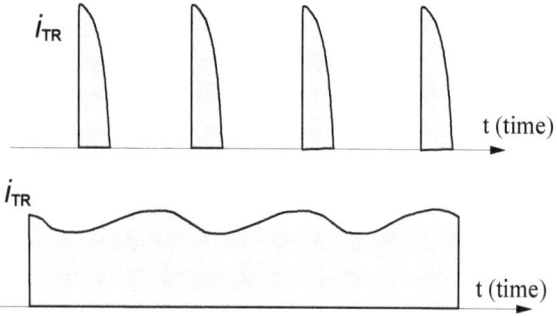

Fig. 4.30: The stark contrast between current pulses drawn from the mains by the power transformer of an amp or preamp with CLC (capacitive) filtering and the smooth & continuous transformer current of an amp or preamp with LC (inductive) filtering.

How to add up the power consumption of your audio components

When buying a power filter or conditioner or when building your own, you must make sure that the power rating of such a device exceeds the total power draw (consumption) of all audio components connected to it.

All audio components should have a "nameplate," a sticker, metal or plastic plate glued or screwed to its back, or the bottom cover. It should specify, among other things, the total power consumption of the device. Basic ratings, such as the mains voltage and the power consumption, are often even screen-printed on the chassis itself, as illustrated on the right.

Sometimes the power in Watts is not specified, but the mains voltage and current draw are, for instance, 120 V, 2 A. In that case, for estimation purposes, just multiply the two to get the power consumption P = V*I = 120*2 = 240 Watts!

Since most power boards and conditioners are rated in terms of maximum allowable current, the power consumption figure of each audio component needs to be converted into its nominal current draw. Then all such figures must be added up to determine the total current load.

The Polk subwoofer is rated at 80 Watts, and its switch-mode power supply is universal, meaning it can operate on any mains voltage between 100 and 240 V. In Japan, with its 100 V mains voltage, the current draw will be the highest, I= P/V = 80/100 = 0.8 A (Ampere). On Australia's 240 V mains voltage, the subwoofer's current draw will be only I = P/V = 80/240 = 0.33 A.

To add to the confusion, the power consumption of some audio components is specified in Watts (W), while for others it is in VA (Volt Amperes). Watts indicate active power, VA (Volt Amperes) are the unit for apparent (total) power, which is the vector sum of active and reactive power, the unit of which is VAR (Volt Amperes Reactive). The active power is apparent (total) power times cos F (the "power factor"). Most audio components have a power factor close to 1 (maximum), usually 0.8 or 0.9, so there isn't that much difference between the watts and volt-amperes they pull from the power grid.

So, either convert all power consumption figures into Watts and calculate the amperes by dividing the Watts by the nominal mains voltage in your country, or add up the amperes (the currents drawn) and size the filter - conditioner - regenerator that way.

Fig. 4.31. A typical fused IEC power inlet socket of a double-insulated (the double-square symbol) audio component. The associated specifications (voltage range, frequency and power consumption) are screen printed. The fuse type (T), current (2.5 A) and voltage ratings are also specified.

Fig. 4.32. The "power triangle", illustrating the phase relationship between the active, reactive and total or apparent power.

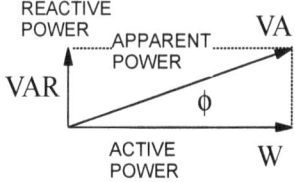

REDUCING ELECTROMAGNETIC AND RADIO FREQUENCY INTERFERENCE

EMI/RFI filters

EMI (Electromagnetic Interference) and RFI (Radio Frequency Interference) mains filters come as stand-alone units or integrated with an IEC socket (and often as combo units with a fuse and on-off switch). Most are single filters comprising of three film capacitors and two differential chokes in a symmetrical PI (Π) configuration. There are even double filters, as illustrated in the circuit diagram below.

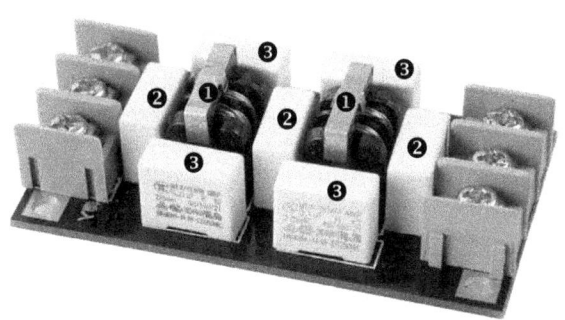

Fig. 4.33. DIY 2-stage RFI filter sold as a kit on eBay, 1) Filtering choke 2) X-type capacitors between L (Line or Live) and N (Neutral) 3) Y-type capacitors between L/N and GND

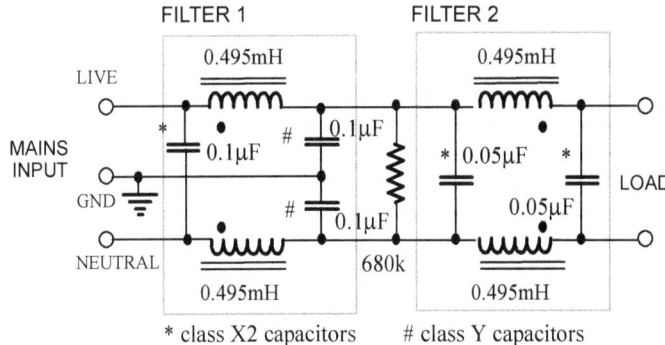

Fig. 4.34. Circuit diagram of a typical double (2-stage) mains filter

The differential-mode choke is directly inserted in the L (live) and N (neutral) supply lines, and thus all the load current passes through it. This affects the current delivery by restricting the instantaneous current drawn by high power amplifiers.

One consequence of such limitation is dynamic compression and anemic, lackluster performance of power amplifiers. Many audiophiles claim that such filters inside audio amplifiers and preamplifiers negatively affect their sonics and I tend to agree. We'd used IEC mains power sockets with single-stage RFI filters (illustrated on the right), but only on the first batch of our tube amplifiers and never again.

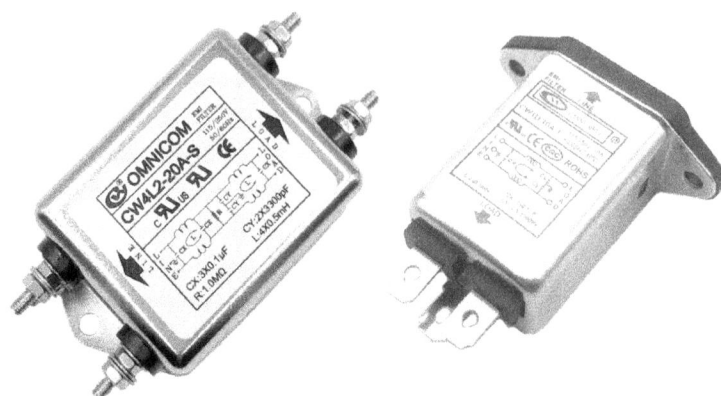

Fig. 4.35. 2-stage (1st order) RFI filter for mounting inside the equipment (amplifier, preamplifier) enclosure (exposed terminals)

Fig. 4.36. Chassis-mount IEC mains power socket with an integral single-stage RFI filter

How RFI enters audio equipment

Mains filters only act on the radio frequency interference present on the AC power lines. However, most such interference travels as electromagnetic waves through air and can enter audio equipment through the power and audio cabling and wiring. All wires (hookup wires inside audio equipment) and external cables (interconnects, power cables, and speaker cables) act as antennas and will pick up all kinds of airborne interference.

That is why effective electromagnetic shielding of both audio equipment and audio cables is of paramount importance. These are highly technical issues and beyond the scope and level of this book, but luckily, there are some simple steps and solutions you as an audiophile can take or implement in that regard.

How to eliminate RFI using small ferrite rings

A very effective RF interference measure is to wrap cable or wire around a ferrite ring or bead a few times. This creates a tightly coupled in-line high-frequency transformer. Since the impedance of a choke raises linearly with frequency, such added series inductance increases impedance for high radio interference frequencies (the common-mode signal) without attenuating the lower frequencies (the differential audio signal).

The added impedance is negligible at low 20 Hz-20 kHz audio frequencies (when used on interconnects or speaker cables) or 50/60 Hz mains frequency (if used on a power cable).

The ferrite rings come in various shapes and sizes (2). There are even clip-on ones that don't require threading the wire through a hole; you simply clip them onto power or signal (interconnect or speaker) cable (1).

The photo shows a complete wiring loom from a CD player, starting with an IEC power socket (3). The earth wire is threaded a few times through a ferrite ring (5) and terminated with an eyelet to secure it to the metal chassis (6), while the L and N conductors are fed together through another ferrite bead (4).

Ferrite rings or beads are especially effective in damping high-frequency oscillations and "ringing" caused by digital amplifiers and switch-mode power supplies, such as those used in "wall-warts" powering up many turntables phono stages and digital sources such as CD players, DACs, and music streamers.

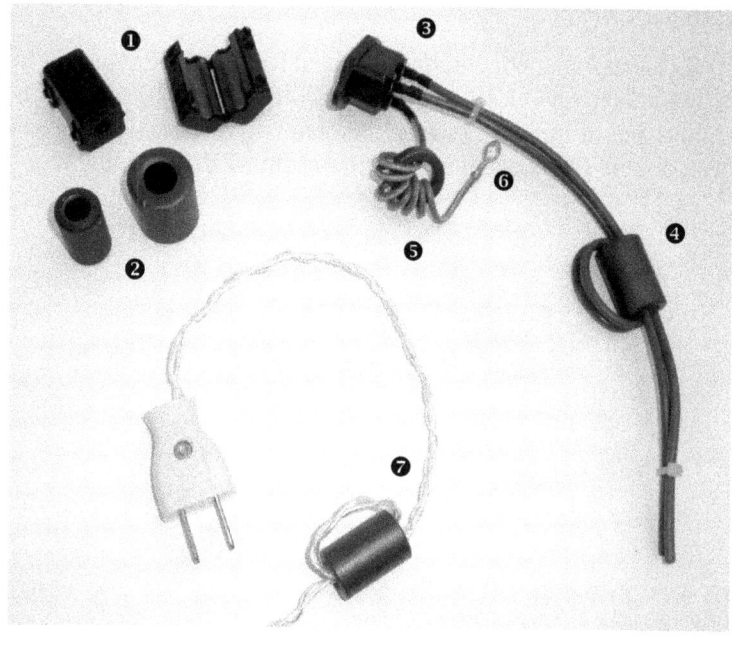

Fig. 4.37. DIY two-core power cord for double insulated components such as CD players, DACs and music streamers. We used Teflon-encased silver conductors and fed it twice through a large ferrite bead (7).

Noise reduction technique: Double-shielded phono and interconnect cables with DC biasing

Since their voltage level is the lowest in the signal chain, thus closest to the noise levels, phono connections are the most susceptible to noise and interference. Instead of using a cheap shielded cable, invest in a double-shielded coaxial cable between the turntable and phono stage. The inner shield is the return conductor, while the outer shield does not conduct the signal (as on single-shielded unbalanced RCA cables) and is purely for shielding purposes

There is usually no need to go to this much trouble in line-level connections, where signals are around 2 V_{AC}. The outer shield must be grounded (earthed) on one side only, preferably at the phono stage input! Connecting both sides of the shield to the ground could create a ground loop and introduce a hum.

This technique is limited to preamplifier stages with tubes only and requires a degree of modification of existing equipment and the technical knowledge and skills involved. However, even paying a technician to perform this modification (should not take more than an hour or so) could result in a highly cost-effective sonic improvement.

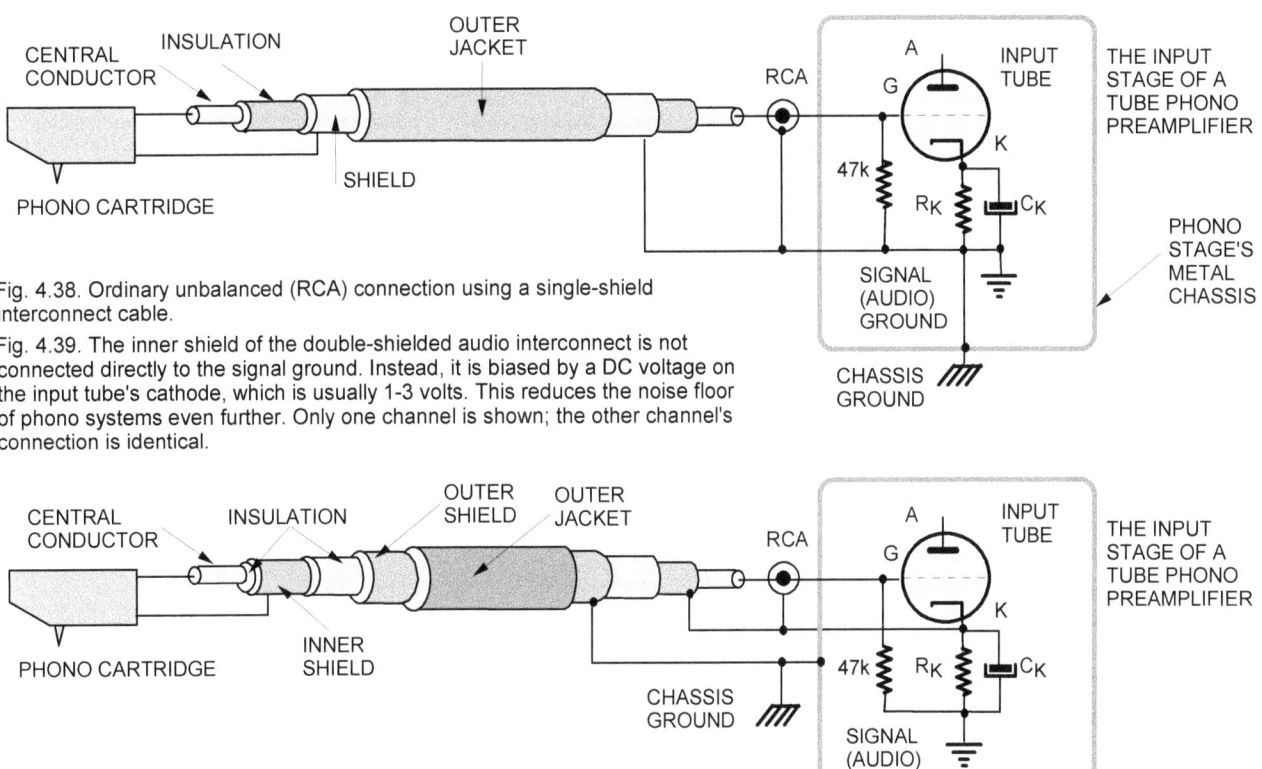

Fig. 4.38. Ordinary unbalanced (RCA) connection using a single-shield interconnect cable.

Fig. 4.39. The inner shield of the double-shielded audio interconnect is not connected directly to the signal ground. Instead, it is biased by a DC voltage on the input tube's cathode, which is usually 1-3 volts. This reduces the noise floor of phono systems even further. Only one channel is shown; the other channel's connection is identical.

How ground loops are formed and how to break them

Apart from the noise and RFI interference, ground loops are another scourge that plagues many audiophile systems. This issue affects both turntable-based systems (a ground loop between the phono stage and a preamplifier) and line-level connections, between a line-level preamp or a CD player and the power amp.

Although both audio components in the diagram (next page) are grounded, for various reasons, there may be a certain resistance between the two grounding points, represented by R_G. Usually, such resistance would be zero or so low that it is negligible.

Suppose an interference or hum (AC voltage signal V_L) is externally induced in the earth loop. In that case, the loop current I_L will flow and cause a voltage drop on the series resistance of the shield around the interconnecting cable R_S, which will appear as an AC signal at the input of the amplifier.

Even a slight potential difference (voltage) between two grounding points causes the loop current to flow. This AC signal of mains frequency (50 or 60 Hz) enters the amplifier and causes an audible hum or buzz. Remember, the mains frequency is an audio frequency and can be heard!

There are two types of "ground" in an amp or preamp. One is the signal ground, which is the negative side of the signal chain (the outer ring on the RCA plug or socket). The other is chassis or safety ground, which must always be connected to the earth pin on the mains cable. The only exceptions to this legal rule are double-insulated power supplies and appliances. Common double-insulated hi-fi components are CD players and network audio streamers, which use two-pin power cords. Their metal chassis is not earthed in any way.

To avoid internal ground loops, properly designed and constructed amps and preamps have their signal ground connected to the chassis (safety ground) at one point only. That point is often called a star-ground since all the points in the circuit that need to be grounded come to it, like a star. There could be four, five and more wires joined together at that point.

Both of the components illustrated comply with that requirement. However, notice that when the two devices are linked by an RCA interconnect cable (only one channel is shown), the negative side of the signal between the two components is now grounded in two points (A and B), which could be far apart.

The preamp may be plugged into one power point, while the power amp could be plugged into a totally different power point, two or even three meters away. The earth wire between these two mains sockets and the RCA interconnect cable are now forming a closed loop.

Every loop is an antenna that picks up all sorts of interference and electromagnetic fields. These manifest themselves as hum, if their frequency is low, such as the mains 50 or 60 Hz frequency, or as hiss, which is an unwanted signal of a higher, albeit still audible frequency.

The loop voltage and current are proportional to the physical area of the loop, so the first hum-reducing measure is to break the loop. This is usually done by lifting one of the signal grounds from the chassis ground.

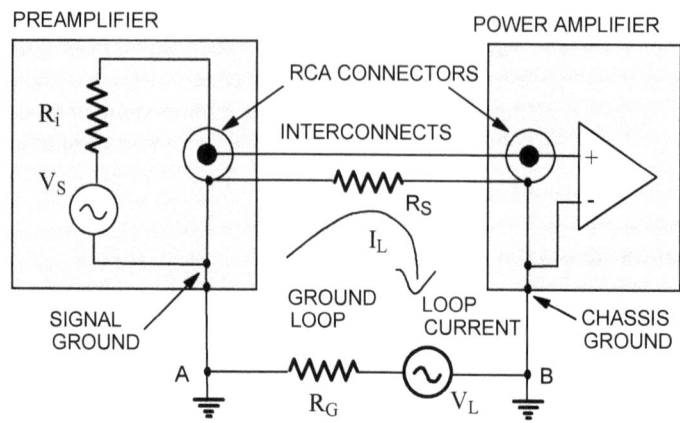

Fig. 4.40. Due to a small resistance R_G between two ground points (A & B), a ground loop is formed causing hum.

Fig. 4.41. Disconnecting the audio (signal) ground from the chassis of one component (mains ground) breaks the ground loop and eliminates mains-caused hum.

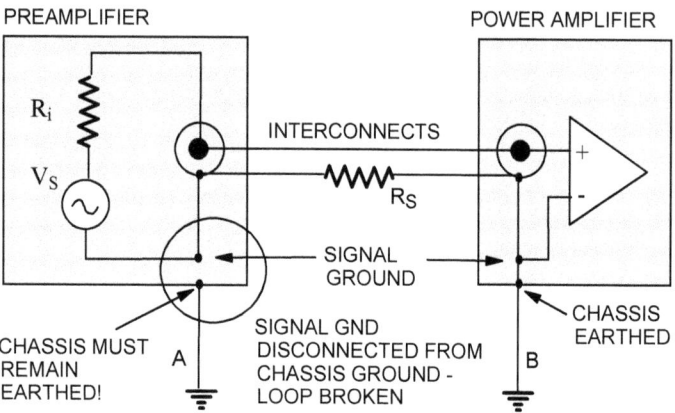

Notice that the signal ground of the preamplifier is still connected to the chassis ground of the power amp, but this time only through the interconnect cable. In a sense, we have removed an unnecessary double-ground.

Of course, the metal chassis of the preamp must stay earthed or grounded (at point A) at all times for safety and legal reasons. Never, ever use ground-lift adapters ("cheater" plugs). They disconnect the powered device entirely from the mains earth and are not just illegal but deadly dangerous!

Sometimes disconnecting the audio ground from the chassis/mains ground is too difficult or, for some reason, not possible (when the chassis is used as a ground bus, conducting the audio signal). The solution is to disconnect the earthing pin from all but one component in the audio chain (usually the power amplifier). However, those chassis must then be externally earthed via the power amplifier's chassis, as illustrated below.

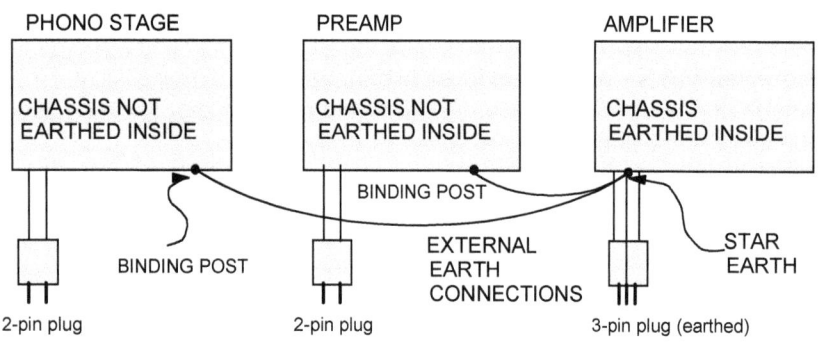

Fig. 4.42. This external DIY earthing arrangement is safe, but may still be illegal in many countries!

While this earthing scheme ensures safety, since you are tampering with the earthing of mains-approved equipment, it may still be illegal in many countries.

Suppose someone gets killed or a fire is started. In that case, the insurance company may reject your claim, and you may even be prosecuted, so treat this idea as an educational example and not a recommended practice!

CLEANING UP THE POWER SUPPLY TO REDUCE NOISE, HUM, AND INTERFERENCE

SAFETY WARNING

To eliminate hum, many "repairman," guitarists, and audiophiles (even some ignorant manufacturers) install "ground lift" switches in amplifiers in such a way that both the audio common and the chassis are disconnected from the mains earth. Not just illegal but potentially deadly. Any short circuit or malfunction can make the chassis "live," and not even RCD circuit breakers or ground-fault interrupters will operate if the earth connection to the equipment is lost or disconnected.

A few guitar players have been electrocuted on stage for this very reason. The same applies to the so-called "cheater plugs." They disconnect the mains cable's ground pin from the wall socket and thus "float" the previously grounded AND safe piece of equipment (amp, preamp, etc.) and make it unsafe and deadly dangerous.

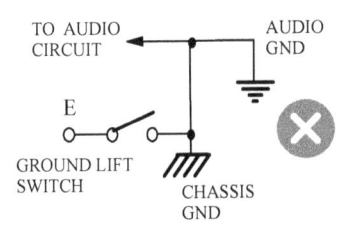

LEFT: Mains earth (ground) disconnected from the chassis. It kills the hum, but it may also kill you!

RIGHT: Mains earth E stays connected to the chassis; only the audio "common" or audio ground is lifted (disconnected) from it.

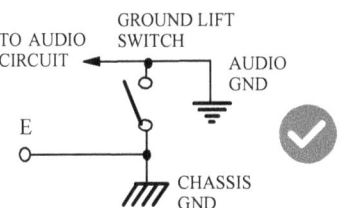

UPGRADING AUDIO EQUIPMENT AC & DC POWER SUPPLIES

Replace noisy switch-mode AC/AC converters with analog AC power supplies

Switch-mode power supplies, be they of AC/AC or AC/DC type, involve rectification and high-frequency oscillators ("choppers") operating at ultrasonic frequencies, usually 40 kHz or higher. Such frequencies may not be directly audible, but such oscillators and controlled rectifiers send high-frequency pulses back into the mains network, which then finds its way to other audio components through their power supplies. They also emit significant levels of RFI (radio frequency interference).

Replace switched-mode power supplies with quieter, better performing (lower ripple and better regulation) analog power supplies. Since they use large and costly low frequency (50 or 60 Hz) transformers, linear power supplies are more expensive and physically larger, but that is a small price to pay for a significant sonic improvement.

A "conventional" or analog AC power supply is fundamentally simple, just a transformer that reduces the mains voltage to the required level. EI, C-core, R-core, or toroidal transformers may be used; they all perform the same function. The closer the transformer's core approaches the doughnut shape, the lower its electromagnetic radiation. Toroidal and R-core transformers radiate less than C-core, while those of EI construction radiate the most.

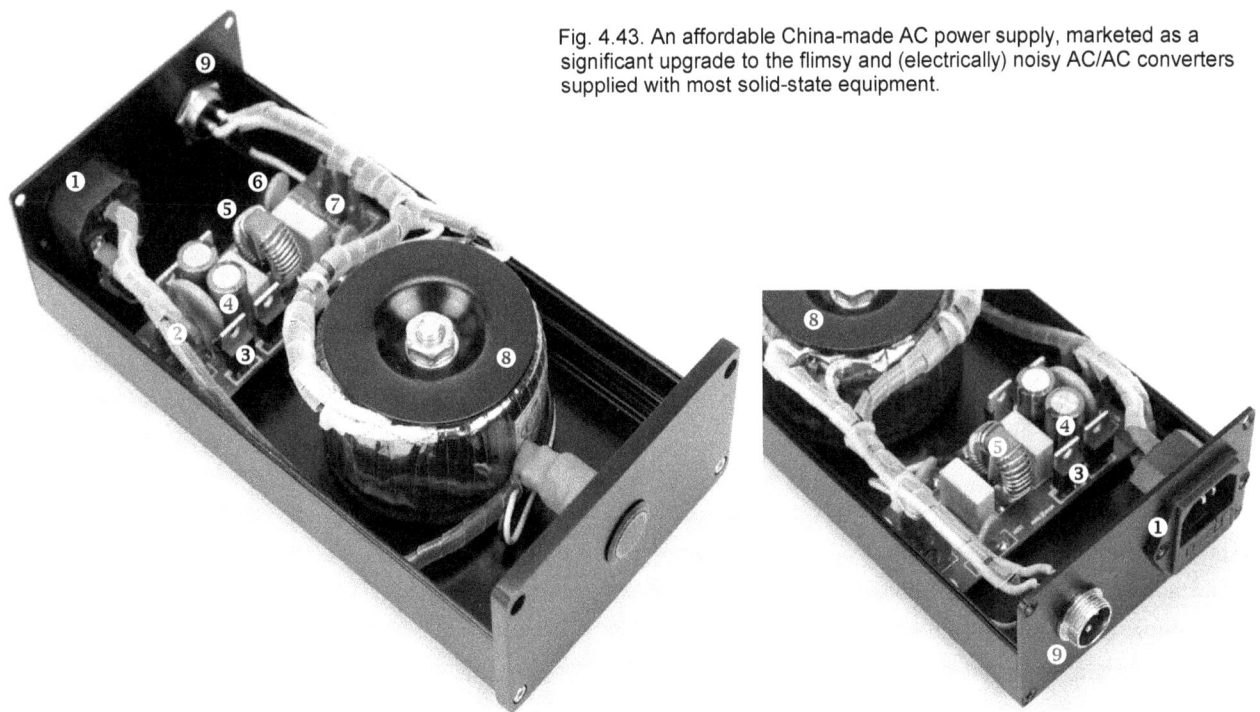

Fig. 4.43. An affordable China-made AC power supply, marketed as a significant upgrade to the flimsy and (electrically) noisy AC/AC converters supplied with most solid-state equipment.

As a rule, to minimize the possibility of such radiation being picked up by the rest of your audio installation (usually through cables acting as antennas), always keep power supplies as far from the sensitive electronics as possible.

This particular AC power supply has an on-off switch at the front, and a toroidal transformer, and an RFI filter inside. For the technically-minded, one look under the cover says it all.

The IEC mains socket (1) at the back has an integral fuse. The current enters the PCB (2), flows through a DC eliminator, consisting of four discrete diodes (3) and two electrolytic capacitors (4), and then through a simple RF filter whose job is to suppress radio frequency noise and interference.

You will recognize a dual toroidal choke (5), flanked by two film capacitors (white boxes) and two ceramic capacitors (6). The filtered voltage exits the PCB on the other side (7), feeds a toroidal step-down transformer (8), and exits the unit through a three-pin socket (9). The large disc ahead of the DC eliminator (4) is a varistor ("voltage-dependent resistor"); it provides overvoltage protection by clipping off the voltage spikes. The transformer in this model provides 16 V_{AC} output voltage at 1 A nominal current, up to a maximum of 1.5 A.

Upgrading your turntable's or phono stage's power supply

We got our Pro-Ject RPM 3 turntable at a clearance sale at half of its retail price of AU$1,199.-

Its outboard power supply is a switch-mode AC-DC converter with the following specs: Input: 100-240 V_{AC}, 0.5 A max., frequency range: 47-63 Hz, output 15 V_{DC}. Googling the model number yields its datasheet, specifying the ripple noise of 200 mV_{PP} (0.2 Volts peak-to-peak), the AC component riding on top of the DC voltage.

Even more important is "Line regulation" of +/-5%, meaning the +15V_{DC} output voltage could go up to 1.05*15 =15.75 V_{DC} and down to 0.95*15V = 14.25 V_{DC}. Since the speed of a DC motor varies proportionally with the DC voltage supply, the Project RPM3's speed will also shift +/-5%, which is quite a lot. A better quality DC supply with tighter regulation (say +/- 1%) would minimize that problem.

One of the smartest practices you as an audiophile could adopt is to read specifications and reviews of various pieces of audio gear, especially those of advanced design (read "expensive"). That could give you a valuable insight or two, ideas that, if technically minded, you could implement in your audio system at a much lower cost.

For instance, Acoustic Signature makes turntables at around US$8,000 price point (Storm Mk II, for example). Its motor is driven by an electronic controller, a switching power supply. The AC mains voltage is first rectified and filtered to DC voltage and then converted by a precision DC/AC converter into a low distortion and stable frequency sine wave at 12V_{AC} to run the synchronous AC motor.

Such power supplies are not affected by the fluctuations of the mains voltage. They also insulate the motor from the distortions (higher harmonics) and noise on the AC supply line, two imperfections of the mains supply with a detrimental effect on the turntable performance and their sound quality.

There is a 5-pin XLR input for an external DC supply, so their AC-1, a 24 V_{DC} linear power supply can be added for €299.00 (presumably superior to its internal power supply), or upgraded to PS-1 for €649.

Of course, this kind of power supply upgrade is only applicable if your turntable uses an AC motor. Ours uses a DC motor, and in those cases, your upgrade task is much simpler and cheaper.

Fig. 4.44. Pro-Ject RPM 3 Carbon turntable's outboard power supply

Commercial DC power supplies

There are many affordable DC power supplies available, either generic ones or those aimed at specific devices. Still, they could be perfect for your turntable and phono stage as well. Study their specifications and if they are superior to your existing power supply, go for it.

The specs to consider are:

1. Input (mains) AC voltage and frequency range
2. The nominal output DC voltage
3. The maximum output (load) current
4. AC ripple at the output (the lower, the better), and
5. "Line regulation"- again, the lower, the better, so +/-1% is much better than +/-5%

CLEANING UP THE POWER SUPPLY TO REDUCE NOISE, HUM, AND INTERFERENCE

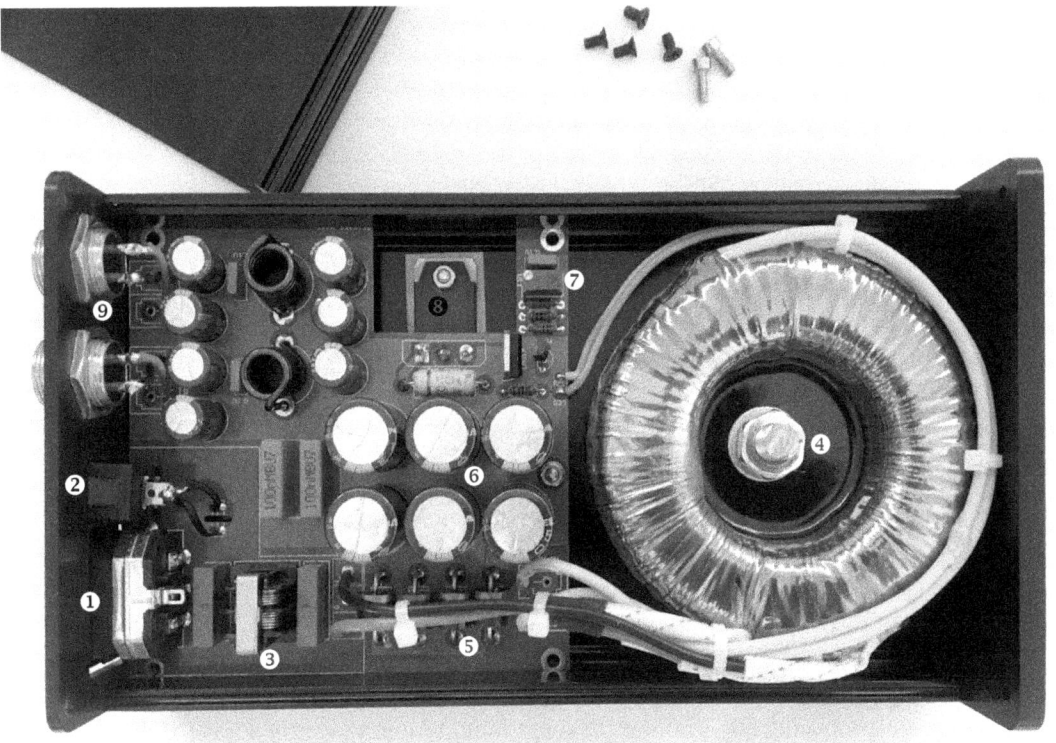

Fig. 4.45. The internal view of the generic made-in-China DC power supply
1. Mains IEC socket (power in)
2. fuse holder
3. single-stage RFI filter
4. toroidal step-down transformer
5. four discrete diodes (bridge rectifier)
6. filtering capacitors
7. output voltage adjustment (trimmer potentiometer)
8. voltage regulator integrated circuit
9. dual DC outputs

Although we ordered a 240 V mains version, the power supply had a "230V$_{AC}$" sticker at the back. On our 247-250 V mains AC voltage, the output DC voltage was a bit high at 15.26 volts.

The trimmer potentiometer (7) made it easy to adjust it precisely to 15.00 volts, but it required the unit to be opened up.

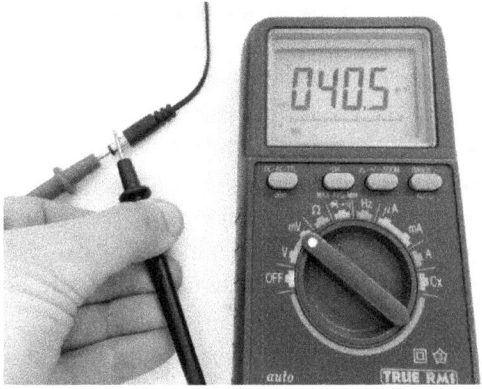

Fig. 4.46. Measuring the AC ripple of Pro-Ject RPM 3 Carbon turntable's outboard power supply

How to measure DC power supply's output voltage and AC ripple

With an auto-ranging digital multimeter on the "DC Volts" range, connect the red lead (positive) to the inside of the connector and the black lead (negative) to the outside. The display should read the DC voltage produced by the power supply, in our case, around 15.0 volts. This voltage is without any load connected, so the operating voltage (turntable motor or phono stage connected) may be slightly lower.

To measure AC ripple (the small unwanted AC component riding on top of the DC output), change the multimeter range to AC volts or mV. In our case (Pro-Ject RPM 3 Carbon turntable's outboard power supply), we got 40.5 mV or 0.0405 V, meaning the ripple was 0.0405/15 = 0.0027, or 0.27 % of the DC voltage. The lower the AC ripple, the better the DC power supply.

A simple and cost-effective DIY variable DC power supply

For under AU$10 on eBay, this assembled 80 x 40mm PCB module seemed a good value (next page). One could not buy the parts for that money in Australia, let alone a printed circuit board and labor for its assembly.

The design is simple, two integrated circuits (ICs), 3-terminal voltage regulators, LM317 (3), providing the positive, and LM337 (4) supplying the negative voltage. Each output voltage can be separately controlled (varied) by its own 5k trimmer potentiometer, (1) and (2). The two regulators are heatsinked and can easily provide up to 1.5 A of load current.

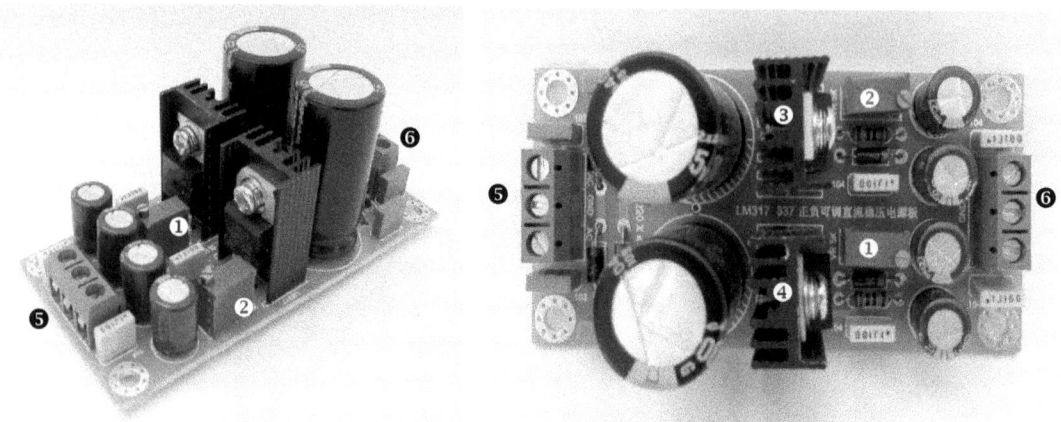

Fig. 4.47. This regulated DC power supply module (and many similar ones commercially available) can be used to power up various audio components (turntables, phono stages, etc.). You could even use it as a DIY field control module for electrodynamic speakers - just replace trimmer pots (1) & (2) with externally-mounted potentiometers for voltage and field current regulation.

LM317 and LM337 can provide up to 37 V_{DC} on their outputs, but with the effective doubling of that voltage due to dual polarity outputs, this design can provide up to a maximum of +/-37 V or 74 Volts.

An external power transformer with a symmetrical secondary winding is required, 2 x 12 V_{AC}, 2 x 15 V_{AC}, etc., depending on the desired DC voltage at the power supply's output, the three terminals marked (6). The three transformer secondary terminals are connected straight into the three input terminals on the PCB (5).

BATTERY-POWERED PHONO STAGES, PREAMPLIFIERS AND POWER AMPLIFIERS

The advantages of battery-based power supplies

Linear power supplies have a limited capacity to deliver power when needed by sudden increases of demand by the audio circuitry, resulting in distortion and dynamic compression at higher volume levels. Although quality amplifiers have large capacitor banks for energy storage, the energy stored in a large lead-acid battery is many times greater. The ultimate power source in hi-fi is a battery bank of a suitable voltage and current capability. Look at it this way: if such a battery can crank over a car or truck engine, it can undoubtedly supply enough power even during the loudest crescendos in music reproduction.

When battery-powered, the audio circuitry is disconnected from the mains, making any distortions and fluctuations in the mains voltage or frequency irrelevant. It also means that mains-propagated spikes, surges, and noise ("grunge") cannot reach the audio circuitry. This benefit on its own is enough for us audio fanatics to stop dreaming about breasts and buttocks and start dreaming about batteries of all shapes, sizes, and technologies.

Transformers radiate electromagnetic waves. This radiation is picked up by the audio circuitry and amplified as a signal since it is of an audio frequency (50/60 Hz for the mains sine voltage or 100/120Hz for a full-wave rectified voltage pulses). This is what we call "hum."

Transformers can also vibrate mechanically and produce their own sound, called a buzz. Hum and buzz are nasty twins, the biggest enemies of high fidelity. There is nothing in your audio system to hum or buzz with battery power and nothing to generate RFI (Radio Frequency Interference).

The more technically savvy amongst you will now scream, "Wait a minute, Igor, you've forgotten one of the biggest benefits of battery supplies, their low internal impedance!" Dead right.

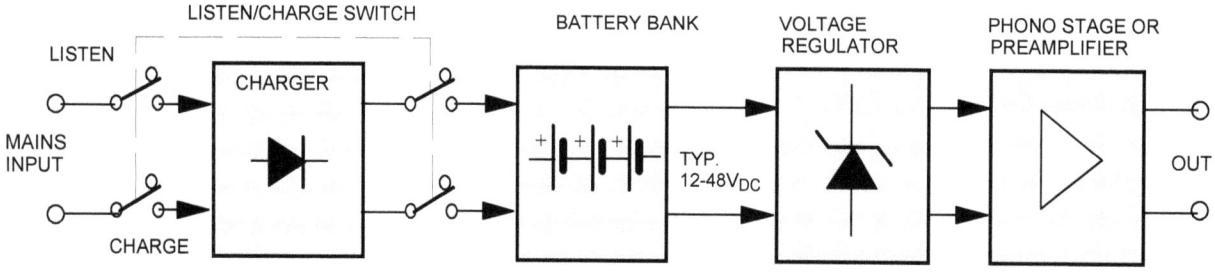

Fig. 4.48. Block diagram of a battery-powered low voltage phono stage or line preamplifier (with the LISTEN/CHARGE switch shown in the "LISTEN" position (battery bank discounnected from the charger and the mains input).

An ideal voltage source has a zero internal impedance, so all the power is transferred to the load, with nothing wasted on its internal impedance. Batteries have a very low internal impedance, and arguably, voltage regulation in battery power supplies may not even be required. The internal resistance of a 12 Volt 7.0 Ah rechargeable sealed lead-acid battery (photo below) is typically 20-30 mΩ, at least 50 times lower than the typical 1Ω (or higher) output impedance of the best regulated power supplies!

Sealed lead-acid batteries for preamplifiers and phono stages

How would we go about designing and building a battery-based audio power supply? Well, it all depends on to which camp you belong. Those of you of the solid-state persuasion can stop reading right now and start stuffing your black boxes with batteries. Due to their low current draw (consumption), battery-powered solid-state phono stages and line preamplifiers are relatively easy to design and build - semiconductors work on low voltages anyway.

"Low voltage" in audio power supplies means DC voltages under 48 V. Battery-powered tube phono stages and preamps are a much more complex undertaking, but not impossible. We published a few such designs in my book *Audiophile Vacuum Tube Amplifiers* (Volume 1), so there's no need to repeat all that here.

Phono stages benefit the most from being battery powered since batteries are noise-free, at least the noise produced and propagated through the ordinary, mains powered power supplies. In terms of dynamics and the avoidance of compression and distortion, power amplifiers benefit the most from the low internal impedance and high instantaneous current capacity of batteries, something no conventional power supply can achieve.

Tube phono stages usually work with high anode voltages (200-300 Volts), but we have successfully designed various 24 V_{DC} tube phono stages using two 12V rechargeable sealed lead-acid batteries in series. One such circuit is described on page 138.

PS-1270 is a 12 Volt 7.0 Ah rechargeable sealed lead-acid battery, typical of this technology, Absorbent Glass Mat (AGM). Its internal resistance is 23 mΩ, which is as close to an ideal voltage source (zero internal resistance) as we can get. It sells for AU$27, so quite affordable.

The discharge depth of a battery will affect the number of cycles you will get out of it. With 30% discharge, more than a thousand cycles are possible, while the 50% discharge depth halves that to less than 500 cycles. A 100% discharge is a no-no; it would reduce the battery life to only 200-250 cycles!

Let's say you have a tube preamp with four ECC88 tubes, two in the phono stage and two in the line stage. With their 6.3 V heaters in series (for 24 V operation), the heater current of 0.365 A will discharge two batteries in about 18 hours. Thus, it would be prudent to limit the listening time to about 30% of 18 hours or about 5 hours.

If that isn't enough for you, use three batteries in series for 36 V or four for 48 V operation, or use higher current capacity batteries.

We have neglected a few milliAmps of anode current that such a phono stage would draw, compared to 365 mA that tube heaters consume; even 10-20 mA of some higher current phono stages is minuscule.

So far, so good, very few audiophiles would listen to their systems for more than 5 hours in one "sitting." The use of voltage regulators would lower the output voltage (due to the voltage drop on the regulator itself). It would affect the sonics since a regulator is directly in the signal path. Luckily, their use is entirely optional. If you limit your listening time to an hour or two, the output voltage of a high capacity 4-battery bank will drop from say 4*12.7V = 50.8V to 4*11.5V = 46 V, which would have a minimal impact on the sonics so that you can power up your phono stage directly from the battery bank.

Fig. 4.49. Power Sonic PDC-1275F2 is a12 Volt 7.5 Ah sealed lead-acid battery with F2 (1/4") blade terminals. For better contact, the link connecting the two batteries was soldered permanently.

Battery-powered power amplifiers: extreme audio fanaticism, but possible

There are two ways to run a tube power amplifier from batteries. We can use a single battery and then convert its low DC voltage (12 or 24 Volts) to a much higher DC voltage needed (200-400 V) using a switch-mode DC-DC converter. An affordable option is to use a 12 V car battery (or a sealed lead-acid battery), or, even better, a 24 V truck battery. 24 V is better since it is easier to design DC/DC converters with lower step-up ratios, for instance from 24 V to 200 V instead of 12V to 200 V.

The second route is much better sonically but a nightmare from a practicality point-of-view. We would need to buy sixteen 12 V batteries and connect them all in series (to get around 200V), or 32 batteries for 400 V!

The financial aspect of the whole nightmare is that sixteen good-quality car batteries would cost around US$1,500 or more (AU$3,000 in Australia). Double that for an amplifier working at 400 V.

The batteries' life span, depending on their quality, the discharge currents, and depth levels, would be only one to two years, so you'd spend US$1,000-2,000 annually just on battery replacements. Your local battery shop owner would become your best friend, and you'd be paying for his new luxury car.

The health and safety issue is of even more concern. During charging, lead-acid batteries release toxic and flammable gasses, so to keep sixteen of those in your listening room is out of question. You don't want to endanger your life or the lives of your loved ones. An elaborate charging circuit with automatic voltage monitoring and switching is also needed. Batteries would have to be charged one-by-one using a smart 12 V charger, and the management circuitry would need to switch over the charger to the sixteen or so batteries sequentially (one-by-one) until they are fully charged.

Why don't we develop a high voltage charger that could charge the whole bank as it is, in series? Sure, we could, but that is another device someone has to design and build.

High power transistor amps would also need quite a few batteries, depending on their rail voltages. For example, with +/-75 Volts power rails, we'd need two banks of seven 12 V batteries, 14 in total.

CASE STUDY: Battery-powered SET amplifier

Due to their high anode voltages and high power levels, battery-powered tube power amplifiers are a much more complex undertaking than preamps or even solid-state power amps.

Never being the ones who surrender easily, we decided to design and make one. The sonic results were worth the enormous effort. The amplifier sounded smooth, dynamic, yet open and detailed, with incredible microdynamics.

To power this amp, we used two 12 V car batteries connected in series. One look at the PL508 power tube specs reveals the reasons for our choice. Its 17 V heater voltage is lower than the 24 V battery voltage, so no issues there. 17 V plus 6.3 V for preamp tubes' heaters is 23.3 V, an ideal match to a 24 V battery.

However, preamp tubes usually draw 300 mA, and very few power tubes draw such a low heater current. Tubes with standard 6.3 V heaters, such as 6L6, EL34, KT88, and others, draw from 1.2 to 1.9 A each, so they are out. Luck would have it that PL508 power tube draws only 300 mA, the same heater current draw as preamp tubes of the 12AU7, 12AT7, and 12BH7 variety! In fact, that tube was designed so its heaters can be connected in series with all other tubes in vintage TV receivers. Thus, the 300mA was a fixed requirement, and 17 V was the consequence, the voltage needed to develop the required anode power.

The heaters of its bigger brothers, PL509 and PL519, also draw only 300 mA, but they need 40 V of heater voltage due to their higher anode power dissipation (30 W and 35 W, respectively).

PL509 and PL519 tubes could also be used for this project (to get higher output power levels), but we would need four 12 V batteries for a total of 48 V.

We knew from our previous mains-powered amps that triode-strapped PL508 had a refined treble, great microdynamics, and fast and tight bass.

Initially, we were concerned about the sonic impact of the switch-mode power supply, but such concerns proved to be unfounded. The power supply was very quiet, and the amp sounded very "musical" and engaging.

Fig. 4.50. The rear view of the battery-powered SET amplifier with PL508 tubes. Notice the two large output transformers with no power transformer. The AC/DC switch-mode converter was under the chassis.

5 | CABLES, FUSES, CONTACTS, AND CONNECTIONS

In my primary school days, I wanted to become an architect. In a way, I did become one; only instead of designing buildings, bridges, or highways, I design audio components such as amps, preamps, and power supplies. If audio components were cities, cables would be the equivalent of roads, railways, and rivers (waterways) that connect them.

In other words, both internal "hookup" wires (inside amps and preamps) and external cables between audio components are sonically critical. "Do cables make a difference?" is a wrong question. The right question is "How much difference (to the sound, not to your financial situation) do audio cables make?" and the answer is - a lot!

Talking about the financial side of things, the prices of many hi-end cables are so preposterous they are arguably the main reason many audiophiles view hi-end cables with a mix of emotions, from suspicion and derision to open hostility.

Luckily, DIY once again comes to the rescue. Providing you choose the right materials, superior-sounding audio cables can be made quickly and cheaply; this chapter will show you how.

- HOW AUDIO SIGNALS PROPAGATE THROUGH CABLES
- INTERCONNECTS
- POWER CABLES
- SPEAKER CABLES
- HEADPHONE CABLES
- COPPER, SILVER, GOLD, NICKEL, RHODIUM, BERYLLIUM AND TELLURIUM: A CRASH COURSE IN AUDIO METALLURGY
- CABLE TERMINATIONS AND CONTACTS
- FUSES

> "People admire complexity. They don't trust something that looks too simple. But only the simple ideas will work. The more powerful ideas have an elegant simplicity about them. Less is more."
> Al Ries and Jack Trout, *Horse Sense (The key to success is finding a horse to ride)*

HOW AUDIO SIGNALS PROPAGATE THROUGH CABLES

The "cable as a water pipe" model: simple, easy to understand, and wrong

Ask any audiophile or even an electrical engineer to explain in a few sentences how electric signals travel through cables, and you will get the usual spiel of electrons moving up and down the cable, just as water molecules travel through a water pipe

The simplest electrical circuit, illustrated on the right, would have only three components or parts, a voltage source (DC battery in this case), a load (light bulb or a resistor), and interconnecting cables or wires. The electrochemical energy stored in the battery is converted into light energy and heat, both electromagnetic waves of different wavelengths and frequencies.

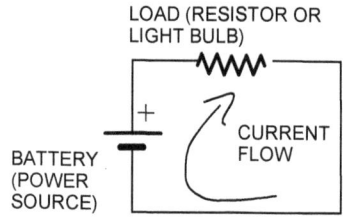

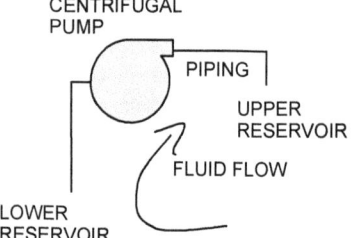

Fig. 5.1. The simplest electric circuit (ABOVE) and its hydraulic equivalent (BELOW)

In mechanical or hydraulic terms, the equivalent would be a pump (equivalent to the battery), pipes (cables) that transport a fluid such as water from a lower level (a pond, a river, or tank) to a higher tank. That requires work, expanding energy, just as in the electrical circuit equivalent.

This "cable as a water pipe" model is simple and easy to comprehend, but for our purposes, understanding audio signal propagation, it's dismally inadequate. How little Johnny imagines things is far from the complex reality.

Electromagnetic waves and cables as their "guides"

Despite their relatively simple anatomy, cables are very complex beasts from an engineering perspective, so the following summary will, out of necessity, be very simplified but still of sufficient depth to provide some answers as to their behavior.

An electromagnetic wave (audio signal) propagating through a cable is a transversal wave, meaning the vectors ("arrows") of the magnetic field and the electric field are perpendicular to each other and to the axial or "travel" direction of the signal (along the cable). This assumes an ideal conductor, with no resistance, acting as a waveguide. In that case, the electromagnetic signals travel entirely on the outside, in the dielectric, and the surface of the conductors.

In real, lossy conductors (with resistance), the electric field is not exactly perpendicular to their length anymore. A loss field is present, with its maximum on the surface and decaying exponentially towards the center of the conductor.

The first issue is that the speed of the internal loss field is much slower than the speed of the outside electromagnetic wave. Plus, the signals traveling closer to the cable's center travel slower than those near the surface. It is believed that this causes audible "time smear" since different frequencies (the harmonics of a complex musical signal) arrive at slightly different times. This "time dispersion" becomes more prominent in thicker (larger diameter) cables.

The second issue is that such trapped energy remains stored inside the cable even when the audio signal ceases. It cannot collapse or dissipate instantaneously, so the cable itself becomes a source feeding this unwanted delayed and time-smeared artifact into the load.

The electromagnetic energy wave ("the audio signal") will undergo multiple reflections at the cable's ends, bouncing back and forth between the source and the load multiple times. To minimize such "ringing" and multiple reflections, it is essential to terminate the cable at one end (the generator end is usually better) with an impedance equal to or close to its characteristic impedance.

Stranded cables introduce another unwanted phenomenon. Since the loss component propagates radially across the strands ("jumping" between them), it crosses random air/copper discontinuities, which slow it down again and cause multiple reflections at each such interface.

Crystal boundaries within copper can be viewed in the same manner as tiny discontinuities that cause subtle yet audible signal degradation. Some cable manufacturers produce single crystal copper wire (purity 99.9999 %) without such discontinuities, resulting in 8-13 % lower resistance and superior sonic properties.

Finally, mathematical analysis shows that as the wire's radius approaches the skin depth at a particular frequency (more on the frequency-dependent skin effect in the next section), the incident (incoming) and reflected fields start to cancel out, and the energy storage is minimized. The wire behaves almost as a purely resistive conductor, and its reactive, energy-storing inductive and capacitive parameters are significantly reduced.

These are the main scientific arguments in favor of the claims that cables should be solid, as thin as practical, with as few discontinuities and contacts as possible, and using low loss, superior-sounding dielectrics.

CABLES, FUSES, CONTACTS, AND CONNECTIONS

What is wrong with stranded cables?

Stranded cables are made up of many thin wires (strands) twisted together. The only benefit from such construction is increased flexibility, so regular twisting, bending, and kinking of the cable does not result in any damage or breakage, as it is likely to happen with solid core cables. Stranded cables still suffer from the skin effect, just as solid wires do. The skin effect is a tendency of current to flow closer to the outside surface or "skin" of the conductor.

DC current (frequency = zero) does not suffer from skin effect and flows across the whole cylindrical cross-section. As the signal frequencies increase, the skin effect becomes more and more prominent, so the skin depth continues to decrease, and the current path "hollows out" and becomes more and more concentrated towards to outer surface of the cylinder. This means that different audio frequencies travel through differently-shaped conductive paths.

This is indicated in the cross-sectional diagrams below as different shades of gray. The darker the wire's shading, the more current flows through it. At the lower limit of human hearing, at an audio frequency of 20 Hz, the skin depth is 9.35 mm, quickly dropping to 6.6 mm at 100 Hz and only 2.1 mm at 1 kHz (the midrange). At the upper-frequency limit, 20 kHz, the skin depth in copper is only 0.47 mm!

Stranded cables also suffer from the proximity effect. The signal current, flowing in multiple parallelled conductors, produces a varying magnetic field around each strand, which then, through electromagnetic induction, generates longitudinal eddy currents in its neighboring strands. This "pulls" their current towards the adjacent side, resulting in further current crowding, especially at higher frequencies, just as the skin effect does!

Let's start with a single strand as an example; at one end of the cable, it is in position #1, close to the center of the bunch. Due to the twisting action, a few millimeters further down the length of the cable, that strand will occupy position #2, then move to location #3, and so forth. The same applies to all other strands.

Strands aren't pressed against each other along their whole length; they only touch each other at a few points. The strands aren't all at the same potential, so the flowing current jumps between them almost randomly. When combined with the natural and inevitable oxidation of the surface of the strands, this micro-arcing results in gradually increasing resistance and degradation in sonic performance.

The male spade terminals of the original tweeters of our Opera speakers were soldered to the hookup wires. To audition various tweeters (read about it on page 227), we unsoldered the hookup wires and crimped and soldered female spade terminals onto them, thus making tweeter "rolling" quicker and easier.

When we stripped off an inch of the outer insulation from the stranded hookup cables the manufacturer used, all the strands were black, totally oxidized!

The humidity in Perth, Western Australia, is very low for most of the year, so in most other, much more humid parts of Australia and the world, such oxidation would happen much more rapidly. Tinned copper strands oxidize less, but even silver-plated cables still oxidize; the extent of oxidation depends on the cable's manufacture quality.

It was then decided to rewire the speakers with solid core, silver-plated, Teflon®-insulated hookup wire.

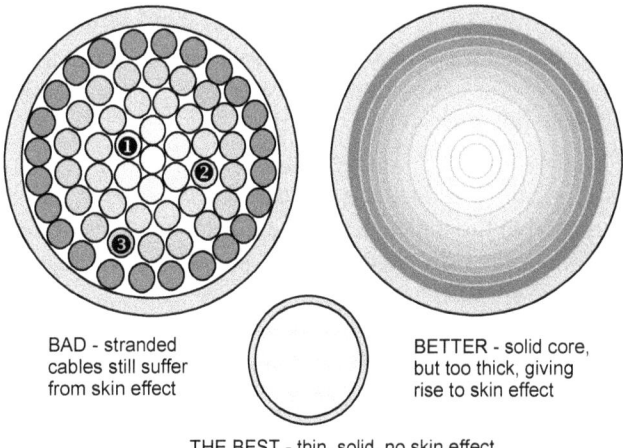

BAD - stranded cables still suffer from skin effect

BETTER - solid core, but too thick, giving rise to skin effect

THE BEST - thin, solid, no skin effect in the audio range of frequencies

Fig. 5.2. Unless elaborate (and expensive!) measures and fancy construction geometries are used to mitigate the proximity and skin effects, stranded cables are generally sonically inferior to the solid core alternative, especially to the thinner (smaller diameter) solid core audio cables.

Litz cables?

Just in case you've missed an essential point from the skin-effect perspective, if bare wire strands are used in a multi-stranded cable, the whole bundle shorts out and thus behaves like a single large diameter wire, as illustrated by various shades of gray in the cross-section above.

Litz wires (from German "litzendraht," Litze meaning a braid, cord, lace, or strand, and Draht meaning wire) address this problem by using individually insulated wires, twisted or woven together. The diameter of each insulated strand is smaller than the skin depth at the frequencies of interest, thus minimizing skin effect losses. All such parallel conductors' ends must be stripped of insulation at both ends and soldered together. This labor-intensive process (must be done manually) further increases the manufacturing cost of such cables.

Until recently, when hi-end cable manufacturers adopted it in their products, Litz wires were only used in radio receivers and other high-frequency devices.

Dielectric is key!

The metal used in audio cables (usually copper or silver), its purity, and crystal structure aren't the only factors determining the cable's behavior and, ultimately, sound at audio frequencies. Geometry and construction of a cable, primarily the number and the arrangement of its cores and strands (if multi-stranded) impact the sonics, but so does the material used as the dielectric. The internal insulation around individual strands and/or cores and the overall jacket aren't there just for structural integrity and protection; the electromagnetic field produced by the power and audio signals propagates mainly in the dielectric.

Of all physical properties, the dielectric absorption coefficient of the material used seems to be the critical factor determining its sonic performance. It goes by many pejorative or popular names, dielectric soak, soakage, dielectric hysteresis, residual charge, or even "battery action."

The phenomenon is most pronounced in capacitors. Discharging them fully removes the voltage only as long as their leads are shorted. Once the capacitors sit disconnected with their leads "open," some of the original voltage reappears. Mylar® (polyester film) has a relatively high dielectric absorption coefficient (0.5 %), while Teflon® and polypropylene show low levels, below 0.01 % and 0.005 %, respectively.

Dielectric absorption causes energy retention and delayed-release, which could explain why some cables (Teflon- or air-insulated, for instance) sound "cleaner." In contrast, others smudge or smear the fine details and introduce a frosty or hazy layer over the sound.

Cables with DC bias

We have talked about the DC biasing concept a few times in this book, namely in magnetic circuits (open-reel tape decks) and electronic circuits (single-ended tube output stages). A similar DC-biasing principle also applies to dielectric materials used as an insulator between conductors in cables.

United States patent US7126055B1, titled "Apparatus and methods for dielectric bias system," was filed on Nov. 3, 2003, on behalf of AudioQuest, a USA-based audio cable manufacturer. They call it the "Dielectric Bias System" (DBS for short) and use it in their interconnects and speaker cables. The patent expires in 2023, meaning other manufacturers and even do-it-yourself cable constructors will be able to use it in their cables without a license.

In its basic form, DBS involves at least two wire electrodes in contact with the cable's dielectric along the entire length of the cable, to which a DC voltage source (an externally attached battery pack) is connected and left permanently energized.

This low-level DC voltage establishes a weak electrostatic field between such two electrodes, which, it's claimed, "significantly reduces nonlinear phase errors," or, in AudioQuest's description, "... saturates and polarizes (organizes) the molecules of the insulation. This minimizes both energy storage in the insulation and the multiple nonlinear time-delays that occur. Sound appears from a surprisingly black background with unexpected detail and dynamic contrast."

The audio cable axioms

Cables are one of the least understood and (perhaps for that reason) most controversial and hotly debated audio components. The following list is my summary of audio cables in general. In the section on speaker cables (page 108), we will add a few more axioms to the list.

THE AUDIO CABLE AXIOMS

Axiom #1: All wires and cables in a circuit affect its sound. This applies to internal ("hook-up") wiring (inside amps, preamps, speakers, and other audio components) and external cables (between audio components). Even power cables, seemingly outside the signal path, will change the sonics of an audio system.

Axiom #2: Every discontinuity and interface (contact, mechanical or soldered joint, discontinuity in the conductor's crystal structure, etc.) in a cable negatively affects its sound.

Axiom #3: The more different metals are present in a cable or connector, the more its sonic performance will be compromised.

Axiom #4: The cable's dielectric (the material used to manufacture its insulating layers or sheaths) significantly affects its sound.

Axiom #5: Audio cables (all of them, power, speaker, interconnects, even digital cables) should be thin (between 0.5 and 1mm diameter) and solid core.

Axiom #6: The closer the cable's characteristic impedance is to the source or the load impedance, the better the cable will sound.

INTERCONNECTS

Since they transmit the lowest level signals in the audio chain, and because their sonic signature (a nice euphemism for all kinds of distortion & delay effects) is amplified "downstream," through the rest of the system, interconnect cables are arguably the most important of all in the audio chain. Their inductance, capacitance, and resistance cause both phase shifts and signal losses that will impact high-frequency reproduction and the overall tonal balance of the whole audio system.

> "I think that any 1m interconnect pair that sells for $1000 or more ought to sound amazingly, obviously good - something that, in my experience, very few do - and I think that any 1m interconnect pair that sells for $5000 or more ought to sound amazingly, obviously good *and* increase the size and functionality of one's penis. Which is, of course, the primary reason men buy such things."
> Art Dudley in Stereophile, June 2015

Audio cables as filters

Every shielded cable is, in its essence, a low-pass filter. From a cross-section of a shielded (coaxial) interconnect cable, it is evident that the whole assembly is one giant capacitor: two metal structures (the inner conductor and the shield or braid), separated by a dielectric, its inner insulation.

In a simple signal chain (illustrated below), the cable will shunt higher frequencies away from the load. That source can be a CD player, while the load can be a preamp or an amp. Or, it may be the shielded cable used inside an amplifier, from the RCA inputs at the back across the chassis to the potentiometer (volume control) at the front.

Once you measure the series resistance R_C and shunt capacitance C_C, you can estimate the frequency at which the cable's capacitance will attenuate the signal by -3dB. At this frequency, called the "half-power" or "cutoff" frequency, half of the power will be lost in transmission between the source and the load).

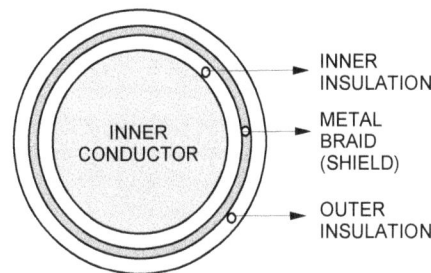

Fig. 5.3. The cross-section of a shielded audio cable (interconnect) illustrating the fact that it is a cylindrical capacitor, with two metal electrodes (the inner conductor and the conductive shield), and dielectric (insulation) sandwiched in between.

Single-core shielded cables

One of our audio cable axioms states that cables should be thin and solid core, so does that mean that single-core shielded unbalanced cables are fine to use in hi-end audio? The answer lies in the cross-section illustrated above. The inner conductor may be solid, but the braided shield is used for signal return!

The different structures and geometry of the two "legs" result in their markedly different electrical and sonic behavior. Secondly, the shield is multi-stranded, and these strands will oxidize with time, just like all multi-stranded cables do, further degrading its performance.

Last but by no means least, the outer braid carries the signal but has to perform its original function as well, to act as a shield from radio frequency interference, thus mixing the audio signal with all high-frequency grunge it picks up.

The solution? Replace coaxial (shielded) audio cables with twisted pairs of thin, solid-core conductors. See our DIY project on the next page.

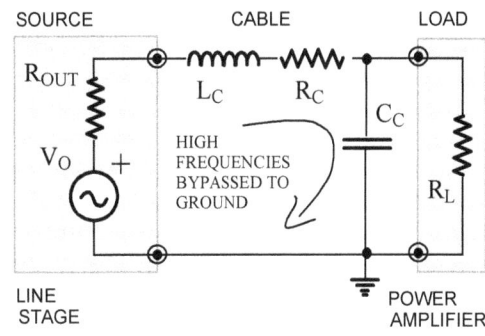

Fig. 5.4. A high capacitance cable is a low-pass filter that will shunt higher signal frequencies to the ground. The effect of the series cable inductance LC is the same, further contributing to high-frequency audio signal attenuation.

Quick capacitance and inductance measurements of a few interconnects

To show you how to test your audio cables, we took a few interconnects and performed a basic LCR-meter test, the capacitance between the active pin and shield (or common conductor on the one cable that was unshielded), and the series inductance of the cable.

Ideally, a cable should have zero capacitance, inductance, and resistance. Since that is physically impossible, the rule is obvious - the lower the L, C, and R of a cable, the better it will perform in every aspect.

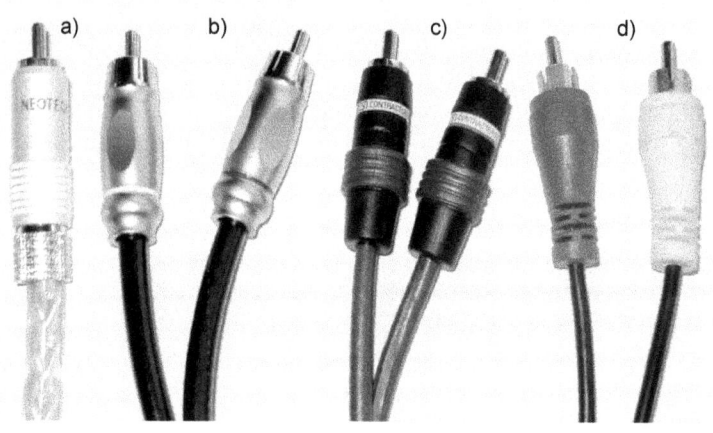

INTERCONNECT CABLE (L-R)	C_C [pF/m]	L_C [µH]	Z_0 [Ω]
Neotech "Element"	42.5	1.2	168
Neotech "Origin"	70.0	1.7	155
Ultralink "CS1"	87.0	0.9	101
Budget	83.0	0.3	60

Fig. 5.5. The two measured paramaters, capacitance and inductance, and the calculated characteristic impedance of a small selection of commercial interconnect audio cables

We didn't measure the DC resistance of these cables since multimeters cannot measure precisely below 1 ohm; the error when the multimeter displays fractional values (0.2 ohms, for instance) is so significant that such measurements are meaningless.

Once the capacitance between the positive and negative connectors has been measured, we can calculate the characteristic impedance of each cable, which is defined as $Z_C = \sqrt{(L_C/C_C)}$.

Cheaper cables, Ultralink CS1 (c), and the generic or "budget" interconnect (d) had the highest capacitance but, also, interestingly, the lowest series inductance. The most expensive cable, Neotech "Element," had the highest inductance and the lowest capacitance (a). This could be expected due to its twisted construction; it is not a coaxial cable like all the others.

From this simple and thus limited experiment, the two parameters, cable's capacitance, and inductance seem to be inversely related.

DIY PROJECT: Pure silver Teflon®-insulated interconnects

Many years ago, we bought various lengths of pure silver and silver-plated copper cables from a USA-based eBay seller who claimed they were used in avionics and other industrial applications. These were all single-core shielded cables, black, brown/bronze, or white outer jackets, some with a solid core, others stranded, all an incredible value-for-money. We've made many interconnects and speaker cables out of them, all superb sounding.

The capacitance between the central conductor and the shield of this 1.05 m homemade interconnect was only 25 pF, an excellent result. As a comparison, Neotech "KHS152 Balanced Interconnect", also 1.05 m long, was measured at 116.4 pF, a 4.66 times higher capacitance!

The first few steps of the termination procedure for RCA interconnects are always the same. Cut off 15 mm of outer insulation. Gently separate the outer braided shield (don't cut through it and then twist!) and twist it together (1). Remove the insulation from 5mm of the inner conductor and twist (if stranded). These RCA connectors use a screw-in terminal (3) for the central conductor; most others require it to be soldered in.

Likewise, some designs also need the screen to be soldered to the outer casing; with this one, you simply pull it back and crimp it in place (4); no soldering is required at all.

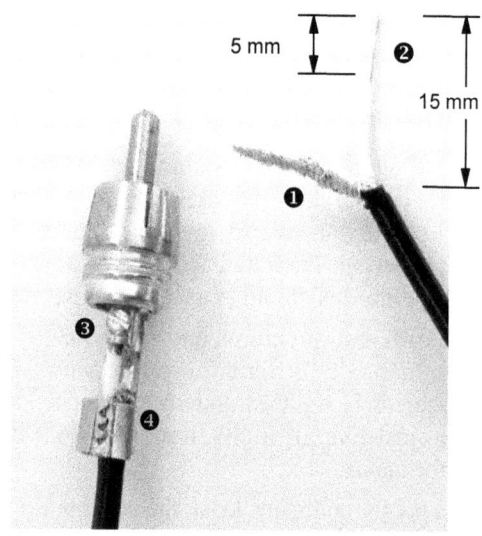

Fig. 5.6. A few termination tips if you decide to make your own interconnects

DIY PROJECT: Solid-core UHP copper Teflon®-insulated interconnects

Bought from an eBay seller in China, 10 meters of this 17AWG (1.12mm) wire cost us US$11.50 (around AU$15.50 at the time), including international airmail postage. Rated at 300 °C and 600 V, this solid core hookup wire is allegedly made of 6N or UHP (Ultra High Purity) copper with Teflon® insulation.

Teflon® is a registered trade name of DuPont; its proper name is polytetrafluoroethylene or PTFE. The silver-plated single-core Teflon®-insulated UHP copper wire of the same size (18 AWG) from the same seller was much more expensive at around AU$4.40 a meter, so AU$44.- for 10 meters.

Alas, the insulation was not PTFE at all. PTFE is very hard to strip with wire cutters, and it doesn't melt at normal soldering iron temperatures, so if you touch it with a hot soldering iron, nothing should happen. This insulation melted very easily! It also made us wonder what else had the seller lied about, the voltage rating, the copper being ultra-high purity, or both?

At 1kHz test frequency, the capacitance between the two cores of the finished cable was 67.2 pF or 39 pF/m (the cables were 172 cm long). Famous brand 3-core braided interconnects, 151 cm long, measured 116.3 pF or 77 pF/m, almost double ours.

Our cable's inductive reactance between the two cores was 2,312,840 ohms minus 2,368,377 of capacitive reactance or -55,537 ohms in total (the - sign indicating a capacitive nature), equivalent to the effective capacitance of 2.87 nF.

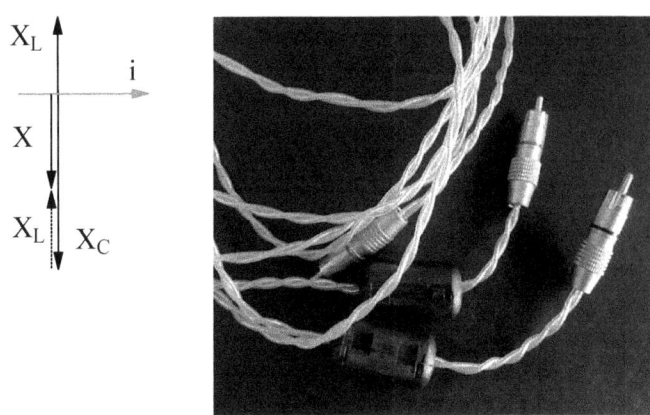

Fig. 5.7. Vector diagram for a cable whose capacitive reactance X_C is bigger than X_L, its overall reactance X is capacitive. It voltage, in phase with X, is lagging the signal current i.

Fig. 5.8. DIY single-core UHP copper interconnects

The branded cable had equivalent effective capacitance of 7.45 nF, 260% higher. Despite its longer length, our cable was superior in every sense, not just in terms of these measurements but also sonically; the listening impressions were very positive.

POWER CABLES

Power cables affecting the sound of audio components - fact or fiction?

Since they are so removed from the "signal path," power cables should have no or minimal impact on the sound of an audio component. However, from my personal experience, they usually affect the sound of an audio system more than interconnects or speaker cables! Again, that makes no technical sense whatsoever. Neat engineering models and simple measurements fail to fully explain the complex and messy audio reality.

In the early days of my audio journey, the engineer in me dismissed those claims about different "sounding" power cords as ridiculous until I tried a few and heard noticeable differences, sometimes subtle, other times significant.

The first time we took part in an audio show was at one of Perth's 5-star hotels, under a local hi-end dealer - only dealers could participate, not manufacturers; he was kind enough to invite us into one of his rooms.

He also helped us with a loan of a high-end CD player and various cables. We kept trying different interconnects and speaker cables, but neither he nor we were happy with the sound. Then, he brought in a pair of power cables that he was making and selling at the time.

His power cables immediately improved everything! The bass was cleaner, tighter, and better defined. The tonal balance became better balanced, the sonic presentation more engaging.

He nodded, walked out, and left me sitting there with my jaw dropped. Luckily, we were in a hotel room with a bathroom, so I went there for a cold shower. I thought I imagined it, but the improvements were still there upon my return. I was not delusional after all; it was real!

I bought a pair of those power cables without hesitation, and they'd served us for many years afterward in our sonic evaluations of amps and preamps.

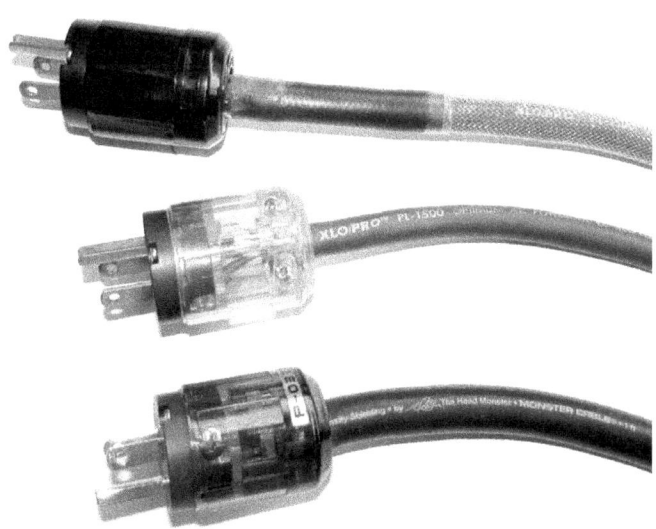

Fig. 5.9. Some "audiophile" power cables sold online use genuine plugs, sockets, and wire by Furutech, Monster, XLO, and other reputable brands. Many, however, are cheap fakes.

Where does an audio system start?

Power (mains) transformers in audio equipment are bi-directional components. They reflect impedances and signals from their primary into secondary circuits and vice versa. Any changes on the power supply side will affect the circuitry on the secondary side (the load side). Some changes on the power supply side will be audible; others may not be. Some will be significant, others subtle; some will improve the sonics, others will make it worse.

Your audio component does not start at its IEC power socket; it doesn't even start at the power outlet on your listening room's wall. Changing a supply cable (installing a dedicated supply) from a house's switchboard to the listening room significantly improved the sonics of many audio systems.

Many audiophiles report significant sonic changes (usually negative) after a changeover switch or a solar power generation system was installed, and inverters were added to their homes' wiring. In houses with a 3-phase mains supply (as ours is), changing the power supply to the listening room from one phase to another usually makes a noticeable sonic improvement.

Should power cables have a wide bandwidth?

An advertisement in a well-known audiophile magazine claims that "Our high resolution, wide-bandwidth power cables will ..." First of all, the term "resolution" applied to power cables is meaningless. Unlike interconnects or speaker cables, power cables transmit a single frequency, simple sinusoidal waveform AC power "signal," so there is no other "detail" to "resolve." The wide bandwidth of such cables is actually detrimental.

Ideally, a power cable should behave as a perfect low pass filter, meaning that it should not transmit any of the harmonics of the fundamental power 50 or 60 Hz sine wave into audio equipment. As we have seen in Chapter 4, these harmonics are created when the relatively low distortion sine wave produced by the generators in a power plant is distorted along the way, first by its transmission & distribution (power lines, switchgear, transformers, etc.), and then by the various loads on the power grid.

The anatomy of a few commercial power cables

Quality power cables come in all sizes, materials, and types of construction. An audiophile-quality power cable should not just have superior conductor cores but also a braided shield and/or an aluminium sheet wrap, usually used in conjunction with a copper drain wire.

The drain wire and, indirectly (since it is in continuous contact with the drain wire inside the cable), the braid should be connected to the earth wire (usually with green or yellow/green colored insulation) and the earth pin at the mains connector plug. The other end of the shield (at the IEC plug or "the equipment end") is left unconnected.

Audioquest AC-12 cable is not shielded. Its cross-section shows the symmetrical arrangement of eight connectors (four for the phase and four for the neutral) around the central earth conductor. The earth is a bundle of 19 wires (AWG #25 size), while each "live" connector is a solid AWG #18 wire made of long-grain copper.

The primary and secondary insulation is PVC with a thin internal Mylar jacket. The cable's outside diameter is 11mm.

XLOs website describes the UltraPLUS™ cables as "designed with classic XLO Integrated Field-Balanced Surface/Diving Winding Geometry™ so topologically speaking, the cable has no surface or center, and the frequency-related phase shift is effectively canceled. Very low capacitance and inductance mean XLO UltraPLUS™ cables work with a wide variety of components."

For more information, see U.S. Patent Number 5,110,999. Notice how each of the L, N & E conductors (claimed to be monocrystal copper) is a bundle of a large diameter central core, surrounded by multiple smaller diameter wires (black dots), equidistantly spaced from the center.

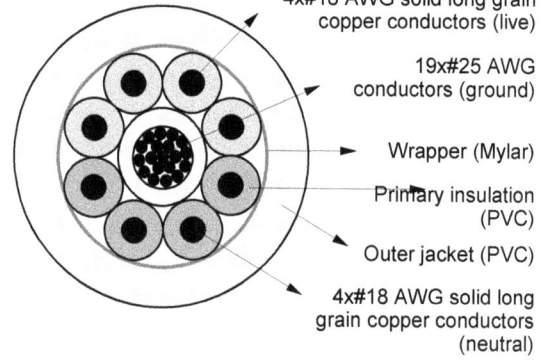

Fig. 5.10. The internal construction of *Audioquest AC-12* power cable. NOTE: Conductors (copper) are black, insulation and shields are various shades of gray

Fig. 5.11. The internal construction of *UltraPlus U10* power cable 1) L (Live) conductors 2) N (Neutral) conductors 3) E (earth) conductors 4) drain wire 5) shield foil 6) filler 7) insulation

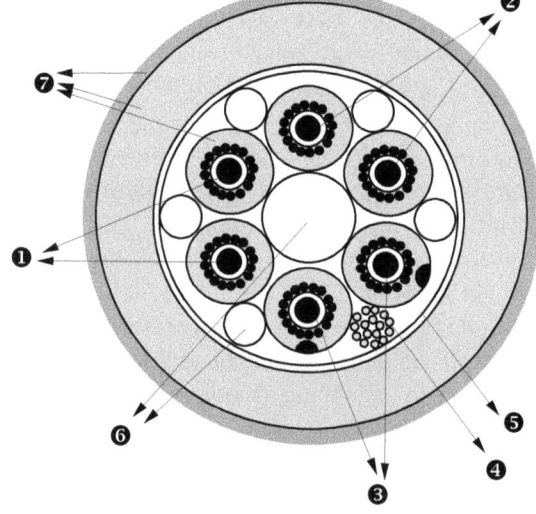

DIY power cables: many benefits but also some risks

The audiophile community has debated the claim that solid core cables sound better than braided (multicore) ones for years. Our experiments with solid core interconnect and speaker cables (some also featured in this chapter) have confirmed their sonic superiority. However, it remained to be seen if power cables of braided construction (most commercially available ones) would also be inferior to the solid core type.

One type of approved solid core (PVC V90 insulated) wire (1 mm^2) is the type used for switchboard wiring in residential premises. Available in commonly used colors, it is rated at 600 V_{AC}. However, solid core mains cables are only approved for "stationary" applications inside walls and switchboards. Any other use (if used as power cords, for instance) may be illegal due to the flexing issues - the solid core may break after continuous twisting.

That is only one reason why you should proceed with caution. If you get the wiring wrong or if the insulation gets damaged, there are serious safety issues. In general, homemade mains cables have not been approved by statutory authorities and are thus illegal. If a fire starts in your listening room or anywhere else in the house and the insurance or safety assessors determine that the cause of the fire was your DIY power cable, your insurance claim will be denied. Sure, it is an unlikely event, but why risk it?

Obtaining safety approvals and certification is an expensive process, so many "boutique" or audiophile-grade power cables (made in small quantities) may not have been approved either, making them illegal as well.

DIY PROJECT: Solid core copper power cable

The 2-core (red + white) mains cable used for lighting and some alarm circuits has a conductor cross-section of 1mm^2 and sells for AU$1.17 per meter. The manufacturer specified the inductive reactance at 50 Hz as 0.184 Ω/km. The DC resistance at 20°C is 18.1 Ω/km or 0.0362 Ω for our length, and the conductor's AC resistance at 50 Hz is 25.8 Ω/km or 0.0516 Ω for our length.

Stripping or rather peeling away the outer PVC sheath was very easy and took only a few seconds. Braiding the two cores together took about ten minutes, plus another five minutes to add a green/yellow earth connector.

Terminating the ends took another ten or so minutes, plus a few minutes for ohmmeter tests, to make sure no catastrophic error had been made in connecting the L (live), N (neutral), and E (earth) pins at both ends. All in all, less than half an hour of time investment.

The sonic improvement over the ordinary power cords (and even many audiophile-grade ones) was enormous. This power cable made the sound cleaner, the bass tighter, the treble less sibilant and more refined, and the backgrounds quieter (blacker).

Fig. 5.12. The power cable braided and ready for the final assembly
Fig. 5.13. The finished cable

Shielding of power cables - is it necessary?

Our solid core power cable wasn't shielded in any way, which may raise some eyebrows. Some audiophiles have noticed a sonic improvement when shielded power cables are substituted for the unshielded ones. Such improvements are more likely to result from other aspects of such cables - different dielectric, superior conductor arrangement and geometry, etc. Shielding may lower the noise floor by reducing RFI and EMI, especially if unshielded interconnects or speaker cables are run close to such power cables. That situation should be avoided, but it may sometimes be inevitable due to space and other restrictions.

However, shields increase cables' capacitance and change their sonics. Luckily, in most listening rooms, the level of RF and EM interference is so low that even the shielding of interconnects isn't necessary. The only possible exceptions are the interconnects between turntables and phono stages, especially when minuscule signals (microVolts!) produced by moving coil cartridges are involved (0.05-0.5 mV).

Subjective evaluations indicate that shielded power cables negatively impact both the dynamics of audio reproduction and the resolution and microdynamics of most systems, possibly because a shield will significantly increase the cable's capacitance. Thus, the shielding of power cables isn't just unnecessary but is generally counterproductive.

Minimize crosstalk and electromagnetic interaction between power and audio cables

1. Separate audio cables from power cables

Power cables should physically be as far from the audio cables as practically possible. Under no circumstances should they run in parallel or be bundled together (some audiophiles even use cable ties to make things neat & tidy)! If they have to cross each other (and that is often inevitable), make sure they cross at the right angle (90 degrees) and don't sit directly on each other but that there's at least an inch or more of physical separation. This is where cable lifters really can make a difference!

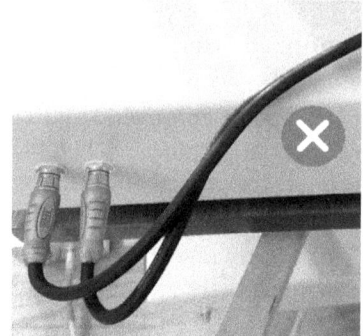

2. Separate power cables for the source components and low power devices (such as preamps and phono stages) from the high current power cables (for power amplifiers and subwoofers).

The lower the audio signal level at a component's input, the more it is susceptible to interference and noise. With that rule in mind, low signal, low-power components such as turntables and phono stages (especially moving coil step-up transformers) should be as far from the power cables and higher current devices (such as power amplifiers and subwoofers) as possible.

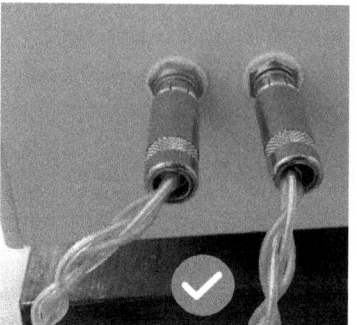

If power filters or conditioners must be used, it is better to use them only on those low-power components. The power amplifiers and subs should be powered up directly from the wall outlet in order not to limit the rate of current delivery (as, unfortunately, many conditioners do).

3. Separate paralleled interconnects and speaker cables

Cables carrying audio signals for the two channels should never run together (joined) or in parallel, close to each other! No matter what type of construction is used (round, flat, braided, etc.), if your interconnects (for the two channels) or speaker cables (the positive and negative wire for each channel) run together, in parallel with each other, separate them if at all possible. That will reduce the electromagnetic interference between them (some call it "print-through"), resulting in reduced crosstalk, improved channel separation, and a better sound stage.

Fig. 5.14. Running two interconnects together (some are even joined by design, as pictured above), even if they are shielded, increases their capacitance and mutual inductance thus reducing channel separation.

Fig. 5.15. To minimize electromagnetic interaction, two DIY cable lifters are used to increase the distance between the crossed interconnects or speaker cables and power cables.

Fig. 5.16. (BELOW LEFT) Physically separate power cables for the source components (1) and low power devices (such as preamps and phono stages) from the high current power cables for power amplifiers and subwoofers (2).

Fig. 5.17. (BELOW RIGHT) Separate the two speaker cable cores and run them as far from each other as possible.

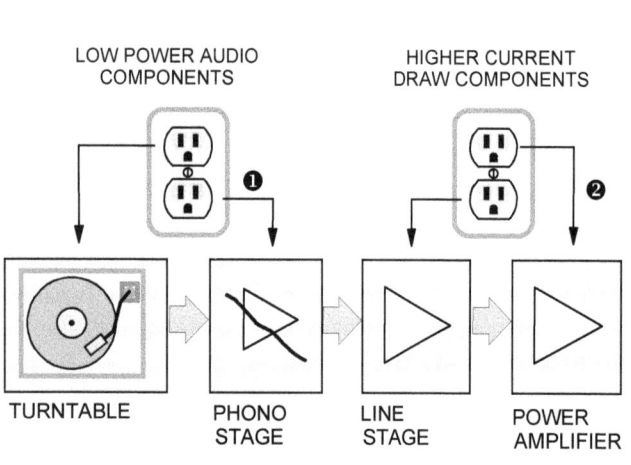
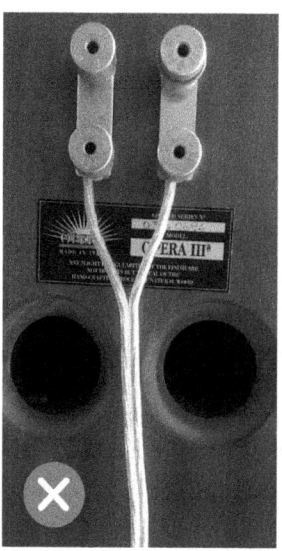
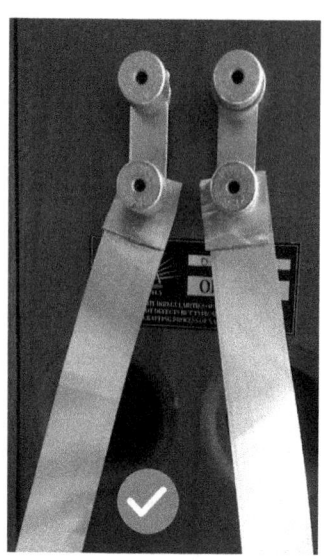

SPEAKER CABLES

DIY PROJECT: Coaxial cables as speaker cables

The RG59/U 75 ohm unbonded coaxial cable sells for AU$1.00 (US$0.72) per meter. The 6 mm outside diameter means the cable isn't too bulky or rigid, and the shield coverage is high, 95%. The conductor area is specified as 0.51 mm^2, meaning it is an AWG20 conductor (AWG stands for the "American Wire Gage"). Its current-carrying capacity will depend on many factors. For instance, a derating factor must be used if conductors are bundled or enclosed in a confined space. Still, in our application as a speaker cable, the conductors are not bundled and are not enclosed in any way (as inside a chassis or duct, for example).

The main factor in determining the current capacity of a wire or cable is the type of insulating material used. With polyethylene, neoprene, or semi-rigid PVC insulation, that AWG20 cable can take 10 A for 80°C temperature rise in the copper. Polypropylene, high-density polyethylene or irradiated PVC insulated cables can take 12 A

Power is related to current and resistance as $P = I^2R$, so with 4-ohm speakers, we could go up to $10^2*4 = 400$ Watts, or 800 Watts with an 8-ohm load. Remember, these are continuous ratings. No domestic speaker can take 800 Watts of continuous power, so this cable is more than large enough for any domestic listening application.

The maximum frequency for 100% skin depth for this size solid copper conductor is 27 kHz, meaning the skin effect will not happen within the 20 kHz audio range. The audio signal will be distributed across the whole cross-sectional area and not concentrated on the surface (skin) of the central core.

Beware, many coaxial cables use copper-clad steel or copper-clad aluminium as central conductor material; they don't sound as good as cables with pure copper conductors. Also, many use a much thinner central conductor, typically with a 0.27mm^2 conductor cross-section (close to AWG27 size).

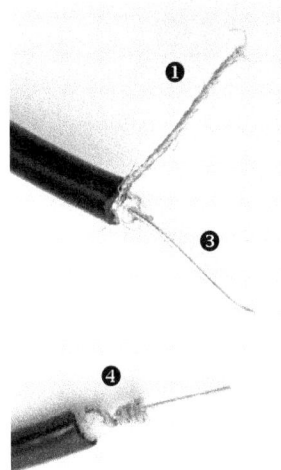
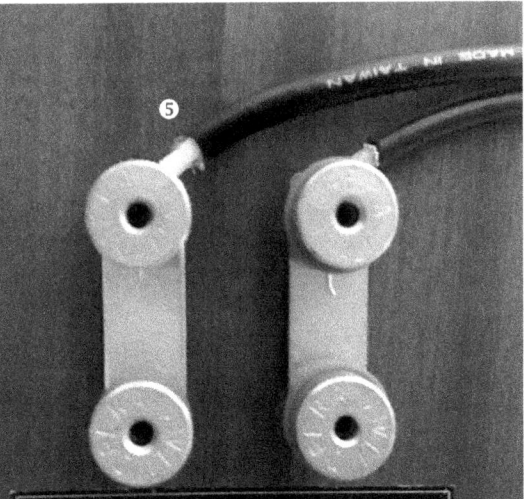
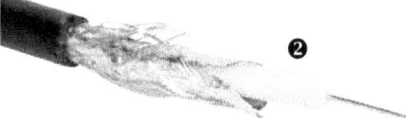

Fig. 5.18. Remove the outer jacket from about 2.5 cm. Remove the aluminium foil shield and twist the braided shield (1). Remove the inner dielectric, the insulation (2). Scrape off the insulation coating from the copper wire along the whole exposed length (3). Wrap the braided shield tightly around the central conductor (4). Solder them together.

Fig. 5.19. The braided shield should only be connected (soldered) to the central conductor at one end, and that end should be connected to the amplifier's binding posts. The photo shows the speaker's connection. Notice the screen cut right back at the speaker end (5).

With each speaker cable 4 meters long, we needed two lengths for each speaker, or 16 meters in total (AU$16.-). With a resistance of 33.3 ohms/km, the 8 m run to each speaker will have a resistance of $33.3*8/1,000 = 0.267$ ohms, which is low enough and can be neglected when used with high output impedance tube amplifiers. Still, it will lower the damping factor somewhat when used with very low output impedance solid-state power amps.

The listening tests were very positive. Compared to our standard "audiophile quality" speaker cables, priced at around AU$500, this coax cable improved the bass, becoming tighter and better defined. The treble was also more precise, with a slight improvement in microdynamics. We didn't terminate the cables with spades or banana plugs; instead, we fed them directly through the holes in the binding posts at both ends (5). If you opt for this type of connection, make sure you tighten them as much as possible. Use a contact protector/enhancer as well.

The 15 minutes of cable preparation time (for end terminations) and less than $20 in financial investment is well justified, so make and keep one set, if not for permanent listening, then as a reference. If the $500, $1,000, or even higher-priced speaker cables you are considering buying don't sound significantly better than these babies, ask yourself (or the audio dealer) some hard questions.

DIY PROJECT: Copper foil loudspeaker cables

Copper or silver foil satisfies all the requirements for superior sounding cables: they are solid (not stranded) and thin enough to be smaller than the skin depth at high frequencies, thus, if not eliminating, at least minimizing the detrimental consequences of the skin effect.

The idea of using thin conductive foils with a rectangular cross-section in audio isn't new. In 1989, Madrigal Audio laboratories marketed their HPS (Helical Planar Copper) interconnects and CPC (Co-Planar Copper) speaker cables. The co-planar means the two speaker cables are placed side-by-side in a joint plastic sleeve, while the two much narrower strips are wound around each other in a spiral manner in the interconnects.

Nordost, with their Flatline range, Alpha Core with the Goertz range, and Tara Labs (RSC-rectangular Solid Core), continued the flat cable trend in the early to mid-1990s.

Copper foil tapes are used mainly for RFI shielding of equipment cabinets and sold in various lengths and widths. For this experiment, you'll need a roll at least 4x longer than your required speaker cable length.

For instance, for 4 m cables, the roll will need to be at least 16 m long (since you need four equal cable lengths).

Instead of buying RFI foil, we unwound four lengths of 4 meters each from these Taiwan-made DIY Zone copper foil inductors (leftovers from speaker crossover upgrades). The logic was, if the signal from a power amp runs through a few of these coils inside speakers' crossovers (up to 100 meters of copper foil in each), 3-4 meters of such cable outside the speakers will be sonically innocuous!

The foil was 1" wide and fit perfectly within our DIY cable raisers (3). The ends can be rolled and soldered to spades or banana plugs, but we opted for bare ends.

Sonically, the best banana plug or spade is no plug or spade at all. Fold 1" ends and then fold again to get 4-fold 1/2" termination (1).

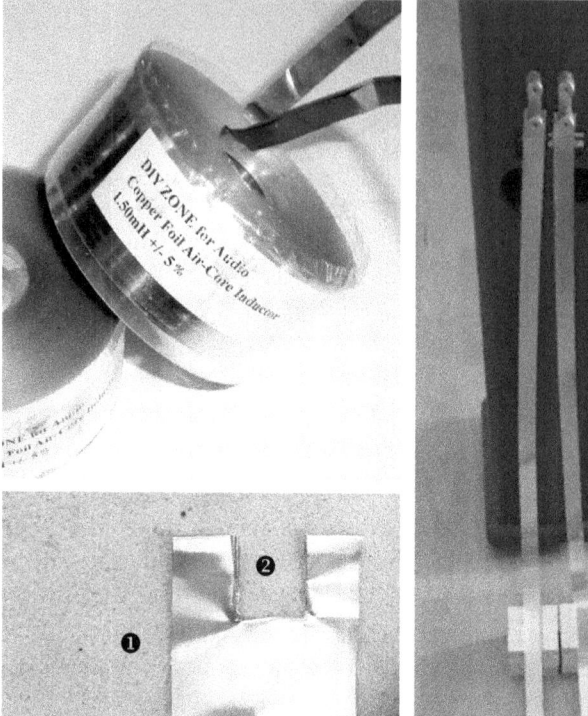

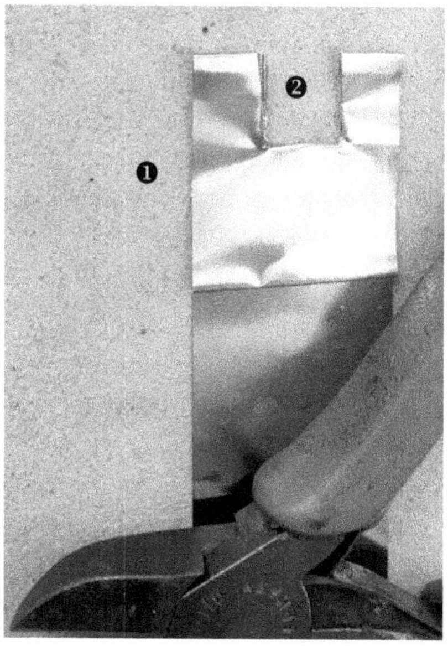

Fig. 5.20. One copper foil inductor can make quite a few speaker cables!

Fig. 5.21. How to prepare the ends - no spades or banana plugs are needed

Fig. 5.22. Make sure the unterminated ends are tightened at both speaker and amplifier binding posts as much as possible (4). Every few months, disconnect them, treat them with a contact enhancer, and re-tighten.

Using sharp scissors or wire cutters, cut out a rectangular notch, large enough to accommodate the center threaded shaft of your amplifier's and speakers' binding posts (2).

Copper foil isn't lacquered or insulated in any way, so caution is required. Make sure the positive and negative runs don't touch each other; otherwise, an instant short circuit at your amplifier's output will be created. They should not be touching any other metal surface, either!

For that reason, and also to prevent oxidation, you should buy a can of polyurethane coating and spray both sides of the cable. For a thicker insulating film, apply two or three coats. This is also a protection against abrasion and atmospheric influences such as moisture, oxidation, dust, or corrosive vapors. You can buy urethane sprays in hardware shops, electronic parts stores, and artist supply places (artists use them for a transparent protective film over paintings).

The positive and negative feeds are separate, run in parallel (photo above). Assuming you haven't coiled them up, the cable's inductance will be very low, as will be its capacitance. Furthermore, the dielectric between the feeds is air, so electrically (and sonically), this is the most optimal arrangement of all.

It took a dozen or so hours of break-in time for the cables to "settle and relax." The sound was fine to start with, but it resulted in cleaner and tighter bass, more refined treble, and smoother midrange when it opened up. Sonically, this should become your reference cable, so each time you are contemplating spending heaps of money on "hi-end" speaker cables, run a comparison test with this baby. You may save yourself a fortune.

DIY PROJECT: The sandwich-type copper foil loudspeaker cables

Keeping the two copper strips separate, as in the previous case, may not always be easy. If bi-wiring or bi-amping is used, four of them will exacerbate this problem, so sandwiching two of them (+ and -) together is a natural progression. Electrically, that will create a capacitor (two metal plates separated by a thin insulating strip), so we will have to make them first, measure them later, and then do a listening test.

Our copper foil strip (1) was 1" (25.4 mm wide), so we needed a wider double-sided tape (2), 1.5" (38 mm) was a perfect width. An electrical insulating tape (3) was used to double insulate the ends to prevent accidental short circuits (4). A knife and scissors were the only tools used.

Cut four copper foil strips; make sure they are oriented the same way by marking the ends (just in case the "cables are directional" crowd is right). The first copper strip was placed on the double-sided tape, followed by another double-sided tape and the second copper strip. So, for each cable, two copper and two tape strips were needed.

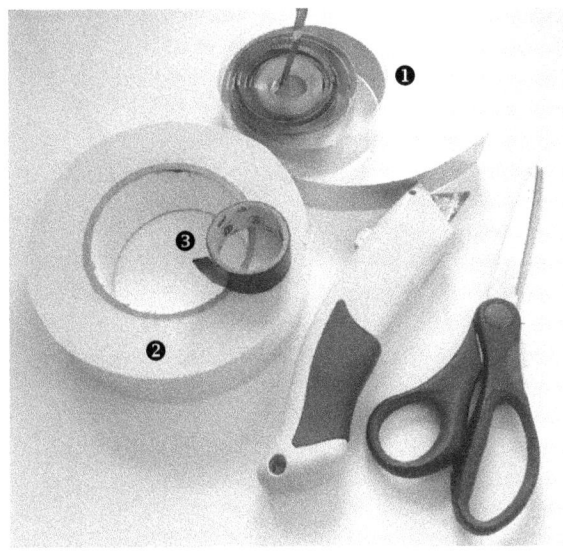

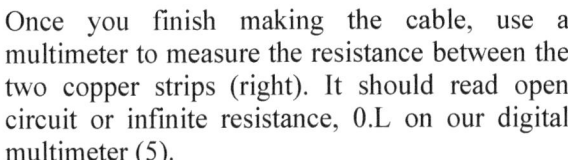

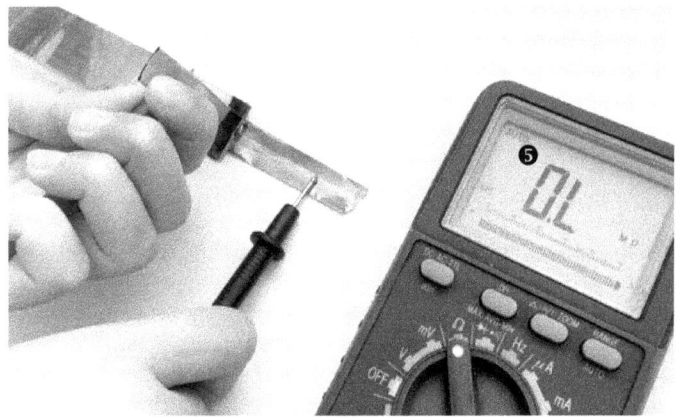

Once you finish making the cable, use a multimeter to measure the resistance between the two copper strips (right). It should read open circuit or infinite resistance, 0.L on our digital multimeter (5).

The resistance measurements were very encouraging, only 83 mΩ or 48.8 mΩ per meter (the cables were 1.7 m long). However, the capacitance was much higher than the 2-core twisted cable (next page), around 5 nF/m.

This cable is, after all, two metal plates with a dielectric between them; in other words, one giant capacitor. This capacitance may result in an noticeable treble roll-off. However, if your speakers sound bright and treble-dominant, it may be just what the doctor ordered.

The resonant frequency of the cable is $f_R = 1/2\pi\sqrt{(L_C C_C)}$ = 5.4 MHz, way above the audio band. The cable's characteristic impedance was $Z_C = \sqrt{(L_C/C_C)} = \sqrt{(100/8.61)} = \sqrt{11.6} = 3.4$ ohms, very close to the impedance of a typical 4-ohm loudspeaker, meaning the impedance matching would be achieved at the load end.

The characteristic impedance of ordinary speaker cables is at least ten times higher, usually in the 50-200 ohms range.

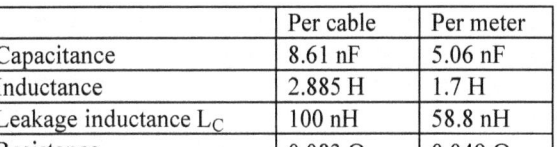

Fig. 5.23. The materials and tools required for this project

Fig. 5.24. One way of terminating the cables. Spades and banana plugs can be used, of course, but we opted for the cheapest, simplest and best solution - no plugs or spades of any kind, direct cable to speaker/amp terminals connection!

Fig. 5.25. To check for shorts, use a multimeter to measure resistance between the two copper strips - it should read open circuit or infinite resistance (0.L on our digital multimeter).

TEST RESULTS at 1kHz		
	Per cable	Per meter
Capacitance	8.61 nF	5.06 nF
Inductance	2.885 H	1.7 H
Leakage inductance L_C	100 nH	58.8 nH
Resistance	0.083 Ω	0.049 Ω

Impedance matching eliminates multiple signal reflections, which sonically improves the transparency, detail, and microdynamics, especially in the upper midrange and treble frequency band.

The cables were used with tube monoblocks placed relatively close to speakers and sounded great, the same as separate foil cables from the previous DIY project. You can see them in the photo of the lounge room used as a listening room on page 270. Highly recommended!

DIY silver-plated copper loudspeaker cables

The same single-core shielded wire (silver-plated copper with Teflon® insulation) we used for making DIY interconnects (page 100) is thick enough to work as a speaker cable. We twisted the two wires together, left the silver braided shield cut off on one and connected it to the central core on the other side, and terminated them with gold-plated spades. Sonic bliss for $20? You bet!

The capacitance between the two cores of this 3.25 m speaker cable was 161 pF, or about 50 pF/m, an excellent result (the lower, the better). The parallel inductance (between the two cores) was 154 Henries, while the short-circuit leakage or series inductance was $L_C = 0.7$ µH.

The cable's characteristic impedance was thus $Z_C = \sqrt{(L_C/C_C)} = 66$ ohms.

Our $500 commercial speaker cables had a capacitance of 150 pF/m, three times higher. Their short-circuit (leakage) inductance was higher as well, 1.852 µH.

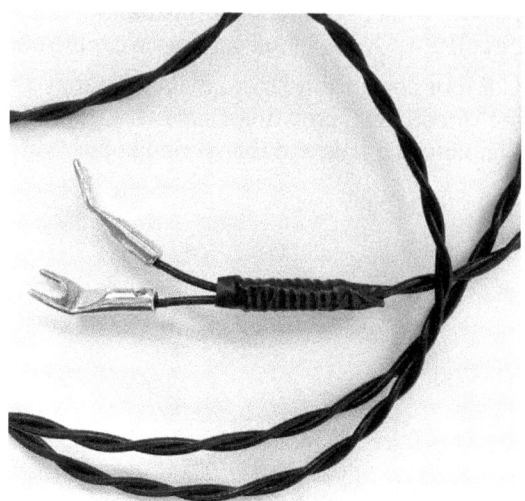

Fig. 5.26. Can $20 DIY silver-plated copper Teflon-insulated speaker cables sound better than many $500.- commercial speaker cables? Yes they can!

> THE SPEAKER CABLE AXIOMS
> 1. The shorter the speaker cable, the less it impacts the sound.
> 2. The higher the output impedance of a power amp or the speaker's impedance, the less speaker cable's parameters matter (the less the speaker cable impacts the sound).
> 3. Both the serial inductance and shunt capacitance of a speaker cable cause attenuation of high frequencies.
> 4. Out of the three cable parameters (RC, LC, CC), the series inductance is the most important.

The "breaking-in" phenomenon in cables

The break-in phenomenon is well known and acknowledged in complex mechanical devices such as machinery and cars, for instance, which include many moving parts. In audio equipment, the issue is a matter of certain controversy. Some audiophiles and "experts" flatly dismiss the possibility that the sound of brand new amplifiers, speakers, and cables improves in the first fifty or hundred hours (sometimes even longer). Our listening experience, however, definitely confirms the existence of the break-in phenomenon.

With audio and electronic components, this process could also be called "run-in." Its nature depends on the type of component in question. The break-in of vacuum tubes, which are also mechanical devices with many interrelated parts, is different from the run-in process of audio cables. In cables, it is primarily a consequence of forming the dielectric, which requires time and continuous signal propagation to adapt to a charged state.

In most cases, the opposite is also true. Once de-energized for a more extended period, the process is reversed; the dielectric slowly reverts to its original state, so it is subject to a new, albeit somewhat shorter, run-in period once the equipment is turned on again.

Fakes and originals

Be they interconnects, speaker or power cables, audiophile cables are expensive, some extremely expensive, so it should come as no surprise that there is a whole global industry producing fakes that to the untrained eye are indistinguishable from the originals.

As with all fakes (watches, jeans, etc.), the first warning sign is their lower or much lower price. If the genuine cable costs $50 per foot and an eBay seller or a website sells it for $10 per foot, you know something's not quite right. Even in a high markup hi-fi industry, a seller cannot stay in business at five times lower prices.

However, while the purity and crystalline structure may not be as good as the original's, the fake can still be better than your garden variety cable.

We opened up a power cable bought on eBay quite a few years back as a matter of interest. Printed on the dark green cable sheath is "XLO/PRO PL-1500 OPTIMUM AC POWER LINE" (3) and "Protected by U.S. Patent No. 5,110,999". The said patent is available for download online, so that we won't repeat it here.

However, the patent description talks about a "twin-axial bundle of insulated conductor strands which are helically and symmetrically wrapped around a dielectric center." This cable is just an ordinary 3-core mains cable with a foil shield (1) and a drain wire (2), totally different from the construction described in the patent.

Fig. 5.27. The alleged "XLO/PRO PL-1500 OPTIMUM AC POWER LINE" cable sold online from China

There is nothing helically wound here, no twin-axial bundles of five separate conductors each, and instead of two separate ground (earth) wires there is only a single one. There are at least two possibilities. Either the manufacturer used the printing related to the patent for marketing purposes (anything "protected by a patent" is bound to be seen by prospective buyers as more valuable and "different"), or, this cable is a total fake.

In any case, the said patent expired in 2010, meaning anyone has been able to use such cable arrangement (design) since, but this cable has nothing to do with such a design and so it seems to be a cheap knockoff.

HEADPHONE CABLES

Upgrading headphone cables

Two detachable cables were supplied with OneOdio Pro-30 dynamic headphones. The shorter (127 cm) straight cable had a capacitance of 339 pF on one and 255 pF on the other channel, while the longer coiled cable had a capacitance of 445 pF on one and 427 pF on the other channel.

The 2.70 m length means the longer cable had a capacitance of 445/2.7 = 165 pF/m, much lower than the 267 pF/m of the shorter black cable, but still way too high! Sure, OneOdio Pro-30 are budget headphones, but you'd be surprised how many headphones priced ten times higher still come with substandard cables, so don't automatically assume that your expensive cans won't benefit from this upgrade.

Our DIY solid-core UHP copper Teflon®-insulated braided interconnects (pages 100-101) had a capacitance of only 39 pF/m; the same wire also makes great H/P cables. Capacitance aside, the wire used must be solid core and thin! Since the return connector (GND or COM) is common to both channels, you will need to braid only three conductors.

Alternatively, if your headphone amp uses two separate mono jacks for the L & R channel, you'll need to braid only two wires together but will need to make and terminate two identical cables.

Many commercial headphone cables use braided construction and are priced between $200 and $4,000. For $200, there's no point mucking around, but a cable in that price bracket isn't likely to bring any improvements.

Fig. 5.28. At 1kHz test frequency the LCR meter shows the 120nF capacitance and 148 µF inductance of OneOdio Pro-30 dynamic headphones.

For much less money, you could buy 15 or so meters of the best solid-core wire, ultra-purity copper, or even pure silver if that is your cup-of-tea, and the best damn connectors (since you need only two or three).

If you suspect that your headphone cables could be the sonic weak point, instead of blindly forking out wads of money, spend a fun afternoon making a few cables of your own. Your DIY project will only take a few hours and save you enough for a nice romantic weekend getaway for two ...

DIY PROJECT: WWW headphone cable

Wire wrapping is a technique of connecting electronic components without soldering by wrapping 4-6 turns of solid core (usually silver-plated copper) wire around a component lead or a socket pin. Quite a few1960s & 1970s made-in-Japan audio components used wire wrapping connections.

Fig. 5.29. Dual rectifier PCB salvaged from 1970's Japanese solid-state amplifier with two pairs of AC inputs (from power transformer) and the white-colored COM or GND wire, all connected using the wire-wrapping method

Wire wrapped connections are more reliable than soldered connections and more immune to vibration and mechanical stresses. Most importantly, the cold weld of silver-plated wire to the pin is gas-tight, preventing oxygen from penetrating and oxidizing the joint. Problems such as cold, dry or cracked soldered joints are avoided.

Perhaps counter-intuitively, wire-wrapped contacts have a lower electrical resistance than soldered joints.

The wire wrapping wire is usually rated up to 300 V_{RMS}, with a wide operating temperature range (-20 ^{O}C to +130 ^{O}C). The solid core 0.254 mm diameter (30 AWG) copper wire is silver plated and insulated with PVDF (polyvinylidene fluoride), better known under one of its trade names, Kynar™.

The cross-sectional area of 0.05 mm^2 results in the current rating of at least 0.4 A (400mA). The wire's outer diameter is only 0.5 mm. The cheapest we could find was a bundle of eight colors, 35 m of each (280 m total length), for AU$13.00, but that wire is tin-plated with PVC insulation. A 30 m spool of silver-plated Kynar™-insulated wire has a retail price of AU$17.

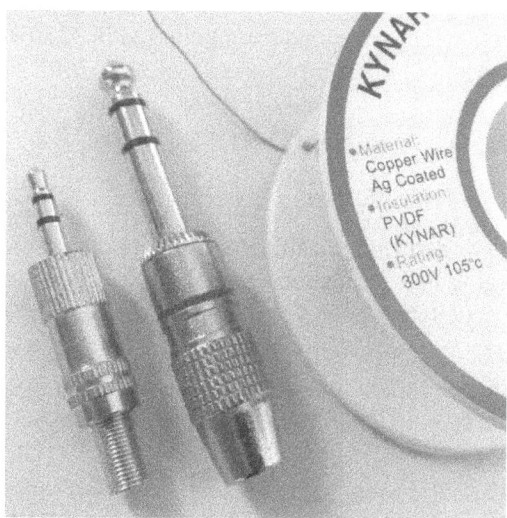

Fig. 5.30. 3.5mm and 6.3 mm jacks next to the spool of silver-plated wire-wrapping wire used to make this DIY headphone cable

Fig. 5.31. The finished headphone cable

Since the wire is so thin, we doubled up. Instead of three strands, we twisted six strands together, paralleling them in pairs.

A 50' (feet) spool (15.24 m) of braided PET (Polyethylene Terephthalate) sleeving sells for US$30, but such braids can be salvaged from your stash of unused audiovisual cables. We reused a 2 m long one (1) from a TDK HDMI cable which was itself used in the next DIY project.

The finished cable was 204 cm long, with a capacitance of 268 pF (131 pF/m), which was a bit high, but still acceptable. Substituting the supplied cables with our WWW (Wire-Wrapping Wire) cable improved the transparency and microdynamics; the DIY cable sounded cleaner, with superior detail retrieval.

DIY PROJECT: Headphone cable using TDK HDMI cable

While looking for a length of braided sleeving (to use in the previous DIY project) in our cable stash, we came across this HDMI (High-Definition Multimedia Interface) cable. The printing said "TDK Life On Record - High Definition Multimedia Interface - 1.25% Silver Cable, " which was promising.

It had four solid core silver-plated conductors (2), just what the doctor ordered (photo on the next page), plus five larger multi-stranded conductors (3), all shielded by aluminium foil and a braided copper screen (4).

CABLES, FUSES, CONTACTS, AND CONNECTIONS

We doubled up two solid wires as the ground (COM) conductor, with the other two as left and right channel "hot" leads. The outside screen was twisted up and connected to the COM (ground) tab on both sides (4) to retain its RFI screening role. The 197 cm cable had a capacitance of only 141 pF (72 pF/m), a great result.

Both DIY cables sounded tonally balanced, with refined details and microdynamics, similar in many respects, probably due to their similar diameters and silver-plated copper metallurgy.

The only drawback was that they were not nearly as flexible as commercial headphone cables, especially the stiff, large outside diameter ex-HDMI cable.

COPPER, SILVER, GOLD, NICKEL, RHODIUM, BERYLLIUM AND TELLURIUM: A CRASH COURSE IN AUDIO METALLURGY

Contacts - the invisible components

When two or more wires or component leads are joined together, a contact is formed. For best audio performance, contacts must be clean and tight. Clean means that the wires and leads must be free from dirt, dust, grime, and oxidization, i.e., that they must be cleaned before they are joined together.

The best electrical contact is mechanically sound. When wires are crimped together or wire-wrapped onto a terminal, the shape of the metal is deformed, forming a "cold-weld," an airtight connection, meaning the contact will not oxidize over time since air cannot get into it. Thus, if you can, twist wires together first or wrap them around terminals before soldering them together.

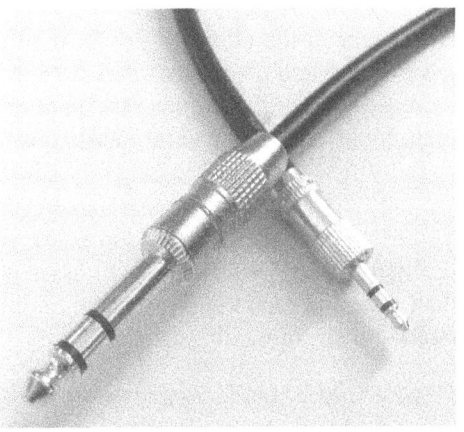

Fig. 5.32. The finished headphone cable using the silver-plated solid cores of a HDMI cable

The lowest grade: Steel or brass base, nickel interlayer, gold or silver plating

As for the pins, spades, and lugs used on speaker cables and RCA plugs and sockets, the aesthetics often interfere with the sonics. The cheapest plugs use steel, an inferior conductor (only 3% of copper's conductivity). Almost all are gold plated, and to make the gold shiny, nickel plating was done first underneath the gold layer. That nickel layer's distortion is irritating and causes listener fatigue.

Get a small magnet (5), such as the button type pictured, to check if your connectors are magnetic. If they are, they'll be attracted by the magnet and stick to it. Ditch them and upgrade to nonmagnetic ones. Beware! Some banana plugs advertised as "pure copper made, nonmagnetic" were magnetic for sure, meaning they were not made of copper, but of steel or some other ferromagnetic material.

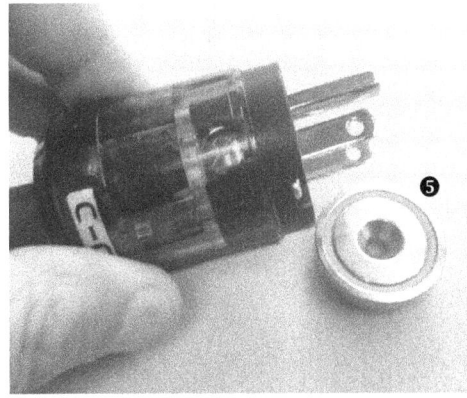

Fig. 5.33. The magnet check on a 3-pin "audiophile grade" power plug. It wasn't magnetic.

Gold

Despite common perception, just because gold is an expensive noble metal and doesn't oxidize, that does not automatically mean that it is a good electrical conductor, not as good as silver, anyway, and not even as good as copper! Silver is around 8% more conductive than copper, gold has only 75% conductivity of copper, and nickel is way at the bottom with only 25% of copper's conductivity.

The negative effect of alloying gold with silver and other materials such as copper is exponential: adding just 1% gold lowers silver's conductivity from 108% to approximately 90%, and 5% gold drastically reduces conductivity to only 60%, approaching that of brasses, bronzes, and steel.

Technically, gold is only used to prevent contact corrosion. Still, since such a thin layer very quickly rubs off by frequent plugging and unplugging, I suspect the main reason is of a marketing nature - anything plated in gold seems more expensive and "hi-end."

Gold-plated beryllium-copper

When added to copper, aluminium, iron, or nickel, beryllium improves their physical properties. Brass and beryllium-copper (BeCu) are cheap, have good mechanical properties, and are much easier to machine than pure copper or silver. Beryllium copper alloys contain only 0.7% beryllium, plus some nickel and cobalt. However, their electrical conductivity is low, depending on the percentage of Be, only 15-30% that of pure copper.

Many audio and power connectors use gold-plated brass for the body and gold-plated beryllium copper for the pins. Beryllium copper alloy is nonmagnetic, does not suffer from metal fatigue, retains the high electrical and thermal conductivity of copper, and is much stronger and harder than copper alone, a soft metal.

Some manufacturers, such as Oyaide, use a beryllium copper base for their mains plugs, first platinum-plated, then plated with palladium.

However, many audiophiles don't like the sonic signature of beryllium copper alloys either, preferring pure copper. One may argue that a concoction of four metals (copper, beryllium, platinum, and palladium) violates the Simplicity Rule. On the other hand, copper connectors oxidize quickly, so some plating is required unless you tighten the hell out of your speaker's copper spades (to create an airtight contact) and never touch them again.

Rhodium-plating

A member of the platinum group, rhodium is a precious metal (much more expensive than gold) used for plating jewelry and audio connectors. A 0.75 to 1.0-micron thick protective layer created by a rhodium "dip" or "flashing" increases durability, scratch resistance and minimizes corrosion. Gold and silver can also be rhodium plated, making them highly lustrous and tarnish-free for a long time.

Tellurium-copper

Tellurium-copper alloy is approx. 99.49% copper, 0.5% tellurium, and the rest is phosphorous, which gives it mechanical properties similar to brass or beryllium-copper and electrical conductivity at 93% of coppers (higher than gold and much higher than aluminium). Most are still plated with silver or gold but without nickel underlay, making them sonically superior.

Copper cables: OCC copper versus OFC

OFC stands for Oxygen-Free Copper and is the most commonly used grade of copper in audio. The minimum copper content (purity) is 99.99 %. Gas impurities are lower than 10ppm (parts per million) for Oxygen (O_2) and lower than 0.5 ppm for Hydrogen (H_2).

OCC (Ohno Continuous Cast) copper is of higher purity (above 99.99998 %), while the gas impurities are about 50 % lower. The fundamental difference is in their crystal structure. The average crystal size in OFC is around 1mm, while in OCC, it is 100 meters or more. To my knowledge, no scientific tests have been done in comparing the sonics of OFC versus OCC copper.

Still, it is claimed that multiple crystal boundaries (potential barriers) introduce a string of nonlinear resistances which the signal current must pass through and that such irregularities negatively affect the sonics of the copper wire and cable. Likewise, many cable manufacturers and audiophiles claim that a single crystal structure of an OCC copper cable results in improved sonics, in particular superior transparency and more faithful reproduction of fast transients.

What exactly is the problem with a mixture of metals in audio plugs and sockets?

When two or more different metals form a contact in a connector, each interface behaves not just as a resistance, but also as a battery and a rectifier diode (P-N junction). Such dissimilar metals form a chemical cell, V_{EC} in the diagram below. The electrochemical or electromotive series lists metals in the order of their standard potentials to the referent hydrogen electrode.

Of all the metals used in audio connectors, beryllium sits at the far negative end of the scale at -1.85 V, followed by tellurium (-1.14 V), iron (-0.44 V), and nickel (-0.24 V). On the positive side are copper (+0.34 V), silver (+0.80 V), platinum (+1.20 V), and gold at +1.50 V.

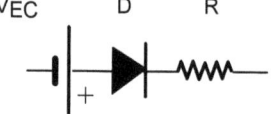

Fig. 5.34. A simplified electrical model of an audio contact - a voltage source (DC battery), a rectifier (diode) D, and contact resistance R.

If you thought having a battery cell and a couple of nonlinear, distorting, and signal-rectifying diodes in series with your audio signal is bad enough, things get even worse. Metals higher up in the series (more positive standard potentials) displace metals lower in the series. In other words, in the joint of two dissimilar metals, the one with the lowest potential corrodes (gets "eaten away") with time and eventually disappears.

Solder is a necessary evil. It is made of at least two, often three of four dissimilar metals and then used on copper or tinned copper leads. When different metals are joined, a parasitic diode is formed, which acts like a micro signal rectifier and an RF (radio frequency) signal detector. Of course, these are subtle effects, but their cumulative effect may be noticeable. Thus, it is wise to minimize them by reducing the number of soldered joints in an amplifier.

CABLE TERMINATIONS AND CONTACTS

What to use, banana plugs, spades or bare wires?

The link between a power amplifier and loudspeakers is electrically simple, just a pair of speaker wires, but there are various options of "interfacing" those wires at both ends, and they aren't all equally good

Speaker cables usually come terminated with either banana plugs or spades. The contact between the banana plug and the internal cavity of the binding post is not uniform or complete around the perimeter of the plug. For that reason, spades are a better option since they provide a much larger contact surface. The contact area can easily be cleaned or enhanced with contact cleaners, while access to the insides of a binding post is much harder.

However, even if spades are used, and even if the contact between them and the binding posts is as tight as possible, the signal must still pass through those mechanical discontinuities. As we have just seen with our battery- diode-resistance model of a typical contact, it is not only the contact resistance that is of importance. What is worrisome is the conductive structure and properties of the materials the signal must pass through.

What happens to complex audio signals when they pass through numerous interfaces of dissimilar metals is not a topic often discussed in audio magazines and their repetitive and relatively superficial reviews. Of course, this issue isn't present just in power and speaker cables; it applies to interconnects as well.

Most binding posts, banana plugs, and spades are made of brass and, even if gold plated, introduce a sonic coloration into the signal chain. Gold, unlike silver, is not a very good conductor of electrical current. Apart from the marketing hype and the ability for the cable/spade/banana plug manufacturer to charge a premium for a minute amount of gold present (only a few microns thin coating), gold plating is used primarily to minimize corrosion and oxidation of contacts, not to minimize the contact resistance or improve the signal flow.

Threading a bare speaker wire through the hole in the binding post is by far the best option. It eliminates one contact from the signal path between the spade or plug and the binding post. However, no matter how (air)tight you make such contact, wires corrode, so if you really want to get serious, permanently solder your speaker wires at both ends. That means inside the amplifier (if it's a tube amp directly to output transformer terminals) and inside the speaker. That way, all questionable connections in between will be eliminated!

Speaker binding posts and shorting links

Due to the higher currents involved, the most critical contacts in the audio chain are between the speaker cable and the binding posts of the speaker at one and the amplifier at the other end.

Suppose your speakers have two pairs of binding posts and you aren't using bi-wiring (driving them through two separate speaker cables). In that case, the signal must pass through two additional mechanical contacts, between the entry binding post (the lower one in the photo) and the shorting link, and the second one between the shorting link and the upper binding post.

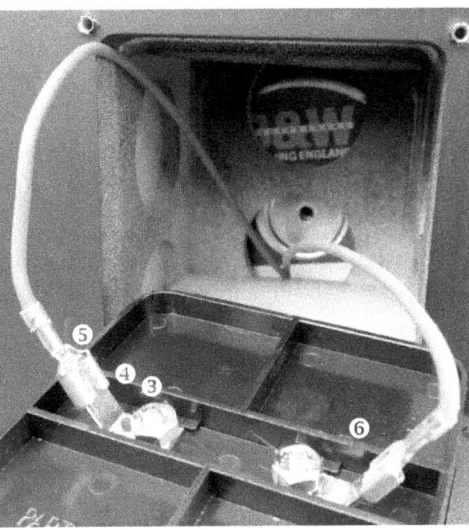

Fig. 5.35. How many interface points (mechanical contacts) can you identify between the red (1) and white (2) banana plug?
Hint: The internal speaker hookup wires are soldered to the other side of the speaker binding posts.

Fig. 5.36. The internal view of binding post connections on B&W FCM8 speakers, using ordinary hookup wire and multiple mechanical internal contacts: the binding post to the male spade terminal (3), then female-to-male spade terminal connection (4), and finally the crimped hookup wire-to-female spade terminal joint (5). The same happens on the other side (6)!

The signal also has a return path back to the amplifier, so the same happens through the other pair of binding posts. That is four additional unnecessary interfaces in your signal chain.

But wait, you may protest, my shorting links are gold-plated, which makes it OK. Well, no, it doesn't; it actually makes it worse! In terms of signal conductivity, gold is inferior to copper. As we have seen in a recent discussion (page 111), the main problem is not the slight increase in contact resistance due to gold plating; it is the material underneath the super-thin gold layer that may be responsible for all kinds of sonic ills.

Thus, since the best contact is no contact, the most prudent course of action is to open up your speakers and solder short internal jumper wires between the two pairs of binding posts, thus ditching those external shorting links.

Such soldering is quite a simple and easy task, but gaining access to the back of the binding posts is usually much harder. In our case (photo on page 222), the woofer had to be unscrewed and removed and the rear of binding posts accessed through its opening in the speaker's front baffle.

Clean the gunk from those contacts!

Contact cleaners are available in electronic parts shops, some hardware shops and auto-parts stores. Some "electronic contact cleaners" can be quite expensive, especially those that leave no residue.

Carburetor cleaners do the same job (can be bought in automotive parts shops) but contain acetone and other chemicals that dissolve plastic, so they are not recommended for audio connectors. Notice how these two products specifically emphasize that they are "safe on most plastics and rubber" (1) and (2).

You'll only need a small quantity, so a spray can will last you many years. Plus, since you will be cleaning and polishing your contacts by hand (after spraying it with the contact cleaner), even cleaners that would leave a residue (and aren't suitable for cleaning printed circuit boards, for example) may be fine.

The two products pictured (350 grams cans) sell for only around US$7.- in Australia but are typical of dozens of chemically similar or even identical ones sold worldwide. The active component is isohexane (above 60% of the total weight), while the propellants are hydrocarbon gasses, butane and propane.

The Hosa D5S6 DeoxIT® contact cleaner is well regarded but very expensive. However, compared to simple contact cleaners just mentioned, it is claimed also to be a "rejuvenator, conductivity enhancer and lubricant" that "dissolves oxidation and corrosion on metal surfaces" and "fills in microscopic gaps and reseals surface for better contact."

Fig. 5.37. Dozens of commercial contact cleaners are available

FUSES

Fuse standards, shapes and sizes

A fuse is an electronic component designed to act as the weakest link in an electrical circuit and burn out ("blow") when the current through the circuit exceeds its rated value. It can be considered a special case of a very short cable, deliberately made the weakest point to protect the cables and components on its load (downstream) side.

In audio applications, the mains-side fuse protects the primary winding of the mains transformer and the wiring inside the amplifier, making sure neither of these burn out before the fuse does. Thus, it is essential not to replace a fuse of a specific current rating with one of a higher rating.

If the current through the primary of a mains transformer is 1 A and we replace a 1.5 A fuse with a 3 A fuse, should a fault develop (a short-circuit on the secondary side), the $300 transformer's winding may burn out before the five-cent fuse!

There are many sizes, types, and standards of fuses. The 3AG (American size) and 2AG or M205 (European size) are the two miniature fuse types most commonly used in audio electronics. Both are so-called "cartridge-type", of cylindrical shape.

AG stands for "automotive glass" since these sizes (2AG, 3AG, etc.) were originally developed for use in cars. Paradoxically, cars today use blade-type fuses, which are easier to change in a hurry by simply unplugging them, while 3AG and M205 are almost always enclosed in a fuse holder whose top cap needs to be unscrewed for a fuse to be changed. 3AG fuses are 31.8mm long and 6.35mm in diameter, while metric-sized M205 fuses are 20mm long and have a diameter of 5mm.

CABLES, FUSES, CONTACTS, AND CONNECTIONS

Fast and slow acting, ceramic and glass fuses

Two main types of fuses are used in electronics, fast and slow-acting (delayed-action fuses). Slow acting or "delay" fuses are often called SLO-BLO(W) and are designed to withstand higher inrush currents that flow when equipment is powered up, but only for a short time.

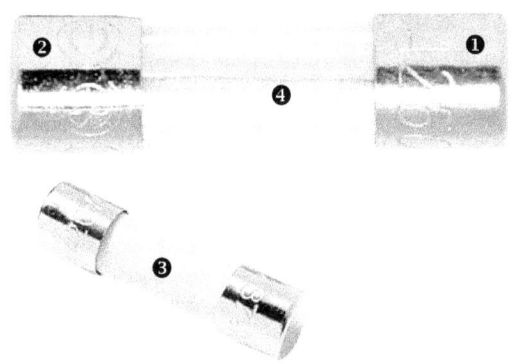

Fig. 5.38 Typical markings on 2AG and 3AG fuses. Ceramic fuses can be recognized by a white opaque body (3), meaning the internal wire (4) cannot be seen as with glass fuses.

If the higher current persists, the fuse eventually melts. They are specified as "T" speed rated and sometimes referred to as "anti-surge."

The naming convention of 3AG fuses is best illustrated with an example: 3AGDA1.25 = 1.25 A 250 V Slow 3AG fuse. M205 fuses a include letter D for "Delayed action" in their model number: M205DA01.6R = 1.6 A 250 V Delayed M205 fuse.

An example of a fast-acting M205 fuse would be M205002R (2A, 250V Fast M205 fuse). The fuse rating is stamped onto one metal end (1), something like "2A250V", while the other cap (2) is embossed with various symbols of approval and certification bodies such as UL (Underwriters Unlimited).

Some amplifiers use ceramic fuses, which are designed for safety and don't explode if there is a fault (as glass fuses can). Thus, do not replace ceramic with glass-type fuses even if they are of the same rating.

Voltage and current rating of fuses

The overload versus time curves for the two types of fuses (illustration on the right) show a few interesting aspects of their design and operation. Notice that fast-acting fuses take 3 seconds to blow at 150% of their rated current (point X) but only 1 second at twice the rated load (200% overload), point Y.

With 200% load (twice their maximum current) the slow blow fuses would take a very long time to react (20 seconds), point Z on their curve, even allowing a 300% load current to flow for ten seconds (point V).

The normal current draw of an audio component should never exceed 75% of the nominal fuse current rating. An amplifier drawing 1A of current from the mains should have a mains fuse of at least 1/0.75 = 1.35 A rating.

However, things aren't that simple. The ambient temperature also plays a role in fuses' behavior. Although audio amplifiers in audiophile setups operate exclusively at room temperatures, depending on the topology and mechanical design of the chassis and element positioning inside an amplifier, temperatures inside the chassis can reach higher levels, especially during hot summer days.

For instance, a 30 °C room temperature and a 30 °C temperature rise next to the mains transformer means a fuse would be at 60 °C. According to the de-rating curve, a slow-blow fuse would need to be de-rated to around 85%. For instance, with a current of 1.5 A, we would need a slow-blow fuse rated at about 1.5/(0.75 x 0.85) = 2.35 A.

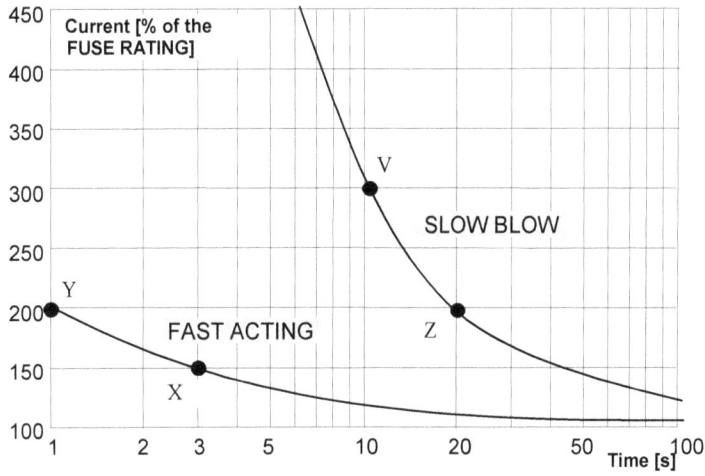

Fig. 5.39. Typical characteristic curves (overload vs. time) for 2AG and 3AG fuses (SLOW BLOW and FAST ACTING)

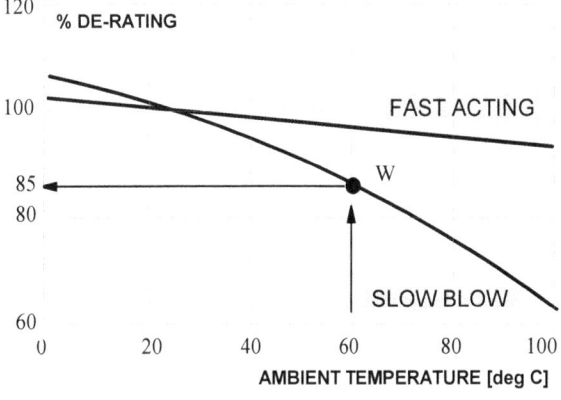

Fig. 5.40. Temperature de-rating curves for fast acting and slow blow fuses

Do fuses affect sound?

A fuse is a short length of wire specially selected and engineered to be the weakest link in a circuit it was designed and chosen to protect. Fuses have two "interfaces," consisting of an internal joint between the wire and the fuse's end-cap and the mechanical contact between the fuse and the fuse holder. Thus, like all cables and hookup wires used in audio, fuses introduce contact issues into the sonic equation.

Despite the fact that their length is negligible compared to the rest of the circuit wiring, some audiophiles claim that fuses nevertheless affect the sonics, that, for instance, ceramic fuses sound better than the glass ones and that even the direction of a fuse inside a fuse holder makes a difference to the sound.

Of course, there's money to be made from such claims, so there are also "audiophile" fuses, usually featuring silver, gold, platinum, and similar metals. While some are cryogenically treated, others, such as HiFi-Tuning's Supreme range of fuses, claim to be "treated with a proprietary quantum level process," whatever that means.

Compared to ordinary fuses that sell for 10-20 cents, these are priced from US$50 to a few hundred dollars each! Fuses can be used in three types of audio applications - as AC fuses on the mains side of audio components, inside such components in their DC supply lines, and as speaker fuses. Depending on their location in the audio system, fuses could have different sonic effects, from none at all to highly noticeable.

Cryogenic treatment of tubes, cables, fuses and other audio & electronic components

Cryogenic treatment involves gradual cooling of the treated part down to temperatures around -300 $^{\circ}$F (or -180 $^{\circ}$C) and then reversing the process and bringing them back up to the room temperature. The indisputable scientific fact is that cryogenic treatment hardens steel and aluminium, so it is plausible that it also hardens other ferrous materials, such as nickel, and the nonferrous ones, too, namely copper. Both are used in vacuum tube construction.

Tubes have no moving parts, so the only foreseeable benefit from hardening would be to change the elasticity (increase the stiffness) of the tube's metal parts and thus their vibrational propensity and resonant frequency. So, again, a reduction in noise and microphony (caused by tube electrodes vibrating when hit by sound waves and moving relative to one another) in tubes as a result of cryo treatment seems plausible. It is interesting that cryo vendors haven't published any such comparison data, namely the noise and microphony before and after figures.

Cryo vendors claim longer tube life, reduced "quantum noise," and improved sonic performance. However, if the gradual cooling changes the molecular structure, what does the re-heating of the tube do? Are the changes permanent, or does the structure changes back after you operate a tube in an amp, where its temperature skyrockets way above the room temperature, 2,200 $^{\circ}$C for tungsten filaments and "only" 700 $^{\circ}$C for oxide-coated heaters.

However, although tube makers pay attention to this fact and try to use materials with similar thermal expansion coefficients, different metals and materials (glass, mica, etc.) in a tube still expand and contract with temperature changes at different rates, so isn't such exposure to extremely low temperatures stressful for the tube? Many audiophiles report tube failures not long after their cryo treatments. Buying cryo-treated currently produced tubes may be OK, but I wouldn't even think about subjecting my precious vintage Amperex, Telefunken, Mullard, or Western Electric tubes to such a risk.

The ultimate skeptics also mention the possibility that you'd pay for a treatment that hadn't been carried out at all. There is no visual indication that a tube has been cryo-treated, and it is impossible to confirm it either way by measuring the tube's parameters in a tube tester.

I could not find even a single video on YouTube demonstrating a cryogenic treatment of a vacuum tube. If I was a tube vendor selling cryo service and/or cryo-treated tubes, to gain credibility, I'd invest a couple of hours into producing a 5-minute video, wouldn't you? Of course, even that wouldn't be much of proof (just like photos, videos can be "faked" or "doctored"), but at least prospective buyers would see the equipment in action and the cryo process time-lapsed into a minute or two.

By the same token, get two identical speaker cables, power cords, or interconnects, have only one cryo-treated, and compare their sonics. Being lazy and skeptical, I could not be bothered. Some sloppy writer I am. Please report your findings to me so we can expand this discussion in future editions of the book.

CAN PAYING MORE FOR CRYO-TREATED TUBES BE JUSTIFIED?

How do you know if paying US$64 cryo-premium for an octet of new power tubes is justified? The only way is to buy two identical sets (same tube, same brand, and batch) from the same vendor, one cryo-treated, the other not, insert them into your amp or preamp and listen. Then, unless you want to keep it as a spare, sell one set and lose money in the resale.

6 | UPGRADING & FINE-TUNING THE SOURCES: OPEN REEL RECORDERS, TURNTABLES, PHONO STAGES, AND CD PLAYERS

This advanced chapter is for audiophiles with technical know-how who aren't afraid to open up their phono stages, DACs or CD-players, and modify their circuits. Alternatively, you could hire the services of a competent audio technician or engineer who will perform such mods for you.

Apart from getting the acoustics of your listening room right, these improvements will result in the most significant improvements of all, so paying someone say $800 to modify your amp and/or CD player will be equivalent to buying a better one for $3,000, $10,000 or even more. That is the order of magnitude of such savings, a tenfold better value for money than upgrading your existing gear!

- THE ARDUOUS SONIC JOURNEY FROM RECORDING MICROPHONES TO YOUR EARS
- REEL-TO-REEL (OPEN REEL) RECORDERS
- TURNTABLES, TONEARMS AND TRACKING ERRORS
- DRIVE MECHANISMS AND VIBRATION MINIMIZATION
- TURNTABLE SETUP & ADJUSTMENTS
- CARTRIDGE AND STYLI TYPES
- UNDERSTANDING PHONO STAGES
- THE CARTRIDGE-CABLE-PHONO STAGE INTERFACE
- MOVING COIL STEP-UP TRANSFORMERS
- SOUND IMPROVEMENT HACKS FOR TURNTABLE-BASED SYSTEMS
- TUBE BUFFERING YOUR CD PLAYER, DAC, OR MUSIC STREAMER
- ADDING AN OUTPUT TRANSFORMER TO A CD PLAYER, DAC, OR MUSIC STREAMER

> "At some point in the not too distant future, once time has granted enough of us the wisdom of hindsight, the virtue of the analog record will be rediscovered. Then the LPs that people so blithely discard today will become treasures in the analog renaissance."
> Laura Dearborn's accurate prophecy from her 1987 book *Good Sound*

THE ARDUOUS SONIC JOURNEY FROM RECORDING MICROPHONES TO YOUR EARS

Depending on the medium and the underlying technology in question (magnetic tape, vinyl disc, or CD), the music production process varies, as does the initial part of the reproduction process. Still, the recording & mastering process is often identical (assuming the use of analog technology). Once you understand the complexities and the sheer number of steps involved in each of these processes, it is indeed a true miracle that analog tapes and vinyl records sound as good as they do!

The sound recording & reproduction miracle

Although digital recording prevails in the 21st century, most musical masterpieces of the past were recorded using analog tape and mixed on analog mixing desks. Even today, quite a few production studios use purely analog recording or mix analog and digital aspects, trying to get the best of both worlds. Let's have a look at the long and arduous journey through various transducers, from the studio or live microphones to your loudspeakers.

The six major physical processes involved in sound recording, mixing, mastering, and reproduction are numbered. It all starts with a microphone and the inverse electromechanical conversion (1) it performs, converting mechanical vibrations (sound waves that hit its membranes) to an electrical signal. In the process of magnetic recording (electromagnetic conversion), this signal is converted by tape deck recording heads into magnetic flux variations, which are then stored on a magnetic tape (2).

Magnetic reproduction (magnetic to electrical conversion) follows (3) when individual tracks are read by a playback or reproduction head, converting back the musical information to electric signals for mixing and mastering purposes. Another magnetic recording conversion (2) follows back onto a master tape.

Tape production (duplication)

The master tape is a precious asset, and since tapes can be so easily damaged, demagnetized, or destroyed (even completely lost due to theft, earthquakes, or fire), the second generation copies are immediately made. Even these "mother" tapes are too valuable and aren't used in the commercial tape duplication process.

Each time a tape is copied, its magnetic performance is degraded. The tapes from which copies are made could be a third or even a fourth-generation from the master tape. So, you could say that commercial music tapes are a copy of a copy of a copy of a copy. The source copies can only be copied a few hundred times and have to be discarded.

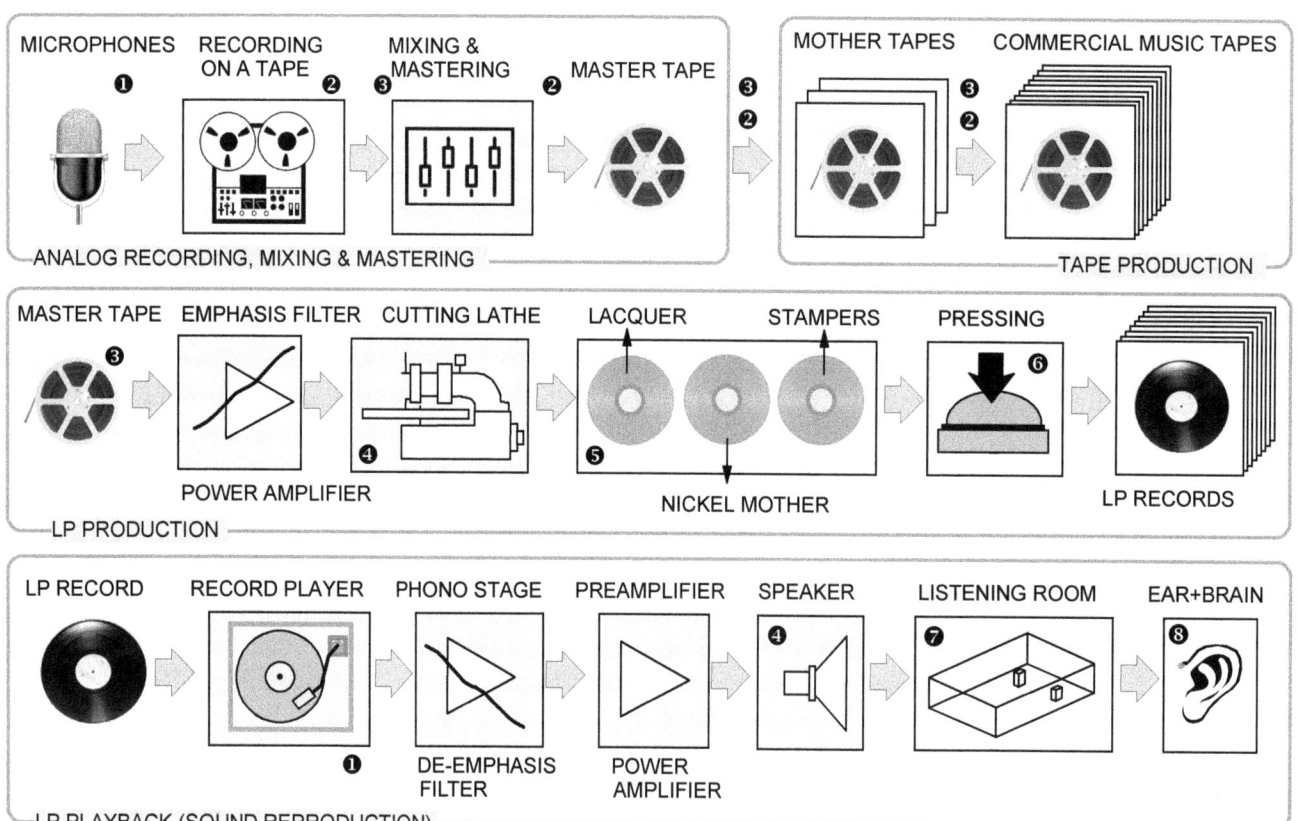

Fig. 6.1. The major steps and equipment involved in the recording, tape and LP production, and LP playback (sound reproduction) processes

UPGRADING & FINE-TUNING THE SOURCES

Unfortunately, each time a tape is copied, its magnetic recording must be read and converted into an electrical signal by the tape deck's reproduction head (3) and then converted back into the magnetic recording (2) by its recording head. Such dual conversion degrades the fidelity of the recorded sound, just as all conversions do.

No matter how good the high-speed, high-quality tapes may sound on a top quality reel-to-reel tape deck, the sound of the master tape and even of the mother tapes is in a different league, the closest we can get to the live sound!

The best quality commercial tapes are produced in a real-time copying process, so a 10" reel tape recorded at 15 IPS (inches per second)) can store just over 30 minutes of music, and, you've guessed it, would take more than 30 minutes to copy. This is a very slow and expensive process from a commercial moneymaking point of view. In addition, blank tapes aren't cheap either, contributing to the astronomical price of superior quality commercial prerecorded tapes.

Although duplicating tapes at 16, 24, or 32 times the normal speed is faster and cheaper (but still relatively slow compared to CD and other digital duplication methods), it requires proportionally higher bias frequencies and smaller gaps between the recording head and the tape.

As a result, consistency suffers (small misalignments have large negative consequences), and the recording quality is seriously reduced, especially at high frequencies. More on tape decks and these performance issues soon.

LP production

The first generation disc is called a lacquer, a blank aluminium disc covered by a surface coating of lacquer material. As the master tape is played back, the signal is processed by an emphasis filter that attenuates bass frequencies (the grooves would be too deep and wide) and then amplified by a special stereo power amplifier that drives the cutting head mounted on a lathe. The cutting lathe moves the cutter head tangentially (radially) across the lacquer.

The cutting head (akin to a moving coil cartridge but working in reverse) has two coils attached to the stylus at 45-degree angles to record two stereo signals. Just as a turntable that reproduces vinyl discs, the cutting lathe also has a rotating platter, to which a suction system tightly holds the lacquer. Another vacuum suction system sucks in and thus removes all the burrs or "chips" cut from the discs as it rotates.

The lacquer is first coated in a two-molecule thick silver coating, making it electrically conductive so it can be electroplated, process number (5) in our diagram, and then nickel plated. This "negative" nickel master is removed from the lacquer and plated again, creating the "positive" nickel mother.

The mother is plated again to make another negative copy called the "stamper," which is used in the stamping press (6) to produce vinyl records.

The pressing stage is the final step. Using hot steam to heat up the vinyl "biscuit", a hydraulic press closes and compresses the hot patty for about 30 seconds to make sure the vinyl particles have filled every nook and cranny. The longer the pressing cycle and the hotter the initial patty temperature (up to a certain limit, of course), the better the quality of the final disc.

Just as with magnetic tapes, a few dozen masters, mothers, and stampers are made, not just one of each, to speed up the disc production process.

Sound reproduction (the LP playback example)

The quality of the recorded medium (LP disc) and the electromechanical transducer (turntable or record player) is crucial. This means attention should be paid that the turntable is set up properly and that all alignments and adjustments (VTA, VTF, anti-skate, etc.) are in the optimal range.

Matching the turntable's tonearm with the right cartridge is a factor that even turntable manufacturers don't pay enough attention to. Most bundle their budget and medium-priced turntables with whatever budget cartridge will keep the total cost down to the predetermined figure and thus make the package competitive on the market.

Most of us are familiar with the rest of the playback chain, two or three electronic components (phono stage, line stage, and power amp), followed by loudspeakers, the final electroacoustic transducer of the audio chain. However, due to its acoustic properties, the listening room can also be considered an acoustic-to-acoustic transducer since it changes the loudspeaker sound considerably.

If you doubt it, compare their sound in a "dead" anechoic chamber, where speaker manufacturers perform their tests, and in a "live" room, full of reflections, standing waves (room modes), and reverberation!

The final transducer is the human ear, which transmits the changes in sound pressure through the eardrum (a vibrating membrane) onto the inner ear and its mechanical parts and finally onto the cochlea and the optical nerve. It is an analog-to-digital converter of sorts since the signal to the brain is not a continuous, analog electrical signal but a series of digital pulses.

REEL-TO-REEL (OPEN REEL) TAPE DECKS

How they work

Reel-to-reel tape decks record and playback audio material on a magnetic tape wound on reels of various diameters. 7" and 10.5" reels are typically used in domestic machines. A full tape on the supply or feed reel (4) is placed on a spindle or hub. The tape is then manually threaded through mechanical guides and a tape head assembly onto the initially empty take-up reel (5).

As a motor (or two or three) drives the reels and the capstan (6) against the pressure roller (8), the tape is tensioned and transported at a constant speed in very close proximity to the erase, record, and playback heads. During reproduction of a prerecorded tape, the erase and record heads are inactive. The playback head detects variations in magnetic flux density caused by the permanently magnetized particles on the tape (residual magnetism). A variable (AC) voltage is produced in the playback head's winding (coil) and fed into a reproduce or playback preamplifier, which equalizes and amplifies that voltage and takes it to the "Line output."

During recording, the process is reversed. A special AC signal is fed to the erase head's coil to erase any previously stored magnetic information on the tape. The recording head is next along the tape's path, which also has a magnetic core and a signal coil.

The input audio signal (from the "LINE IN" input) is processed by the record amplifier, mixed with a high-frequency bias signal, and fed to the coil of the recording head. The changing voltage in the coil induces a changing magnetic flux in its core, which through the gap in the head penetrates the magnetic tape and magnetizes its particles as it travels along (illustration on the next page).

You can say that the magnetization pattern on the tape corresponds to the varying audio signal in a magnetic way, just as the variable depth and width of the grooves in a vinyl record mechanically correspond to the music signal.

The magnetizing curve of all magnetic materials is highly nonlinear (read: "causing serious distortion"). For instance, this applies to laminations of output transformers in tube amps, not just to tape deck heads and magnetic tapes, and has to be dealt with to minimize distortion.

Just as in the DC biasing of a Class A audio amplifier stage, the AC biasing in magnetic recording shifts the audio signal to the relatively linear part of the magnetizing curve. However, in contrast to DC biasing in tube amps, AC magnetic biasing is superior in this application, so DC biasing is not used commercially.

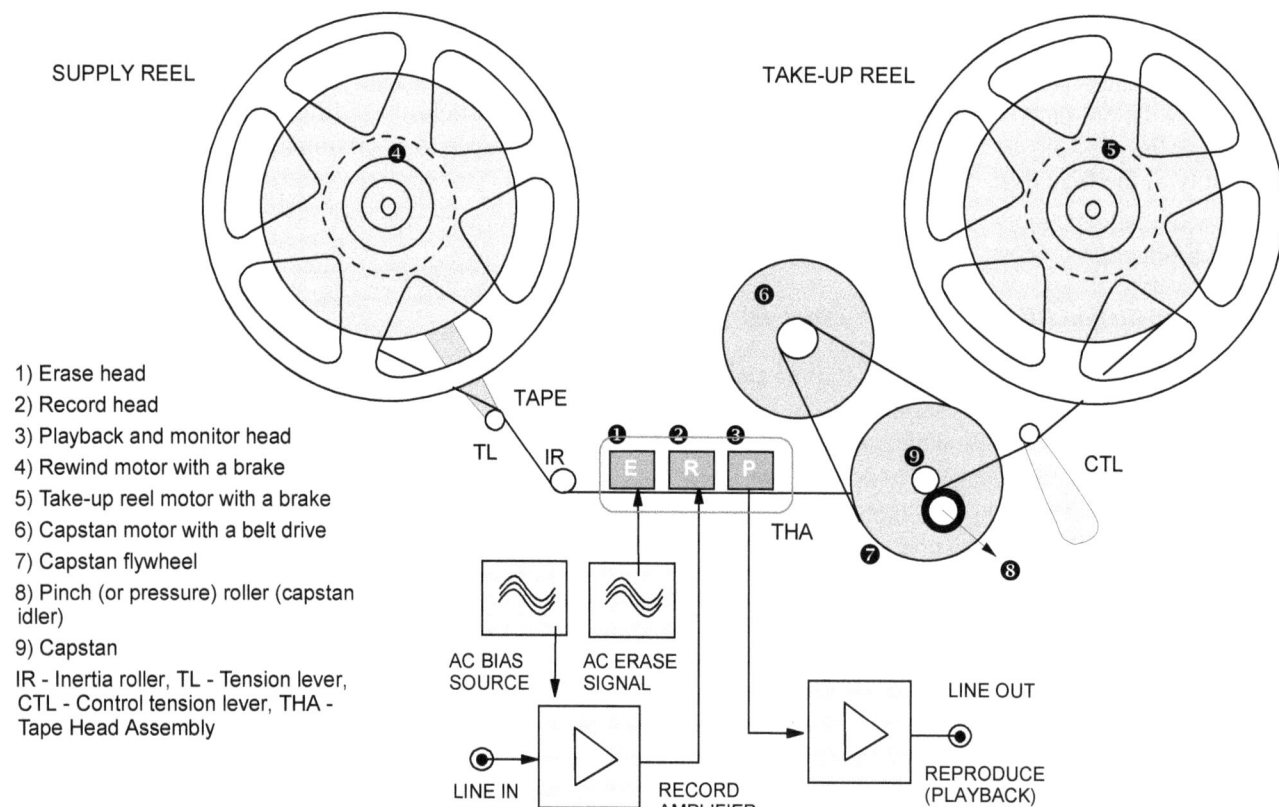

Fig. 6.2. A typical arrangement of a 3-head 3-motor reel-to-reel tape deck. Instead of a belt drive, some machines use a direct drive capstan motor or even dual capstan drive system, which must be precisely microprocessor controlled. The basic block diagram of the amplification chain is also included. The mechanical-electronic interface and logic controls are not shown for clarity.

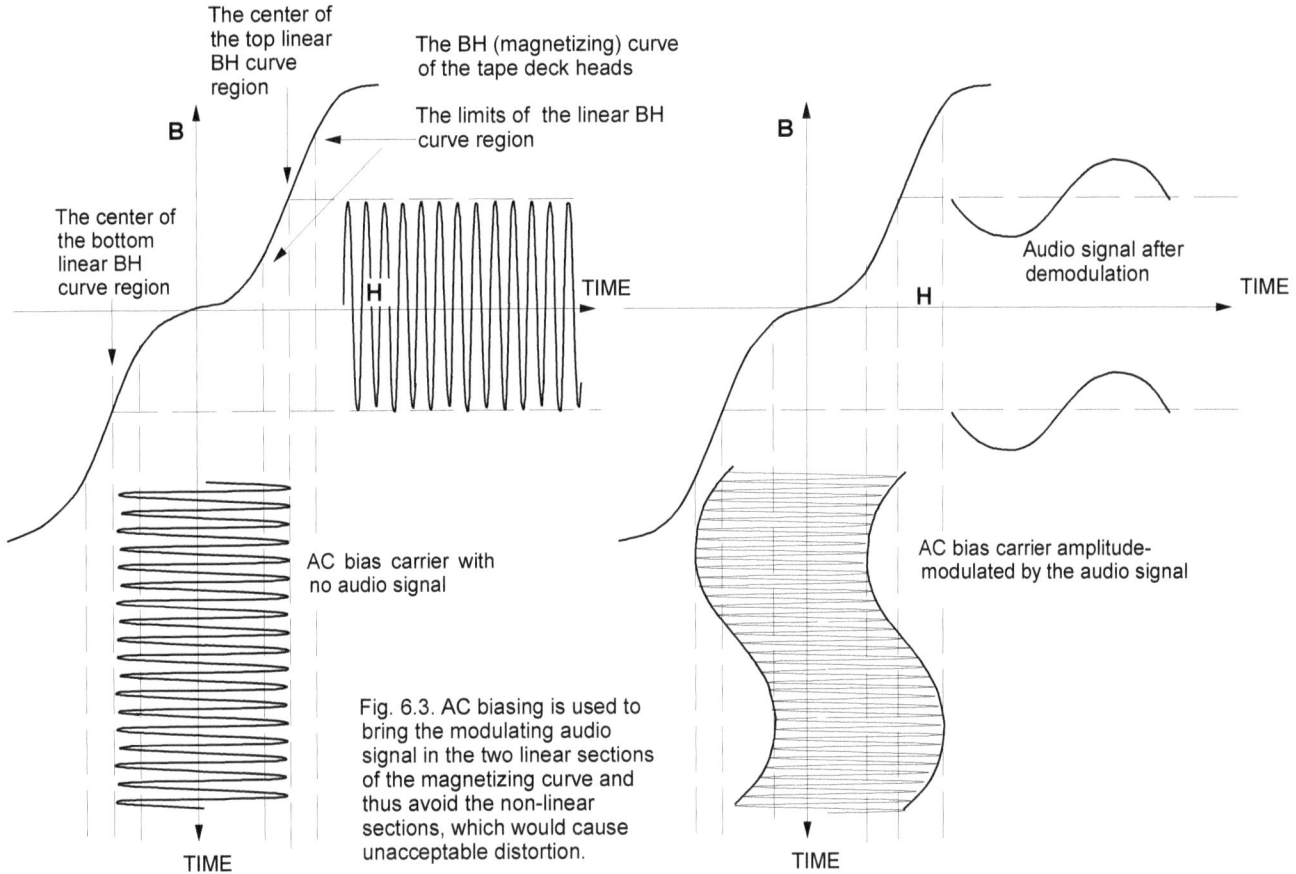

Fig. 6.3. AC biasing is used to bring the modulating audio signal in the two linear sections of the magnetizing curve and thus avoid the non-linear sections, which would cause unacceptable distortion.

The high-frequency supersonic bias signal (cannot be heard by humans) is amplitude modulated by the lower frequency audio signal. The higher the bias frequency, the better the performance; some tape decks use AC bias frequencies up to 160 kHz!

The illustration above shows the B-H (magnetizing) curve of the magnetic material on a tape and the sine wave bias signal without and with the audio signal. The heads are electromagnetic transducers, converting electrical into magnetic signals and vice versa. "B" is the magnetic flux density through the magnetic core, and "H" is the strength of the magnetic field that caused such flux, proportional to signal current through the coil.

While the arrangement on the previous page is of a 3-head 3-motor deck, there are simpler reel-to-reel tape recorders, for instance, with only two heads and one motor (illustration below right). The even more complex auto-reverse machines may have up to six heads. Three are active in one and three in the other direction, which the control logic changes automatically once the end of a tape is detected.

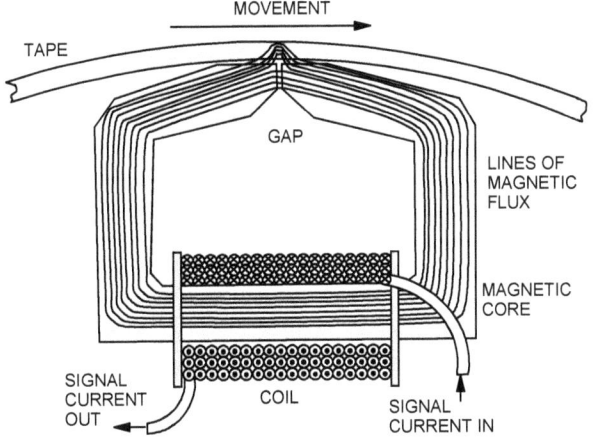

Fig. 6.4. The in-principle operation of a recording head, showing the lines of magnetic flux produced by the flow of signal current through the coil. The dot symbolizes the current flowing out of the page towards you, and the cross symbolizes the flow of current into the page, away from you (like an arrow's tip and cross at its back).

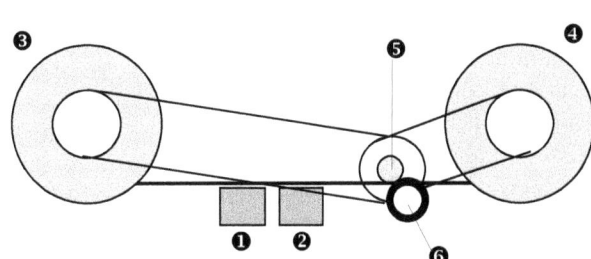

Fig. 6.5. A typical arrangement of a 2-head 1-motor open reel tape recorder. 1) Erase head 2) Record & Playback head 3) Supply reel 4) Take-up reel 5) Capstan with drive motor with belt drives to (3) and (4) 6) Pinch (or pressure) roller (capstan idler)

Closed-loop tape transport

There are many other variations on the tape transport design side of things, for instance, dual capstan designs and closed-loop tape transport systems, which became popular with certain Japanese tape deck brands in the second part of the 1970s.

The isolated loop tape transport mechanism was designed to maintain constant and stable tape tension, thus reducing wow & flutter and modulation noise. Tape speed variations of less than +/-0.1% and 0.05% tape speed-accuracy were claimed.

The large diameter capstan is usually driven by a quartz-controlled PLL (phase-locked loop) direct-drive motor, while the in-going and out-going capstan idlers (acting as pressure rollers against the centrally placed capstan) isolate the heads from the reels.

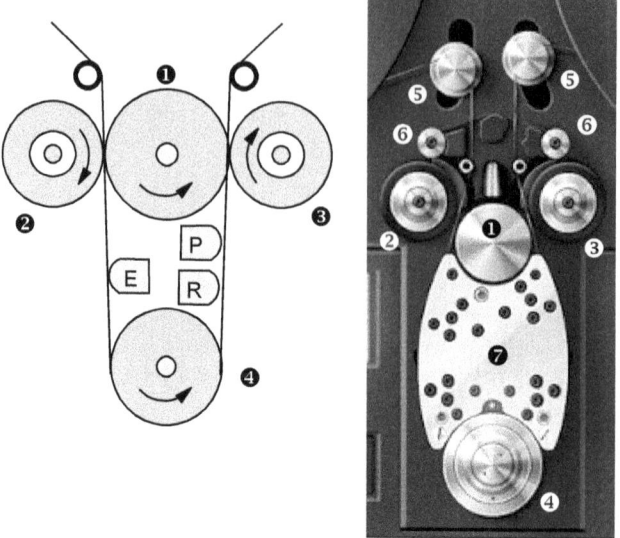

Fig. 6.6. Closed-loop tape transport with 3 heads, and as implemented on Technics reel-to-reel machines, with 4 heads: 1) Capstan 2) In-going capstan idler 3) Outgoing capstan idler 4) Reversing idler 5) Tension rollers 6) Tape edit markers 7) Head block

Servicing and alignment issues

For proper operation, the alignment of the recording and replay heads is critical. There are four such adjustments: height, zenith, wrap, and azimuth. The "height" alignment (1) ensures that the head is right above the tape. It can be checked visually by comparing the distance from the edge of the tape to the top and bottom shield plates, which should be equal.

Zenith or "vertical alignment" (2) is the degree to which the head is parallel to the tape.

Azimuth (3) is the tilt in the vertical plane, while wrap or "horizontal contact angle" (4) is a tilt in the horizontal plane. Wrap only applies to better-quality machines that don't use pressure pads.

The alignment procedure will depend on the design of a particular machine. Usually, one spring-loaded screw adjusts both height and zenith, while two other screws adjust height and azimuth.

The bias and equalization levels have a similar effect on the frequency response curve as misaligned or worn-out heads. If under-biased, the bass will be severely attenuated, and the highs will be lifted, resulting in a thin and harsh sound. Over-biasing is the opposite, characterized by severely rolled-off treble and slightly elevated bass levels. In both cases, the phase errors will be introduced as well.

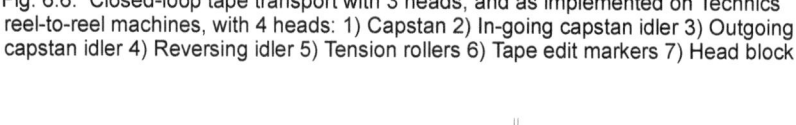
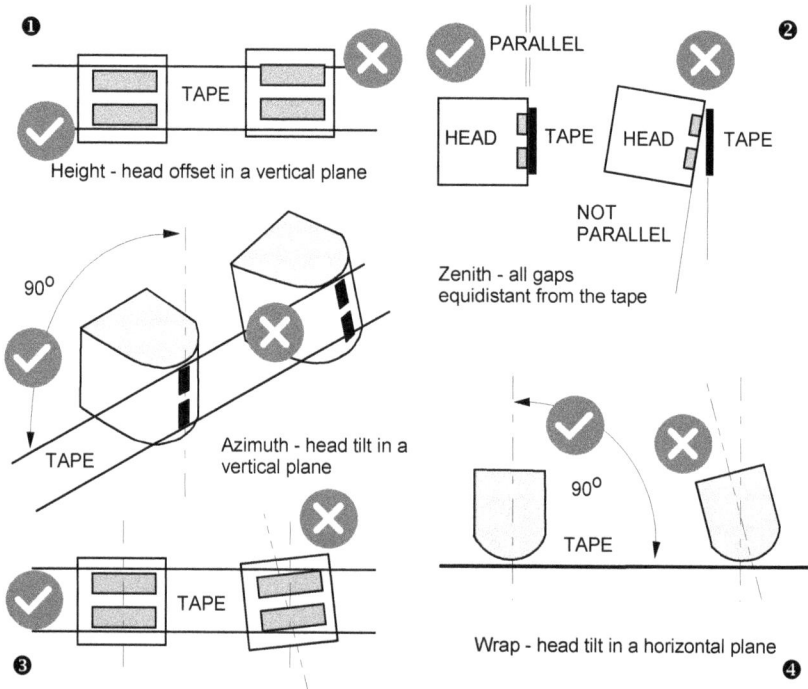

Fig. 6.7. Height, zenith, wrap and azimuth illustrated

Fig. 6.8. Suboptimal bias current levels cause a sever shift (or rather "tilt") of the frequency response curve. Underbiasing diminishes the bass and raises treble, while over-biasing has the opposite effect, slightly elevating the bass frequencies but severely reducing treble levels!

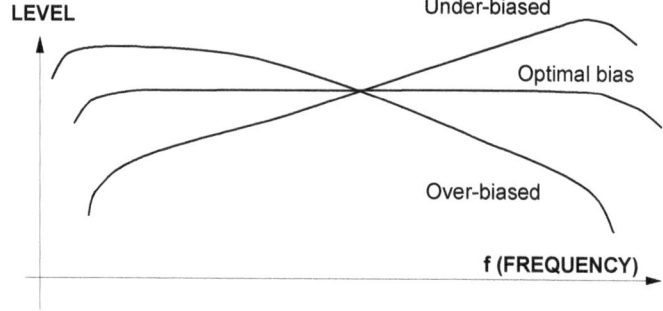

Do all quality tape decks sound the same?

In short - no. Transport problems aside (speed variations, wow & flutter), two major factors determine the sonic character of a tape deck. The input and output signals are both processed (amplified and equalized) by the deck's electronics, and both pass through the recording and playback heads, so their design and construction leave a major imprint on the sonics.

Just as various phono cartridges and turntables sound different, so do tape heads and tape decks. Look how two late 1970s open reel tape decks, Tandberg TD20A (below left) and Akai GX267D (below right), both quality machines, reproduce the same square wave test signal.

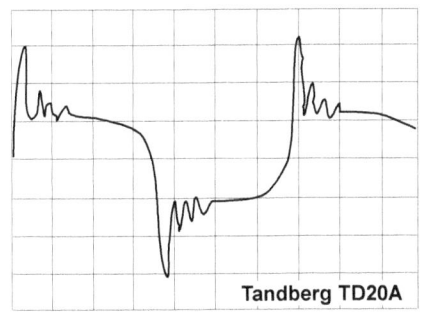
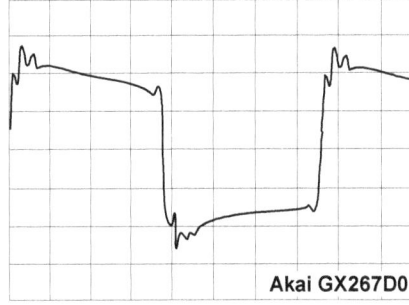

Fig. 6.9. Square wave response of two late 1970s open reel tape decks, Tandberg TD20A (LEFT) and Akai GX267D (RIGHT)

Why tapes sound better than CD players or turntables

As we have just seen, during the analog recording process, individual tracks are recorded on multi-track tape recorders and mastered onto a two-channel master tape. It is that tape that is used in LP production.

A disc mastering facility uses cutting lathes to cut a "lacquer," which is then "silvered," and after that silver coating, it is dipped into a "bath" where it's nickel-plated. The temperature of the plating solution, the quality and purity of the water used, and the speed and skill of the plating operator all impact the sonic quality of the disc. The "master," or the negative of the original lacquer, is plated again, thus forming a positive metal "mother."

The "mother" is now plated itself, forming one or more of the negative "stampers," which are used in the molds of the recording presses to produce final LP records.

LP production involves so many highly critical steps and processes that it is a true miracle LPs sound good at all! No matter how good all that sophisticated equipment may be, each step degrades sound quality, so the master tape sounds infinitely better than any LP can possibly sound. The difference is truly dramatic.

However, as we have seen, the tape multiplication process isn't benign, either. Consumer prerecorded tapes are produced from tapes that are a 3rd or 4th generation from the master tape, a copy-of-a-copy-of-a-copy-of- a-copy if you will, and such tapes only last a few hundred runs, meaning only a few hundred consumer tapes can be made before the source tape deteriorates so much that it must be thrown away.

The fidelity is further reduced by high dubbing speeds, 10X or faster, which severely reduce high frequencies, and by the use of the cheapest consumer-grade tape possible. Welcome to capitalism, where accountants and bankers rule over engineers! However, tape dubbing does not suffer from possibly the major source of sonic degradation LPs suffer from, and that is the issue of bass.

Before cutting the "lacquer," bass frequencies must be highly attenuated; otherwise, the grooves on an LP would be incredibly large and deep. Depth-wise, such deep groves would be impossible to record onto such thin discs; the cutting heads would drill straight through them! Thus, the bass performance of magnetic tape far exceeds that of an LP. Magnetic tape has a wider dynamic range than a vinyl disc, with superior sound, especially at the frequency extremes, the bass, and treble.

The quality of phonograph records' reproduction deteriorates with each play. The surface noise from handling (scratches, dust, lint, oils, and acids from fingerprints) and stylus wear develop and cause a noticeable loss of fidelity. With normal precautions (keeping tapes away from high temperatures and strong magnetic fields), the quality of a magnetic recording is practically permanent.

However, storage conditions affect magnetic tapes much more than LPs. Humidity, mold, and high temperatures cause tapes to become sticky or the binder to separate from the oxide layer, clogging up tape guides and heads.

Both turntables and tape deck suffer from wow and flutter (a low- and high-frequency periodic speed variation, respectively). Wow can be caused by a warped record or by an offset created by imprecisely punched LP's center hole. Flutter, more common in tape decks, can be caused by worn or deformed rotating components such as the capstan or the pinch roller. These objectionable distortion artifacts are quite noticeable, especially on the sustained (prolonged or slow decaying) piano or organ tones.

Are cassette decks hi-fi components?

Many 1980s and 90s mid-fi audio systems featured single or double cassette tape decks. How do "compact cassette" tapes compare to their open reel brethren?

Arguably, cassette tape decks aren't true audiophile components. Compact cassettes were invented for convenience and ease of use, but fidelity-wise, their performance pales in comparison to even consumer-grade open reel machines. Even in their heyday (the 1970s and 80s), cassettes were primarily a handy medium for recording music at home (from radio stations and LPs) and playing them in cars and other vehicles. There were no compact disc players in vehicles in those days, let alone mp3 players or Bluetooth capabilities.

Why do reel-to-reel machines sound better? When it comes to magnetic recording, bigger (wider and thicker tapes) and faster (the tape speed) are better, and compact cassettes, with their thinner and narrower tapes and very slow tape speed, are inferior in every way. Even the modest 7½ IPS speed of consumer open reel decks is four times the speed of a cassette tape, and increased speed widens the frequency range (bandwidth) and reduces the S/N (Signal-to-Noise) ratio. Open reel tapes are double the width of a cassette tape, further improving the S/N ratio and the increased quantity of the oxide coating on the tape retains more magnetism, thus increasing the output level.

What features to look for when buying an open reel machine

The basic consumer-grade reel-to-reel machines will have two speeds, 7½ and 3¾ IPS (inches per second), equivalent to 19 and 9.5 CPS (centimeters per second). Better machines have 15 and 7½ IPS options. However, keep in mind that most prerecorded music tapes were recorded at either 7½ or 3¾ IPS, so 15/7½ IPS machines will not be able to playback 3¾ IPS tapes. A 4-track machine will allow recording in stereo on both sides of the tape.

Most consumer-grade tape decks use a single capstan tape transport system that is prone to wow and flutter. Better ones have a dual capstan system, controlled by a closed feedback loop logic. Dual capstan arrangement isolates the tape in the head region and thus ensures a constant tape-to-head contact, drastically reducing wow and flutter.

Each capstan will have its own high torque motor, so the two maintain a constant tape tension at all times. DC motors minimize changes in motor temperature under different loads, thus maintaining constant torque. Other machines use eddy current motors with circulating oil for cooling.

The auto-reverse feature almost doubles the complexity of the machine, requiring twice the number of checks and adjustments. This is of marginal benefit and not worth paying a price premium for.

For optimal performance, a tape deck should be rebiased for each type or even brand of tape used. . Better decks will have a selector switch for different bias settings, such as "Standard/High" or "Normal/Special. In addition, there are at least two EQ (Equalization) settings, three on better machines.

Older models, released before higher coercivity metal tapes, don't have the sufficient headroom to benefit from higher coercivity tapes released afterward and do not exploit their superior performance parameters fully, so if you intend to use metal tapes, for instance, look for machines that were designed to use them.

The major open reel tape deck makers were Teac (later Tascam), Akai, Pioneer, Technics, Tandberg, Revox, Studer, Uher, and Crown. Sony, Denon, Otari, Kenwood, Roberts, Viking, Magnecord also produced a few notable models. Many others produced cheaper and inferior models below the audiophile quality level.

Before purchasing a vintage unit: the basic operational test and visual inspection

You don't need any tapes for this test, but it is a very basic one; it only tests for the start-stop-run & speed change functions, in other words, the motor(s) and the associated hardware.

Press the "ON-OFF" button to turn the machine on. Lift the CTL lever up and observe the capstan - it should start rotating. Press the "PLAY" button while still holding the CTL up. The pinch roller should snap up towards the capstan, and the reels should start turning.

Press the "SPEED" button to change from "High" to "Low" or the other way around. Notice if the reels change the rotational speed. On some units with three or more speeds, the switch will look different, but in any case, make sure that the machine responds to speed change commands.

As part of the visual inspection, pay particular attention to any visible rust. Tape heads rust easily and may have to be replaced. It doesn't take much rust to seize adjustment screws and smaller mechanical parts inside the machine, which would then be a nightmare to unscrew or take apart. Heads should have a convex face area and thus touch the tape over a relatively small area. Worn heads have been ground to a flat surface.

Minor misalignments may not be visible upon cursory inspection, but major ones can be spotted if you know what to look for, and that is an asymmetrical wear pattern. For instance, a zenith error will result in a wider worn area at the top or bottom of the head, while a wrap error will cause higher wear on the left or the right side.

Many opportunistic sellers sell old decks without any tapes. In fact, most will claim that they don't even have any tapes around. This is a warning sign. Either the seller had not used the machine himself, and thus its condition cannot be described as "working" or "fully operational," or the seller knows the deck does not work properly and is hoping that you will not fully test all of its functions (which you cannot possibly do without any tapes).

So, always have a commercial prerecorded tape and a pair of reels with you, to be played back at 7½ (preferably) or even 3¾ IPS, the speeds most machines should have.

Even if a deck passes the quick operational and even listening tests, you still cannot be sure how well it will ultimately perform. The quality of the tape deck's transport mechanism is its most important performance factor, and, out of all transport components, the most important part of any drive is the capstan assembly. The size, machining tolerances, and the overall quality of the capstan shaft drive impact tape speed and its fluctuations.

The magnetic hysteresis curve is symmetrical, so even harmonic distortion is generally not present in magnetic recorders. If it is present, it points to faults and problems in the tape deck itself - magnetized heads, a DC current in the heads, amplifier overload or fault, or asymmetrical bias signal. This is one of the main reasons why precise biasing, calibration, and alignment are critical in open-reel recorders.

The perils of buying a 40 - 60 years old open reel machine

The prices of used tape decks have gone through the roof lately (the end of the second and the first half of the third decade of the 21st century). Decks that you'd have to pay someone to take away or that you would dump on a verge for council pick up ten years ago are now selling for $1,000-2,000!

The adjustments and calibrations of reel-to-reel tape decks are beyond the knowledge or skill level of most audiophiles and even the majority of audio technicians, who were not even born when the last tape deck was manufactured. Apart from the intimate knowledge of tape decks, special tools, and calibration tapes are needed for such highly specialized work. So don't let just anyone loose on your vintage open-reel deck.

Compared to turntables and later on CD players, tape decks were relatively rare even in their heyday, but now can be truly considered hi-end esoterica. Audio technicians well versed in such work are rare in most parts of the world, and their work doesn't come cheap. Fixing rusted and seized mechanisms, worn-out heads, and electronic components that have drifted way out of their tolerances is a long and laborious process, so such a significant expense should be factored in before you purchase a vintage unit.

After 40 or more years of use & abuse, a tape deck will need new rubber belts and brake pads, and, most likely, replacement pinch rollers as well. On the electronics side, the electrolytic capacitors will have dried out, while the film caps will be "leaky" (not leaking physically, but developing electric leakage, like an added resistor that bypasses the capacitor). Many resistors will have drifted up or down in value and will have to be replaced as well.

Add to that the cost of new heads, all the adjustments (bias, head alignment, etc.), and detailed cleaning, and you can safely assume the final cost to be at least double the purchasing cost. Due to the very high cost of imported parts and exorbitant labor rates in countries like Australia (AU$100-150 per hour for technician's time is the norm in 2021), a $1,000 machine suddenly becomes a $3,000 proposition!

Are the rewards of reel-to-reel decks commensurate to the associated risks?

So, after all is said and done, with all the likely problems, limitations, caveats, significant expenses, and lack of user-friendliness (compared to other audio sources), why would an audiophile want a reel-to-reel machine? Does their natural and "organic" sound justify all that trouble?

In my view, no. You'd have to be a true purist, a nostalgia-ridden sentimentalist who is in awe of the spinning reels and illuminated moving analog meters, or a sadomasochist who perversely enjoys the long and tedious process of setting up, maintaining, and fixing audio equipment more than the musical enjoyment itself.

Then again, if you have a first-class audio technician who specializes in servicing and calibrating tape decks on your list of close friends (no, Facebook "friends" that you have never met in real life do not count!), by all means, ask for their advice (which machine to buy, etc..) and get them to service it for you.

CONFUSED BY SO MANY TYPES AND FEATURES OF TAPE DECKS? PROCEED WITH CAUTION!

When it comes to choosing a tape deck, as if all the transport system design variations and various types of heads (soft iron versus ferrite and "glass and X'tal ferrite" GX heads, plus a few types not discussed, such as cross-field heads) weren't confusing enough, add to that so many tape widths, speeds, types and reel diameters, plus various noise-reduction methods used (DBX, Dolby A, B, C, S and SR), and a you have a true minefield.

TURNTABLES, TONEARMS AND TRACKING ERRORS

The fiddly and flawed technology that works surprisingly well and sounds much better than it should

"At some point in the not too distant future, once time has granted enough of us the wisdom of hindsight, the virtue of the analog record will be rediscovered. Then the LPs that people so blithely discard today will become treasures in the analog renaissance."

That prophecy was made by Laura Dearborn in her 1987 book *Good Sound*. Remember (if you are old enough), that was only five years after the introduction of compact disc technology! In other words, some experts recognized very early on the limitations and sonic inferiority of the (then) new medium, and, to give credit where it's due, many audiophiles refused, even then, to abandon their turntable-based systems in favor of the new fad.

Although the CD's dominance lasted for almost 30 years (1986-2016), it seems to be in its final and terminal decline, with music streaming and LPs taking over. It may be too harsh to call it a fad (a short-term phenomenon), so "temporary sidetracking" or even a "dead end" could be more appropriate.

Some predict a resurgence of open-reel tape decks and analog tapes. However, I would hazard a guess that, due to the limited range and the costs of recorded material on tape, and the cost, complexity, and lack of user-friendliness of the machines they need to be played on, as just discussed, such development is highly unlikely.

Modifying & improving turntables and phono stages versus messing around with CD players, music streamers and other digital gear

Sonic merits aside, from a purely DIY modification and improvement perspective, the comparison between the mechanical/analog phonograph and an electronic/digital CD player cannot be starker. You can improve a record player's sonics by adding a record clamp or by changing its mat, by substituting arms and cartridges, by changing the VTA (vertical tracking angle), tracking force, and anti-skating settings, by careful and precise cartridge alignment, and many other relatively simple interventions. Even the most technically challenged audiophile should be able to manage those after reading this chapter or watching a few YouTube videos.

Other mods, such as adding an isolation platform or vibration-reducing feet and upgrading the usually flimsy stock motor power supplies, are so simple all of us should be able to do them. Others, such as upgrading tonearm wiring or replacing bearings, are more involved but not beyond the capability of most technically-minded tweakers.

By comparison, the range of possible *simple and easy* improvements to a CD player, DAC, or music server is very limited. For most upgrades you'll require the services of an audio expert.

Not concerned with future upgrade potential or possibilities, most manufacturers solder ICs ("chips") directly to the printed circuit board (instead of using IC sockets and simply plugging the chips in), so even this simplest possible upgrade is a frustration supreme. Without special soldering iron tools, have you tried desoldering 8, 12, 16, or more pins at the same time? Practically impossible!

> **THE STEAM TRAIN ANALOGY**
>
> A steam train is a multi-sensory, romantic, and nostalgic experience; an electric car is but a quirky utilitarian vehicle. Turntables, reel-to-reel tape decks, and tube amplifiers are warm, sexy, rotating, glowing works of art, objects of beauty and awe. Most CD players, DACs streamers, and solid-state amplifiers are cold and impersonal black boxes.

The geometry of a phonograph

The correct name for a record player is the phonograph, but somehow a more colloquial yet incorrect term turntable became the norm. Strictly speaking, a turntable is what its name says, a table that turns, i.e., the platter, together with its base or chassis (plinth) and the motor that powers it. A phonograph is a turntable plus a tonearm and a cartridge. In this book, just as in many others, we will use the less formal naming convention, turntable = plinth + platter + tonearm + cartridge! We'll also use the acronym TT for a turntable.

A turntable is a precision instrument that detects minute vibrations of a needle (stylus), which is transmitted through a cantilever to the cartridge, an electromechanical transducer that converts such vibrations into an electrical signal. A cartridge is mounted on a tonearm, apart from the platter, the only other moving part of a turntable.

A tangentially moving arm operates in the same manner as the cutting lathe that produced the master, but very few turntables use it nowadays due to certain engineering problems, the discussion of which would be beyond the scope of this book.

Since most modern turntables use the pivoted or "radial" tonearm, the pivot point being the point of its contact with the bearing on which it balances (10), it pays to understand its basic parameters and operating principles.

Ideally, the stylus should always stay aligned in such a way that its tip (2), as it travels across the record, remains at a tangent to all record grooves. However, while the cutting head moved in a straight line across the record, a stylus of a pivoted tonearm moves in an arc, which introduces a tracking error. The stylus can only be tangent to the grooves in a maximum of two points, which are known as "null" points.

The two-point protractors used in the alignment of lateral tracking angle (tangency and overhang) are usually supplied with a turntable and have such points marked. More on that soon (see page 132).

Ultimately, the shape of the tonearm does not matter from the geometrical point-of-view; what matters is "L" or the effective tonearm length, the distance between the pivot point or bearing (10), and the tip of the stylus (2). Most tonearms are in the 210-230 mm range. The effective length is always longer than "D," the mounting distance between the spindle (6) and the bearing (10), and that difference is called an overhang (O=L-D).

The longer the effective length L, the smaller the curvature of the arc (stylus path) and the lower the tracking distortion. This has led to a renewed interest in longer (12") arms.

However, longer arms have higher mass and inertia, increased internal resonances, and a greater possibility and consequences of misalignment.

The distortion doesn't just vary with the tracking error; it is also a function of the groove speed, which is the slowest at the outer rim of the LP and increases as the distance from the spindle reduces. Thus, the distortion is much higher in the innermost grooves, almost double the distortion at the outer grooves!

A properly designed, correctly installed & aligned tonearm and cartridge can achieve a tracking error of under +/-1 %, which isn't bad at all, compared to much higher distortion levels of most tube amplifiers and all loudspeakers.

Also, the tracking distortion is predominately a 2nd harmonic distortion, which is pleasing to the ear and sonically "benign" or "innocuous," as magazine reviewers like to say. Is that the secret behind the "natural" turntable sound?

Going back to the tonearm geometry, notice that the headshell (and the cartridge attached to it) is angled in relation to the tonearm tube (1). In modern 9" tonearms, that offset angle is typically in the 22-25° range, the optimum when it comes to reducing tracking distortion, which is the reason behind such arrangement.

The alignment shown in the graph (right) and on page 132 has null points 130 mm and 250 mm from the spindle.

There are many other possible alignment geometries named after their proponents, such as Baerwald, Loëfgren, Stevenson, and UNI-DIN. They all position the two null points differently along the arc, resulting in different distortion "profiles," so it's obvious that there is no optimal choice in this game.

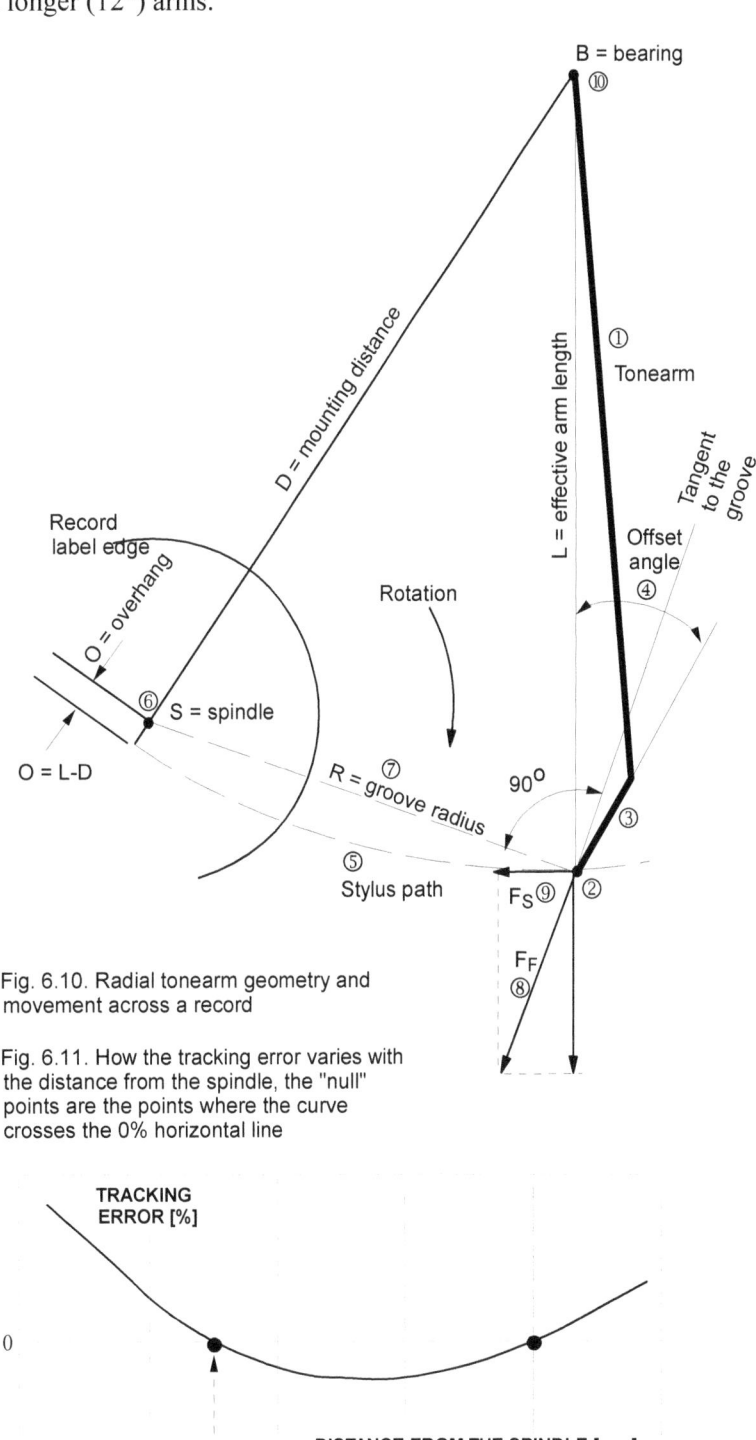

Fig. 6.10. Radial tonearm geometry and movement across a record

Fig. 6.11. How the tracking error varies with the distance from the spindle, the "null" points are the points where the curve crosses the 0% horizontal line

However, due to the offset, the tangential or frictional force F_F (8) that opposes the movement of the stylus (2) now has a perpendicular or radial component F_S referred to as the "skating" force (9). This force causes the tonearm to skate or slide towards the spindle (6), thus favoring the inner grove wall (left channel signal) while losing contact with the outer wall (the right channel signal). Special "anti-skating" measures are thus required, which will be outlined in the "Turntable setup & adjustments" section a few pages ahead.

DRIVE MECHANISMS AND VIBRATION MINIMIZATION

Turntables have been in use for so many decades, and yet their designers and manufacturers still cannot agree on many of the fundamental aspects of their design, among them the best or most optimal materials to be used and the best arrangement and mechanical (de)coupling between their main parts, namely the plinth, the platter, the motor, and the tonearm.

Although, strictly speaking, the type of drive and the vibration-mitigation are two distinct topics, since they are mutually interactive aspects of turntable design, we'll address them together in this short overview.

A phonograph is a very sensitive vibration detector since its very essence is to as faithfully as possible detect and reproduce mechanical vibrations that the stylus experiences in a modulated groove. However, since such a seismograph cannot distinguish between wanted (the music etched into the modulated grooves of an LP) and unwanted vibrations (from the motor, acoustic feedback from loudspeakers, structural vibrations), it is up to turntable designers to address that problem, and they do it in various ways and with varied success.

Belt drive

Since the motor driving the platter is considered the main offender (vibration-wise), its mechanical decoupling from the platter and tonearm is the underlying theme of all designs. The contemporary leanings are towards a rigid or solid plinth on which the platter and tonearm are both mounted, with a motor totally separate in its own cylindrical pod.

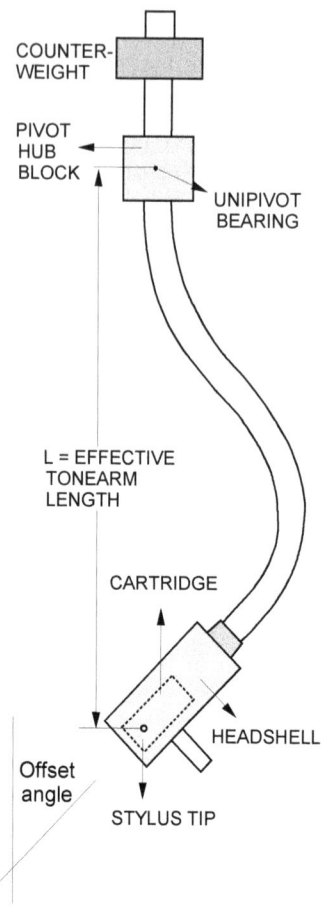

Fig. 6.12. How the effective tonearm length is defined and measured for S-shaped tonearms

Even some cheaper, entry-to-mid level turntables such as Shiit Audio Sol and Pro-Ject RPM 3 have isolated motors (separate from the plinth). However, the motor's base (pod) and the plinth still sit on the same support shelf, so some vibration transmission is inevitable; at least they aren't directly coupled as with models where the motor is directly bolted onto the plinth.

The only option of coupling an isolated motor to a plinth is via a belt drive, and since belt drive is the most common way of rotating a platter, it pays to understand its pros and cons.

Another decoupling option is to rigidly fasten the turntable's motor on the main chassis or plinth (1) but mount its tonearm and platter on a separate rigid platform or "sub-chassis" (2), which is then suspended on springs (3) from its main chassis or base. The spring suspension is akin to the use of a spring on a car's shock absorber.

The underlying assumption behind both the rigid mounting and the suspended sub-chassis design is that vibrations that affect the platter will also equally affect the tonearm, so there'd be no relative movement between the two. This, however, isn't strictly true. LPs are mechanically imperfect and vibrate (wobble) in unpredictable ways, and so do platters, whose bearing quality and performance are of paramount importance. Tonearms are a different story again, with longitudinal vibrations traveling across them in both directions.

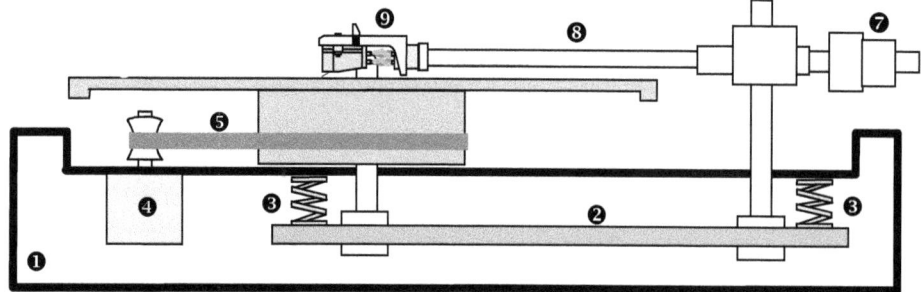

1) Main chassis (the base or plinth)
2) Suspended sub-chassis
3) Suspension springs
4) Motor
5) Pulley and belt
6) The platter
7) Counterweight
8) Tonearm
9) Headshell and cartridge

Fig. 6.13. A belt-driven turntable with a suspended sub-chassis

Thus, some TT designers have adopted the philosophy of isolating all major TT components from each other, but more complex solutions inevitably cost more, so only pricey TTs will be in that category.

In belt drive turntables, the belt acts both as a primitive shock absorber, minimizing the transfer of vibrations between the motor and the platter, and also as a speed regulator, its elasticity dampening small changes in the motor's rotational speed. The rotational inertia of the platter (which is in effect a large flywheel) also helps all but eliminate the "cogging" problem, those small jerky motions (jumps) in the motor's rotor position caused by the finite number of poles in its stator. Due to such motor design, the stator's rotation suffers from "cogging". You could call it mechanical or rotational jitter if you wish.

To improve the motor-platter isolation even further, some designs include two motors working in unison (dual drives), while others interpose an idler wheel between the motor's pulley and the platter.

Belts are made of soft rubber compounds, and such materials stretch, age, and deteriorate depending on temperature, humidity, and many other factors. Belts have a finite life, and regular replacements are necessary.

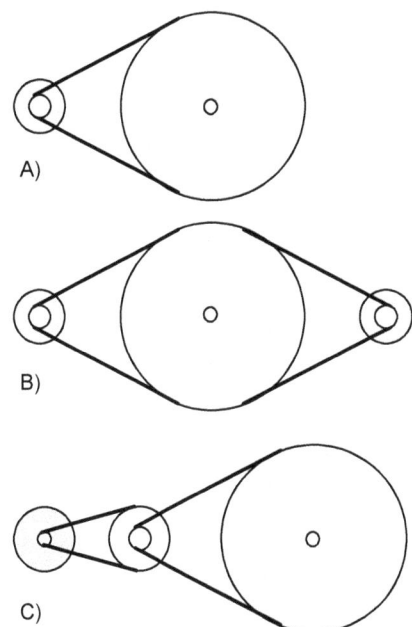

Fig. 6.14. A few variations on the belt-drive theme: A) Single motor belt drive B) Dual motor belt drive C) Single motor belt drive with an intermediate idler wheel

Idler wheel drive

An early method involved a rubber-rimmed idler wheel (4) between the stepped shaft or boss (3) and the inside rim of the platter (5). Speed changes (6) were provided by raising and lowering the idler to engage a different diameter boss. The stepped shaft is belt-driven (2) by a motor (1).

Idler-wheel drive turntables suffered from poor rumble performance due to the lack of decoupling (direct mechanical contact between the motor and platter via the idler wheel).

Rumble is a very low-frequency noise (picked up by the phono cartridge and reproduced by the rest of the system), caused by a loose or wobbly platter, motor, or idler-wheel bearing, or by any such mechanical irregularity that results in a relative movement between the platter and the cartridge.

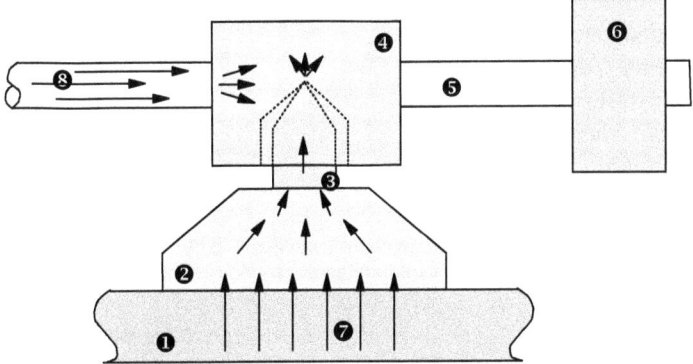

Fig. 6.15. How vibrations propagate through and get dampened by the major mechanical blocks of the tonearm and turntable: 1) Turntable base (plinth) 2) Tonearm base mount 3) Unipivot 4) Unipivot hub 5) Tonearm 6) Counterweight 7) Motor and base vibrations 8) Longitudinal tonearm vibrations

Fig. 6.16. The idler-wheel drive mechanism with two speeds

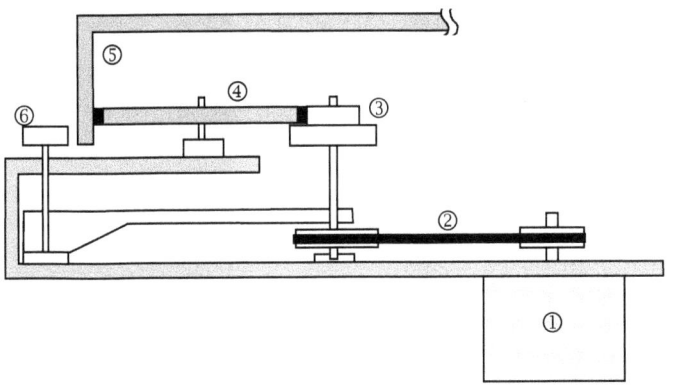

Recently there's been a revival of interest in this type of drive, most notably in Thorens TD-124 (as probably the best of the Garrard-Thorens-Lenco bunch), but also even budget vintage idler-drive turntables by Dual and Elac.

Direct drive

Marantz, Luxman, Audio Technica, Technics, Denon, Aiwa, JVC, Torrens, Onkyo, Hitachi, Pioneer, Sony, Toshiba. The list of direct-drive turntable manufacturers reads like "Who's who in the Japanese audio industry."

In direct-drive turntables, the motor (usually a brushless DC motor) is integrated with the platter (the drive shaft is the spindle), so it rotates at the exact record speed (33½ RPM or 45 RPM).

Their impressive technical specs include a very low wow & flutter, high torque, and a fast, well-damped response to dynamic stylus drag. Modern coreless (without an iron core) motor designs and advanced control systems (PLL - Phase Locked Loops) have, if not eliminated, significantly reduced cogging.

TURNTABLE SETUP & ADJUSTMENTS

Turntables differ in design, features, and setup procedures, so for best results, you should study your turntable's user's manual and any information and/or video available online for that specific model. Some manuals include excellent, detailed, and well-explained steps; others outline only the basics, leaving you with more questions than answers. Here, to illustrate a few important points, we'll use Pro-Ject RPM 3 turntable as an example.

Step #1: Leveling the turntable

Some turntables have adjustable (screw-in and -out) feet; others don't. For this adjustment, you will need a small spirit level or a record clamp (1) with an integral level indication (as illustrated). These days they are so cheap there is no excuse for not using one.

While the normal spirit level checks in only one direction, these use a single bubble to indicate in both dimensions, thus making the adjustment so much faster and easier. The bubble should be right in the center, marked with a small circle and a dot in its center (2).

Pro-Ject RPM 3 has three conical feet, two of which are adjustable. If your turntable's feet aren't adjustable, you have two options. You could place "shims" under one or more of them. These are thin wafers of any hard material. I find samples from kitchen and bathroom vanity tops great for this purpose, lice "Formica" and similar brands. Others use old credit cards or even paper business cards.

Alternatively, upgrade the stock feet (usually not the best sonically, since most turntables are built to compete on price, so costs are cut everywhere) with adjustable ones.

Fig. 6.17. A record clamp (solid machined aluminium, available in different color finishes) with an integral level indication

Step #2: LTA - the lateral tracking angle (tangency and overhang)

The only reference to this adjustment in Pro-Ject's "Instructions for use" document is "The full sound quality of the record player can only be achieved if the cartridge is correctly adjusted. Particular tools like the Pro-Ject Alignment Tool are required to accomplish this job properly. If you are not well acquainted with the adjustment of cartridges, you are advised to call upon the willing help of your Pro-Ject dealer to accomplish this task for you."

In this era of online buying, the dealer you purchased from could be thousands of kilometers away, and you have to perform such alignment yourself. Expecting you to return the whole turntable to a dealer every time you change a cartridge, paying high transport costs and his professional fees while subjecting such a precise machine to vibration and possible damage in transport? Time for a serious reality check!

Most protractors used for setting the LTA are based on the two-point cartridge alignment principle and are usually supplied with your turntable. This alignment sets both the specified stylus overhang (the distance the stylus tip extends beyond the spindle when the headshell is positioned directly above it) and the tangency.

This two-point protractor, supplied with Pro-Ject RPM 3 Carbon turntable, has two small crosses with a stylus symbol above it, an outer (5) and an inner one (6). These are for the stylus tip to rest on. The arm in the photo was moved to the space between the two points so they can be seen; that is not its proper resting position, of course. In our case, the inner dot is 65 mm (130 mm is Φ - the diameter) while the outer one is 125 mm from the spindle (3). These distances are not universal; they depend on the effective length of each tonearm.

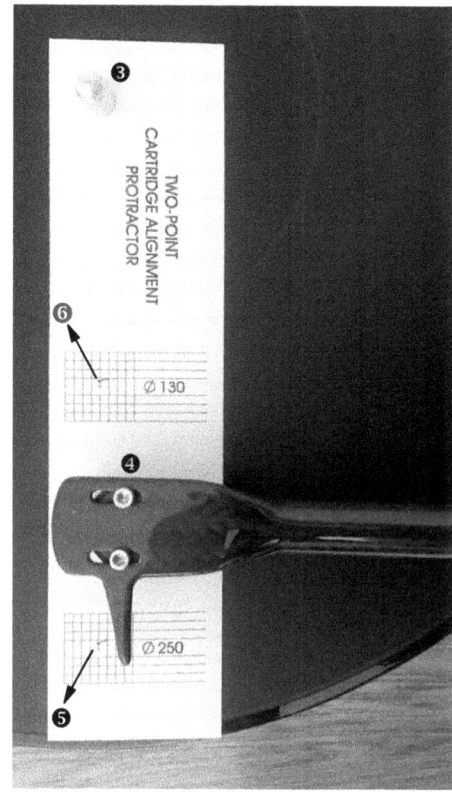

Fig. 6.18. The two-point paper protractor supplied with Pro-Ject RPM 3 Carbon turntable

There are also two grids to which the stylus cantilever should be aligned. If the cantilever cannot be seen, it is usually satisfactory to align the sides of the cartridge body to the lines on the grid instead. Such an approach assumes that the cantilever is centered, straight, and parallel to the sides of the cartridge, which is usually the case (but not always).

The procedure is simple. With the cartridge mounting bolts (4) slightly loose, twist the cartridge and slide it along one and then the other headshell's slot until the stylus cantilever is parallel with the tangential lines on the protractor.

The outer alignment is usually adjusted first, and this will set the correct stylus overhang. Then, if you are lucky, the inner alignment, at point (5), will also be perfect.

If it isn't, or if, for some reason, you cannot achieve the optimal alignment at both points, it is more important to achieve the perfect alignment at the inner point since it is in that region where the arm/cartridge has a more difficult tracking task.

The LTA of cartridges mounted on a removable (twist-lock type) headshell can be adjusted at its base (7).

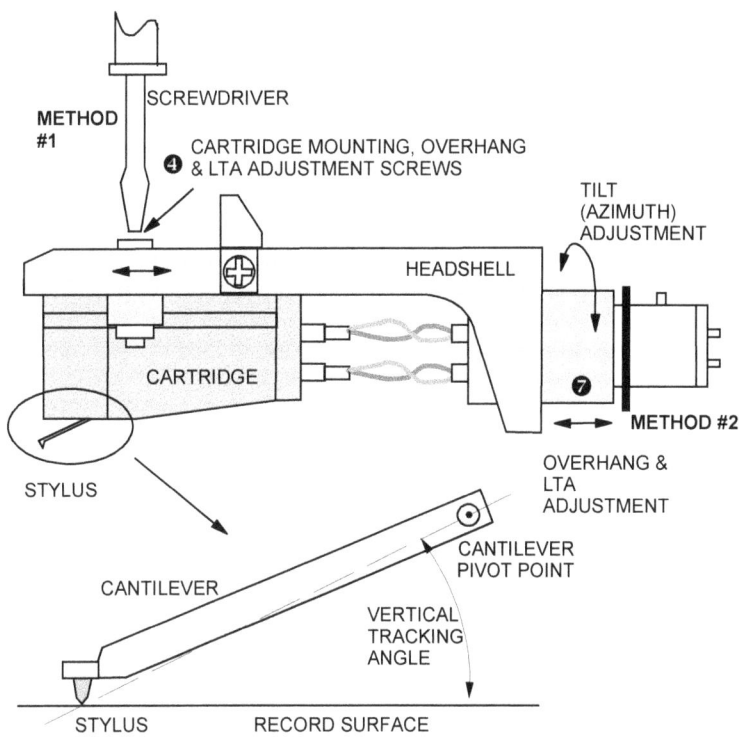

Fig. 6.19. Two methods of adjusting the overhang and the lateral tracking angle (LTA). Method (1) is for fixed headshells with two cartridge mounting slots; method (2) is for cartridges mounted on a removable headshell. The enlarged detail of the stylus and cantilever shows the definition of the Vertical Tracking Angle (VTA).

Step #3: Adjusting VTF - the vertical tracking force

VTF is downward pressure the tonearm and cartridge make on the stylus and the record grooves. Too much force will result in accelerated record wear since the stylus will then permanently damage the grooves. However, perhaps counter-intuitively, too little downward pressure can damage the records even more due to the stylus bouncing around in the groove and chipping it away instead of sliding along as smoothly as possible.

A simple mechanical tracking force gauge is included with many turntables, but I find these fiddly and time-consuming to use. They aren't very precise either. So invest in a digital pressure gauge; they cost very little these days.

To perform this adjustment, set anti-skating to zero (more on that in a moment) and balance the tonearm by sliding the counterweight (CW) in either direction until the balance is achieved and the arm is suspended in the air. At this stage, there is no vertical tracking force. Then slide the CW towards the cartridge, gently drop the needle to the designated spot on the digital pressure gauge and read the tracking force. Adjust the CW in small increments until the desired force is achieved.

Fig. 6.20. Digital pressure gauges have become so cheap, you'd have to be mad to use the plastic ones supplied by turntable manufacturers, who should be supplying such digital gauges with their turntables (especially the expensive ones). This one reads 1.68 grams of vertical tracking force.

Fig. 6.21. ACF = arm + cartridge force (weight), CWF = counterweight force (weight). The counterweight is moved until the balance is achieved, L*ACF = X*CWF. In Step 2, slide CW towards the cartridge.

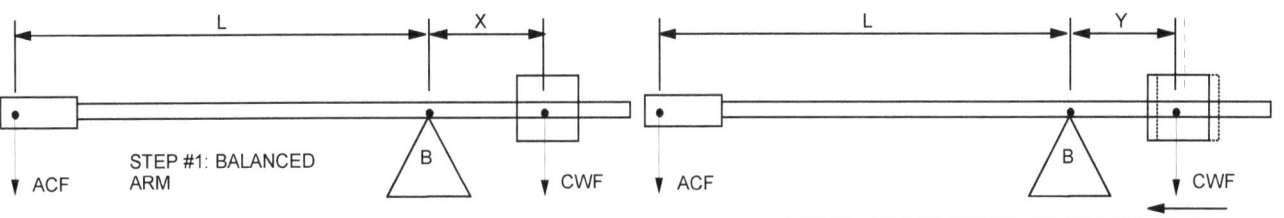

Adjusting the tracking force on Pro-Ject RPM 3 turntable is difficult. Once the single grub screw that holds the counterweight in place is loosened, it is impossible to smoothly slide the weight along the shaft. Even budget turntables from the 1970s had calibrated dials on their counterweights that were so easy to turn and adjust (no grub screws there); they were precisely machined pieces of engineering. This counterweight design is crude and unfriendly.

Step #4: Adjusting VTA - the vertical tracking angle

With a record on a platter but with your turntable turned off (not rotating), lower the tonearm onto the record. The tube of the tonearm should be parallel to the surface of the record.

If a spirit level shows an incline either way, loosen the two screws (1) in the tonearm base or collar (2) and slide the arm pillar assembly (3) up or down until the tube of the tonearm is parallel to the surface of the record (4). Re-tighten the screws without using excessive force, which could deform the arm pillar.

On some turntables, the arm pillar will drop as soon as you loosen its screws, so it pays to support the pillar using thin shims - thin rectangular objects such as business or credit cards. Assuming the pillar is too low, keep adding the shims until the arm is horizontal and then re-tighten the set screws.

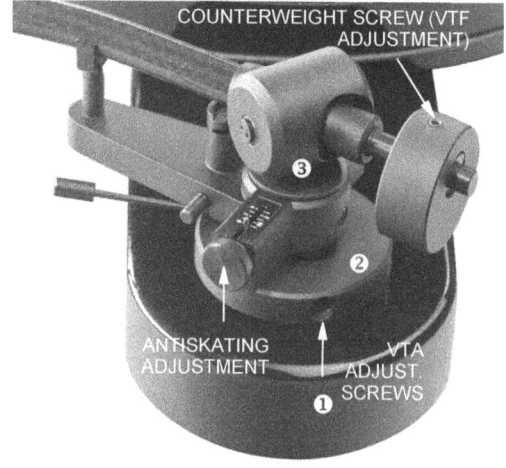

This method of VTA adjustment obviously assumes that once the tonearm is horizontal and the optimal VTF is dialed in, the manufacturer of the cartridge assembled it correctly so that the cantilever and the rubber or elastomer fulcrum it is pivoting on are providing the exact amount of compliance (or "elasticity") as specified, which would then automatically result in the cantilever assuming the correct tracking angle.

If a tonearm does not offer VTA adjustment, cartridge and platter mats changes may result in a suboptimal vertical tracking angle. Likewise, this would make platter upgrades a questionable proposition unless the replacement platter is of exactly the same height (or thickness)!

Fig. 6.22. The location and design of various adjustments on Pro-Ject RPM 3 Carbon turntable

Fig. 6.23. The principle behind the VTA adjustment

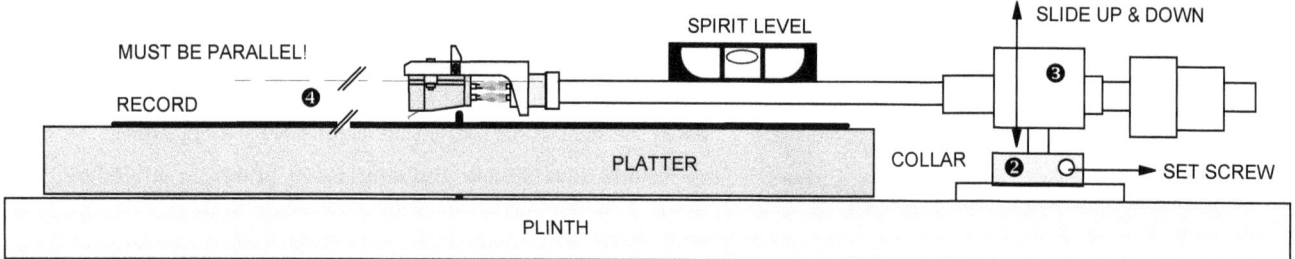

Step #5: The "antiskating" adjustment

The adjustment of the "anti-skating" force varies between turntables since various methods can be used to generate such force. On Pro-Ject RPM 3, it is done by rotating a knob and observing the indication on a dial, marked in grams of force and calibrated from 0 to 5g. The manufacturer recommends the set force to be equal to the "downforce" (VTF) set in step #3. Since our VTF was set at 1.68 g, we set the anti-skating force to 1.7g (more precise graduations aren't available on such a small dial).

Another method is to find an LP with a wide run-out (a segment with grooves but without any music) or a totally blank disc since even the "silent" grooves offer much more resistance to the stylus than the blank discs, but still less than when playing the heavily modulated grooves.

In that situation, the arm should slowly drift outwards (away from the spindle or the label). The fine-tuning should then be done by ear. There should be no mistracking, distortion, or loss of the right channel signal. Without anti-skate, the outer groove wall (the right channel, closer to the rim of the LP) may be lost due to the skating force acting towards the inner wall or the left channel.

Listen for a soundstage not quite centered (but make sure that is not due to other factors such as volume control potentiometers with large imbalance or suboptimal speaker placement, for instance).

Step #6: *Adjusting the azimuth (axial tilt)*

Just as with the open reel tape decks, the azimuth in turntables is the vertical alignment of the arm - headshell - cartridge - stylus assembly in relation to the record's surface, or "axial tilt" if you will. Viewing the cartridge "head-on," as illustrated below, the stylus should be perpendicular (at 90°) to the groove walls.

This ensures that the stylus traces the groove wall modulations correctly and does not favor or emphasize one groove wall over the other; in other words, it maximizes channel separation and minimizes the crosstalk between channels.

Even one or two degrees of azimuth error drastically reduces channel separation and increases crosstalk between channels, as illustrated. Notice that right-to-left and left-to-right channel crosstalk figures may not be identical, except at the zero degrees reference point X, in this case around -31 dB.

Since a small and thin stylus is hard to see with a naked eye, some audiophiles assume that the stylus is by definition (proper manufacturing and mounting in the factory) perpendicular to the cartridge's body, so they align the top surface of the cartridge, so it's parallel to the record surface.

That assumption is usually correct (unless the cartridge is faulty), but it pays to check with a magnifying glass that the stylus really is at the right angle to the record.

If not, most turntables have a small screw at the bearing end of the arm for this alignment. Release the screw (some are hexagonal, so you will need the right size Allen key), but don't remove it completely.

Gently rotate the arm tube in the direction opposite of the needle's tilt angle until the needle (stylus) is vertical in the groove (i.e., perpendicular to the record's surface) and re-tighten the screw.

In the drawing (right), the cartridge is tilted to the left, so you'd rotate the arm tube to the right (clockwise).

Make sure the stylus is not in the record groove during this adjustment, or irreparable damage may be caused to the cantilever suspension! You must lift the arm, make a small adjustment, lower it back onto a record, and if still not perfectly perpendicular, repeat the adjustment sequence until it is.

Adjusting azimuth with a naked eye is questionable, and even with a magnifying glass, it is open to interpretation and highly dependent on your skills.

Electronic solutions are available, for instance, a Fozgometer or "Azimuth range meter." Azimuth testers don't actually measure the azimuth angle directly (that would make them a very precise and thus hugely expensive class of optical instruments!) but measure crosstalk between the channels instead.

However, the discrepancy between crosstalk figures can be caused not just by an improperly set azimuth angle but also by an imperfect assembly of the cartridge.

A test record that features tracks for azimuth adjustment is used in conjunction with these electronic testers, the aim being to get identical readings from the left and right channel test signals, thus "matching" the signal level from both channels. These could be signals either directly from the turntable (the tonearm cables) or from the output of a phono stage.

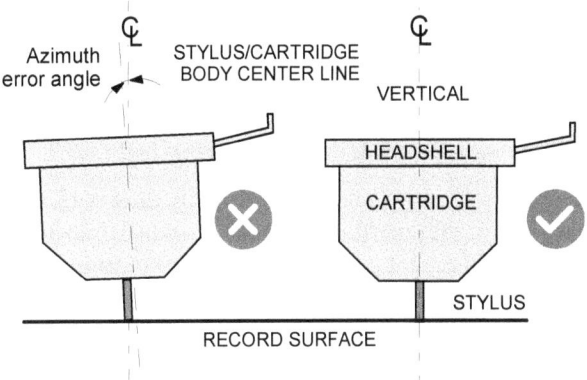

Fig. 6.24. The azimuth angle (in reference to the vertical center line CL) should be zero.

Fig. 6.25. How even one or two degrees of azimuth error drastically reduces channel separation and increases crosstalk (right-to-left and left-to-right channel)

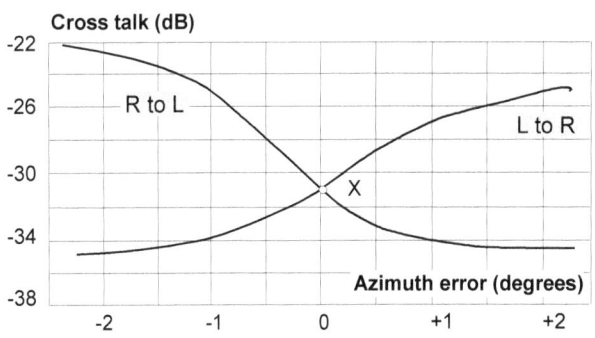

Fig. 6.26. The location of the azimuth adjustment screw on Pro-Ject RPM 3 turntable.

CARTRIDGE AND STYLI TYPES

Electromagnetic cartridges

When a coil (a spool with a number of windings of conductive wire) moves in a fixed & uniform magnetic field, a changing (alternating) voltage is induced at its ends, and, if a load (for instance, a resistor) is connected across such terminals (thus completing the circuit), an AC current will flow through the said coil and the electrical power will be generated and dissipated on the load.

Moving coil cartridges are just such transducers. The stylus has miniature coils attached to its cantilever, and as the groove modulations cause the stylus to track them, the coils move in unison, and an audio signal is generate.

The situation is reversed in moving magnet cartridges. The coils are stationary but two tiny magnets (attached to the stylus) move in their vicinity. There must be some relative movement or change between the magnetic field (magnets) and the coils. In principle, it does not matter which one moves and which one is stationary.

There is yet another way a vibrating stylus can generate electrical signals even if both the magnets and the coils are stationary within the body of the cartridge, and that principle is exploited in moving iron cartridges. The movement of the cantilever changes the distance between a moving piece of iron attached to it and the stationary magnets, thus changing or modulating the strength of their magnetic field, and such changes induce tiny electrical currents in the stationary coils.

Moving magnet (MM) versus moving coil (MC)

Magnets are heavy, and that is the only drawback of MM cartridges. Compared to the MC type, their higher moving mass and higher inertia reduce the response time and the ability to reproduce fast transients. That has been somewhat mitigated in recent years with the use of extremely powerful neodymium magnets, which reduce the overall weight attached to the cantilever.

The low number of turns means MC cartridges have a negligible inductance, which is significant in MM designs (500-800 mH), and they behave like current generators while MM engines are closer to voltage sources.

To keep the moving mass as low as possible, the coils in moving coil cartridges have a much smaller number of turns (compared to MM designs) and thus produce much lower signal amplitudes. Since their signal is 10-20 times lower, additional amplification stages or step-up (voltage step-up, that is, meaning current step-down) transformers must be used, and they introduce a whole gamut of new problems and challenges, noise, distortion, etc.

MC cartridges are generally much more expensive (some skeptics would say grossly overpriced for what they are) To add insult to injury, in contrast to the more user-friendly MM variety, the styli in MC cartridges are non-replaceable, a major drawback that significantly increases their cost of ownership and cost-per-hour played.

So, is the prize worth the price? Most audiophiles believe so, but I am not convinced. MC transducers generally (not always) sound more detailed and more transparent, but to me, also less "musical" or "euphonic," kinda like electrostatic versus electrodynamic speaker comparison. In many systems that I've heard, they sound too bright, with artificially "etched" outlines of notes and transients and a somewhat fatiguing character.

In addition, the extended frequency range of MC engines is meaningless since no such HF signals will be found in the grooves, thus "reproducing" or rather "introducing" only pops, clicks, dust, and vinyl imperfections.

Moving iron (MI)

All advantages of MC cartridges over their MM brethren also apply in the comparison of MI versus MC types. The moving mass is reduced even further, and with its lower inertia, even faster response times are achieved. Superior channel separation (lower crosstalk) figures are perhaps even more important. MC cartridges typically achieve only up to 25-30 dB, compared to 35-45 dB for the best MIs.

MI cartridges can be made with a higher degree of manufacturing consistency and, since both the coils and the magnets are fixed within the body, it is much easier to change their parameters and thus customize them to achieve specific loading and performance parameters.

Finally, the stylus replacement of MI cartridges is much cheaper than the "rebuilding" of the MC type since "rebuilding" usually means simply the manufacturer replacing the whole MC cartridge at a reduced price.

Optical cartridges - the idea whose time has finally come (after 40 years in the wilderness)

The optical cartridge idea isn't new; the first such cartridges were released more than 50 years ago by Toshiba and other Japanese manufacturers. However, they suffered from rapid aging and deterioration. Standard incandescent bulbs were used as light sources, and these produced heat, which aged the rubber in the cartridge and changed its compliance parameters.

Also, by then, the attention of hi-fi makers had already switched to digital reproduction, namely compact disk technology, and turntable-related developments fell by the wayside.

Modern optical cartridges use LEDs (Light Emitting Diodes), which don't produce heat, and highly sensitive photo-diode as detectors, matching the spectral sensitivity characteristics of the two. This results in very high voltage outputs in the order of 40 mV, 10 and 100 times higher than MM and MC cartridges, respectively.

Both MM and MC cartridges work on the magnetic induction principle, where counter-EMF (electromotive force, or "voltage") develops, which tries to resist any changes (kinda like a "choke" or inductor) in the magnetic field.

In optical cartridges, there is no magnetic induction and counter-electromagnetic force; the moving systems are much smaller and lighter, making the speed of the response the primary advantage of optical cartridge technology.

The phono stages for optical cartridges are completely different from MM or MC preamplifiers. Such phono stages must provide a supply voltage to the LED in the optical cartridge, but their amplification and filtering circuits are simpler. The MM and MC cartridges are velocity sensors; their output signals are proportional to speed, with the output increasing with rising speed (at higher frequencies).

In contrast, the output voltage of optical cartridges is proportional to the amplitude of the stylus vibrations (yes, optical cartridges still need a stylus as a sensing device), meaning their frequency response is flat throughout the audio range of frequencies. Thus, optical phono stages are simpler, needing less than 20 dB of gain, with less signal processing, another sonic advantage, since simpler circuits usually sound better.

Styli types

Conical styli start as rods machined into a cone shape with a rounded tip. However, due to their special (circular) cross-section, they are very different from the cutting styli used in recording lathes and cannot accurately track the grooves. This results in tracing distortion and the pinch effect, where the stylus is pushed upwards out of the groove. Conical styli are the cheapest to make, used only on budget cartridges.

Elliptical styli are machined further, with two polished sides. The oval cross-section has a larger contact area with the grooves, thus reducing record wear, reducing distortion, and eliminating the pinch effect.

The hyper-elliptical variety is ground even further and more precisely, so they are closer to "line contact" styli than to the ellipticals. These "fine line" styli have all four sides machined, thus making the tip even narrower so it can sink a bit deeper into grooves to lengthen the area of contact with the side walls.

The line contacts come under various names, "Shibata," after the guy who invented it in 1971 working at JVC in Japan, Fineline, Microline, Paratrace, Sas, and other trade names. While both the conicals and ellipticals come in nude and bonded varieties, the line contact type is almost exclusively of "nude" variety to minimize the mass.

"Nude" is cartridge-speak means that the diamond tips are mounted directly onto the cantilever, without the usual metal shank (a steel pin) as an interface, so the tip mass and inertia are reduced, resulting in faster response and better reproduction of fast transients.

CASE STUDY: Who is who in Ortofon 2M line of moving magnet cartridges

As Yogi Berra once famously said, "You can observe a lot by just watching," you can learn a lot by just reading equipment specs if you know what to look for, which specs really matter and which don't.

Let's use Ortophon's 2M lineup of MM cartridges to illustrate a few points. The Australian RRP (recommended retail prices) as of this writing are AU$189 for Red and AU$369 for Blue, while the Bronze and Black will set you back AU$659 and AU$1,099, respectively.

There's also 2M Silver, an OEM version of 2M Red, with an upgraded suspension, and coils wound with silver-plated copper wire (as used in 2M Bronze and Black).

Since the Red and Blue have identical coils and magnets configuration, and so do the Bronze and Black, and since the styli are interchangeable between the Red, Blue, and Silver, and again between Bronze and Black, that effectively means that there are two quality levels in the series, Red-Silver-Blue and Bronze-Black.

The primary difference is in the stylus shape, elliptical (Red and Silver), nude elliptical (Blue), nude fine line (Bronze), and nude Shibata (Black), and in the frequency range extension at the upper end, 22 kHz, 25 kHz, 29 kHz, and 31 kHz respectively.

Two different "engines" are used; the Red-Blue-Silver has 700 mH internal inductance and 1.3 kΩ DC resistance, versus 630 mH and 1.2 kΩ for Bronze-Black. Channel separation and channel balance figures are also slightly better for the Bronze-Black level but so marginal as to be inaudible (1 dB better and 0.5 dB lower respectively at 1 kHz). The recommended load resistance (47 kΩ) and capacitance (150-300 pF) are the same across the line.

UNDERSTANDING PHONO STAGES

Why are phono preamplifiers needed?

While the master tape is played back, its output signal goes through an "emphasis" filter which attenuates low frequencies (since they would require very wide and deep grooves in a record). During the reproduction process, a "de-emphasis" filter is required to boost the low frequencies, which must be a "mirror image" of the emphasis filter used during the mastering process.

This "RIAA" filter (named after the Recording Industry Association of America that specified its levels and frequencies), is thus a critical part of every phono preamplifier. A phono stage is thus a frequency-selective audio preamplifier since its amplitude response varies with frequency according to the RIAA de-emphasis curve.

Moving magnet (MM) cartridges provide a low-level voltage signal (1-5 mV), which is directly amplified by a phono stage before further amplification and volume control by a line-level preamplifier.

However, since moving coil (MC) cartridges produce very low voltages (10-20 times lower than the MM variety), they require an additional voltage amplification stage, usually a step-up transformer, which transforms high current-low voltage MC cartridge's signal to a higher voltage, lower current signal, which is then fed into a MM phono stage. Of course, "high" current is a relative term, meaning simply that the current through the primary winding of the transformers is higher than its secondary current.

Some phono stages include an additional amplification stage instead of a step-up transformer. While avoiding all the difficulties in designing a transformer of superior characteristics, that introduces a different set of problems, primarily the issue of signal-to-noise ratio, which can be compromised by a noisy amplification circuit.

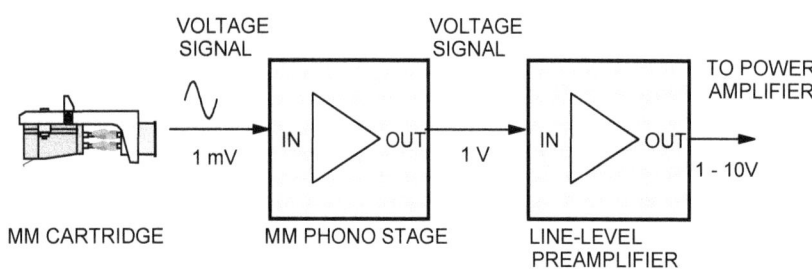

Fig. 6.27. Moving magnet cartridges provide a low-level voltage signal (1-5 mV), which is directly amplified by a phono stage before further amplification and volume control by a line-level preamplifier.

Fig. 6.28. Moving coil (MC) cartridges are current sources. A step-up transformer transforms a high current/low voltage input signal to a higher voltage, lower current signal for the MM phono stage.

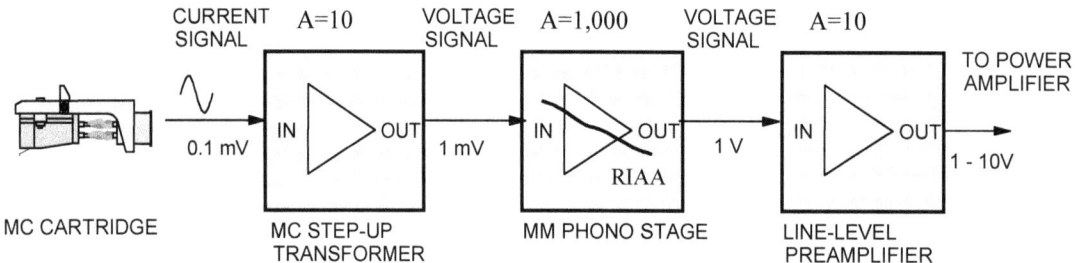

The RIAA de-emphasis curve

The RIAA de-emphasis curve essentially boosts bass frequencies from 500 Hz down to 50 Hz by 20dB and attenuates treble (high frequencies) by the same amount, 20dB, from 2,120 Hz to 21,200 Hz. The referent or 0dB level is exactly at the central frequency (f_C) of 1 kHz!

There are three original time constants corresponding to turnover frequencies f_3, f_4, and f_5: 3,180 μs (50 Hz), 318 μs (500 Hz), and 75μs (2,120 Hz). The newly introduced f_2 =20 Hz gives as the time constant of T_2 =7,950*10^{-6} s.

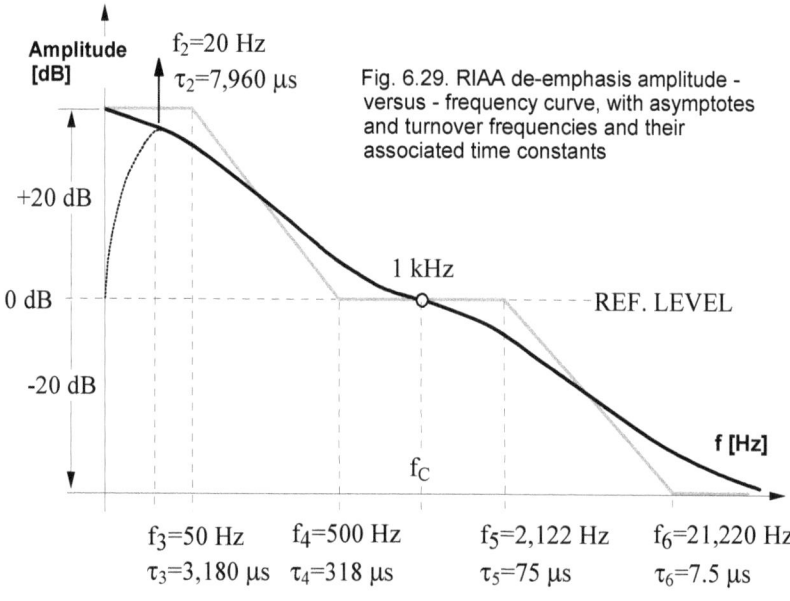

Fig. 6.29. RIAA de-emphasis amplitude - versus - frequency curve, with asymptotes and turnover frequencies and their associated time constants

UPGRADING & FINE-TUNING THE SOURCES

Some designers argue that frequencies over f_6=21,200 Hz should not be continued to be attenuated indefinitely, so they introduce another time constant at that frequency, with time constant τ_6= 7.5*10-6 s!

Another issue is the optional corner frequency f_3. Sometimes called the "New Orthophonic" standard, it is based on the recommendation of 1953 by NARTB, of 1955 by IEC No.98, and B.S. (British Standard) No. 128.

The original RIAA de-emphasis curve is shown in full-line, and the later (1976) low-frequency modification to the curve is in the dotted line. This modification attenuates (or rather doesn't allow for the amplification of) frequencies below 20 Hz so as not to amplify the turntable rumble and to avoid the tonearm resonance frequency range.

The Magnificent Seven: Design & build quality criteria for phono stages

Phono stages are more complex audio components than line stages (line preamplifiers) since they involve precise filters and must amplify much lower voltages, emphasizing the need for their superior noise immunity. The following factors play a crucial role in the sonics of a phono stage:

- Low hum and noise, resulting in high signal-to-noise ratio
- Accurate frequency response curve (correctly designed RIAA de-emphasis filter)
- Adequate gain (especially at frequency extremes)
- Low distortion (high linearity)
- Optimized input impedance (cartridge matching) and low output impedance
- Its performance must not be affected by the aging of tubes, the thermal drift of transistors, or the drift of passive component values

Phono stage design choices

When it comes to active components used, filtering types and topologies, audio circuit and power supply options and configurations, voltage regulation, interstage coupling, the use (or not) of negative feedback, and a myriad of other factors, a phono stage designer has so many choices, more than in any other audio component:

- Passive or active (in the negative feedback circuit) equalization, or a combination of both
- CR (capacitor-resistor), LR (inductor-resistor) or LCR (inductor-capacitor-resistor) RIAA equalization
- Discrete components (tubes, FETs, bipolar transistors) or ICs (integrated circuits)
- Bipolar, FET transistors or tubes (or a combination of two or all three)
- Mains or battery power supply
- Regulated or unregulated power supply
- Circuit topology: common cathode (or emitter or source), cascode, SRPP, µ-follower, differential amplifier
- Capacitive, direct, or transformer-coupling
- High impedance or low impedance output (cathode follower or output transformer)

A few common phono stage topologies

While a detailed technical analysis of phono stages is beyond the scope of this book, if you belong to a turntable connoisseur club, you should understand at least the basics of phono stages. Regardless of the technology used (tubes, transistors, or ICs), there are only a few ways RIAA filtering may be implemented, so let's illustrate them using simplified block diagrams.

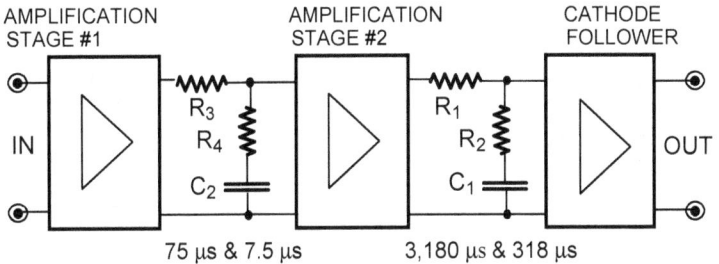

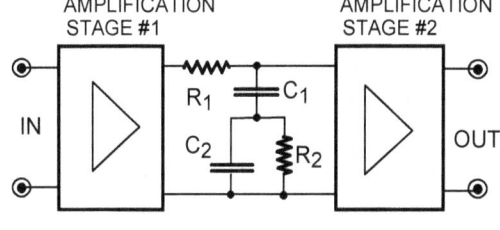

Fig. 6.30. The block diagram of a 3-stage phono preamplifier with RIAA filter split into two separate passive filters. The time constants of each filter are indicated.

Fig. 6.31. The block diagram of a 2-stage phono preamplifier with a single passive RIAA filter between them. Notice that the whole RIAA filter consists of only four passive components, Resistors R_1 and R_2 and film capacitors* C_1 and C_2!

Passive filtering is a simple filtering network between two amplification stages. RC filtering is illustrated, but chokes (inductors) can be used instead of capacitors, although that is a much less common approach. Since there are four turnover frequencies (and corresponding time constants) in the RIAA filtering curve, all four can be implemented in a single filter, or two separate filters can be used.

Active equalization consists of a single filter in the feedback loop, taken from the preamp's output (1) back to its input (2).

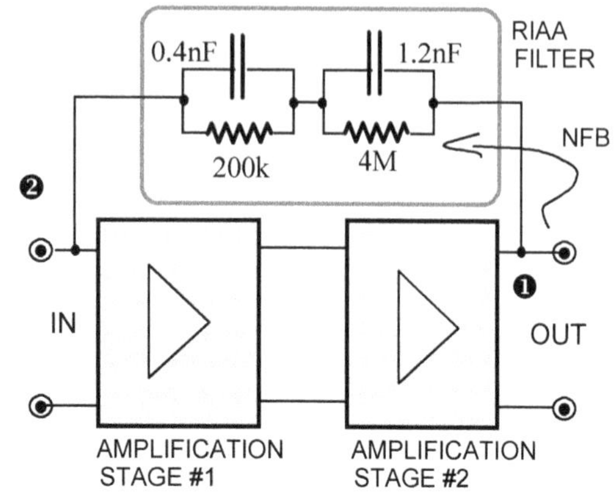

Fig. 6.32. The most common implementation of the RIAA filter (in both solid-state and tube versions) is in a negative feedback circuit. It is generally considered to be an inferior approach, both technically and sonically.

The anatomy of a high-gain MM and MC tube phono stage

This design uses a passive RIAA filtering (3), but instead of one input stage with a high gain ECC83 triode, two triode stages are needed, (4) and (5), since the gain of ECC88 is three times lower than ECC83. The total voltage gain of the first two stages is 16*18.75 = 300 times (49.4 dB). DC voltages are marked next to significant points.

After 20 dB of attenuation at 1 kHz of the passive RIAA filter, this leaves us with a midrange gain of 16x18.7x0.1x16 = 480 or 53.6 dB, which is quite high. For 0.5 mV input from an MC cartridge, the output should be around 0.24 V. If followed by a line stage preamp with a gain of 7 or higher, that should be plenty for typical MC cartridges. Simply add a 10-47 ohm resistor between the input and ground for proper loading of your low impedance MC cartridge (6), and you are in business. For MM cartridges, omit the 10-ohm input termination resistor.

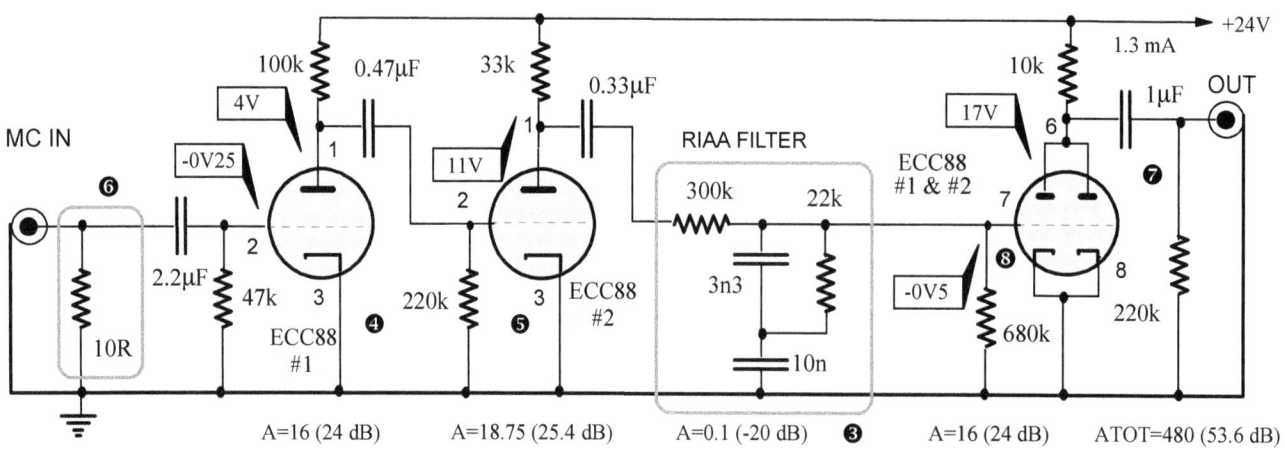

Fig. 6.33. Circuit diagram of the audio section (one channel only shown) of a high gain MM or low gain MC phono stage. Three class A triode amplification stages, single-stage passive RIAA filter and low voltage operation.

The input 2.2 µF capacitor, together with the 47 kΩ input resistance, forms a high pass filter and determines the low cutoff frequency (1.5 Hz in this case). The same applies to the coupling capacitor between the 1st and 2nd stages, 0.47 µF and 220 kΩ grid resistor, which results in f_L of 1.8 Hz. Finally, the 1 µF output capacitor with 220 kΩ loading resistor forms another high pass filter (7) with the lower -3 dB frequency f_L = 0.72 Hz.

The input resistance of the following preamp or power amp will be added in parallel to this resistance and will lower it unless there is a series capacitor at its input. Say, for instance, the input resistor of the following amplifier is 100 kΩ. That would lower the parallel resistance to 68.75 kΩ and raise the f_L to 2.3 Hz. So far, so good.

Even if a solid-state preamp with a typical 10 kΩ input impedance is used, the f_L would shoot up to 16.6 Hz, which is still below 20 Hz and thus fine.

Notice that there are no cathode resistors; all three stages use contact bias (also called grid-leak bias) where the small grid leakage current (0.73 microAmperes in stage #3) flowing through a large value resistance to ground (680 kΩ in stage #3) creates such a voltage drop, -0.5 V on the grid (pin 7) in stage three (8).

The beauty of this tried & tested design is that it works on the extra-low supply voltage of only 24 V_{DC}. For the battery aficionados, the power draw is only 1.3 mA per channel, so even a smallish battery would power this phono stage for many hundreds of hours without recharging.

Channel imbalance and frequency response characteristics of phono stages

The frequency response (amplification factor versus frequency) of an ideal phono stage should faithfully follow the RIAA equalization curve, so one criterion used to evaluate the fidelity of phono preamplifiers is to measure the deviation from that ideal or the "RIAA error." Obviously, the smaller such error, the better.

Among other factors such as the input and output impedances or channel separation (or its opposite, the crosstalk), the shape of the error curve (often published in audiophile magazines as part of their equipment reviews), gives us valuable insights and enable us to predict with higher certainty how such a phono stage will perform and sound as part of an audio system. This "sonic signature" (below) belongs to an expensive, currently produced vacuum tube phono stage. How should an audiophile analyze such a graph?

First, starting with the midrange and lower treble frequencies (1), between say 1 and 10,000 Hz (Hertz) or often written as 1k - 10k (k or "kilo" signifying a factor of 1,000x), we see a slight imbalance between the two channels, but only around 0.4 dB, which isn't serious enough and should not be audible.

What will be audible, though, is the significant rise (2) between 70 and 300 Hz (upper bass, lower midrange), its peak reaching +4 dB at around 120 Hz. A positive error means a boost of those frequencies.

We see a sharp drop in the lower bass frequency range; the error becomes negative. Those frequencies will be attenuated or, more precisely, will not be boosted by the phono stage as much as they should have been.

From the +3 dB peak (2), the curve drops to almost -4 dB at 50 Hz, an attenuation of 7 dB, which is significant. Apart from the artificial boost between 70 and 300 Hz, this phono stage may sound too lean and lacking in bass.

The channel imbalance increases in the bass region; for instance, at 50 Hz, the difference between the two channels is 3 dB, which will be noticeable by a critical ear (3).

Finally, look at the rise of the RIAA error at high frequencies (4). It happens in the ultrasonic region (above 20 kHz) and should not be directly audible. Notice the word "directly"; it may still be indirectly perceived as glare or harshness of the highest audible notes.

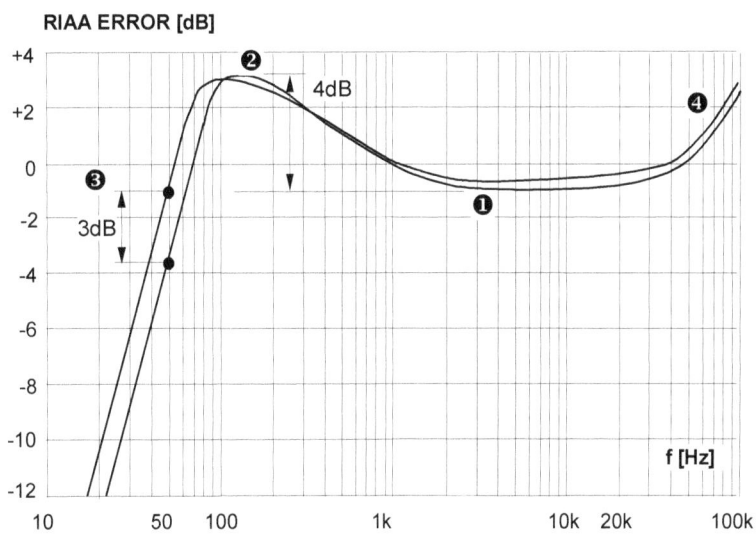

Fig. 6.34. The RIAA error for the two channels of an expensive currently produced vacuum tube phono stage

Tweaking the frequency response of a phono stage

The graph shows the measured values of our Accurus MM phono stage (one channel only). Notice the positive error at low frequencies, peaking around 50-60 Hz (5). That was done deliberately to provide a slight boost to the low bass.

In the midrange and lower treble, up to 14 kHz, the error is around -0.15 dB (6), which is negligible. It then increases in a linear fashion, reaching -1 dB at 20 kHz (7).

Slightly attenuating the treble, as in this case, produces a smoother top end and helps in reducing most of the remaining traces of harshness from bright-sounding cartridges.

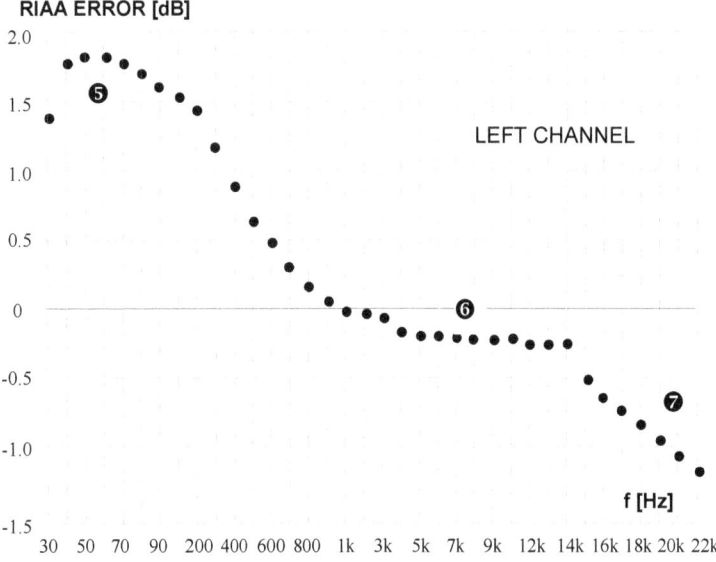

Fig. 6.35. An example of the equalizing error versus frequency of a moving magnet (MM) tube phono stage. The dots are the measured values.

THE CARTRIDGE-CABLE-PHONO STAGE INTERFACE

Moving magnet and moving iron cartridge interfacing

A moving magnet cartridge can be modeled by a signal source in series with its internal resistance and inductance. The shielded cable has its distributed resistance R_C and capacitance C_C, and the input stage of the phono preamp is represented by the parallel RC combination of its input resistance R_{IN} and capacitance C_{IN}. R_G (pickup's resistance) and cable's resistance can be lumped together into one equivalent resistance R, as can the cable's lumped capacitance C_C and the preamps input capacitance C_{IN}.

This constitutes a series LRC resonant circuit, which is a 2nd order low pass filter, whose response to a square wave excitation depends on the three parameters, R, L, and C, and the frequency of the signal.

The square wave response of a phono cartridge

As with power amplifiers, the response of the cartridge - cable - phono stage chain to a square wave test signal (measured at the output of the phono stage) is the best indicator of its character.

The underdamped, oscillatory response with high initial overshot and prolonged "ringing" is easy to detect by ear; it sounds harsh, the voices have a nasal, shrill character. At the other extreme, the overdamped response makes the system sound slow, lacking dynamics and top-end sparkle. The musical transients are severely curtailed or not reproduced at all.

Your aim is to optimize or correct the square wave response, from the underdamped to the critically damped curve, or, if preferred, to mildly underdamped response.

This optimization can be performed on a test bench, using a square wave source (function generator) and observing the waveform on an oscilloscope, or, as most audiophiles do, simply by listening and substituting various values for R_C and C_C (by changing interconnect cables) and/or R_{IN} and C_{IN}, by changing the input terminating resistors and capacitors of the phono stage.

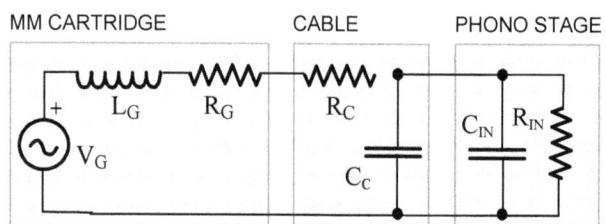

Fig. 6.36. The simplified circuit model of the MM cartridge - cable - phono stage chain

Fig. 6.37. Underdamped, critically damped, and overdamped response of a cartridge - transformer - phono stage chain

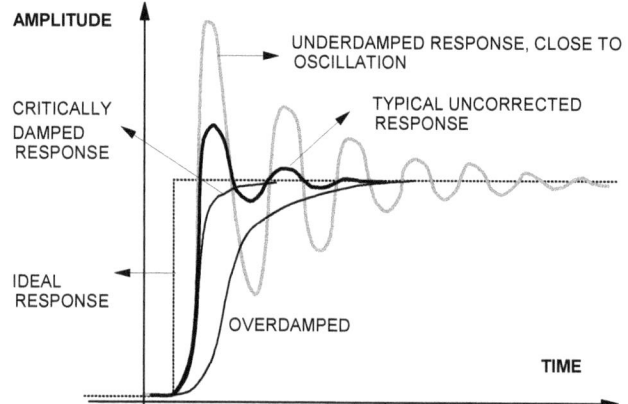

The interconnects between your turntable and the phono stage

The capacitance the cartridge "sees" is not just the input capacitance of the phono stage, but the total capacitance between the cartridge and the phono stage, which comprises of the capacitance of the tonearm wiring, the input capacitance of the phono stage, and the capacitance of the interconnect cable between them.

For instance, the Ortofon 2M Silver cartridge has the following specs: DC Resistance: 1kΩ, inductance: 550 mH, recommended load resistance: 47 kΩ, recommended load capacitance: 150 - 300 pF. However, the capacitance of the RCA cable (no model, manufacturer, or descriptor of any kind) supplied with the Pro-Ject RPM 3 Carbon turntable was measured as 419 pF (at 1 kHz test frequency) or 335 pF/m, an incredibly high figure! If you perform such a measurement, of course, make sure the other end is unplugged from the turntable. In other words, the cable must be unplugged on both sides.

As just the cable's capacitance (419 pF) had already exceeded the recommended maximum for that cartridge (300 pF), no wonder the turntable initially sounded dull, lacking dynamics and "sparkle." After ditching that shocker of a cable and replacing it with our DIY solid core twisted interconnects (page 100), whose capacitance was only around 39 pF/m, everything came to life.

Another example of why one should never blindly trust manufacturers and audio dealers. "Oh, since this cable came with my $1,000 turntable, then it must be a good match; the manufacturer must have selected it for a good reason!" Noooo! Most manufacturers select such "acessories" because they are cheap, not because they sound good!

In the March 2021 issue of the *Stereophile* magazine, Alex Halberstadt describes his experience with the interconnect cable supplied with Pro-ject Debut Carbon EVO turntable sent to him for a review.

After experiencing a "distractingly loud hum" using the supplied cable, he replaced it with another cable, and all was well again. "... I suspect that I received a faulty cable, and I left the good enough alone," he wrote. We cannot generalize based on two instances only, but the cables supplied with our two turntables may point to a wider issue with their cable suppliers, something Pro-Ject would be wise to look into.

Before you buy or use any audio cable, but especially the interconnects between your turntable and a phono stage, measure their inductance and capacitance. Borrow or buy a digital LCR meter. It is a very useful test instrument for any audiophile to have, and the handheld ones only cost a couple of hundred bucks new and even less if used.

How to determine the optimal input capacitance of a phono stage

The shunt capacitance works with the source inductance and resistance (cartridge+cable) to form a low pass filter. Too much capacitance will shunt higher frequencies (treble) away from the phono stage's input, resulting in the loss of sparkle, reduced transparency, and microdynamics, and a dull muffled sound.

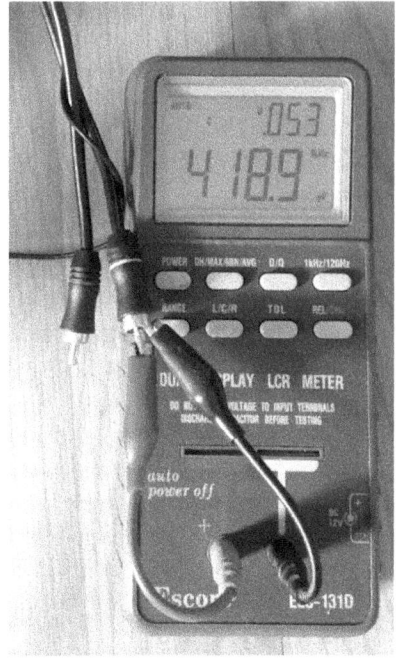

Fig. 6.38. The capacitance of the 125 cm long interconnect cable supplied with Pro-Ject RPM 3 Carbon turntable was measured as 419 pF (at 1kHz test frequency) or 335 pF/m, an incredibly high figure!

A very low total capacitance, on the other hand, may result in a bright and edgy sound, with oscillations of the kind illustrated by the underdamped response to the square wave signal on the previous page.

There are two ways to determine the optimal input capacitance of a phono stage. You can use a square wave test signal from a function generator and observe the system's response on an oscilloscope; or, since most audiophiles don't have the necessary equipment or knowledge to perform such a test, the alternative is to do the fine-tuning by ear.

Use various interconnects between the turntable and the phone stage (their capacitance will vary a lot), and decide which results in the best treble (smoothest, best balanced with the bass and midrange).

Alternatively, get a few film capacitors of different values, say 22, 33, and 47 pF, and keep adding them in parallel to the input of your phono stage. Start with 22 pF, and if the sound is still too bright or harsh, change to 47 pF. Initially, use crocodile clips to clip them to the circuit, and when you find one that results in the most optimal sound, solder it permanently to the circuit.

The drawback of this approach is obvious - you need to open up your phono stage and mess around with its circuit, which may void your warranty.

For that reason, better phono stages have a selector switch or an array of DIL (dual-in-line) switches, so various levels of input capacitance can be selected by the user. Such a feature is one of the most useful, and if you are in a dilemma about which of the two phono stages to buy, all other things being equal or similar, always buy the one with user-selectable input capacitance.

Phono stages with user-selectable parameters

Some phono stages give the user the flexibility of selecting a few values of input resistance and capacitance. Some do it internally, using tiny DIP (dual in-line package) switches; others make such controls accessible from the outside, either using mechanical switches or electronic selection.

For instance, Pro-Ject Phono Box DS2 is a solid-state phono stage whose two cartridge loading parameters (the input capacitance and the input loading resistance) and voltage gain can be selected on the front panel.

The three capacitance values (100, 220, and 320 pF) are only relevant for MM cartridges; MC cartridges are not sensitive to input capacitance. The lower gain values of 40, 45, and 50 dB are used for MM cartridges, while the MC variety requires much higher gains, 60 or 65 dB.

The 47 kohms loading is standard for MM cartridges, the other four (low resistance) options (10, 50, 100, and 1,000 ohms) are obviously aimed at matching MC cartridges' low internal impedance.

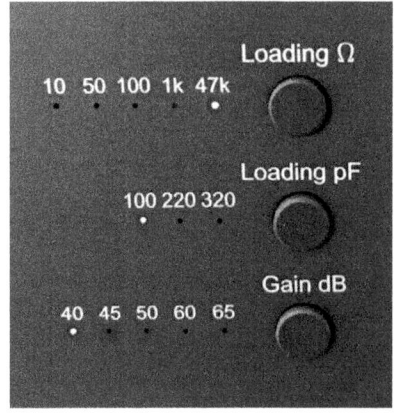

Fig. 6.39. The loading and gain choices of Pro-Ject Phono Box DS2 phono preamplifier

Fine-tuning the capacitive loading by ear

Adding discrete capacitors in parallel with your MM phono stage input terminals isn't difficult, but it only gives you a few jumps in the value of capacitance. This alternative would enable you to continuously and most precisely adjust the value of the capacitive loading on your MM cartridge.

Small size variable air trimmer capacitors are cheap and widely available, including surplus electronics sales. They have a relatively wide range of capacitance (50 - 200 pF, 30 - 150 pF, etc.), which is achieved by rotating a shaft (like a trimmer potentiometer, usually with a screwdriver, but you could install a knob). This rotates a bank of parallel metal plates, changing their overlap, and since the capacitance is proportional to that area, the total capacitance.

You could drill a hole at the back of your phono stage and install the trimmer capacitor there since you don't need to adjust it on a regular basis, only when you change cartridges or interconnect cables.

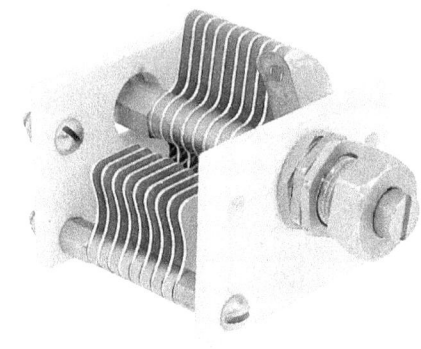

Fig. 6.39. Small size variable air trimmer capacitors are cheap and widely available, including surplus electronics sales.

This particular one has a range of 10-100pF and is rated at 525 Volts.

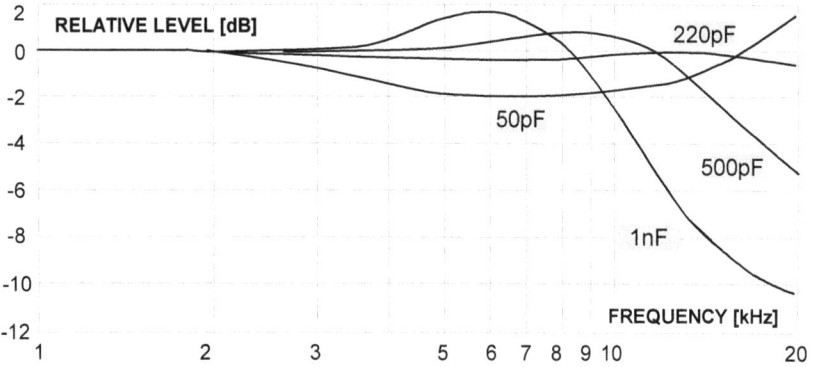

Fig. 6.40. Typical frequency response curves of a MM cartridge for various values of the loading capacitance.

The higher the impedance of the cartridge, the lower the peaks and dips. Lower impedance MM (moving magnet) and MI (moving iron) cartridges are more sensitive to improper capacitive loading.

How to foretell cartridge's sound by its square wave response

Since a square wave contains a large number of odd harmonics (1st, 3rd, 5th, 7th, 9th, etc.), and since the rising and falling times of its edges are very short (almost vertical leading and trailing edges), a square wave response is one of the most useful measurements on audio components, especially turntable cartridges, CD players, preamplifiers and power amplifiers.

Compare how these two moving coil cartridges reproduce the same 1 kHz square wave signal. Cartridge A has a very high initial overshot (1) and an oscillatory response whose amplitude drops exponentially (2) towards the horizontal top of the waveform but never quite settles down.

Cartridge B's initial overshot is practically nonexistent, but its response also oscillates around the horizontal tops and bottoms of the waveform. Notice a drop just before the falling edge of the signal (3), which is somewhat unusual; the response of cartridge A is certainly more common.

These two cartridges will sound very different. Usually, high overshoots and less dampened oscillations, as in the case of cartridge A, result in a crispier but often harsher and more strident sound.

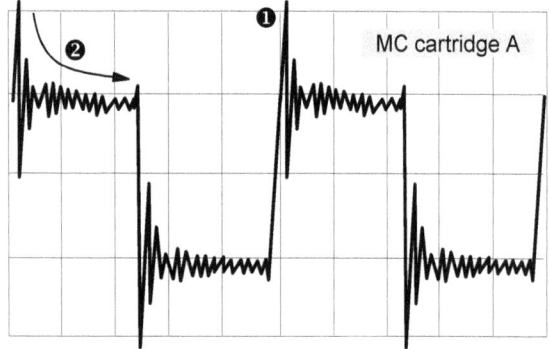 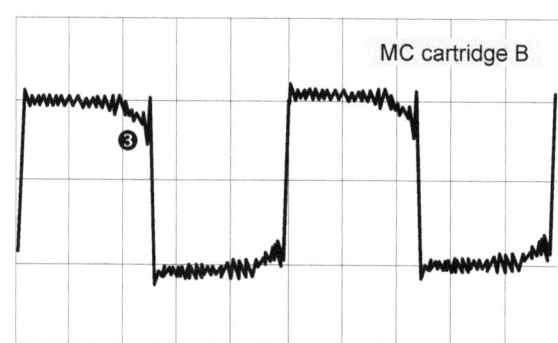

Fig. 6.41. Which of these two MC cartridges do you think will sound better?

UPGRADING & FINE-TUNING THE SOURCES

Checking how closely matched your cartridge's two channels are

Once you know the DC resistance and inductance of your cartridge, as specified by its manufacturer, assuming a stereo cartridge, of course, you can check how closely matched the cartridge's two systems are (left and right).

A word of warning: never measure a cartridge's DC resistance using an analog multimeter! Some push such high current through the cartridge's coil (which is wound with very fine wires) that it may fry the coil. Even if the coil does not burn out, such a high current may magnetize the cartridge and change its properties. Digital multimeters and LCR meters are generally safe to use, although a few also use higher than usual test voltages.

Differences in DC resistance cause an imbalance in the two channels' output levels, while coils of unmatched inductance will have varying frequency response curves, affecting the tonal balance between the channels. Some cartridge manufacturers grade specimens of the same cartridge design in terms of how matched the two mechanisms are and even name them differently, with the better balanced or matched ones attracting a premium price in comparison to those of inferior balance and other parameters.

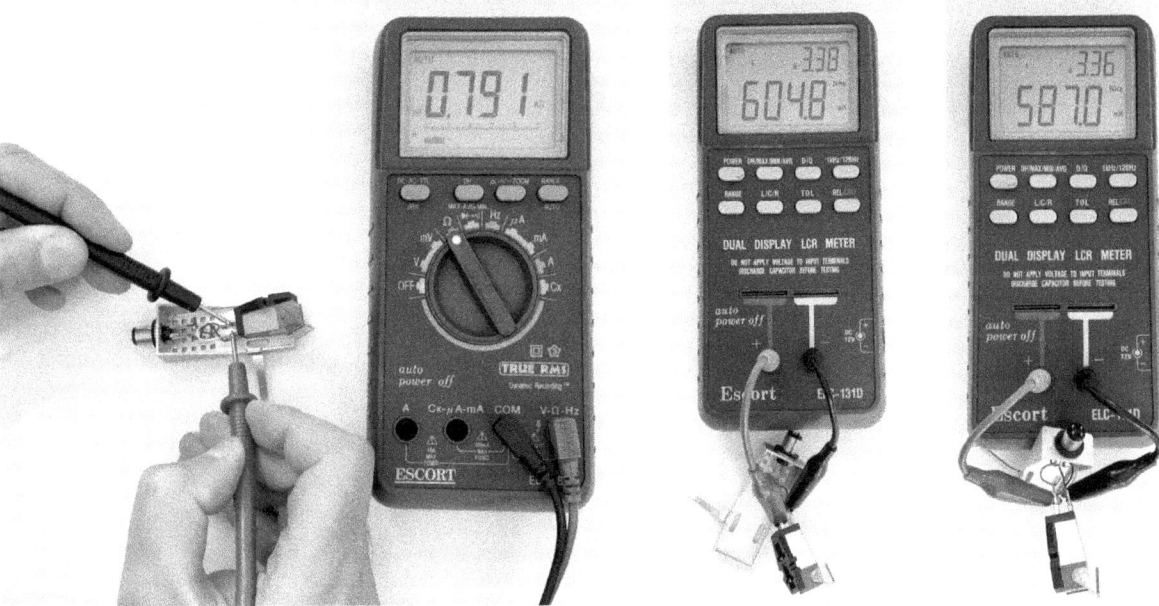

Fig. 6.42. Measuring Ortofon MM cartridge with a multimeter on "Ohms." The DC resistance of the two systems were 791 and 786 ohms, quite well-matched channels. The inductance at 1 kHz was 604.8 mH and 587 mH, around 3% relative error.

MOVING COIL STEP-UP TRANSFORMERS

Moving coil cartridge interfacing

Moving coil cartridges have very low inductance, so the equivalent circuit is simpler than the model for MM cartridges, just an AC signal source in series with cartridge's internal resistance R_G ("G" for generator), which is typically in the order of 10-50 Ω.

To bring the low voltage signal from an MC cartridge up to the level required by MM phono stages, voltage step-up transformers (SUT) are used. The most important parameter (for us audiophiles) of a SUT is its voltage ratio (equal to its turns ratio), which is usually expressed as 1 : N, meaning that for each primary winding turn, the secondary winding has "N" turns. In reality, the primary winding has N_1 number of turns, and the secondary N_2.

As a result of MCC's very low inductance, the input capacitance of a phono stage is irrelevant since their resonant frequency is in the MHz (megahertz) region, in the spectrum of radio waves, way above the audible range. This is why the diagram here has only R_{IN} as a relevant factor.

The turns ratio TR is equal to the voltage ratio VR; a transformer with a turns ratio of 1:10 will have a voltage a step-up ratio of 10. For 1 mV signal at its input (from the MC cartridge) at terminals X & Z, its output voltage at terminals A & B will be 10 mV.

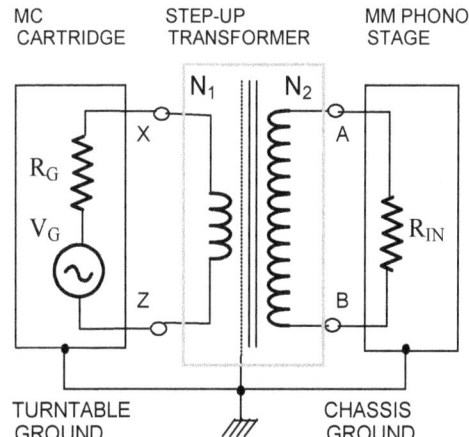

Fig. 6.43. A simplified circuit diagram (model) of an MC cartridge connected to a phono stage via a step-up transformer (SUT).

However, a transformer doesn't just "transform" voltage and current, but impedance as well (since the impedance is a product of voltage and current). The impedance the cartridge "sees," reflected by SUT from its secondary to its primary, is N^2 times less than the actual resistance or impedance connected to the secondary.

An SUT with 1:10 voltage ratio or $VR^2 = 100$ impedance ratio, will reflect 47,000 ohms from its secondary as $47,000/100 = 470$ ohms. Likewise, assuming a say 10 Ω resistance of a certain MC cartridge, the SUT will reflect that impedance as 100 times higher to the secondary side, or $10 \times 100 = 1,000$ ohms.

Commercial example: Ortofon ST-80 moving coil transformer

Different MC transformer makers specify them in different ways, taking considerable liberty, making the specs inconsistent and often confusing. It's time to debunk them!

Ortofon ST-80 MC transformer is specified as "Gain 27 dB at 1 kHz" and also "Input impedance: 2 - 6 Ω, output impedance: 47 kΩ ". On other websites, its recommended cartridge impedance is specified as less than 10 ohms. How can the output impedance be a fixed 47 kΩ if the input impedance varies from 2 to 6 Ω? Of course, it can not. What they are actually saying is that the resistance of the cartridge should be in the 2 - 6 Ω range and that the transformer should be terminated by 47 kΩ (the input resistance R_{IN} of a phono stage).

What is its voltage or "step-up" ratio VR? Ortofon did not specify it, but, from the specified 27 dB "gain" *(voltage gain)* we know $dB = 20 \log VR$, so we can calculate $\log VR = 27/20 = 1.35$ and $VR = 10^{1.35} = 22.39$!

So, the voltage step-up ratio is 1 : 22.4, meaning the 47 kΩ secondary load will be reflected to the primary as $47k/VR^2 = 47,000/22.4^2 = 47,000/501.8 = 93.7$ ohms.

Impedance matching the cartridge with the step-up transformer

Apart from the mismatch between the power amplifier and loudspeakers, the mismatch between the impedance of a phono cartridge and the phono stage has the potential for the largest negative impact on the whole audio chain.

Since a picture is truly worth a thousand words, consider two MC cartridges connected to the same step-up transformer. Using the 30 Ω cartridge as a referent 0 dB level in the midband, the -3 dB point ("cutoff frequency") of that cartridge is at around 14 Hz (1).

The output of the 100 Ω cartridge, due to its higher internal impedance, is 1 dB lower, which isn't a significant issue.

However, instead of being flat down to almost 50 Hz as is the case with the 30-ohm cartridge, the output of the higher impedance (100 ohms) cartridge starts drooping at around 100 Hz.

Its -3 dB limit (actually -4 dB in comparison with the first cartridge, since its reference level is 1 dB lower) is at a relatively high frequency of 34 Hz. As a consequence, its reproduction of the lower bass frequencies will be affected.

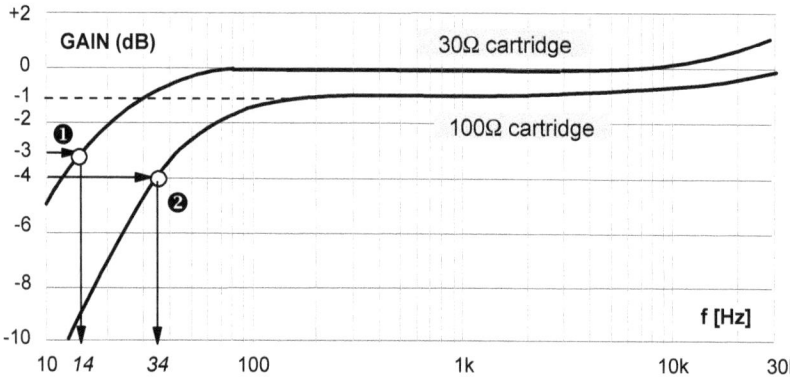

Fig. 6.44. Frequency response curves of a 30 ohm and 100 ohm MC cartridges connected to the same step-up transformer.

How to determine the required terminating resistance for MC cartridges

The problem with a step-up transformer (SUT) is that its voltage or turns ratio automatically determines its impedance reflection ratio. You cannot independently change both! If the voltage step-up ratio is 1 : N, the impedance ratio IR is 1 : N^2!

So, a 1:10 SUT, which steps the voltage up ten times, reflects the secondary impedance back to the primary side as $10^2 = 100$ times lower! So, a standard 47,000 ohms input resistance of the MM phono stage will be reflected as $47,000/100 = 470$ ohms. That is the load that the MC cartridge will "see." But what if the MC cartridge manufacturer specifies a different optimal load for its cartridge?

The required primary side resistance (the resistive load on the cartridge) can be precisely manipulated by adding a terminating resistor R_X at SUT's secondary side, in parallel with the input of the MM phono stage. I call it R_X because its exact resistance is the unknown value that we have to calculate here.

The secondary resistance R_S is now $R_S = R_X \| R_{IN}$ ($\|$ means "in parallel") or $R_S = (R_X \ast R_{IN})/(R_X + R_{IN})$.

UPGRADING & FINE-TUNING THE SOURCES

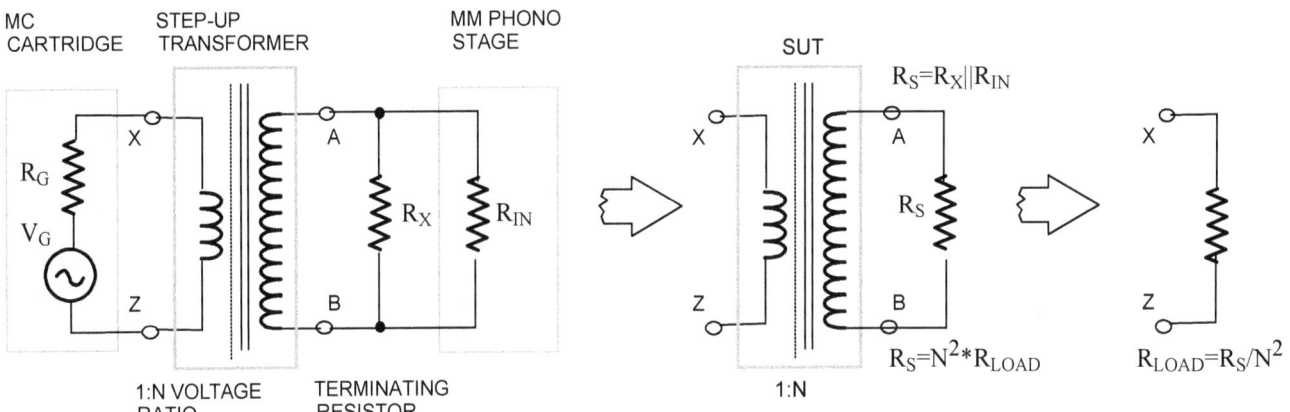

Fig. 6.45. The total secondary resistance is the input resistance of the phono stage R_{IN} in parallel with the terminating resistor R_X, whose value is unknown and has to be calculated. The SUT reflects this resistance to the primary side as $R_{LOAD}=R_S/N^2$!

We know that the required secondary resistance is $R_S = R_{LOAD}*N$, where R_{LOAD} is specified or recommended by the manufacturer and N is the voltage ratio of the step-up transformer (SUT). So, we have two equations with R_S on one side, meaning that the other sides must be equal, too: $R_{LOAD}*N = (R_X*R_{IN})/(R_X+R_{IN})$, and from this equation, we have to express R_X. Here's the 3-step procedure:

HOW TO CALCULATE THE REQUIRED TERMINATING RESISTANCE FOR MOVING COIL CARTRIDGES

1: Determine the required (recommended) load resistance (specified by cartridge manufacturer): R_{LOAD}
2: Calculate the required secondary resistance $R_S=N^2*R_{LOAD}$
3: Calculate the value of the terminating resistor to be used: $R_X = \dfrac{R_S*R_{IN}}{R_{IN} - R_S}$

As an example, let's say you have a Lyra Atlas Lambda SL cartridge, whose recommended load via step-up transformer is 5 to 15 ohms. The manufacturer says "step-up transformer's output must be connected to 10kohm ~ 47 kohm MM-level RIAA input, preferably via short, low-capacitance cable", which makes us none wiser. You'd imagine after forking out a small fortune for it (AU$16,900.- in 2021, a cost of a brand new small car), they'd tell you *exactly* what to do? Anyway, let's do it here together.

Let's say that our SUT has a voltage ratio of 1:15 (meaning our N is 15), and we choose a load of 10 ohms, in the middle of the 5 to 15 ohms recommended range. The input resistance of our MM phono stage is the standard 47,000 ohms (47 kΩ). First, we calculate the required secondary resistance $R_S=N^2*R_{LOAD} = 15^2*10+225*10 = 2,250$ ohms or 2.25 kΩ. Now we can calculate R_X as

$$R_X = \dfrac{R_S*R_{IN}}{R_{IN} - R_S} = \dfrac{2.25*47}{47 - 2.25} = 2.363 \text{ kΩ or } 2,363 \text{ ohms}$$

Resistor color chart and standard values

The issue now is that you cannot just walk into an electronic parts store and ask for a resistor with 2,363 ohms resistance; you'd make their day and be laughed out of the store.

Resistors come in certain standard values, which follow geometric progressions, such as E12, E24, and E48 (10%, 5%, and 2% tolerance, respectively). The values between 1 and 10 are shown below. All other values are obtained by multiplying those with 10, 100, 1000, and so on.

E12 series (+/- 10% tolerance) has the values in multiples of 1.00, 1.20, 1.50, 1.80, 2.20, 2.70, 3.30, 3.90, 4.70, 5.60, 6.80, and 8.20. We need 2.36; the closest standard value is 2.2, meaning we would buy a 2,200 ohms (abbreviated as 2k2 or 2.2 kΩ) resistor, whose color bands would be red (2), red (2), red (multiplier 100), and silver (+/- 10% tolerance). Alternatively, a 2.7 kΩ resistor would also be OK, red-violet-red-silver.

Color	Band 1 1st digit	Band 2 2nd digit	Band 3 Multiplier	Band 4 Tolerance
Black	0	0	1	
Brown	1	1	10	+/- 1%
Red	2	2	100	+/- 2%
Orange	3	3	1000	
Yellow	4	4	10,000	
Green	5	5	100k	
Blue	6	6	1M	
Violet	7	7	10M	
Gray	8	8	100M	
White	9	9	1G	
Gold			0.1	+/- 5%
Silver			0.01	+/-10%
None				+/- 20%

Fig. 6.46. Resistor color chart and standard values

E24 series of resistors is the 5% tolerance progression (more precise due to more choices), in the following values: 1.00, 1.10, 1.20, 1.30, 1.50, 1.60, 1.80, 2.00, 2.20, 2.40, 2.70, 3.00, 3.30, 3.60, 3.90, 4.30, 4.70, 5.10, 5.60, 6.20, 6.80, 7.50, 8.20, and 9.10.

The 2.40 is perfect for our needs, so we would need a 2,400-ohm resistor, whose color bands would be red (2), yellow (4), red (multiplier 100), and gold (+/- 5% tolerance).

DIY MC step-up transformer

Say you want to get a superior MM phono stage but haven't got much money left for the purchase of a quality moving coil (MC) step-up transformer. These can cost $3,000 or even (much) more. There are also well-priced vintage SUTs such as Denon AU-320, Ortofon ST-80, and Ortofon MCA-76, which can be found for $300-400 used.

Many types of professional audio gear used (and some still do) step-up transformers, not just for MC cartridges but also for microphones, and such individual transformers can be salvaged or bought very cheaply online. All you need to do is install them in a suitable enclosure, solder a few wires to a pair of input and output RCA sockets, and you are laughing all the way to the bank. This DIY project can literally save you hundreds if not thousands of your hard-earned dollars!

Before we have a look at some DIY SUT examples (pictured below), let's analyze a few examples of step-up transformers (pictured on the right). This could give you some valuable clues as to what to look for before you buy step-up transformers for your own project, should you decide to go the DIY route.

A pair of vintage Russian microphone transformers, made in the Soviet Union in the early 1980s, with the selectable step-up ratios of 1:2, 1:10, and 1:20 are enclosed in a magnetic shield.

Fig. 6.47. The four step-up transformers tested. Notice the apparent relative size difference. The rectangular-shaped NO233BK is not shielded at all, so it is most likely the same size as the Chinese Raphaelite and Russian transformers, which are shielded in large cans.

Raphaelite PM30Ω :7k is of current production, made in China, with 30 Ω primary and 7 kΩ secondary impedance.

However, once we got them, their actual weight was only 207 g each, not 1 kg as listed. Plus, with 7 kΩ and 30 Ω, the impedance ratio IR would be 7,000/30 = 233, and the voltage ratio is thus 15.3, as was indeed measured, not 13.5 as specified.

The Sennheiser and NO233BK were made in Western Europe in the 1960s for professional studio and cinema equipment. The Sennheisers are encased in a mumetal housing; the NO233BK is in a plastic case and thus not shielded at all.

The primary-to-secondary parasitic capacitance of the Russian transformer was roughly half of the Raphaelite ones and their leakage inductance was 9 times lower. Newer = better? I don't think so.

The primary inductance at 120 Hz (important for bass reproduction) of the Russian transformers was 15 times higher!

Fig. 6.48. Three DIY SUT projects. Sennheiser microphone transformers (1) and the rectangular-shaped unshielded NO233BK (2), salvaged from old studio gear, were mounted in extruded aluminium enclosures sold online. The vintage made-in-USSR (Soviet Union) shielded transformers (3) were installed in an old printer switch enclosure (6).

UPGRADING & FINE-TUNING THE SOURCES

Despite all that, which would predict that Russian SUT will have a much wider frequency range than the Chinese one, that was not the case; the results were very close.

NO233BK had by far the highest primary inductance of 35 H at 120 Hz, yet it had the worst f_L (40 Hz). It had the lowest parasitic primary-secondary capacitance, and as a result, had the highest f_U of all, by far, a whopping 90 kHz! Subjectively speaking, the Russian transformers had the best sonics, were not necessarily most transparent or precise, but were undoubtedly most musical and emotionally engaging.

The bottom two SUTs (photo on the previous page) use identical enclosures made from an extruded aluminium profile, the left one with Sennheiser (1) and the right one with NO233BK transformers (2), which are simply glued to the bottom. Both transformers had dual secondaries for two different voltage ratios. With Sennheiser, we used only one ratio, but with NO233BK different ratios can be selected using a side-mounted sliding switch (5).

Since the Russian transformers were bigger (3), we had to use a larger steel enclosure from an old printer switch. The two pairs of RCA sockets are on the same side, at the back of the unit, one above the other (4).

SOUND IMPROVEMENT HACKS FOR TURNTABLE-BASED SYSTEMS

Reverse one channel's connections on a turntable cartridge

This quick and easy trick can easily be done in turntable-based systems since it only requires unplugging two wires (one channel) on a cartridge and swapping them around. So, unplug white and blue wires (left channel's plus and minus) and swap them around, plug the blue wire to the top pin (L+) and the white wire to the bottom pin (L-). Alternatively, as illustrated below, swap around the green and red wires.

That reverses the polarity of one channel's signal, so it will propagate with the opposite phase through the rest of the system (phono stage, preamp, power amp). Of course, you must also swap around the speaker wires at the back of your power amp or at the back of the speaker on that channel to bring it back in phase with the other channel.

This modification can also be done in CD-player or digital streamer-based systems, but it requires opening one end of the interconnect cable, unsoldering the center connector and the screen and swapping them around there), or the same minor soldering-iron surgery inside your source (one channel's output on your DAC or music streamer).

Most of the energy of music signals is in the bass region (low frequencies). The bass frequencies are usually present in both channels of a stereo amplifier (this applies to both tube and solid-state amps) and are often identical. A strong simultaneous signal in both channels may overload the amp's power supply, which is common to both channels in most stereo amps. This drains the energy from the power supply capacitors (discharges them significantly) and temporarily brings down its voltage.

As a result, the operating point of all amplification stages (and the output stage in particular) is altered, causing compression of the signal and significant distortion.

By reversing the polarity of one channel, the bass signals now appear to the amplifier out of phase, which significantly decreases the instantaneous loading of the power supply. A positive peak in one channel is balanced out by a negative peak in the other. Distortion is reduced, the amplifier recovery time is shortened, and the whole audio system sounds cleaner, faster, and more dynamic. Truly amazing!

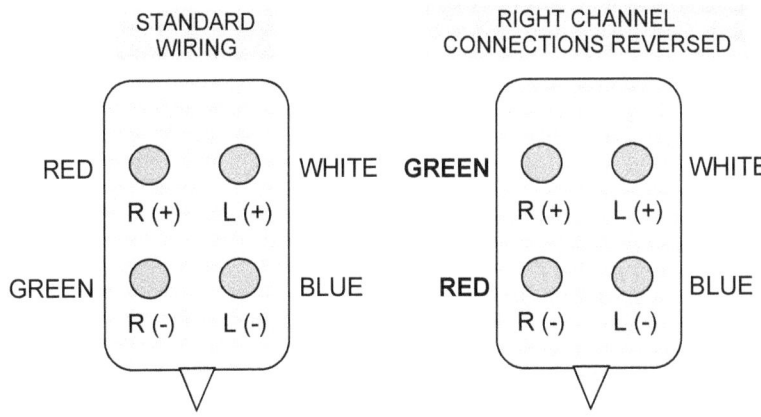
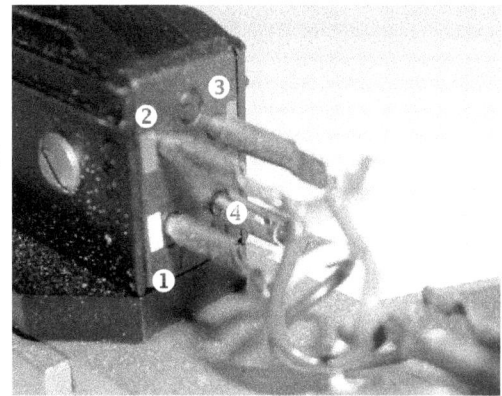

Fig. 6.49.1 Standard wiring on a turntable cartridge, viewed from the cartridge's connections side, stylus down.

Fig. 6.49.2 Swap wires on one channel only, either WHITE and BLUE or RED and GREEN (illustrated). Then swap speaker wires on your amp or at the back of the speaker on that channel (right channel in this case)

Fig. 6.49.3 The Ortofon cartridge (stylus facing upward) connections, white (1) & blue (2) for the left channel and red (3) and green (4) for the right channel.

This intervention isn't necessary with amplifiers of dual mono construction or with separate monoblocks. In those cases, each channel has its power supply, and there is no interaction between the channels, which are (electrically speaking) independent.

The optimal positioning of turntables and phono stages

The mid-fi console systems of the 1950s and 60s have become chic retro items, in hot demand, especially those with tube amps and tuners and well-preserved cabinets. These were, as a rule, made using real timber veneers and furniture- grade finish and have been fetching high prices lately.

Topology-wise, all had horizontally arranged components, with the tuner and controls (volume, tone, selector, etc.) to the far left and the power supply and amplifier underneath, with the turntable (or record changer) to the far right. This was done to place the sensitive turntable pickup as far as possible from the power transformer, power supply choke, and the rest of the power amplifier as possible, so it doesn't pick up stray electromagnetic fields (EMI).

While most hi-fi racks are still of the vertical variety, the sleek horizontal "entertainment" units are gaining in popularity, usually positioned below a large TV or a projection screen. Since they allow two or even three audio components to be placed side-by-side, it's worth reiterating the main lesson learned from the old consoles.

The turntable should be placed so as to maximize the distance between its cartridge and the power transformer inside a phono stage, preamplifier, or power amplifier. The same applies if outboard power supplies are used; place them as far as possible from the rest of the electronics, especially away from the cartridge.

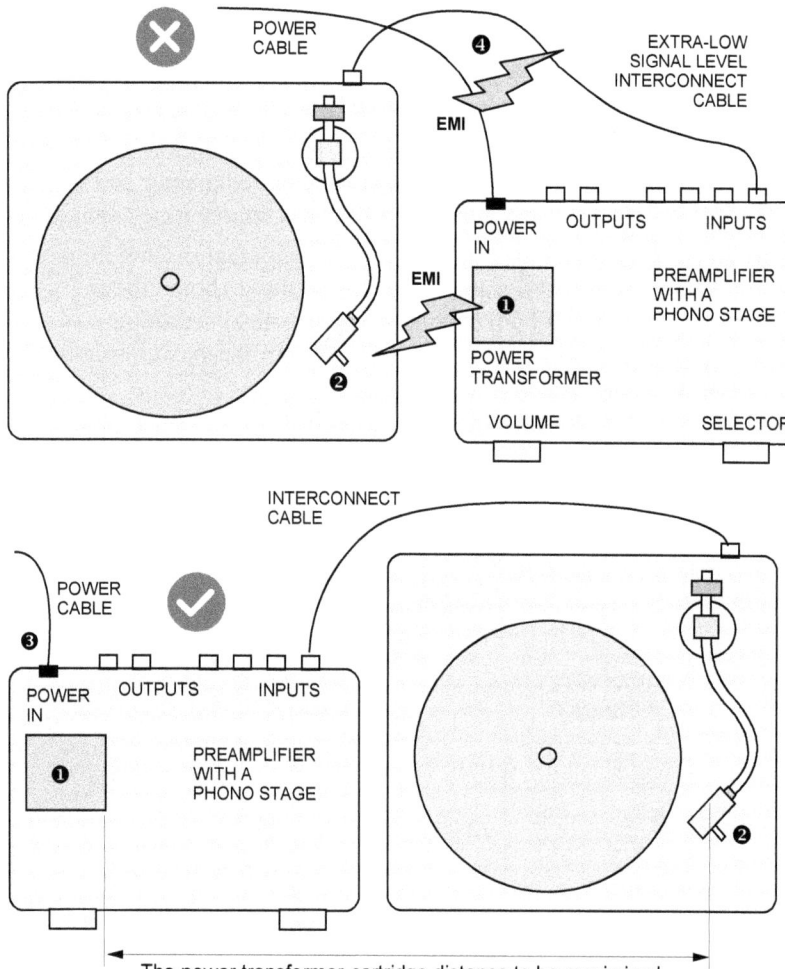

Fig. 6.50. Identify the location of the power transformer inside your phono stage, preamplifier, or integrated amplifier (1) and position your components in such a way as to maximize their distance from the turntable cartridge (2).

This will, if not eliminate, then certainly minimize possible EMI (electromagnetic interference) problems.

Usually, it isn't necessary to open up such electronic components to figure out where the power transformer is positioned; it is almost always right next to the AC inlet into the component (3).

Also, try to avoid situations where the interconnect between the turntable and the phono preamp crosses one or more power cables, as in (4).

THE VIBRATION & INTERFERENCE SOURCE - SINK RULE

To minimize microphony, vibrations, hum, his, electromagnetic and radio-frequency interference, physically separate the sources and "sinks" (or receivers) and position them as far apart as possible. Transformers and phono cartridges, tubes and loudspeakers, power and interconnect cables, dirty loads (dimmers, inverter air conditioners, motors, and such) and audio components, should not be placed close to each other.

UPGRADING & FINE-TUNING THE SOURCES

TUBE BUFFERING YOUR CD PLAYER, DAC, OR MUSIC STREAMER

The goal? To eliminate (bypass) cheap and nasty sounding solid state output buffer circuits

In most CD players and DACs, the output from the D/A converter goes through one or more solid-state devices, either discrete bipolar or FET transistors, or, more commonly, though numerous OP amp (operational amplifier) integrated circuits ("chips"). In Marantz CD40 (partial circuit diagram below), two 8-pin LM833 chips are used, one per channel, marked (3) on the diagram. Thus, such output amplifiers or buffers have a significant impact on your DAC's or CD player's sound.

If you thought having two OP amps in the signal path is bad enough (keep in mind that each operational amplifier has hundreds of transistors "integrated" inside the chip), what would you say about "high-end" CD players such as Kenwood 1000D, which has six of them?

How to do it

Adding a tube buffer stage to a DAC or CD player isn't difficult but should not be attempted unless you have solid electronics knowledge and skills. The first problem could be the physical space inside the chassis. You have to fit at least one duo-triode into it, often two and a reasonably sized power supply. Tubes require high anode (plate) voltages, so a transformer is needed to provide the heater and anode voltage, plus the rectifying and filtering circuit to convert the AC voltage on the transformer's secondary to a DC voltage 150-350V required by the tube.

If you are using a tube preamplifier, you don't need to add a tube buffer stage since the preamp's input tube stage will perform that function.

The exact steps of performing this improvement differ between various DACs, notably those with voltage and current outputs. Let's outline only the major steps involved, using the Marantz CD40 CD player as an example. It uses a very common TDA1541 DAC chip of the current output type. A circuit diagram of your DAC or CD player is helpful but not always necessary.

Once you open your DAC or CD player up, you can usually quickly identify the DAC chip, and a few minutes of googling around will yield its pinout arrangement. Identify its analog output pins; in this case, they are pin number 6, marked AOR (Analog Output Right) and pin 25, marked AOL (Analog Output Left).

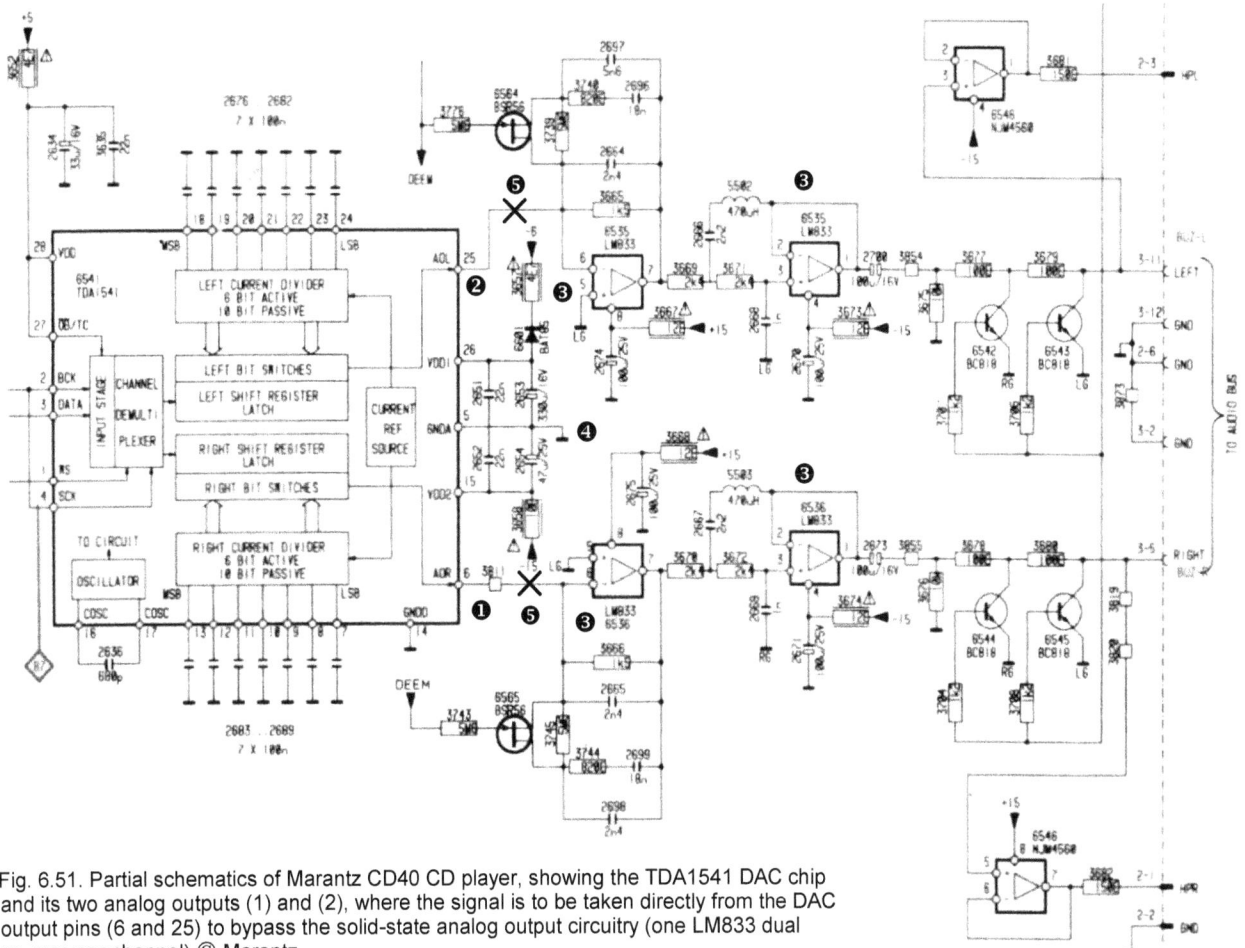

Fig. 6.51. Partial schematics of Marantz CD40 CD player, showing the TDA1541 DAC chip and its two analog outputs (1) and (2), where the signal is to be taken directly from the DAC output pins (6 and 25) to bypass the solid-state analog output circuitry (one LM833 dual op-amp per channel) © Marantz

Voltage output DACs produce audio voltage signals that can be taken straight out to the grid of the tube buffer or preamplifier. DACs with current outputs (as in our example) need a precision resistor added (usually 60-100 Ω) between the analog out and the analog or signal ground (GND), marked (4) on the schematics. This will convert the output current into a voltage; the voltage drop the current signal will make on the precision resistor.

That resistor will also act as a grid resistor for the input tube, referencing the grid to the ground. Usually, there is no need for a coupling capacitor between the DAC output and the grid, so DC coupling is an additional benefit of this simple arrangement.

Solder wires to those points on the circuit board but not directly on the chip itself, so you don't damage it by excessive heat from the soldering iron. Take these wires to a new pair of RCA connectors that you've installed at the back of your DAC or CD player.

The common ground requires only a single wire soldered to point (4), the analog ground, so you take only three wires to the back connections. The existing analog circuitry connected to the two current outputs must be disconnected; cut the PCB tracks (5) at X.

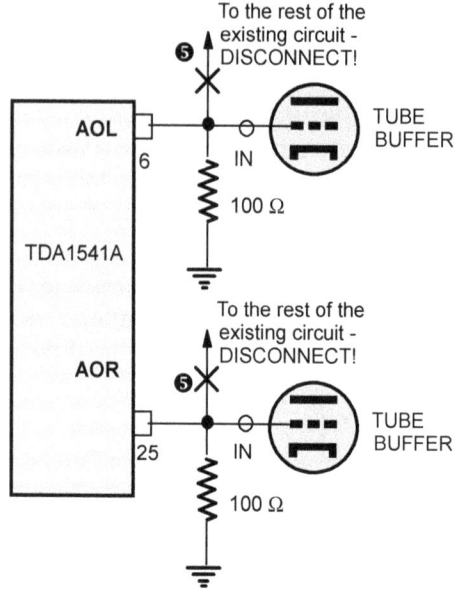

Fig. 6.52. Compared to voltage output DACs, interfacing current output DACs is only slightly more involved, requiring a pair of precision resistors to be added, to perform the current-to-voltage conversion!

Circuit option: Cathode follower (no voltage amplification)

The next choice is between a simple buffer (no voltage gain) or a proper amplification stage, with voltage gain. A cathode follower has a voltage gain below 1, meaning it actually reduces or attenuates the input signal. The anode A is directly connected to the stable DC voltage (+V_{BB}), the input is brought into the control grid G and the output is taken from the cathode K.

One duo-triode such as ECC40, ECC 82 or 6SN7 is sufficient for both channels. In the illustrated case (ECC40), one triode has pins 2, 3 & 4 and the other pins 5, 6 & 7. Pins 1 and 8 are the common heater circuit for both triodes in one glass bulb.

If cathode followers (CF) don't amplify the signal, why would we used them at all? The first benefit is increased input impedance. Since the whole output signal of the CF is fed back into the input circuit, their *feedback ratio* is β = -1 (100% negative feedback). This makes the *feedback factor* (1+Aβ) = 1-A.

All input impedances are multiplied by this feedback factor, so the grid resistor's new effective value is now $R_G(1+Aβ)$.

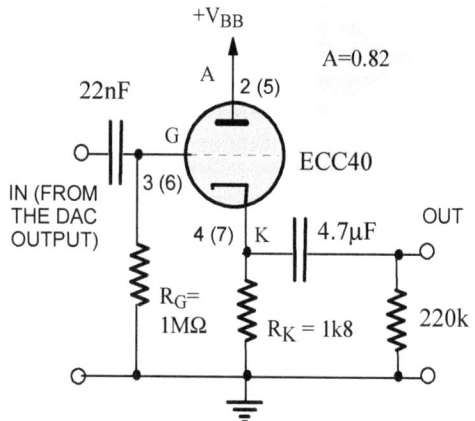

Fig. 6.53. Practical cathode follower buffer stage with ECC40 duo-triode (one triode used per channel)

Since voltage gain of this ECC40 cathode follower stage is A=0.82 and 1+Aβ= 1−0.82 = 0.18, instead of the 1 MΩ resistance of the grid resistor R_G, the input resistance of the cathode follower is 5.5 times higher: 1M/0.18 =1M*5.55 = 5.55 MΩ! This high input impedance means that the tube buffer will not load the output of the DAC at all, thus minimizing distortion.

Since the capacitive impedance between the grid and cathode of the tube is also multiplied by the factor (1+Aβ), this circuit also lowers the input capacitance of the cathode follower way below that of a standard common cathode tube stage. Low input capacitance means faster transients and wider frequency range, since the high frequencies are not bypassed to ground by the negligible input capacitance.

Finally, disregarding for a moment the output CR circuit (the 4.7 μF coupling capacitor and the 220 kΩ output resistor) the output impedance of this buffer stage would be the internal impedance of the tube R_I = 11,000 Ω in parallel with R_K = 1,800 Ω or 1,550 Ω. NFB reduces that figure by the factor (1+Aβ) or 5.55 times to arrive at the effective output impedance Z_{OUT} = 1,550/5.5 = 281 Ω, a very low figure for a tube stage.

In tube audio, anything below 1 kΩ is considered a very low output impedance. Such buffer stage would be able to drive long interconnect cables and low input impedance solid state amplifiers with ease and without distortion.

For comparison, the output impedance of a common cathode stage with bypassed cathode resistor is its internal resistance R_I in parallel with the load resistance R_L, $Z_{OUT} = R_I \| R_L = R_L R_I (R_L + R_I)$.

For our amplification stage with ECC40, that would be Z_{OUT} = 11k || 47k = 8,900 Ω or almost 32 times higher than Z_{OUT} of a cathode follower! With a tube buffer with 10 kΩ output impedance and a solid-state power amplifier with 10 kΩ input impedance, half of the output voltage would be lost; only half would reach the power amplifier. Remember, ideally, we want an infinite input and zero output impedance of an amplifier or an amplifying stage, and the cathode follower gets very close to it.

Circuit option: Common cathode tube stage with voltage gain

Should you need a modest voltage gain of your buffer stage, the common cathode triode stage is just what the doctor ordered. However, as we have just seen, while the 1 MΩ or so input impedance is fine, the price to be paid is much higher output impedance, which may not be an issue if a tube power amplifier is to be driven by it, but will not work well with low input impedance solid-state amplifiers.

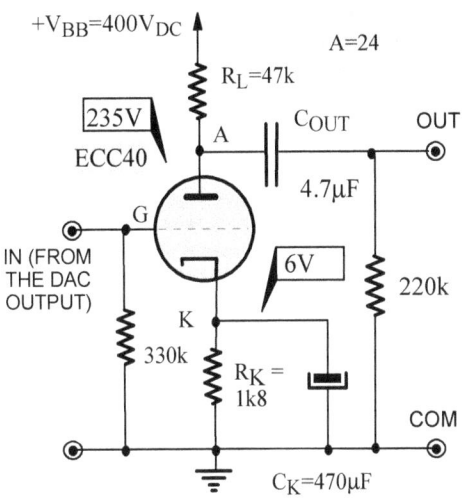

Fig. 6.54. A common cathode triode stage

The input signal is brought to the grid G while the output is taken from the triode's anode A.

Apart from high output impedance, another issue is this stage's high voltage amplification factor or "gain," in this case, 24 times. Power amps need only 1-3 Volts for full output power. For a 1 Volt signal at the output of your DAC, this stage will feed a 24 Volt signal into your power amplifier, which is way too high. One way to solve this problem is to use transformer loading in the anode (instead of the anode resistor).

The first benefit is that a lower DC voltage is needed in the power supply (not shown), in the illustrated case, 250 V compared to 400 V. Also, there is no need for the output capacitor C_{OUT} since the output transformer provides the required decoupling and isolates the high DC voltage from the input of the power amplifier.

Depending on the turns ratio of the transformers, various voltage gains are possible but look for a "step-down" transformer with, say, 10:1 turns and voltage ratio.

Thus, for a 10 Volts signal at the primary terminals, you will only get 1 Volt out at the secondary.

With our buffer's voltage gain of 24 times, we will still have some voltage gain left, 24/10 = 2.4, which is usually sufficient for most audio systems. For each Volt of the audio signal at the DAC's output, you will get 2.4 Volts at the output of your buffer.

If the transformer has a center-tapped secondary winding, as in this case, two output signals are available, one in-phase with the input signal (non-inverting output) and the other out-of-phase (inverting output); this can help you maintain the absolute phase throughout your system. Since the two secondary signals are out-of-phase, it is also possible to wire this buffer to an XLR socket to get a balanced output signal.

How useful and elegant is that! Usually, only costly CD players or preamps have a balanced topology or provide a balanced output signal.

Most audiophiles agree that transformer coupling sounds better than capacitive coupling. We have already discussed some possible reasons for such a claim, so instead of spending hundreds of dollars on upgrading your coupling capacitors, try transformer coupling in your system and make up your mind.

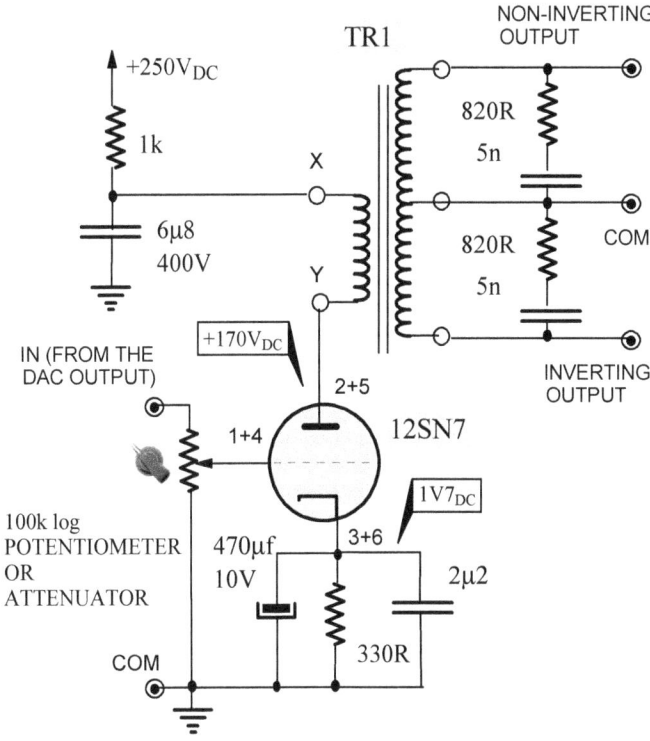

Fig. 6.55. A simple parallel triode line-level preamplifier stage with transformer output

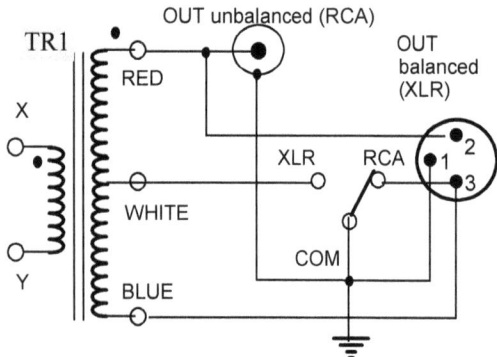

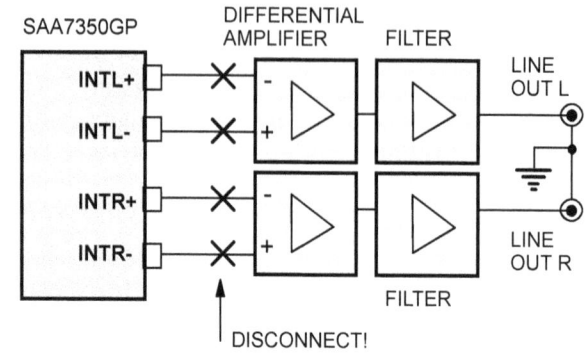

Fig. 6.56. The variation of the switchable output circuit providing both balanced (XLR) and unbalanced (RCA) outputs.

Fig. 6.57. The block diagram of the analog out solid state circuit (Marantz CD52SE CD player).

Interfacing DACs with balanced outputs

DACs with balanced (floating) analog outputs require a more complex tube buffer/amplifier circuit, a differential amplifier. This stage has two inputs and two outputs and amplifies only the difference in the voltage levels between the two inputs, while any signal common to both inputs is not amplified. These common-mode signals are usually interference, hum, and other induced nasties, so a differential amplifier has a high CMRR (Common Mode Rejection Ratio), making it a superior sounding topology.

Instead of a common cathode resistor to ground, better circuits use a constant current source (CCS), stabilizing the operating point and making it rock solid. Noise and distortion are also minimized, and the overall performance is improved. There's no need for elaborate complexities; we've used a simple CCS with one JFET in this example.

A negative DC voltage is needed at the tail end of the constant current source, in this case, provided by a simple power supply that will work with any AC voltage between 7.5 V and 12 V at its input. A half-wave rectifier (sufficient due to the minuscule current draw, so full-wave rectification isn't necessary) is followed by a simple series voltage regulator with a bipolar transistor.

The trimmer potentiometer R_S in the FET's source circuit is adjusted until the required current through the two tubes (FET's drain current) is achieved, in this case, 2 mA per triode (4 mA total).

For a balanced output, connect the two anodes to pins 1 and 3 of the XLR connector, and for a single-ended output, chose either OUT1 or OUT2 and take it out to the RCA socket's center pin. OUT1 is an inverting output, while OUT 2 is a non-inverting output (has the same phase as the INTL+ signal).

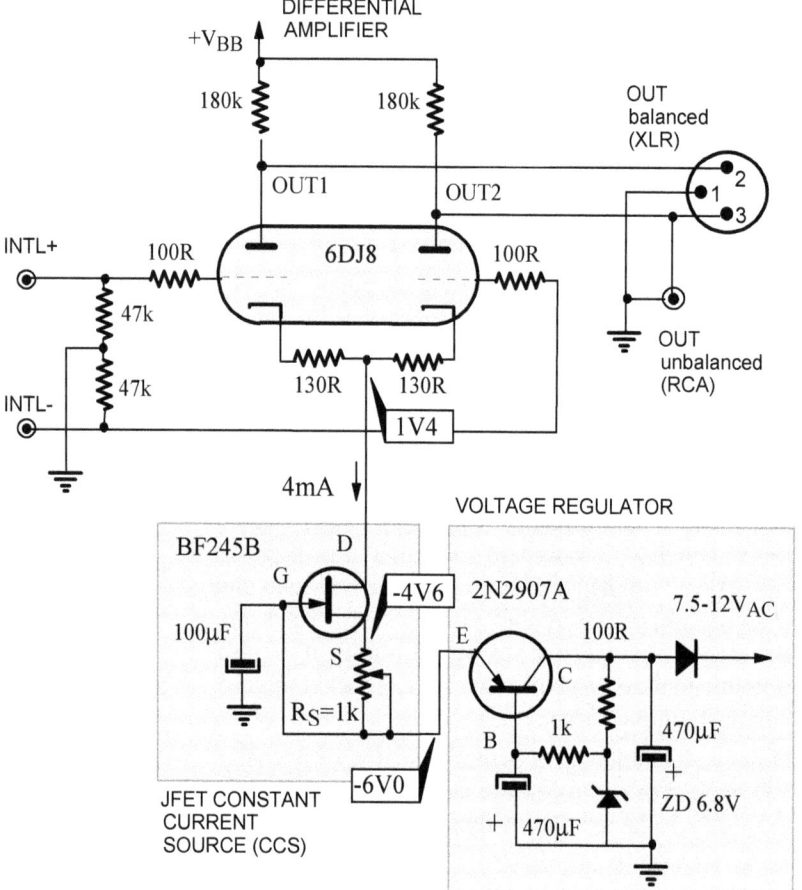

Fig. 6.58. One possible circuit (one channel only shown) of a differential buffer amplifier with a 6DJ8 (ECC88) duo-triode. One tube is needed per channel.

CASE STUDY: Tubing Marantz CD52 - MkII

If you are getting an audio technician to modify your existing DAC or CD player (the "host"), you have a choice between a hand-wired (point-to-point) circuit and a commercial kit on a PCB. The second option should be (much) cheaper; such kits can be built (the PCB can be populated with components) in a couple of hours.

Building circuits in a point-to-point fashion is much slower. Still, the technician will have total flexibility in all aspects, such as choice of the terminal strips or terminal boards (the substrate that will carry the tube sockets and all components), their positioning within the host, the choice and arrangement of the components, and so on.

If you are buying a used DAC or CD player with a view of tubing its output, keep in mind that many aren't suitable for that purpose; there's physically no space inside for the three main add-ons, the power transformer, the power supply, and the audio board.

Google a few models on your shortlist and try to find photos of their internals to assess the size and location of the free space (if any).

In our case, the host (Marantz CD52-SE) had a perfectly positioned ample space, right next to the DAC (1) and to the backspace where a new pair of RCA sockets would be added (2). Buying a cheaper CD player to practice on is a wise move if you choose the DIY option. We purchased this baby for AU$20 from a Salvation Army store. The CD tray would not open, so nobody wanted it. After the broken plastic cog was replaced, everything worked perfectly.

Fig. 6.59. A partial internal view of Marantz CD52-SE once the bottom cover was removed. The pinout of SAA7350 DAC (1) is in its datasheet (available for download online), so the four pins of interest can easily be identified and the associated PCB tracks that need to be cut.

Notice plenty of space at the back, where a new pair of RCA sockets will be mounted (2), and a large space on one side of the CD player (3), approx. 80 x 190 mm, where the tube buffer and its power supply can be installed. Some of the plastic crossbars (4) can be removed without compromising the integrity of the chassis.

All components on this side of the board are the fiddly SMD (Surface Mounted Devices), which is bad news if you have to replace them, but in this case, you can leave the unused solid-state output circuitry on the board, there is no need to remove them.

ADDING AN OUTPUT TRANSFORMER TO A CD PLAYER, DAC OR MUSIC STREAMER

The operational principle and practical implementation

We have already established that the weakest link inside CD players and DACs is the output stage with cheap integrated circuits. However, many "tubed" hi-end players still have such a solid-state output stage in the signal path, onto which a single triode is "tacked on," often a cathode follower to reduce the output impedance and add some tube "warmth." Although manufacturers charge a hefty premium of many hundreds of dollars for such a tube output stage, in many cases, there is no sonic improvement at all, just added tube coloration.

Completely replacing the semiconductor output stage with a quality transformer will significantly improve its sound. Clarity, microdynamics, the level of detail, and "musicality" will be considerably better. The principle is the same for DAC with balanced and DAC with unbalanced outputs. Compared to the complex differential amplifier circuit on page 154, the whole setup is straightforward, and, as you know by now, simple circuits sound better.

The transformer's secondary current will feed the load resistor R_X and perform the current-to-voltage (I/V) conversion. You only need two such resistors, one for each channel) so buy the best and most expensive ones you can find or afford (Kiwame, Holco, Shinkoh, Allen Bradley, Audio Note, TKD, Caddock, Takman, ..., the choice seems endless). Of course, they have to be perfectly matched (have the same resistance), so forget ordinary 10% or even 5% tolerance resistors; go for +/- 1% tolerance precision resistors.

There are additional benefits of this conversion. Distortion and noise can be reduced by isolating the digital (on the primary side) and analog ground (on the secondary side). Also, the transformer's limited upper-frequency range will filter out lots of higher harmonics generated as quantization noise. The transformer would not only serve as an I/V converter and analog filter but, depending on its turns ratio, can also provide signal (voltage) amplification!

As an example, Burr-Brown PCM1738 produces 2.48 mA$_{PP}$ of balanced output current. The effective value is 2.92 times lower than that peak-to-peak figure, so I_1 = 2.48mA/2.82 = 0.88 mA.

If the turns and voltage ratio of the transformer is, say 1:5, the secondary voltage will be five times the primary voltage, while the secondary current will be five times lower, so $I_2 = I_1 * TR = 0.88 * 0.2 = 0.176$ mA. Assuming that all that current will pass through resistor R_X and that none will enter the preamplifier's input stage (connected to the secondary), let's say we need the voltage signal at the RCA output to have a value of $1V_{RMS}$.

$V_{OUT} = I_2 * R_X$, so $R_X = V_{OUT}/I_2 = 1/0.176 * 10^{-3} = 5.68$ kΩ. You could use the standard value of 5k6, but we have assumed an ideal transformer. Real transformers have a significant insertion loss, so we should increase the value of R_X up to 20%, or $1.2 * 5.6 = 6.72$ kΩ, which would make the standard value of R_X = 6k8 perfect for the job.

If the preamp's input impedance that follows (R_{IN}) is not infinite (or very high), then it needs to be added in parallel with R_X and such a value used in calculations.

With R_{IN} = 47 kΩ, the total parallel resistance must be R_P = 6k8, so $R_X = (R_P * R_{IN})/(R_{IN} - R_P) = 6.8 * 47/(47 - 6.8) = 7.95$ kΩ. Use the first higher standard value, 8k2.

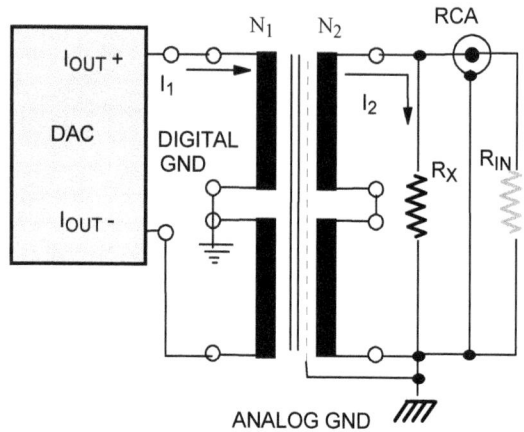

Fig. 6.61. Transformer output stage for a DAC with a balanced current output

A commercial benchmark and a typical DIY option

Using multi-section winding techniques, 1465 and other Sowter DAC transformers have mumetal (76% Nickel) cores and are enclosed (shielded) in mumetal enclosures.

Its twin bifilar primary windings can be connected in parallel, in series, or driven by twin DACs supplying signals of the opposite phase. Its electrostatic shield should be connected to analog ground only, never to digital ground.

Instead of spending £210 plus transport on a pair of commercial cans, we installed a couple of those vintage Russian microphone transformers (from page 148) into our experimental host, the Marantz CD52-SE CD player.

While their 1:10 and 1:20 ratios would be too high for this application, the lowest one, the 1:2 ratio, was perfect. The sonic improvement was significant.

SOWTER 1465 SPECIFICATIONS:

- Dimensions 45 mm (dia.) x 52.5 mm high
- Primary inductance (each coil) 5.0 H typ.
- Primary DC resistance (each coil) 6.7Ω typ.
- Sec. resistance (two coils in series) 1.29 kΩ
- Voltage ratio 1+1 : 5+5
- Frequency response: 5 Hz to 100 kHz (+/- 1.5 dB, paralleled secondary windings), 5 Hz to 50 kHz (+/- 1.5 dB, secondary windings in series)
- £105.89 each (2018)

7 AUDIO AMPLIFIERS - HOW THEY WORK AND HOW TO IMPROVE THEIR SOUND

Since audio amps and preamps are more complex than loudspeakers crossovers, modifying and improving their performance is more demanding. A much deeper understanding of their operational principles is needed. However, there are some quick, cheap, and easy modifications, the low-hanging fruit, so to speak, that will result in noticeable sonic improvements.

Although we use tube amps and preamps as examples in this chapter, the same or similar principles apply to solid-state ones. In general (there are some exceptions, as always), tube amps have fewer amplification stages, use simpler topologies, and only a handful of electronic components. Thus, they are much easier to understand and modify.

- UNDERSTANDING AUDIO AMPLIFIERS
- HOW TO EVALUATE AMPLIFIER SPECIFICATIONS
- THE SONIC ANATOMY OF AMPLIFICATION: RULES FOR BETTER SOUND
- PASSIVE PREAMPLIFIERS
- CLASS D AMPLIFIERS
- OTL TUBE AMPLIFIERS
- TUBE ROLLING: CHANGING AND UPGRADING VACUUM TUBES IN AMPS & PREAMPS
- MODIFYING AND UPGRADING AMPLIFIERS AND PREAMPLIFIERS
- BRIDGING AND PARALLELING: TURNING STEREO POWER AMPS INTO MONOBLOCKS

> "What this country needs is a good five-watt amplifier."
> Paul W. Klipsch, American audio pioneer, designer of Klipshorn, a seminal folded horn loudspeaker

UNDERSTANDING AUDIO AMPLIFIERS

An audio amplifier as a power supply modulator

A variable resistor, with its wiper or "slider" controlled by the music signal, is in a simple circuit with its own internal DC (direct current) power supply and a loudspeaker, connected to the amplifier's output terminals X and Y. Since the circuit is closed, electrical current will flow from (and through!) the power supply, through the part of the resistor in the circuit (between points "U" and "V") and through the speaker (its voice coil).

As the wiper slides to the right ("DOWN"), more and more of the resistor's resistance is in the circuit, and less and less current is flowing through the speaker. As the slider moves to the left ("UP"), less and less of the resistor's resistance is in the circuit, and more and more current is flowing through the speaker.

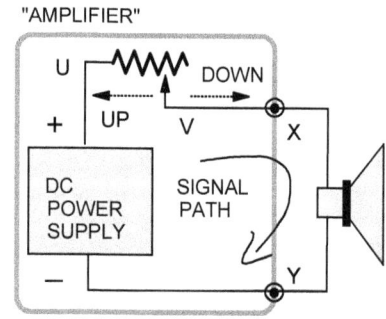

Fig. 7.1. The in-principle operation of an audio amplifier as a modulator of its internal power supply

If we could find an electronic device that could act as such a fast and accurate variable resistor and change its resistance in accordance with the input music signal, we'd be in business. Indeed, such electronic components exist, vacuum tubes ("tubes") and solid-state transistors. We can define an audio amplifier as an electronic device in which a small amplitude (low power, low current, and/or voltage) signal (from a music source) controls large amplitude (high power, current, and voltage) output signals.

Notice a couple of obvious crucial points even from such a simple, "in principle" circuit diagram. First, an amplifier is simply a modulator of its power supply, modulating (converting DC into AC) the current and voltage provided by the power supply before sending it to the loudspeaker (load).

Notice that the signal path passes right through the power supply, or, in other words, that the power supply is in the signal path. Thus, an amplifier is only as good as its power supply, its heart. It is impossible to have a good-sounding amplifier with a lousy power supply.

It also means that the power supply's design (topology) and the components chosen to build such a power supply directly and profoundly impact the sonic properties of the whole amplifier!

You may think, come on, Igor, this is basic stuff, I know all that, but you'd be surprised how many audio designers and other supposed "experts" in audio lose sight of these fundamental truths. They concentrate on fancy circuits, microprocessor controls, automatic bias modules, and similar stuff (most of which make minimal improvement to the sound of their amps), yet pay very little attention to the heart of any design, its power supply.

THE POWER SUPPLY AXIOMS

1. An amplifier's power supply is in the signal path, meaning it will significantly impact its sonics.
2. An audio amplifier is only as good as its power supply. You cannot have a good-sounding amplifier with a lousy power supply.
3. A power amplifier can be considered a signal-modulated DC power supply; thus, the power supply is the most critical part of any amplifier!

Vacuum tube as a variable resistor controlled by the input signal

Vacuum tubes, as they are called in the USA, are also known as "electronic tubes" or "thermionic valves." A vacuum tube is a device in which conduction by electrons takes place in a vacuum within an airtight envelope. The term "valve" is used mainly in the UK and Australian English. While the term electron "tube" probably comes from the glass bulb's tubular shape, the term valve most likely originated in the mechanical-electrical analogy that explains the operating principle of a triode.

Just as turning the handle on a mechanical valve controls the flow of water (or any gas or liquid) through it, changing the voltage (or bias) on the triode's grid controls the flow of electrons through it, between its anode (the positive electrode) and cathode (the negative electrode).

The Americans use the colloquial term "plate" instead of the proper term "anode" since the anode is a large metal plate on the outside of its structure (the closest to the glass bulb) and can easily be seen.

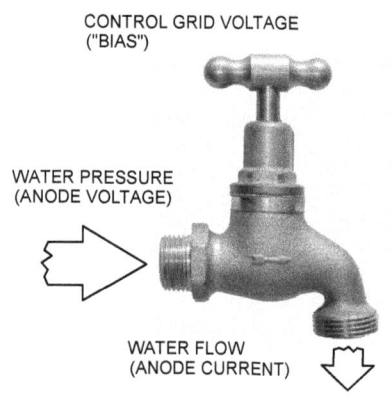

Fig. 7.2. The water tap analogy of a triode explains why vacuum tubes are also called "valves"

AUDIO AMPLIFIERS - HOW THEY WORK AND HOW TO IMPROVE THEIR SOUND

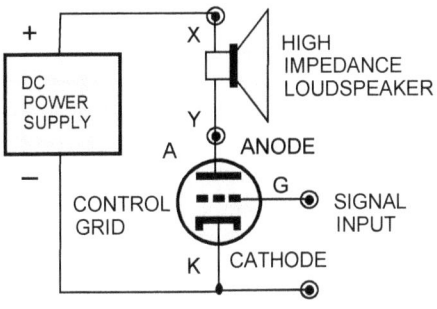

The simplest amplifying vacuum tube is a triode, with three electrodes, anode (A), Cathode (K), and the control grid (CG). There are even tubes with only two electrodes (anode and cathode), called diodes, used to rectify AC signals in power supplies.

With a negative voltage on the grid ("the bias", not shown on this diagram for simplicity), the electrons (negatively charged particles) energized by the heated cathode don't have enough energy to pass through it. Even if they leave the cathode and start flowing through the vacuum, the negative voltage on the grid repels them back to the cathode.

Fig. 7.3. A triode in the audio circuit instead of a variable resistor

Fig. 7.4. A more detailed and practical triode single-ended amplification stage, including the bias power supply and the output transformer OT, transforming the high primary to a low secondary impedance for a better match with low impedance speakers.

However, with a music signal applied to the grid (a varying or AC voltage between its grid and common cathode), electrons start flowing and reach the anode or "plate," which collects them. If the circuit is closed through a load (a speaker), the current will flow.

With such varying currents, the voltage across the speaker also varies. Due to the high DC voltage used (300 - 1,200 Volts), such variations are much higher than the grid voltage variations, and thus, voltage amplification is achieved.

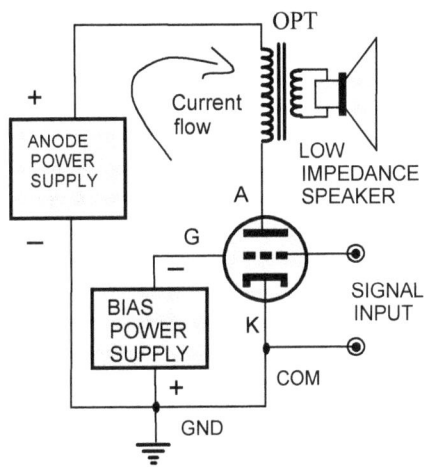

However, this in-principle circuit would not work well. Apart from the AC signal, a DC current would also flow through the speaker, which would need to be wound with a very thick wire. Also, only high impedance speakers (800 - 1,000 ohms) could be used. An output transformer (OPT) would solve both problems - eliminate the speaker's DC current and act as an impedance matcher between the tube and the speaker. Due to their high internal impedance, tubes cannot drive low impedance speakers directly, so output transformers are used to "transform" the low speaker impedance on their secondaries (4 - 8 ohms) into high impedances on their primary sides (1-10 kΩ).

Both the anode power supply and the output transformer are in the signal path, so their type and quality impact a tube amplifier's sonics.

Notice that the standard convention for current flow is the opposite of the flow of electrons (from cathode to anode).

The outside anatomy of a vacuum tube audiophile amplifier

Since almost all solid-state amplifiers look like boxes with heatsinks, meaning there are no outside clues to their design or topology, we'll again use tube technology for illustrative purposes. A "typical" stereo single-ended amplifier is pictured here.

Most tube amps have tubes and transformers on top of a chassis, clearly identifiable. The rectifier tube (1) is in front of two power supply chokes (8) and the power transformer (9).

There are two preamp tubes (2), one per channel, and two output tubes (3). Two RCA input sockets (4) and two pairs of loudspeaker binding posts at the back (7) are the only audio connections, apart from the IEC power socket and a fuse holder behind the power transformer at the back panel (not visible).

The rotary on-off switch (5) and volume control (6) are at the front. Signal flows are indicated by arrows.

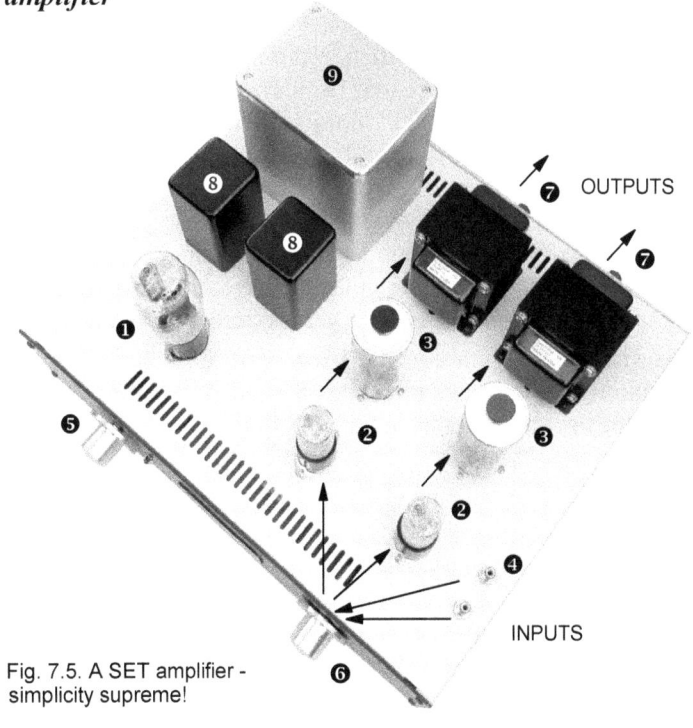

Fig. 7.5. A SET amplifier - simplicity supreme!

Single-ended versus push-pull operation

Both tube and solid-state amplifiers can be designed using single-ended (SE) or push-pull (PP) topology. PP requires an even number of tubes per channel, a minimum of two, but often, to get more power, four, six, or eight identical tubes are used, half-and-half connected in parallel.

SE topology implies only a single output tube per channel, but the name can be misleading, since again, two, three or more tubes can be paralleled. However, they still operate as one "super" tube, so the topology is still single-ended. SE output stages are simpler (fewer parts), that one tube amplifies the whole audio signal.

In push-pull stages, each tube amplifies only part of the signal, and the pair work in opposition, one "pushes" (amplifies the positive part of the signal), the other "pulls" (amplifies only the negative part of the signal). Thus, the preceding stage, called "phase inverter" or PI for short, must split the signal into two signals of the inverted or opposite phase (polarity).

There are two types of bipolar transistors, the NPN and PNP type, and they can be made to be "complementary," close to (but never completely) a mirror image of each other, so when used in a PP circuit, they require no phase inversion. Ditto for MOSFETS and JFETs (N- and P-type).

However, there are no complementary tubes; tubes always conduct the same way, thus requiring the phase inverter driver stage. This additional stage adds to the complexity and cost and introduces various problems that manifest themselves as sonic degradation. The phase inverter stage is one of the most significant weaknesses of push-pull topologies.

In short, the two halves of the push-pull stage (the two tubes and the two halves of the output transformer's primary) are never entirely identical or "balanced" at all audio frequencies, introducing gain imbalance, phase shift, and distortion.

The most commonly used phase inverters are split-load (also called "cathodyne" or "concertina"), the paraphase, and the cathode-coupled inverter or "long-tailed pair." Some more expensive amps use interstage transformers with a center-tapped secondary (or two identical secondaries).

Each of these phase inverter options has its own strengths but also particular problems and shortcomings.

It pays to study the simplified diagram of a push-pull tube stage (below right). Notice that DC anode currents (I_{DC}) of the two output tubes flow in opposite directions through the primary winding of the output transformer and thus cancel each other. Since they magnetize the output transformer's core the opposite ways, there is no danger of saturation. Push-pull output transformers do not need an air gap between their laminations as SE transformers must have; thus, smaller and cheaper cores/lamination stacks can be used.

The signal (AC) currents $i_1(t)$ and $i_2(t)$ flow in the same direction and thus get recombined (added together by the action of the output transformer.

By convention used in electrical engineering, capital letters are used for DC currents and voltages, and lower case letters for signals or AC voltages and currents, (t) signifying that such signals change with (are a function of) time.

Do SE and PP amps (even when the same preamp and power tubes are used in both) sound different? Yes, they do. The split into SE aficionados and PP brigade is another fault line that divides audiophiles.

Fig. 7.6. Can you tell just by looking at an amplifier if it's of SE or PP kind? Usually, no. This amp uses one preamp tube and two 6V6 power tubes per channel in parallel SE topology. Generally, the smaller the size of the output transformer and the more preamp/driver tubes, the more likely that the amp is of a PP kind.

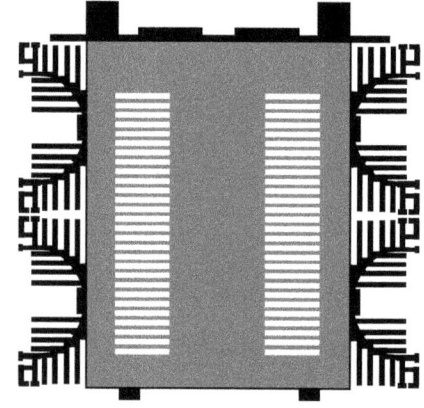

Fig. 7.7. When the only visual appeal of an amplifier comes from its mass-produced extruded heatsinks or a VU-meter that looks like an expensive Swiss watch but is immaterial to sound quality ...

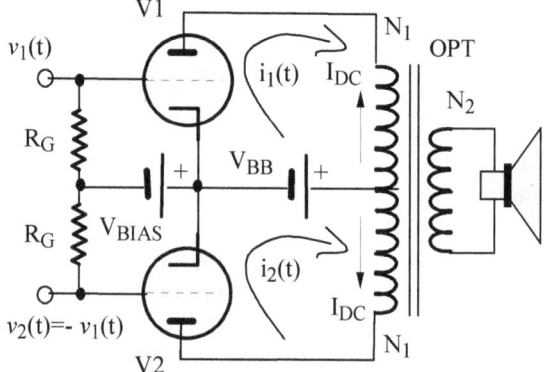

Fig. 7.8. Triode push-pull output stage with a fixed bias applied to the control grids. Signal voltages v2(t) = -v1(t) are of equal amplitude but opposite phase.

Nothing comes even close to the naturalness, liquidity, and magic of a single-ended amplifiers' midrange. However, push-pull amps often have better, tighter bass and sometimes more extended top end. PP amps are usually much more powerful, making them easier to match with inefficient, hard-to-drive speakers.

Push-pull amps usually produce a more forward sound stage. However, their imaging is highly dependent on the quality of the phase inverter (the symmetry of its outputs) and the balance of the output stage. The main factor is how matched the output tubes are. Output tubes are seldom completely matched - they may be matched in one point only but not along the whole transfer curve. To make matters worse, tube aging changes their parameters (gain, mutual conductance, and internal resistance) unevenly, so the imbalance becomes more and more pronounced, and the sonic performance of the output stage deteriorates with time.

I have already declared my membership in the SE camp, especially its SET (triode) "branch." Now, there are many great PP tube amps, which to me still sound superior to most solid-state amps. However, despite their shortcomings (low power, low damping factor, low efficiency), most of which aren't that important to sound quality anyway or can be minimized and compensated for, single-ended tube amps are in their own sonic class.

Ultralinear tube amplifiers

Pentodes and beam power tubes have a screen grid (1), which can be connected in various ways. When held at some stable DC voltage (2), the tube will work as a pentode, and when strapped together with the anode, as a triode. When in the "TR" position (as illustrated), the screen is connected to the anode (4) via a low-value resistor.

In the "UL" (ultralinear) position, the screen is connected to a tap on the primary winding of the output transformer (5). The screen's voltage now follows the audio signal, which means that local negative feedback is applied to the screen grid.

This will "straighten" or "linearize" (make more linear) the curved characteristics of a pentode (anode current versus anode-to-cathode voltage), leading to its name. The higher the percentage of the anode signal that we bring back to the screen grid, the more the pentode curves resemble triode ones, as in the sketch below.

The optimal position of the UL tap on the OPT's primary varies for different output tubes, for instance, between 23% of primary turns for 6V6 and EL84, 33% for EL34, and 43% for 6L6.

In this output stage, a switch (3) enables all three modes of operation. Other amplifying devices, such as the "solid-state" ones, bipolar transistors, and FETs, don't have such capability or flexibility; this is inherent only to vacuum tubes.

The UL connection results in a host of negative feedback benefits:

- output impedance is reduced to about half of the triode's circuit
- linearity is improved, resulting in reduced distortion
- the power output is reduced only slightly compared to the pentode connection, but it's still much higher than in the triode connection
- the power output is more constant because the output stage behaves in-between the voltage (triode) and current (pentode) source mode.

If your amp works in a pentode or UL mode, adding the triode mode capability is relatively easy. If you don't feel confident in performing such surgery yourself, it should not cost much. However, in the triode mode, the same power tube needs a much higher drive voltage, so the gain of the whole power amp will be significantly reduced. You will have to crank up the volume control on your preamp.

Unless the output transformers in your amp have the UL tap, converting it to ultra-linear operation will not be possible.

Despite its wider bandwidth, lower distortion, and higher damping factor, resulting in "cleaner" and more neutral sound, the ultra-linear connection is usually sonically inferior. It does not sound as "organic," relaxed ("free-flowing"), or natural as the triode mode.

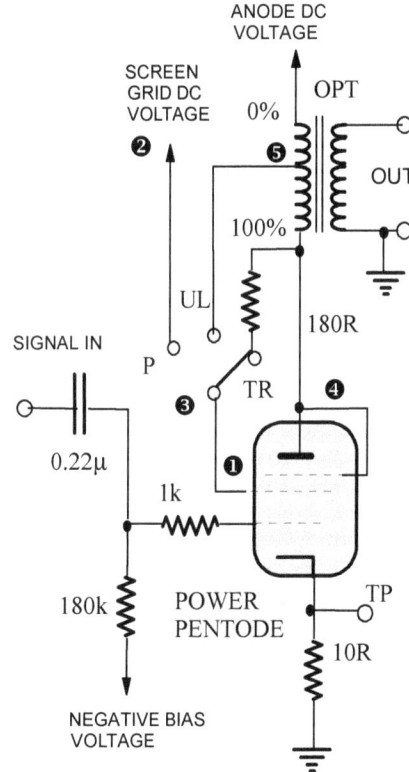

Fig. 7.9. A single-ended output stage using a pentode, whose screen grid connection will determine the tubes operating regime, as a pentode, in an ultra-linear mode, or strapped as a triode.

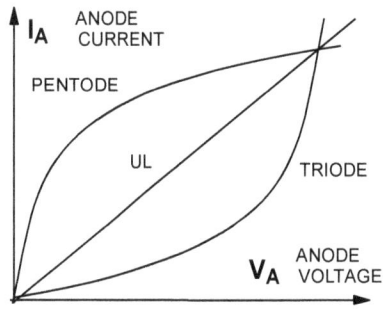

Fig. 7.10. The ultra-linear (UL) connection is an example of the "Linearizing Principle": Convex + Concave = Straight!

Class A or Class AB?

Push-pull amplifying stages usually work in Class A or Class AB. They always work in Class A at low output power levels, where both tubes or transistors conduct the whole signal current. Then, at a certain level, depending on their bias voltage, they cross into Class AB, each tube or transistor conducting only a portion of the output signal (between 50 and 100 %). By definition, single-ended stages must work in Class A; otherwise, a part of the music signal would be cut off.

Very few push-pull amplifiers work in Class B, where each output tube conducts precisely 50% (half) of the signal. Typically, such stages suffer from unacceptably high crossover distortion (when one tube stops and the other starts conducting), primarily due to the imperfect matching between output tubes and other parts of the circuit. The most notable are McIntosh Class B amps. To minimize crossover distortion, they use incredibly complex output transformers and rely on various types and multiple loops of negative feedback (to screen, cathode, and preamp stages), all derived from different primary windings of the output transformer.

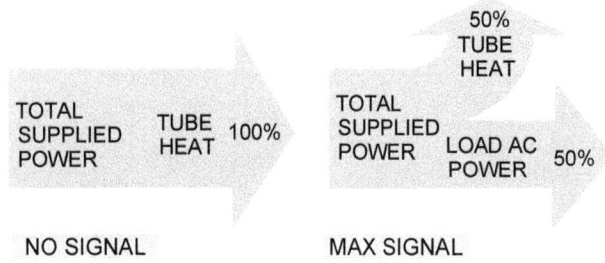

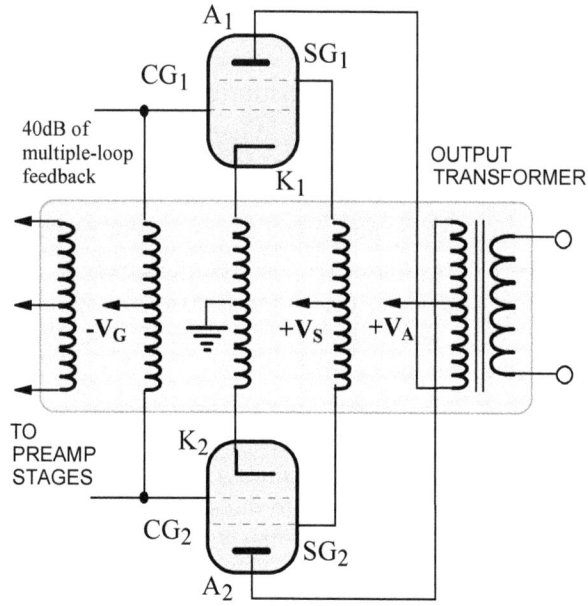

Fig. 7.11. Perhaps paradoxically and certainly counter-intuitively, in Class A amplifiers, the power tubes run cooler with AC (music) signal than when idle. At the amplifier's maximum output power, the output tubes run coolest.

Fig. 7.12. The most complex McIntosh output stage with pentafilar windings (five primary windings!) as used in their MC3500 amplifier. V_A, V_S, and V_G are the anode, screen, and bias DC supply voltages.

Why amplifiers don't like reactive loads (loudspeakers)

An audio output stage with purely resistive load and a straight load line (A-B in the graph below) is a designer's dream that never comes true. An ellipse is the path of operation for a partially inductive or capacitive load such as a typical dynamic loudspeaker.

The actual loads such as dynamic and electrostatic loudspeakers aren't purely resistive. Dynamic speakers are mostly inductive, and in the first approximation, can be modeled by a series RL circuit, the impedance being a complex number $Z_L = R_L + jX_L$. Electrostatic speakers are predominately a capacitive load, modeled by a resistor R and capacitor C in parallel.

In both cases, instead of the operating point moving up and down along the straight load line as with a purely resistive load, the operating point follows an elliptical trajectory. For capacitive loads, it rotates counter-clockwise; for inductive loads, it moves in the clockwise direction.

The larger the reactive component of the load, the wider or "fatter" the ellipse. The smaller the reactance, the flatter the ellipse, and the more such an ellipse will approach a straight line.

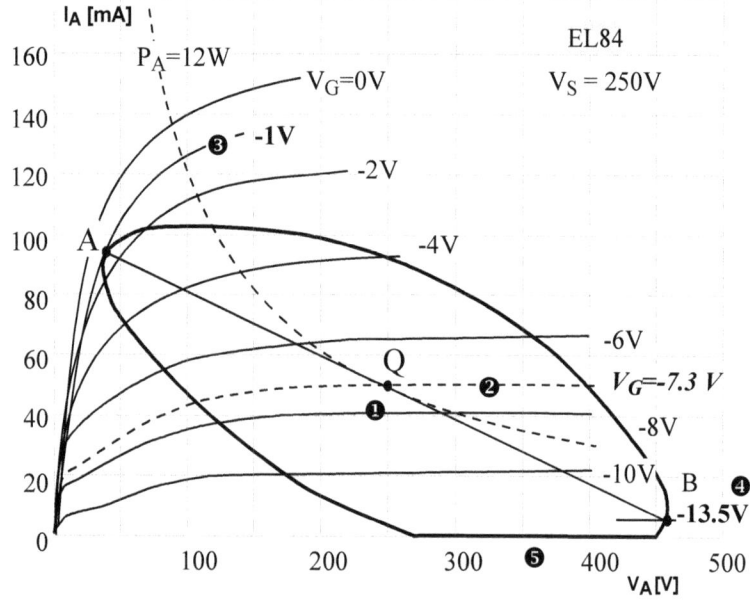

Fig. 7.13. The load line of a single-ended pentode stage (EL84) with a purely resistive load will expand into an ellipse with reactive loudspeaker loads. Since the shape of the ellipse varies significantly with signal's frequency, so will the amplifier's output power and distortion.

The illustration shows the pentode curves of the EL84 power tube (anode current versus anode-to-cathode voltage) working as a single-ended power amplifier. The load line and the quiescent (without signal) operating point Q are indicated (1). The tube is biased at -7.3 V (2). This is the worst-case scenario, for maximum power output. The input signal (grid voltage V_G) swings between anode curves for -1 V bias (3) and -13.5 V bias (4) peak-to-peak. Obviously, the tube is in a cutoff for a large part of the operating cycle (5) because the anode current I_A is zero, resulting in significant distortion.

To make matters worse, the reactive component of the loudspeaker will change with frequency. The higher the signal frequency, the larger the inductive reactance $X_L = 2\pi f L$ will be and the fatter the ellipse.

So, with a complex signal comprising many different components of various frequencies, one ellipse will turn into dozens. If you observe that on an oscilloscope, the load "line" will be a messy, fuzzy, filled "ellipse" of ever-changing size and shape.

Luckily, at lower power levels, the ellipse stays undistorted. We can also conclude that the "fatter" the ellipse (the more inductive or capacitive the loudspeaker load), the earlier such a cutoff situation will happen. For that reason, when looking for speakers to pair with your amplifier, your main criterion should be the loudspeaker's impedance curve. Look for speakers whose impedance curve is as flat as possible.

Can the amplifier drive electrostatic speakers? The 2 μF capacitive load amplifier stability test

Many power amplifiers, such as solid-state ones and those that use high levels of negative feedback, become unstable when driving highly capacitive loads such as electrostatic loudspeakers. However, even the ordinarily inductive dynamic speakers become capacitive in specific frequency bands, where their impedance curve drops or falls between the curve's resonant peaks and dips.

"Unstable" in this situation means the amplifier will go into uncontrolled oscillation, producing high output voltages, as high as their power supplies will allow. The amp will draw a very high current from its power supply and the power outlet. If its protection fuse is properly dimensioned (current rated), it will blow; otherwise, the power transformer's winding wire may melt (burn out), resulting in a costly repair.

If you have electrostatic speakers and are considering purchasing a power or integrated amplifier to drive them, perform this quick & easy test using any dynamic speaker.

Make a test module by cutting a crocodile-clip terminated wire jumper in half and soldering it to a 1.0 or 2.2 μF film capacitor (rated at a minimum of 200 V_{DC}). Connect it in parallel with one or both speakers (you will need a pair of these test modules) by clipping it to the speaker terminals (polarity doesn't matter). Power the amp up and play music.

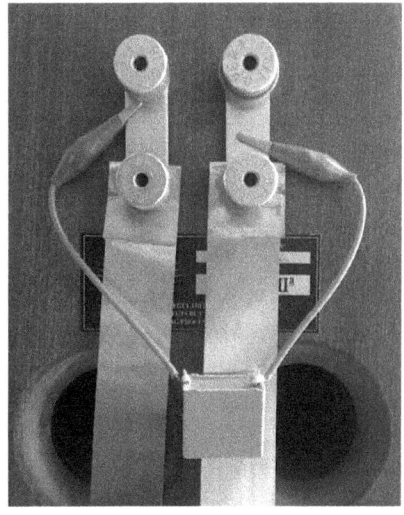

If the amp blows its fuse, it most likely wouldn't be able to drive highly capacitive electrostatic speakers.

Fig. 7.14. 2μF film capacitor clipped onto speaker terminals

Outboard capacitor banks - are they worth the extra expense?

Some tube amp manufacturers make it possible to increase their amps' power supplies' energy by hooking up an external capacitor bank. For an Octave Audio's V40SE amplifier, retailing for US$5,300 (in 2014), that is an add-on expense of US$1,200 for their Black Box, which quadruples the capacitance, presumably in the high voltage power supply.

Interestingly, in his "Measurements" section for a review in *Stereophile*, the editor, John Atkinson, noted that the Black Box had no impact on voltage gain, output impedance, or maximum output power. However, the reviewer (Art Dudley) heard "a distinct improvement in the sound."

Of course, if your tube amplifier is out-of-warranty and if you are reasonably handy with a soldering iron, you can add a few capacitors yourself to its power supply filtering circuit and save a small fortune. This assumes that some real estate (space) is available under the hood, another reason to buy amplifiers with as large a chassis (enclosure) as possible. The smaller its chassis, the more cramped the amplifier will be inside.

Older amps will have large-sized (physically) capacitors, so replacing them entirely with more modern units of the same physical size but much higher capacitance is one way to go if the space is at a premium.

For instance, an amp may have four 220 μF electrolytic caps (total capacitance 880 μF), which can be replaced by the same physical size modern 470 μF caps (or even 680 μF). Such an increase in total capacitance (to 1,880 μF) and the proportional increase in stored energy should result in better dynamics and reduced dynamic compression.

CRITICAL QUESTION: Are output tubes manually biased and if they are, how easy is biasing to do?

A cathode- or self-biased output stage is the cheapest and easiest to implement, a simple resistor in the cathode circuit, on which cathode current will create a voltage drop which will thus lift the cathode's DC voltage above the zero (ground) level of the control grid CG, thus making the CG negative with respect to the cathode.

Remember, a bias must make the control grid negative vis-a-vis the cathode to control the flow of electrons through a tube, like a brake that will prevent the tube from overheating and self-destruction.

"Fixed" biasing is an unfortunate term. It means that the bias voltage is fixed at a specific value and the tube itself has no say in determining that value. In some amps, this negative voltage brought to the CG is set (cannot be adjusted), but the user can adjust it on others. If adjustable, how it is done is a significant issue to be considered by the prospective buyer.

In all cases, the amp must be powered up but without any input signal. Some amps must be turned upside down, the bottom cover removed, and a multimeter used to measure the bias voltage while adjusting it with a small screwdriver (trimmer potentiometers). This exposes you to dangerously high voltages inside. Some amp manufacturers even demand that you return the amp to an authorized dealer and pay them a few hundred dollars to bias the amp. That is ridiculous, not just expensive (transport costs both ways must be added), but also makes transport damage very likely. You will have to be without your amp (and music) for weeks! Stay away from such amps and brands.

The more user-friendly models will have the trimmer potentiometers and test terminals (to plug the multimeter into) accessible from the top. The best designs are those where you don't directly measure the current but a voltage drop on a small resistor between the cathode and ground (1 or 10 ohms are the common values in use). See the schematics on page 176, where this test point (to connect a DC voltmeter or multimeter) is marked with "TP1".

With a specified bias of say 0.6 Volts (bias is always a DC voltage) on a 10-ohm resistor, the cathode current of that tube will be 60 mA (Ohm's Law: $I = V/R = 600$ mV$/10 = 60$ mA).

What does "dual-mono" mean?

It seems that many audio reviewers don't even understand the bare fundamentals of the terms they use willy-nilly. For instance, a commercial "dual-mono" phono stage was reviewed. It had a single power transformer with two identical secondary windings, each supplying power to one channel's power supply.

It's very simple: if it has a single transformer, it cannot be dual-mono! "Dual-mono" means each channel must have everything separate, starting with individual power transformers, except the common chassis and perhaps a shared power cord, on-off switch & protective fuse.

Compared to stereo amps, dual-mono amplifiers or preamps have better channel separation (since they don't share a power supply) and lower crosstalk.

Pushing tubes to their limits

In many commercial tube amps on our test bench, the designer was either ignorant or deliberately chose to push tubes beyond their power dissipation limits. That is done to claim higher power than their competitors or to hot-bias the output tubes. During listening evaluations, hot-biased tube stages (high current flowing) usually sound better than cold-biased ones.

Every tube has its maximum anode current and anode dissipation. The anode collects electrons that bombard it and becomes very hot in that process. Such heat must be dissipated through the surrounding vacuum and the glass envelope into the air around the amp. This is the main reason not to cover your tube amp with anything and to place it topmost on a hi-fi rack, with no shelf above it that could impede the upwards flow of hot air.

There are other voltage and current limitations, such as the minimum and maximum heater voltage, the maximum grid current, the maximum heater-to-cathode voltage (due to the very thin insulation between them), and many others. Unfortunately, unless you get an amp on a test bench before you buy it or have its detailed circuit diagram with voltages marked and get an expert to analyze it for you, there's no way to tell how well or how badly an amp (or any other piece of audio gear) was designed or built.

We'll talk about cases of amplifiers that "eat" tubes (tubes' life is very short, weeks and months instead of years) and the reasons behind this problem.

> Tube amps' undeserved reputation of being unreliable is due to the poorly designed ones just discussed! I believe tube amps are more resilient and reliable than solid-state ones. For instance, making a brief short circuit at solid-state amps' speaker terminals (while fiddling with speaker connections) will instantly destroy all output transistors, and there are dozens of them in a typical amp. Incredibly, many "hi-end" high-power (200-400 Watts) transistor amps (I won't name them here for legal reasons) don't have any output protection. This would never happen with a tube amp - the output tubes would survive even prolonged short circuits at the output!

HOW TO EVALUATE AMPLIFIER SPECIFICATIONS

Bandwidth and -3dB frequencies

These test results (right) are of one of our single-ended tube amplifiers. The bandwidth is usually measured at a low power level (usually 1 Watt) and a higher power level, most commonly at the rated or nominal output power (12 Watts in this case). The bandwidth at higher power levels ("power bandwidth") is always narrower than the frequency range at 1 Watt. In tube amps, that is primarily due to the limitations of their output transformers.

MEASURED RESULTS:
- BW: 5 Hz - 57 kHz (-3 dB@ 1W)
- BW: 10 Hz- 49 kHz (-3 dB@ 12W)
- Class A1: P_{NOM} = 12W
- Class A1: P_{MAX} = 16W
- Class A2: P_{MAX} = 28W

The amplitude of the input signal is held constant, and as its frequency is reduced, the drop in the output voltage starts happening at some point. Once that drop reaches -3 dB (or reduction to 0.707 or approx. 71% of the midrange value), that is the lower -3 dB frequency or half-power frequency.

Decibel or "dB" is a unit for a ratio between two levels, which can be electrical (power, voltage, current) or acoustic (sound pressure level, for instance). It is handy when a sensor or transducer (such as the human ear) responds to stimulus not in linear but in a logarithmic manner.

In terms of power ratio $P = 10*\log(P/P_0)$ [dB], where P is the power produced by an amplifier, P_0 is the referent power chosen for a particular purpose, comparison or measurement. If we express power gain of an amplifier in terms of output voltage we get a simplified formula $P = 20\log(V/V_0)$. Since power fed to the load is $P = V^2/R_L$, when voltage drops to 0.707 V_0 we have $P = 20\log(0.707V_0/V_0) = 20\log 0.707 = 20*(-0.15) = -3dB$.

The results above are excellent; most tube amps do not have such wide bandwidths. Also, the difference between the lower -3 dB frequencies at 1 Watt and 12 Watts isn't significant, which indicates an excellent design and superior quality of output transformers. The same applies to the upper -3 dB frequencies, a drop from 57 to 49 kHz, which is a very moderate reduction. The power bandwidth of inferior amps shrinks much more (for instance, dropping from 37 kHz at the upper end to 21 kHz at higher power levels).

Output power and headroom

There are various definitions of "headroom." The "clipping headroom" relates the nominal output power to the output power at the onset of clipping. For a 100 Watt amplifier that starts clipping at 112 Watts, the clipping headroom is $\Delta P=10\log (112/100) = 0.49$ dB.

Our case (a 300B amp) is not typical since the amplifier could cross into class A_2 operation, which is when the grid current starts flowing. Thus, the maximum output power before clipping was not 16 Watts (Class A_1) but a whopping 28 watts, meaning the clipping headroom was quite high, $\Delta P_{CP}=10\log (28/14) = 3$ dB.

Dynamic headroom is a ratio of dynamic power output, using the IHF tone-burst signal and the rated output power declared by the manufacturer. In our case, the IHF tone-burst test yielded an output power of 22 watts, slightly lower than the maximum output of 28 Watts, so $\Delta P_{DYN}=10\log (22/14) = 1.96$ dB, still relatively high.

Power supply "energy-per-output-watt" as one of amplifier comparison factors

Since every power amplifier, in its essence, is a modulator of its DC power supply, we could argue that an amplifier is only as good as its power supply is, well, powerful! So, the higher the stored energy in its power supply, the better the amplifier's dynamic performance. Thus, the power supply's stored energy per watt of output power is one quick & easy way to evaluate a tube amp's energy reserve.

For instance, Audio Research Corporation's REF150 SE amplifier's specs mention the power supply energy storage of 1,040 J (Joules). The photos show one bank of a dozen 470 µF capacitors and another PCB with twelve more. In this case, with two channels rated at 150 Watts each (300 W total), the EPW factor (Energy-Per-Watt) is 1,040/300 = 3.47 J/W, which is a very high figure.

This criterion can be used for solid-state amplifiers as well. Obviously, they work with much lower voltages, thus the need for a much higher capacitance.

Output impedance

When a spec sheet for a tube amplifier proclaims "Output impedance: 4 and 8 Ohms", that is almost meaningless. All it means is that the amplifier has binding posts marked 4 and 8 ohms (plus the one marked "0" or "COM"). The output impedance is more likely to be 0.2-1 ohm for a push-pull tube amplifier, depending on negative feedback's strength (if it's been used at all), or 1-3 ohms for a single-ended triode amplifier with no feedback. Solid-state amps will generally have a much lower output impedance than tubed ones. More on this issue very soon.

Input impedance

Since measuring the input impedance isn't difficult, you'd think that manufacturers would at least get that right, but no, by my count, a significant percentage of amplifiers reviewed in *Stereophile* magazine have a significantly lower input impedance than their specified figures. Instead of say, 100 kΩ declared impedance, *Stereophile's* equipment tests would return results of 13 kΩ, almost an order of magnitude lower. In other words, don't place too much trust in such manufacturers' specifications!

Keep in mind that by "impedance," everyone refers only to an amplifier's input resistance; the input capacitance (the other component that determines its input reactance and contributes to its overall complex impedance) is never tested (by audio magazines) or specified by manufacturers. Thus, such a figure is both technically and linguistically incorrect, incomplete, and can be misleading.

Generally, the higher the amp's input impedance, the better, the less the amp will "load" the output stage of the preceding preamp or line-level source (CD-player, streamer, DAC, etc.), thus minimizing its distortion.

Signal-to-Noise ratio (SNR)

An important parameter of any amplifier, preamplifier, or electronic system in general, is called signal-to-noise ratio, a ratio of the signal voltage V_S and noise voltage V_N at its output: SNR = 20log (V_S/V_N) [dB].

The lower the noise voltage V_N, the higher the S/N ratio SNR and the quieter the amplifier. Say we have a load of 8 Ω. Without any signal at the amplifier's input (inputs grounded), we take a true-RMS multimeter and measure 1 mV_{AC} voltage on the speaker terminals. This is the hum or noise that the amp is generating itself. We want to specify its S/N ratio referenced to a level of 1 Watt at the output. How do we do it?

1 Watt of power dissipated on 8Ω load means the voltage is $V_S = \sqrt{(PR)} = \sqrt{(1*8)}$ = 2.83 V. Substituting 2.83 for V_S and 1 mV for V_N into the formula we get SNR = 20log (2.83/0.001) = 20*log2,830 = 20*3.452 = 69 dB.

Audio equipment manufacturers use every method at their disposal to make their equipment look better and their test figures more impressive. They can declare SNR ratios at a certain "standard" power level, such as 1 Watt, or at the amp's rated power output. If an S/N ratio is declared with reference to either the nominal or maximum power level, the figures will look much better (be significantly higher) than those at 1 Watt.

CASE STUDY: Classic 16.0 tube amp and its specs

The Classic No.16.0 is a push-pull power amplifier with a pair of EL34 pentodes in each channel. The output stage can operate in ultralinear (U/L) or triode mode.

Notice the high input sensitivity, 220 mV of the input signal would take the amp into full power. Since CD players and other sources produce around 2V_{RMS} output signals, to drive this power amplifier, you need to attenuate the source signal more than ten times, not amplify it. This extra gain may be an issue if you are using an active preamp.

SPECIFICATIONS:
- Output power: 40W (UL mode), 20W (triode mode)
- BW: 15 Hz - 42 kHz (-3dB, at 1W into 8Ω, triode mode)
- BW: 8Hz - 82 kHz (at 10W, U/L mode)
- Open-loop gain 45db, gain with NFB 38 dB
- THD: below 1.2% (at 10W out, 1kHz signal, triode mode)
- SNR: 90db (unweighted)
- Output noise: 1.2mV
- Negative feedback: 5.72 dB
- Input sensitivity: 0.22V for full power

Even if truthful (some aren't), many amplifier manufacturers' specifications are vague (incomplete) or even contradicting. For instance, they may specify "THD: below 1.2 %". We don't know at which frequency and, more importantly, at what power level and into what load. At least here, two out of three parameters are specified: "THD below 1.2 % (at 10 W out, 1 kHz signal, triode mode)"

"SNR: 90 dB (unweighted)" does not specify if that is referenced to 1 Watt output power level or the rated power level. A-weighted SNR considers only the audible range frequencies rather than all frequencies that a measuring device can detect. A measuring instrument can measure noise above 20 kHz, but humans can't hear such frequencies. Thus, the A-weighted spec are always better (higher SNR) than unweighted figures.

However, the noise voltage of 1.2mV is specified, so we can calculate SNR_{1W} = 20log (2.83/0.0012) = 20*log2,358.33 = 67.45 dB (referenced to 1 Watt output level into 8 ohms).

Let's try it again referenced at the nominal power (40 Watts into 8 ohms). The signal voltage is $V_S=\sqrt{(PR)} = \sqrt{(40*8)}$ = 17.89 V, so SNR_{40W}= 20log (17.89/0.0012) = 20*log14,908 = 83.47 dB, still far from the claimed 90 dB!

In conclusion, take all specs with reserve. Ultimately, they are only a starting point. The only thing that matters is how the amp sounds with your speakers and in your listening room.

THE SONIC ANATOMY OF AMPLIFICATION: RULES FOR BETTER SOUND

Believers and deniers

I still find it astonishing that "absolutists, rationalists (or whatever you want to call that group of audiophiles and even audio "experts") could deny the possibility of individual electronic components (tubes, transistors, resistors, capacitors, transformers, etc.) affecting the overall sonics of an audio component (amp, preamp, DAC, etc.). Designing, constructing, and listening to just one amp, preamp, phono stage, or even a loudspeaker, and changing around a few key parts would have been enough to make anyone a true believer that each audio component and electronic part has its "sound" or sonic signature.

You change a metallized film coupling capacitor of a particular brand to the same capacitance film & foil capacitor of another brand, and the sound of the whole amp/preamp changes, often significantly? Yep, you better believe it!

In the following examples of a tube amp and preamp designs, I will try to summarize the most important conclusions we've reached after more than 25 years in the audio business. You may agree with some and disagree with others; that's fine. Your experiences may have been different. Or, if you are an audiophile with no technical knowledge or skills, the whole discussion may be too technical and confusing. That's fine too, don't get frustrated; the main thing is to understand the basic premises and my conclusions (in italics), not the technical details.

The chosen examples reflect our background in tube amplification, but similar conclusions are reached by those experienced in using solid-state amplification, DAC design, or loudspeaker building as well.

The power supply: the heart of every amp or preamp

The power supply is the heart of any amp or preamp, and the power (mains) transformer is the heart of any power supply. An electrostatic shield (1) prevents (or minimizes) RFI (Radio Frequency Interference) and other garbage on the power supply lines and the mains frequency (50 or 60 Hz) hum from entering its secondary windings and the rest of the amp/preamp. Its one end is grounded (to the chassis and the earth conductor).

The high voltage rectifier will determine the essential sonic character of the whole amp. We've made many amps with solid-state diodes. They sounded terrific, but, overall, tube rectifiers sound better. The best sounding tube rectifiers are of the mercury vapor kind, such as 83, a vintage USA-made tube (2). The RF chokes in its anodes (3) help clean the sound and prevent possible oscillations. The dot inside a tube symbol indicates a gas-filled tube.

Directly heated rectifier tubes usually sound better than indirectly heated ones. For instance, replace indirectly-heated 5V4 or 5AR4 in your amp or preamp with 5Y3 (assuming its lower current capacity is sufficient for the application, double-check!) or 5U4, and compare the sound.

The best power supply topology is dual mono, where each channel of a stereo amp has its own, completely separate power supply. Here we have the next best thing, a common filtering choke (6 Henry), followed by individual 2nd stage filters, one for each channel, each with its own choke. That results in a high degree of channel separation and minimal crosstalk.

Chokes make better filters than resistors. Cheap and nasty amps use RC filtering with a 1-cent resistor instead of a $200 choke. The biggest improvement you can easily make in your amp is to replace the series resistors in the high voltage power line (would be used instead of chokes #1 and #2 below) with chokes.

LC filtering sounds better than CLC filters. Notice that the first filtering stage below is LC, choke (4) - capacitor (5); there is no capacitor connected between point (4) and ground. LC filters result in smooth continuous current through the power transformers; CLC filters cause the current to flow in short and sharp pulses, which degrade the sonics.

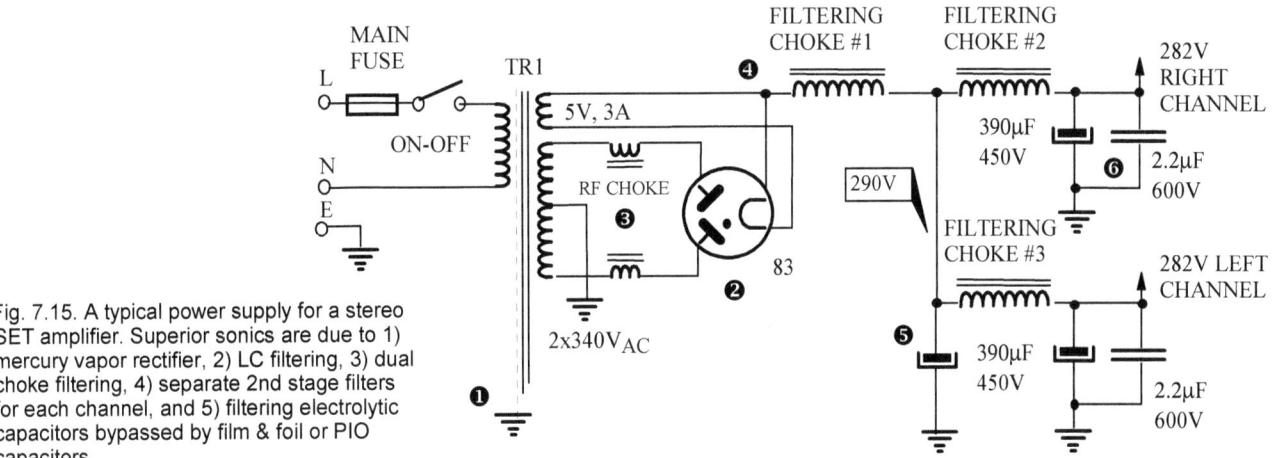

Fig. 7.15. A typical power supply for a stereo SET amplifier. Superior sonics are due to 1) mercury vapor rectifier, 2) LC filtering, 3) dual choke filtering, 4) separate 2nd stage filters for each channel, and 5) filtering electrolytic capacitors bypassed by film & foil or PIO capacitors.

The best electrolytic capacitor (elco) is no electrolytic capacitor. If possible, film caps should be used instead. However, they are large and expensive. The next best solution is to bypass elcos with quality film & foil or PIO (paper-in-oil) capacitors (6). The value is not critical, 0.22 µF, 1 µF, 3.3 µF, whatever you have.

Line level preamplifiers and power amplifiers' driver stages

Directly-heated tubes ("filament type") sound better than indirectly heated ones. Directly-heated means that the filament (heater) also acts as a cathode, so both the signal current and the heater current flow through it. You'd imagine that (mixing the two currents) to be a bad thing, and it is. If AC heating is used, its source connected to terminals H1&H2 (1), the hum is increased. However, for some reason, directly heated tubes, such as 45, 2A3, 300B, AD1, VT25 & VT25A, for instance, sound superior to indirectly heated ones (6L6, EL34, KT88).

AC-heated tubes sound better than DC-heated ones. DC-heating of DHTs reduces hum resulting in quieter amps and preamps, but also negatively affects their sonics. It kills microdynamics, nuances, and "musicality." Thus, it comes down to the compromise, choosing between superior sonics and less hum in an audio system.

High mutual conductance tubes sound more dynamic. ECC88 (6DJ) is a duo-triode (2) of a medium voltage gain (amplification factor), about 33, but with relatively low internal resistance (only 2.6 kΩ) and high mutual conductance (Gm=12.5 mA/V). This means that for each 1 Volt change in input voltage, its anode current jumps by 12.5 mA, which is a huge jump. Such jumps make preamps and amps with such tubes more responsive or "dynamic." Also, when the two triodes in the same glass envelope are paralleled, as they are here, the mutual conductance is doubled to 25 mA/V, and the internal resistance is halved to 1.3 kΩ!, which is of further benefit.

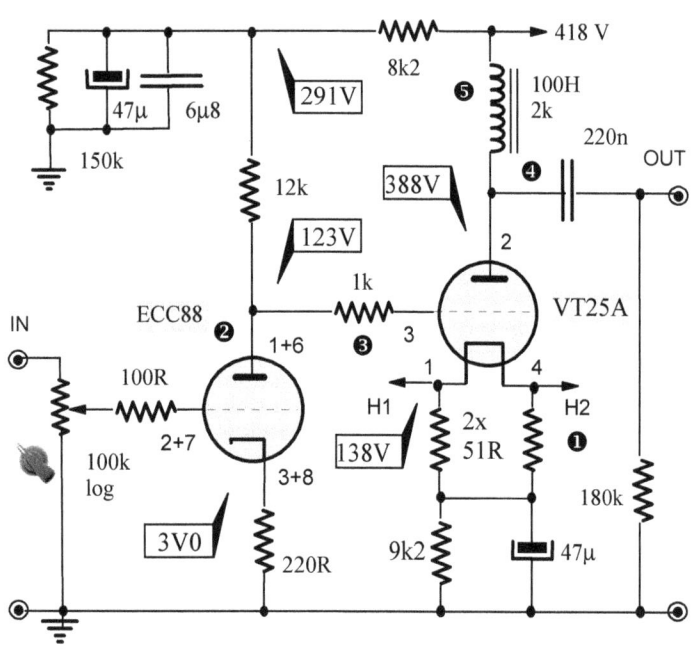

Direct (DC) coupling between stages sounds better than capacitive or transformer (inductive) coupling.

The best coupling capacitor is no capacitor (3). Due to the high DC voltage present at the output of the second stage (4), the use of a series output capacitor cannot be avoided. That capacitor will have a noticeable impact on the sonic character of the whole preamp, so use the best and most expensive type you can afford; you only need one per channel.

Anode chokes (5) or inductive loading sounds better than anode resistors (resistive loading). See the analysis of a commercial headphone amplifier on pages 194-195, where the benefits of inductive loading (a more horizontal load line resulting in less distortion) are discussed.

Fig. 7.16. One of the best sounding line preamplifiers we have ever made or heard, a minimalistic 2-stage design with DC coupling and anode choke.

The power amplifier

Bigger output transformers = bigger & bolder bass!

The nominal power rating of a transformer is proportional to the cross-sectional area of its center leg. That means the bigger its outside dimensions, the better (cleaner, with lower distortion) it will reproduce the low bass notes.

In this SET amplifier, identical GOSS laminations (EI118, meaning the largest dimension is 118 mm) were used for both power (6) and output transformers (7).

The amp's nominal output power is 15 Watts, while the output transformers' cores are capable of 308 Watts (3.9 x 4.5 cm center leg, or 17.55 cm^2, and 17.55^2 = 308). They are 308 W/15 W ≈ 20 times oversized!

Fig. 7.17. If the output transformers of your tube amplifier aren't close in size to its power transformer, you've been short-changed!

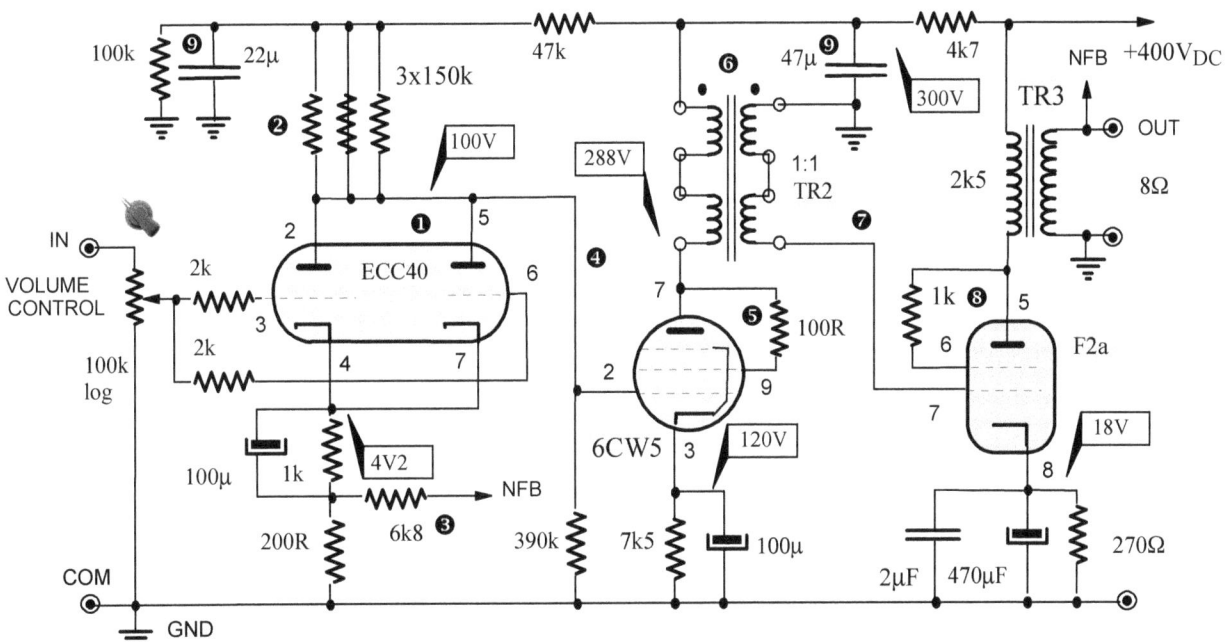

Fig. 7.18. One of the best sounding power amplifiers we have ever made or heard. Audio section only, one channel shown.

Simplicity and minimalism are the key factors. The lower the component count (the simpler the design), the better. Each new component introduces its sonics, adds more hum/hiss/noise, and increases distortion. If you are using a quality tube line stage with a high amplification factor, low output impedance and high current capability, you could even drive the output stages of some tube power amps directly, eliminating the whole input stage.

This amplifier can operate in class A_2, meaning the driver stage (with 6CW5 tube) can supply current into the grid of the output tube through the interstage transformer TR2 (7). This significantly increases the maximum output power available (compared to class A_1) and improves the dynamics.

Notice that the driver tube is a triode-connected pentode (5), as is the power tube (8). The screen grids are strapped to anodes through small value resistors, 100 ohms, and 1,000 ohms, respectively. This leads us to the next claim:

Triodes sound better than pentodes or tetrodes. There is nothing wrong with connecting pentodes or beam power tubes as triodes, despite being called derogatory names such as "pseudo-triodes" or "quasi triodes" by the ill-informed puritans.

For instance, German F2 or F2a11 tubes here sound as good, if not better, than famous triodes such as 300B or 2A3! The driver tube is also strapped as a triode. You can recognize such a trick on circuit diagrams by the presence of a resistor between its screen grid (pin 9 in this case) and its anode (plate), pin 7 (5).

Paralleled tubes sound better. ECC40, one of the best sounding duo-triodes, is used here (1). In its usual operating point, it has a lowish mutual conductance of Gm=3 mA/V, and a highish internal resistance of around 11 kΩ.

When paralleled, the mutual conductance is doubled to 6 mA/V, and the internal resistance is halved to 5.5 kΩ. Apart from improving dynamics, paralleled tubes pass higher currents and supply more current to the load or the next stage, which brings us to the next point.

The higher the current through a tube stage, the better it sounds. A single-ended output stage biased "hotter" at 90 mA will sound better than the one biased "colder" to draw 65 mA. However, the higher the current, the harder the tube has to work and the lower the tube's life expectancy, so don't go crazy when biasing your amp!

DC-coupling (4) between stages is sonically the best, and transformer (inductive) coupling is almost as good (7). Interstage transformers are expensive and bulky, but the results are worth it. Unfortunately, you cannot just replace the coupling capacitors in your amp with interstage transformers; that is a job for true experts. Most amp fixers and audio technicians wouldn't even know where to start.

The best negative feedback (NFB) is no feedback. However, in some cases, low levels of global NFB (taken from the speaker output back to the input stage's cathode) improve things by lowering distortion and increasing the damping factor (3). Experiment by changing the values of the feedback resistor (6k8 shown). A lower resistance (4k7 or 3k3) will increase NFB, a higher value (10 k, 15 k, 18 k, or 22 k) will reduce the strength of the feedback.

Point-to-point wiring is more reliable, better sounding and cheaper & easier to repair than printed circuit boards. PCBs are used by manufacturers not because they sound better (they don't) but because they significantly reduce manufacturing costs.

PASSIVE PREAMPLIFIERS

> "Enough shit already - I'm turning 70 and I'm finally going to say it flat out: *I don't like preamps.*"
> Sam Tellig in his review of *Baby Reference* transformer-coupled passive preamplifier by *Music First Audio*, Stereophile, Oct 2012

How they work

Apart from a few inputs and outputs which are switch-selectable, at its heart, a passive preamp is a volume control potentiometer, attenuator, or transformer, depending on how volume control is achieved. Usually sold on their own, some are also combined with active preamps, enabling the user to chose active or passive operation.

A Promitheus Audio unorthodox line-level preamp is two preamps in one. It can be used as a passive TVC (transformer volume control) passive preamp or an active preamp, a paralleled ECC88 triode stage with a transformer output. The TVC is in the circuit in both cases. The fidelity of the output voltage is low (significant oscillation and waveform distortion).

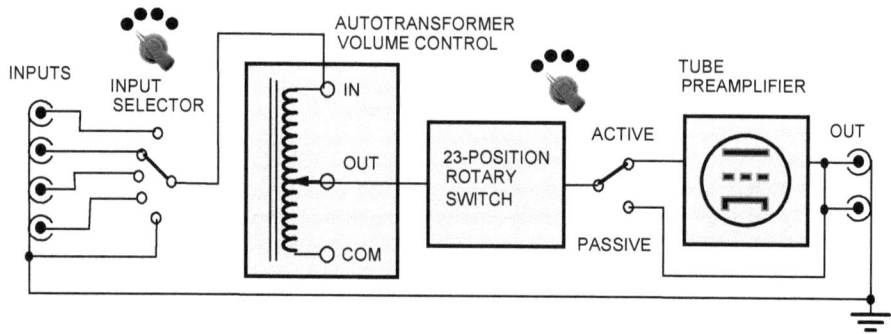

Fig. 7.19. Block-diagram of Promitheus Audio TVC preamplifier (one channel shown)

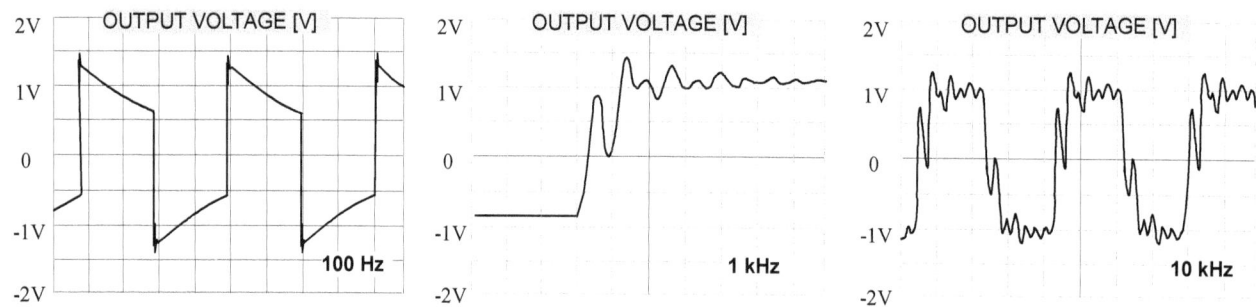

Fig. 7.20. Square wave tests of Promitheus Audio TVC preamplifier at three common frequencies reveal a severe distortion, slew-rate limitation, oscillation and narrow bandwidth.

The impedance problem

Active preamplifiers (those that provide voltage gain) have a fixed or nominal input and output impedance. The input and output resistance and input capacitance are of most interest to us audiophiles. As frequency varies, there is a slight change in output resistance of preamps that use a series output capacitor. Still, for all other intents and purposes, we can consider the output impedance constant.

In contrast, both the input and output resistance of passive preamps vary with their volume control setting. This is undesirable, and here's why.

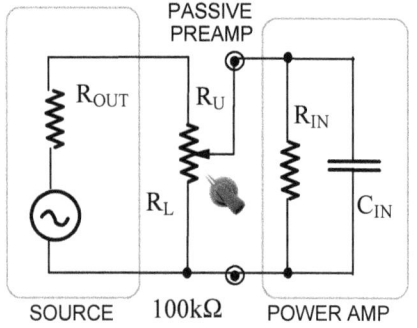 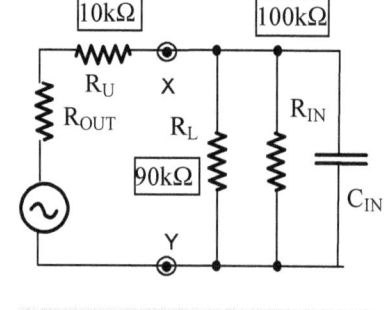 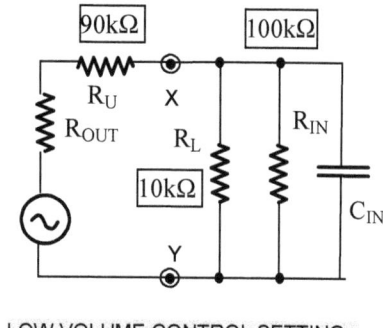

Fig. 7.21. A simplified circuit of a signal source (phono stage, DAC, etc.) driving a power amplifier through a volume control (passive preamp). "Simplified" because the two interconnect cables (one between the source and the volume control and the other between the volume control and the power amp) are not shown, so their resistances and capacitances have not been taken into account.

The secret to analyzing circuits (and many other technical problems) is to split the frequency range into three bands (low, mid, and high audio frequencies) and analyze them separately using various simplifications. Alternatively, as in this case, consider two extreme positions of the volume control potentiometer, one close to the maximum volume and the other close to the minimum or zero volume.

The position of the wiper or slider of the potentiometer (pot) effectively splits the total resistance of the pot in two. Assuming a 100 kΩ pot, in the first case, 90% of the resistance is the lower part (R_L=90 kΩ), and 10% is the upper section (R_U=10 kΩ). The upper resistance effectively adds to the output resistance of the source (they are in series), while the lower resistance is in parallel with the input resistance R_{IN} of the power amp.

Assuming a standard R_{IN} = 100 kΩ input impedance, with a high volume setting, the load on the source changes from 100 kΩ to 90 kΩ in parallel with 100 kΩ, which is 47.4 kΩ, less than a half of the previous input resistance. Unless the output resistance of the source was very high to start with, this should not have any sonic implications. At the chosen low volume setting, the output resistance of the source has increased by 90 kΩ, and the load resistance decreased to 10 kΩ in parallel with 100 kΩ, or 9.09 kΩ! Even if our source had a very low output resistance R_{OUT}, say a 100 Ω, now it behaves as a source with a very high R_{OUT} (above 90 kΩ), which is generally undesirable!

So, at lower volumes, we have an increasingly high impedance source feeding a difficult low impedance load, resulting in an increase in distortion.

Since the source output resistance and the amplifier's input capacitance C_{IN} form a low pass RC filter with the upper -3 dB frequency of $f_U = 1/(2\pi R_{OUT} C_{IN})$, such an increase in R_{OUT} (say from 100 Ω to 90 kΩ) will reduce the frequency bandwidth, causing a noticeable loss of "top end."

To make things worse, the output capacitance of the source (quite high in the case of a capacitor-coupled phono stage), together with the reduced load (passive preamp) impedance, forms a high pass filter. The lower the R_{IN} in the formula for the lower -3 dB frequency, $f_L = 1/(2\pi R_{IN} C_{OUT})$, the more the bass frequency will be attenuated. So, we have a loss of bass *and* treble, plus increased distortion at low volume control settings!

CLASS D AMPLIFIERS

How they work

Class D is a class of switch-mode amplifiers that share the description and the operating principle (and associated problems) with switch-mode power supplies. Pulse-width modulation is commonly but not exclusively employed to derive the two-level waveform. Pulse density modulation, self-oscillating modulation, and delta-sigma modulation are also used.

An audio signal is compared to a high frequency sawtooth waveform in a comparator (block diagram on the next page), whose output is a string of rectangular pulses of various widths. Typically, switching frequencies of at least 300 kHz (15 times the highest audio frequency) are necessary if reasonable fidelity is to be achieved. The higher such frequency, the higher the dynamic range and SNR (signal-to-noise ratio) of the amplifier.

A controller, among other tasks, acts as a phase splitter and produces two out-of-phase signals that drive two complementary MOSFET transistors which work as switches, either fully conducting or being fully turned off.

Finally, a low pass analog filter allows only the fundamental harmonic of these amplified pulses to pass while attenuating all higher ones. It can be shown mathematically that this fundamental harmonic is now an amplified version of the input analog signal.

The inherent problems with and weaknesses of Class D amplifiers

An ideal Class D amplifier would produce no distortion or losses, meaning it would be 100% efficient. However, you can intuitively conclude that any deviation of the HF triangular wave in terms of amplitude (nonlinearity causing distortion) and timing (frequency variations) will distort the sampled and thus amplified audio signal.

Also, MOSFETS are not ideal switches. They have a turn-on and turn-off time, transit time, and other limitations, which vary from one transistor to the next and are also temperature-dependent, which is bad, bad news. "Complementary" transistors of all types, MOSFETs, JFETS, and bipolar, are never a "mirror image" of each other, so they introduce nonlinear distortion in all push-pull stages.

One consequence of such uncertainty is the dead time when both MOSFET switches are off. The designer must address this issue to prevent both switches from being turned on simultaneously, a condition that would effectively short-circuit the power supply and lead to the instantaneous destruction of the MOSFET power transistors.

And you thought vacuum tubes were unreliable? A vacuum tube switching stage would tolerate being short-circuited for quite a few seconds; tubes cannot be destroyed in an instant like solid-state components can.

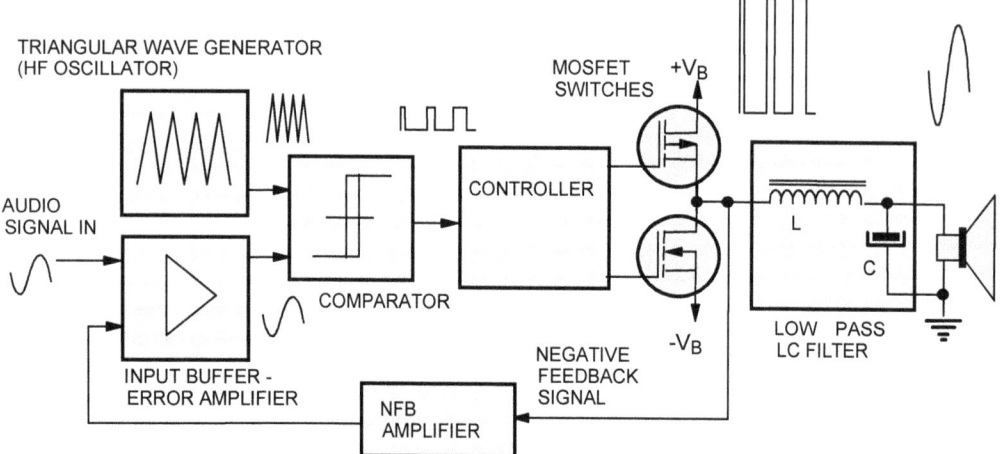

Fig. 7.22. A block diagram of a Class D audio amplifier.

Fig. 7.23. A sinewave audio signal and triangular HF carrier are compared and a string of rectangular pulses is produced. Pulse-width modulation is shown but other modulation methods are also used.

This dead time is the primary cause of nonlinearity and harmonic distortion. That means copious amounts of negative feedback must be used (25-30 dB)!

Although our block diagram shows a simple 2nd order LC low pass filter, its slope would not be sufficient and would allow too much of the higher frequency junk to pass onto the speaker, so more complex, high-slope filters must be used.

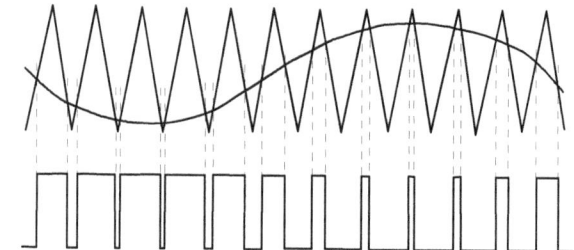

This complicates things since sharp filters suffer from nonlinear amplitude response and wreak havoc with their problematic phase response, creating phase and timing distortion.

The weakest link in this whole story is the output filter. If you imagine a complex LCR network (speaker crossover) connected in parallel with the output capacitor C, you don't have to have an advanced degree in electronics to conclude that the operation of such a filter will be significantly affected by what is connected to it

In other words, its filtering performance will vary with the frequency of the audio signal and the impedance of the loudspeaker; the amp will work and sound differently with different speakers.

Again, with such a small range of tools in their arsenal, audio designers try to solve this problem by, you've guessed it, high levels of negative feedback. Also, without NFB, Class D amplifiers would have a very poor damping factor (high output impedance).

Finally, just like their switch-mode power supplies brethren, Class D amps are generators and hotbeds of high-frequency interference, and we all know how sonically bad RFI is.

Are Class D amplifiers "digital"?

Well, it depends on how you define digital. The "D" in their name does not stand for digital. When they were conceived, letters A, B, and C were taken, so "D" was simply the next available letter in the alphabetical order.

Any signal whose amplitude can take only two levels is a digital signal to me ("di" meaning "two")! What those two levels are in voltage terms, 0 V and 5 V as in TTL logic, or zero and 100 V in Class D amps, is irrelevant. The signal does not vary smoothly and continuously as it does in all analog amplification stages.

A switch-mode power supply is a digital device compared to classic or analog (also called "linear") power supplies. Since Class D amps work in almost identical fashion (in principle, not in detail), they are "digital."

Anything "chopped up" (or "sampled") and suffering from jitter is digital! It is no longer the original, analog (continuous) signal but a series of digital pulses, which are then recreated or cobbled together into an analog signal. A DAC in your CD player works like that, and CD is digital technology. Just like Class D amps, the fact that DACs recreate the analog signal from a pulse stream does not make them analog!

How do they sound?

Even if Class D amps weren't digital, ultimately, does it matter? The ones that I've heard certainly sounded digital to me: cold, uninspiring, lacking emotion and musicality. I agree with Michael Fremer, who calls it a "ringy" character, "a glare or glow around the notes, like a parasitic halo," a "one-dimensional quality."

Professional audio reviewers have to thread lightly not to offend manufacturers who are also advertisers, so they have to be polite and make understated and "qualified" references while avoiding open, in-your-face critique.

I have no such restrictions, so I can express my personal opinions any way I like. Ten or so years after its debut, Audio Research does not feature DS450 amplifier in their product range anymore; that surely tells you something? One of the telltale signs of a product's sonic quality is its longevity, which is the result of its acceptance in the marketplace and customers' demand for it.

> "It's no surprise that this superquiet class-D amplifier excels on bottom. Bass is what class-D was originally built for; at least that's where it did the most good and the least musical damage."
> Michael Fremer in his review of PS Audio Stellar M1200, *Stereophile*, Dec 2020

The future-telling part

Quite a few companies (OEMs) design and manufacture Class D amplifier modules, so most commercial amplifiers use them as the heart of their models. A few other larger audio brands design and make their own modules. Class D amps (or at least the basic modules) can be mass-produced cheaply, ensuring their survival in certain products.

Class D amplifiers have their place in equipment where small size and low weight, cool operation, and super high efficiency are desirable, such as portable audio devices, car audio, etc. However, ultimately, none of these criteria are of any importance in audiophilia.

With the quality of tube electronics on the rise and their prices in constant decline (you can buy a beautiful looking SET 300B stereo amp for under US$800 including transport from China) I see no future for this amplifier class in hi-end audio, except in powered ("active") subwoofers.

In his *Stereophile* article (Jan 2013), Sam Tellig says, "There is so much riding on making class-D work: labor savings, reliability, customer acceptance (smaller size, less heat) and, just maybe, the Green Police." Labor savings are relevant only if manufacturers pass them on to customers in full or substantially; I suspect that most won't. As for the smaller size and less heat being acceptance factors to audiophiles is a serious delusion. The most critical acceptance factor is the quality of sound, which is under a question mark as far as Class D amps are concerned.

OTL TUBE AMPLIFIERS

As amplifying devices, vacuum tubes have a much higher impedance than bipolar transistors. This isn't an issue in preamplifiers, but it becomes a significant problem when tubes are used in the output stages of power amplifiers. Due to the mismatch in impedance, a very high impedance amplifier (typically 3,000-7,000 ohms) driving very low impedance (4-16 ohms) loudspeakers would not work well at all. A similar problem is present in automobiles, where a high revolution engine has to be coupled to slow rotating wheels, so a transmission (gearbox) is needed. An output (impedance) transformer in tube amps is such an electronic gearbox.

However, output transformers were identified as the main culprit or "bottleneck" way before transistor amplifiers (which do not need output transformers) became a serious alternative. Transformers attenuate bass frequencies, introduce their sonic coloration, are large, heavy, and expensive to manufacture.

Fig. 7.24. A stereo OTL amplifier with two triode-connected E130L pentodes per channel. With a 19 Hz - 60 kHz (-3dB,1W into 8Ω) bandwidth, it produced 2 Watts per channel. Two E86C triodes, one per channel, super simple topology and circuit, no negative feedback of any kind. Two 5V4G rectifier tubes are at the back, next to the power transformer.

Fig. 7.25. A push-pull OTL amplifier with two 6C33C-B triodes per channel produced 19 Watts with a very wide bandwidth (5Hz to 140 kHz), partially because of strong negative feedback used.

Many great minds have grappled with the fundamental problem of coupling a high output impedance tube stage directly to a low impedance load, the loudspeaker. The most famous was the Futterman design, which had limited commercial success, but was unreliable. Perhaps that was the reason why OTL amplifiers never moved beyond being a fringe product.

To get a decent output power, almost all commercial OTL amps are of the push-pull type, although it is possible to design and build a single-ended OTL amp, as pictured on the previous page.

How they work

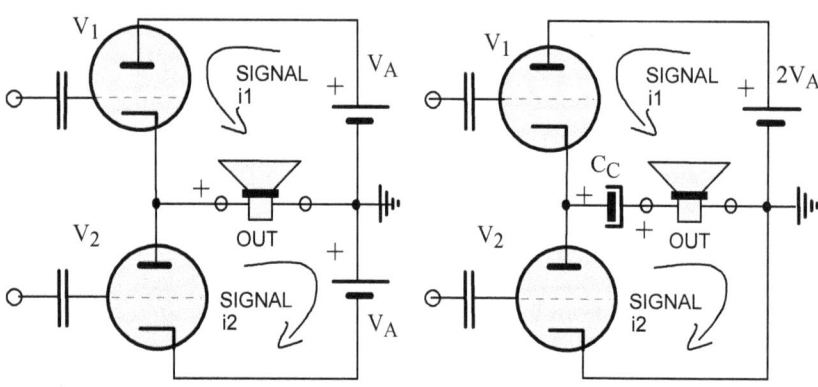

Fig. 7.26. The OTL output stage with symmetrical (bipolar) power supplies (LEFT) and with a single (unipolar) power supply (RIGHT)

When symmetrical (+/-) or bipolar power supplies are used, one end of the load (loudspeaker) can be at the ground potential. If the output stage is properly biased and balanced, there is no DC current flowing through the load; thus, there is no need for a coupling capacitor CC in series with the loudspeaker, as is the case if an asymmetrical (unipolar) power supply is used.

The lower tube V_2 works in a common cathode circuit, but the upper tube V_1 works as a cathode follower, which is where problems start.

Output stage's two "halves" have different voltage amplification factors, different input, and very different output impedances. OTL designers approach this disparity problem between the upper and lower tube in various ways. In 1997 US patent 5,604,461 was awarded to Bruce Rozenblit of Transcendent Sound, Inc. His solution is to boost the amplification of the upper tube by applying positive feedback to its driver.

An electrolytic capacitor in series with signal is bad enough, but even the addition of positive feedback is not the end of troubles. Most OTL amps have such a high distortion that very high levels of negative feedback must be used, and, by now, you should know that NFB is sonically bad, bad, bad.

Are the sonics worth all the trouble and complexity?

The OTL proponents claim that OTL amplifiers don't sound warm and soft like "typical" tube amplifiers, which is true (although well-designed tube amps with output transformers don't sound warm and soft either).

Tube puritans often criticize the OTL sound as being solid-state-like, which is only partially true. A few OTL amps we made and listened to did sound different, but still better than most SS amps I've heard.

Without the high & low-frequency limitations and resonances of output transformers, the bandwidths of OTL amps are impressive, typically from 5 Hz to 140 kHz (-3 dB, 1W into 8 Watts), but we listen to music, not to test data.

The verdict comes down to what kind of sound one likes. If you want the musicality and magic of directly heated triodes (or any triode), OTL will disappoint you - it has neither. Although one of the power tubes we used (6C33C-B) is a triode, it didn't sound like one at all in an OTL circuit.

If you want a fast, punchy, crisp, detailed, and transparent sound, OTL amps may be for you. However, you still get low power and limited speaker driving capability, so OTLs cannot replace solid-state amps driving inefficient speakers, which is their downfall. It is just not clear who they are for.

When you consider the circuit and power supply complexity, the biasing, and power tube matching problems, my personal opinion is no - it's not worth it. Build a 15 Watts to 35 Watts SET amplifier with lovelies such as E130L, EL153, GM70 or 211 (using quality output transformers) and enjoy a true triode sound superior to OTL in almost every respect.

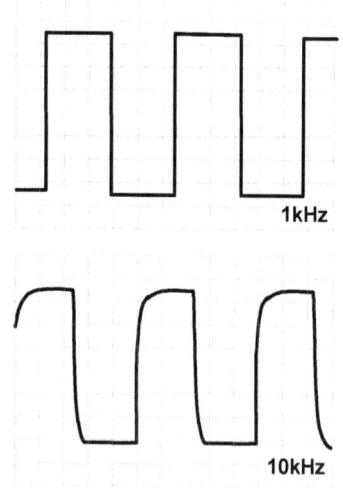

Fig. 7.27. A square wave response of our 6C33C-B OTL amplifier (18 WPC) at two typical test frequencies. Notice the perfect 1kHz result and a very good 10kHz waveform with fast rise time and only a small rounding of the leading edge.

TUBE ROLLING: CHANGING AND UPGRADING VACUUM TUBES IN AMPS & PREAMPS

Experiment: How to plot a tube's transfer characteristics on your tube tester

The worst way to "match" tubes is based on anode current in one steady point only, yet this is what tube sellers and even manufacturers do. The next best option is to match them for both anode current and mutual conductance, albeit again in only one operating point (one anode voltage, one bias voltage). Short of plotting the whole family of anode characteristics (which requires a range of anode voltages), the quickest way to match tubes is to plot their transfer curves (by varying the bias control on a tube tester) and superimpose them on top of each other.

This assumes that the bias voltage on your tube tester is variable. If it isn't, your tester tests tubes in only one point, predetermined by the tester's designer. Even if they belong to the family of mutual conductance testers (much better than the rather primitive emissions testers), these are not serious testers, only "quick tube checkers." A transfer curve shows the relationship between DC plate current I_A (Y-axis) and negative grid bias voltage as an independent variable on X-axis at a fixed plate voltage.

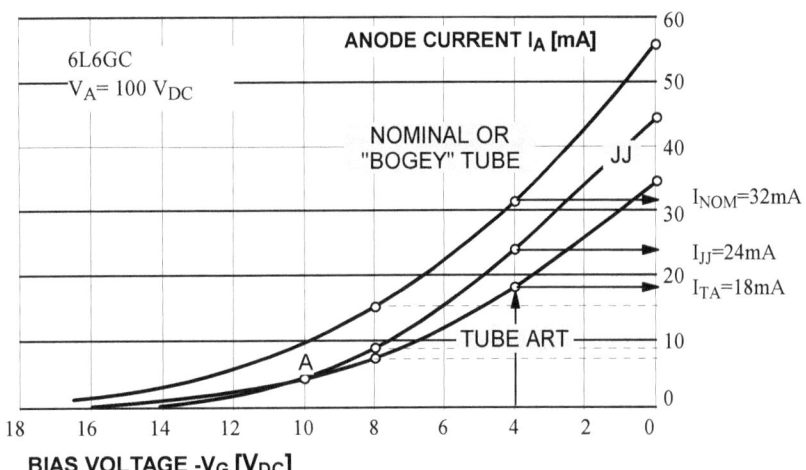

We measured transfer curves on Triplett 3444 tube analyzer of two brands of 6L6 tubes, the current production JJ and NOS Chinese production from the 1960s, labeled "Tube Art" (TA) by an eBay seller.

A 100 V plate/screen voltage was used in order to stay below the 50 mA current measurement limitation of the tester. With 250 V_{DC} anode/screen voltages, the anode currents would approach or even exceed 100 mA. The results were drawn on the graph taken from the GE catalog for a 6L6GC tube.

Notice the difference between the published curve by GE and the two tubes tested. Would these tubes perform differently in an amplifier? Absolutely!

Fig. 7.28. The measured transfer curves of JJ and TA 6L6 tubes compared to the curve published in the 1950s by GE (General Electric). These tubes will behave very differently in the same tube amp and will thus sound significantly different.

Both TA and JJ tubes pull significantly lower currents than the nominal, average, or "bogey" tube from the GE datasheet. Also, below point A (where two transfer curves cross), the TA tube pulls more current than the JJ tube. JJ's transfer curve is steeper in that region than TA's, meaning JJ tube has a higher mutual conductance. To the right of point A, JJ's curve is still steeper (the slope of the tangent at any point) than that of the TA tube, and the anode currents increase faster, so JJ's transconductance is higher than TA's across the whole range of operation.

What would happen if we plugged these tubes into an amplifier? Let's assume that our amp is biased at $V_G = -4\ V_{DC}$. While GE datasheet specifies that their average tube should pass around 32 mA of anode current in that operating point ($V_G = -4\ V_{DC}$ and $V_{AK} = +100\ V_{DC}$), the JJ tube only draws 24 mA and TA pulls even less, 18 mA.

The TA tube will produce the lowest audio power of the three tubes. Assuming identical anode voltage swings, due to its larger current swing JJ's tube will produce a higher power output, but it will still fall significantly short of the vintage GE tube that would produce higher output still.

Since we cannot make meaningful conclusions based on a single tube from each maker, we cannot say that all Tube Art tubes are "weaker" than JJ tubes or that all JJ tubes are "weaker" than the vintage GE tubes. This was simply to illustrate that variations between brands and even tubes made in the same factory can be significant. Swapping such tubes around in your amp will change operating conditions, output power, and the tonal voicing of your amp.

Why most tubes sold as "matched" pairs or quads aren't properly matched at all

The main problem with matching tubes on tube testers is that they do it in one point only. It is like saying you and I both have cute blue eyes and are thus identical humans in all other aspects. That would be laughable at best, or, at worst, you'd end up locked up in a psychiatric ward for a long, long time.

Notice that the two tubes (JJ and TA) in the graph above are matched in point A, with -10 V_{DC} on the control grid (bias voltage) and 100 Volts on the anode and screen. "Matched" in this case are anode currents only; the mutual conductance figures (the slope of the tangent in that point) will be different for the two tubes.

How meaningful are tube test "certificates"?

I put certificates in quotation marks since these flimsy pieces of paper with minimal information could hardly be called certificates. At best, they are more of a marketing tool designed to instill confidence in the buyer. Even genuine ones have a very limited value since only one or two parameters are specified. At worst, they may be forgeries.

A few years back we bought a "matched" pair of current production Shuguang 2A3 tubes from a Hong Kong seller on eBay. Upon receipt, we tested the tubes on Triplett 3444 tube analyzer.

The test "certificates" said I_A = 45 mA at 250 V_{DC} anode voltage and -45 V bias for both tubes. We replicated the test voltages and got 36 mA of anode current for one and 47 mA for the other! Those tubes were not matched in any sense; we returned them and lost $30 on return postage.

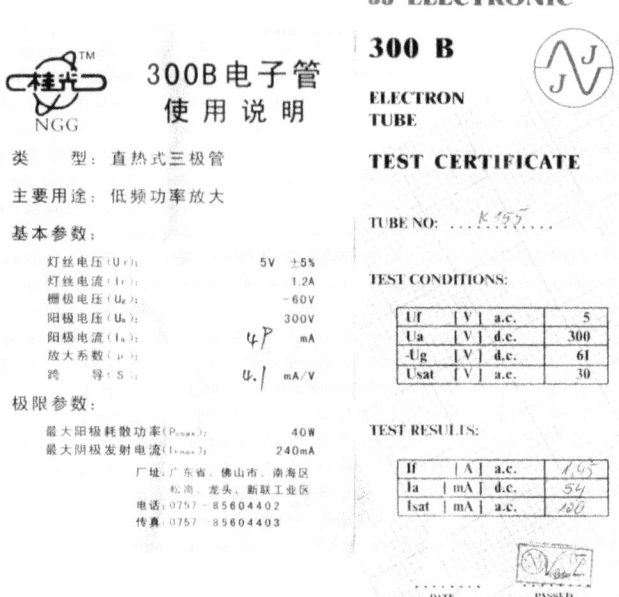

Fig. 7.29. NGG specifies both Gm (4.1 mA/V) and anode current (49 mA) for their 300B. Most other manufacturers only specify one of those. Shuguang, for instance, only lists the anode current. For some strange reason, JJ does not specify Gm either but includes the heater current and I_{SAT} (anode current saturation level) instead, two parameters of no importance to tube buyers.

Mismatched tubes as the cause of harmonic, intermodulation and crossover distortion

Ideally, in an idle push-pull output stage (no signal) the two tubes are pulling the same anode (plate) and screen currents. If the tubes are unbalanced in DC terms (unmatched), their plate currents at the same negative bias voltage will be different, and there will be a resultant DC primary current through the output transformer. This current causes three problems.

Push-pull output transformers are designed on the premise of equal and opposite DC primary currents, so any DC imbalance will reduce the permeability of the magnetic core and may even saturate it. That would have a detrimental impact on the output power, which would drop significantly, especially in the bass or low-frequency region. Also, both the harmonic and intermodulation distortion would dramatically increase.

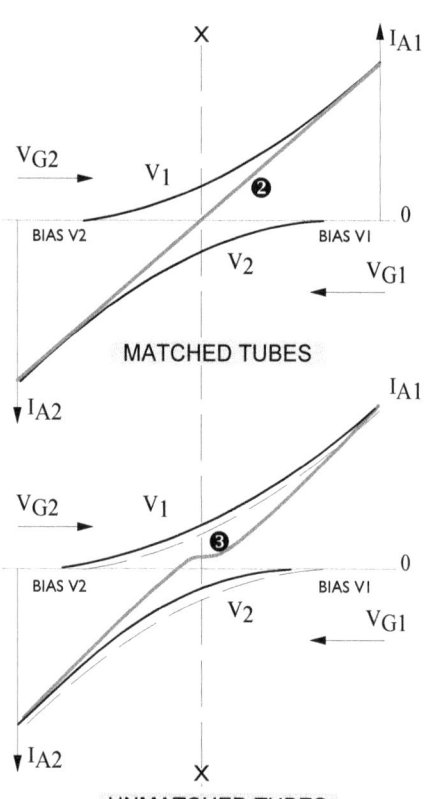

Notice how with only 12 mA of imbalance (1), the total harmonic distortion of a 6L6 push-pull stage increases by 0.2% for a 100 Hz signal, 0.55% for a 50 Hz signal, and a whopping 1.3% for a 20 Hz signal.

The imbalance of anode currents at the same bias voltage indicates the mismatch of transfer characteristics of the two power tubes, and that means that the overall transfer curve of the output stage is not a straight line, as it ideally should be (2), but a kinked curve (3). A curved transfer function means increased distortion.

Fig. 7.30. Harmonic distortion in a U/L push-pull output stage using 6L6 tubes with cathode bias as a result of an imbalance in DC cathode currents. Only one "direction" is shown; the results are identical for imbalance in the other direction.

Fig. 7.31. A matched pair of tubes results in a linear composite characteristic and no distortion in a push-pull output stage (upper graph). The unbalanced case shows stronger V_1 (higher anode current I_A) and weaker V_2 (lower I_A), resulting in a composite curve with a significant "kink" (3) and offset (lower graph).

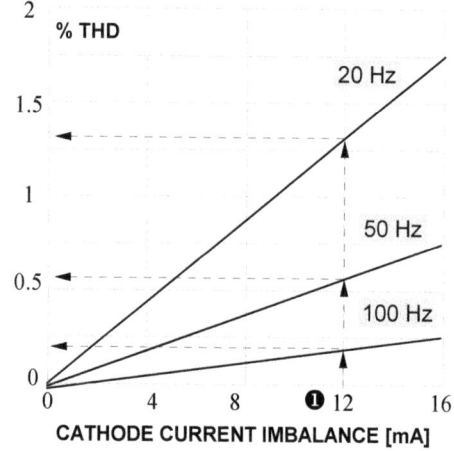

With perfectly matched tubes the transfer curves are mirror images of one another. The grid bias point is in the middle line X-X. The resultant (or composite) dynamic characteristic of the whole push-pull stage is a straight line, resulting in zero distortion.

The unmatched situation shows both transfer curves shifted upwards (one tube is stronger, the other weaker). As a consequence, the resultant transfer characteristic has a pronounced double bend or kink (3), and that kink is the cause of crossover distortion. Paradoxically, the distortion (4) is more significant (pronounced) at low volumes (lover signal amplitudes).

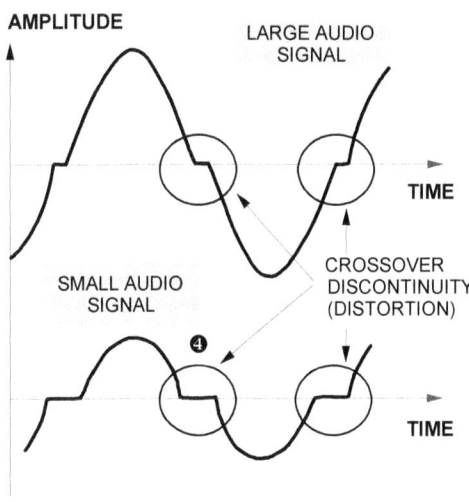

Fig. 7.32. The distorted composite curve caused by mismatched tubes will cause the kink (discontinuity) in the reproduced sine wave signal (4), known as crossover distortion. It happens when the signal crosses over from one tube to the other in a push-pull stage.

Why tubes of the same type sound different?

Most audiophiles will agree that different tubes (of the same type) make the same amplifier sound different. Tube "rolling" makes it easy to compare the sound of various tubes by simply swapping them around in an amplifier and listening critically. If you take a dozen or so tubes of the same type, say 12AX7 (ECC83), from multiple manufacturers, you will notice that anodes or plates are of different shapes and sizes. So are the cathodes and the grids, but these usually cannot be seen from the outside. Some anodes are twice as large as the others (JJ has the smallest anode here, Ei the largest).

Some anodes are "boxed," others flat, some are ribbed, some smooth. Also, they are made of different materials, some shiny silver in color (nickel?), such as Ei, some matte gray (JJ and RCA), others black (Arcturus). Other parts of these tubes will also differ.

Fig. 7.33. All marked 12AX7, but are they all the same tube? L-R: JJ ECC83S (Slovakia), RCA (USA), Ei (Serbia), Arcturus brand (unknown origin)

The material used for cathodes, heaters, grids, and other electrodes, the chemical composition of the glass will vary, the thickness and construction geometry of electrodes will be different. These variations are not apparent, tubes would need to be destroyed and taken apart for further inspection, and only chemical analysis would detect metallurgical differences.

It is indeed a miracle, or rather a long stretch of the imagination, to call all these very different tubes "12AX7".

Sure, they all have a similar amplification factor μ and other basic parameters (transconductance and internal resistance), but they may not behave in the same manner in other respects. No wonder they sound different in the same amplifier.

Tube adapters and substitution of similar, but not identical tubes

While tube rolling limits itself to the same tube type by various manufacturers and results in relatively minor sonic changes, tube substitution is a more drastic move because a different tube from the one specified by the amplifier's designer/maker is plugged in. This usually requires an adapter since the two tube type's plugs/sockets and pin assignments are generally different. The changes of amplifier's sound are usually significant, sometimes for better, other times for worse, depending on various factors, but mainly on the amplifier's design and operating conditions. While buying and installing tube adapters into an amplifier is easy, there is an element of risk.

The two tube types for which the adapter was made usually aren't entirely identical. The design of some amplifiers pushes tubes (especially power tubes) above their maximum voltage, current, and power dissipation limits, so if the replacement tube has lower ratings, it may fail in that amplifier and even cause amplifier damage.

For instance, EL34 has a much higher maximum anode voltage rating than 6L6 (800 Volts versus 400 - 500V for various types versions of 6L6 tubes). Some amplifiers use a very high anode voltage on EL34 tubes, way above the maximum allowed for 6L6 lovelies, which, upon substitution, may work OK for a while but fail prematurely.

The most straightforward adapter is probably for EL34 pentodes to be plugged into amplifiers that use 6L6 beam power tubes. The pinouts of these two output tubes are identical except pin #1. 6L6 does not have pin #1 connected, so on most amps, that pin on 6L6 octal sockets is left unused.

EL34, however, has its suppressor grid connected to pin #1. For those tubes to work in 6L6 amps, pin #1 must be connected to pin 8 (cathode). That can be done inside an adapter, but a more straightforward solution is to open the amp up and bridge (solder) pins #1 and #8 with a short piece of wire. This will not affect the operation of 6L6 tubes but will enable EL34 to be substituted for a (usually significant) sonic change.

EL84 pentodes have a similar power rating and biasing requirements to 6V6 beam power tubes but use Noval sockets. To plug them into amplifiers using 6V6 tubes (with octal sockets), an adapter is needed.

7868 and 7591 both use Octal sockets but with different pin arrangements.

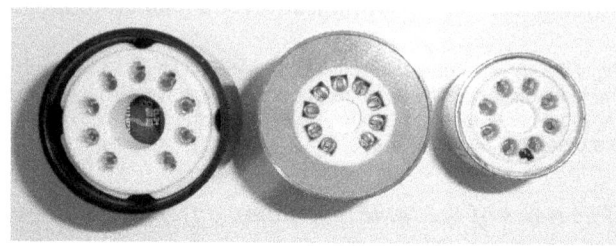

Fig. 7.34. Three currently produced tube adapters, L-R: Octal to Octal 7868 to 7591, Noval to Octal, EL84 (6BQ5) to 6V6, and RimLock to Noval, for instance ECC40 to 12AU7 (ECC82)

CASE STUDY: How tube rolling can result in expensive amplifier repairs

An old client of ours bought a pair of Mullard PL519 tubes on eBay and plugged them into our ValveMark Nimbus SET amplifier. He thought they'd sound better than the stock ones supplied. He noticed a burning smell and smoke coming out of the amp, quickly turned it off, and brought it to us to be checked.

We inspected the Mullard tubes and found that one had a pin missing. It either got broken off in transport, or the dishonest seller sold him a defective tube. After returning the original power tubes in, both channels worked, but one was distorting.

The power tubes were the original PL519 installed by us in 1997, made by Ei in our native Yugoslavia but branded Westinghouse USA. Despite the amplifier being used on average two hours a day, 15 years later, their emission and mutual conductance tested at 90% of the new values.

These great tubes had "clocked" more than 10,000 hours and were still working flawlessly!

Once the amp was opened up, we found that the 10 Ω resistor (1) between the cathode and ground was overheating but did not burn out completely. Its resistance increased to 17 Ω. The PCB started to melt but wasn't irreparably damaged

Once we replaced the resistor, the distortion disappeared. We biased the amp (with old tubes), and it measured like new again. Can you guess which pin was missing on the "new" tube? Consult the circuit diagram.

Something had caused a significant increase in anode current, which could only be a loss of the negative bias voltage. So, it was pin #1, the one connected to the control grid.

The tube had no negative bias voltage; that primary fault caused a secondary fault, a large current through the tube.

Luckily, the primary winding of the output transformer did not burn out, a testament to the quality of Dr. Bob's hand-wound transformers. Replacing it would cost quite a few hundred dollars.

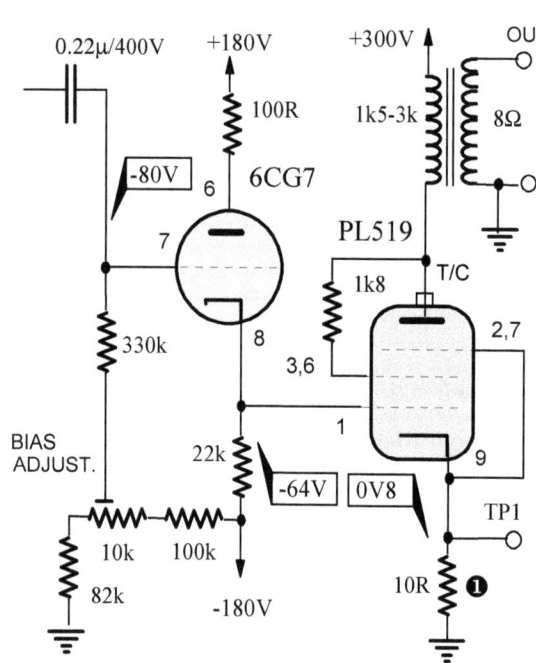

Fig. 7.35. The directly-coupled driver (pin 8, cathode of 6CG7, to pin 1, control grid of PL519) and fixed-biased (-64 V_{DC}) output stage

The tube's current was so high that its cathode resistor, rated at 3 Watts, started to act as a slow-blow fuse. That was its secondary purpose anyway; its main task was to enable us to measure cathode current for biasing purposes.

CASE STUDY: How tube rolling can create problems in preamplifiers

A local Perth audiophile brought in a Ming Da MC-2A3 line level preamplifier which had one channel louder than the other. This preamp is possibly the weirdest audio design I have ever seen, with some seriously poor design decisions (read: mistakes) but that is not our concern here.

The first thing I do when repairing any piece of tube gear is test its tubes on a tube analyzer and see if they are all operational, and if they are, how closely matched or unmatched they are.

One 6SN7 duo-triode is used per channel, in a SRPP configuration, two in total, followed by three 6DJ8 duo-triodes in a cathode follower of some strange topology, one per channel, with the third one shared between the channels.

The 6SN7s' heaters are powered by their own diode bridge and CRC filter (capacitor-resistor-capacitor) with a 13.2 V_{DC} output, meaning the two tubes' heaters are connected in series. One tube had 6.3V on its heater (perfect), the other 6.9 V (too high).

Now you can see why connecting tube heaters in series is a very silly idea, the heaters' resistances vary, and with the same current flowing through them in the series circuit, the voltage drops across individual heaters could be very different. That tube with 6.9 V on its heater will have a short and stressful life.

The three 6922 heaters in series were supplied with 19.8 V_{DC}. Two of them had 7.4 Volts on their heaters (waaaay too high, these tubes would last a month at best) and the third one had only 5.0 V.

Underheating a tube also shortens its life, but not as much as overheating it. However, the amplification factor (mju) of such tube would drop to very low levels, and that's what happened here, that is why one channel was much louder than the other.

That end 6922 was in one channel, the weaker one and was an old Amperex tube of good quality. I solved the problem by swapping that tube with the middle one, which was shared between the channels. After such simple swap, all three tubes had 6.6 V on their heaters (on a high side but acceptable, within allowed tolerances) and the volume levels in both channels were equyilized.

I am not sure if the owner shuffled the 6922 tubes around and created the problem himself, or replaced one of the three Russian-made tubes with that Amperex one (which certainly did not come with the preamp), but he claimed the preamp had recently been "fixed" by a technician and then the fault returned. Well, either that was a very incompetent "technician" or the owner wasn't completely truthful. I've seen many instances of both over the years.

TUBE ROLLING IS RISKY AND SHOULD BE APPROACHED WITH EXTREME CAUTION
Before plugging a replacement or trial tube into your amp or preamp, make sure it is fully tested and 100% operational.

Rolling rectifier tubes in tube amplifiers and preamplifiers - why do they impact the sound?

Most larger amplifiers and even some preamplifiers use 8-pin (Octal) tube rectifiers. Most have a standard (identical) pinout and can be substituted for each other, providing a few rules are followed (below).

5Z3P is a Chinese version of the common 5U4 rectifier; both are directly heated or filament-type rectifiers with 5V/3A heaters. The filament emits electrons and works as a cathode; there is no separate heater and cathode as in indirectly heated tubes.

5AR4/GZ34 is an indirectly heated rectifier with a low voltage drop. Despite its small physical size, its current rating is 225 mA, much higher than the 165 mA maximum for 5R4GA. However, 5R4G has a much higher voltage rating. Hence, it is physically larger (the distances between electrodes must be increased). GZ34 is used in many amps, so the prices of the vintage Telefunken (pictured on the next page), Mullard, or Amperex lovelies have gone through the roof. Currently produced Sovtek, Tungsol, JJ, Shuguang are OK but not as well regarded sonically.

TUBE RECTIFIER SUBSTITUTION RULES
1. Both must use the same socket and have the same pinout
2. The replacement rectifier can be of higher voltage & current rating, but not the other way around.
3. The replacement rectifier must have the same heater voltage and identical or lower current draw.

5Z4P is a smaller China-made directly-heated rectifier, similar to 5Z4G or GZ30. Its maximum ratings are identical to 5Y3GT (not pictured), another popular 5V/2A filament rectifier, but its voltage (350V max.) and current (125mA) ratings are low.

5Y3GT has a high internal resistance and thus a high voltage drop in a power supply. If your amp or preamp uses it, you can substitute a more powerful rectifier such as 5AR4 or 5V4 for a lower voltage drop and higher anode voltages in the amp.

Compared to 5Y3GT, 5V4 is a more modern indirectly-heated 5V/2A rectifier rated at 175 mA.

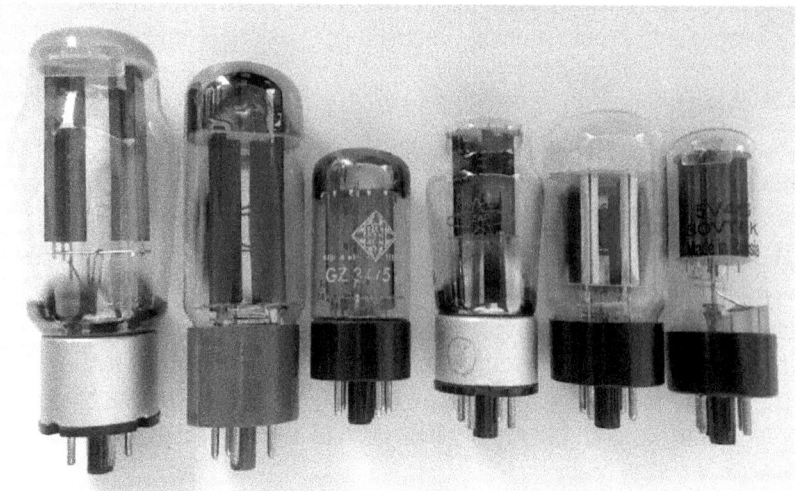

Fig. 7.36. Octal rectifier tubes of various power ratings, origins, and vintages. They can be interchanged (substituted) for one another in many but not all amplifiers. Study their ratings carefully before making any substitutions!
L-R: 5Z3P (old stock Chinese military), 5R4G (NOS Philips), GZ34 (5AR4), vintage German (Telefunken), 5Z4P old stock Chinese military), 5V4GA (vintage USA-made by GE) and 5V4G, current Russian production (Sovtek)

Mercury vapor tube rectifiers - the best sounding of all

Mercury vapor (MV) rectifiers feature an almost constant voltage drop, which does not increase with the load current, as is the case with vacuum rectifiers. The regulation of the power supplies with MV rectifiers is significantly improved, resulting in lower distortion, lower dynamic compression, and better sonics.

However, such tubes aren't in production anymore. Mercury is toxic and classified as a hazardous material. As long as such tubes are intact, it's contained inside the glass bulb and is harmless. The only dangerous situation would be a broken glass envelope while the tube was hot, i.e., operating in an amplifier. Mercury vapors would spread into the surrounding air, so breathing them in would be unavoidable.

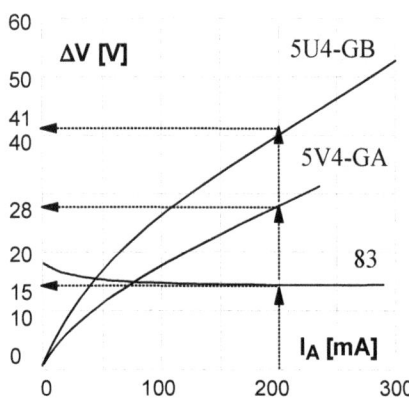

Fig. 7.37. Voltage drop on 5U4-GB, 5V4-GA, and mercury vapor 83 rectifiers against current. Notice a much lower internal voltage drop on 5V4 tube and an even lower drop on the 83 tubes (above 50 and 75 mA, respectively).

Ultimately, the decision is yours. If a few droplets of mercury inside a rectifier tube make you uneasy, by all means, stick to other types.

One reason different tube rectifiers will change the sonics of an amplifier is the difference in their internal voltage drops. For instance, a stereo amplifier with a 5U4 rectifier, pulling 200 mA from the high voltage power supply, works with 350 V. The voltage drop on the rectifier tube is 41 V.

Substituting a 5V4 rectifier with a voltage drop of only 28 V (13 Volts lower) would raise the anode voltage of the amplifier from 350 V to 363 V, enough to change the operating conditions, distortion, and sonics.

Although type 83 tube cannot be directly plugged into an octal socket for 5U4 and 5V4, an adapter could be used, and the voltage would rise even further, 26 Volts, to 376 V!

However, the increase in anode voltage cannot be the only reason mercury vapor rectifiers sound better than any other. There are so many factors that impact the sonics of audio components that we don't even know what they are, let alone why they do what they do.

MODIFYING AND UPGRADING AMPLIFIERS AND PREAMPLIFIERS

Many commercial amps and preamps are old, dated, and deteriorated, requiring a significant upgrade of the compromised parts. Others, including many "modern" currently-produced ones, have been poorly designed or built (constructed) and sound terrible. However, providing their fundamentals are sound (a large, solid, non-resonant chassis, a decent power supply, and good quality power and output transformers), everything else can be fixed.

To give you an idea of what to look for and what is involved, let's look at a few typical case studies.

CASE STUDY: Music Angel preamplifier

A budget made-in-China tube preamplifier ($200-300 + transport) with solid timber fascia, three line-level inputs and two paralleled outputs. The circuit is a copy of the vintage Marantz 7 preamplifier, seemingly adopted without any critical analysis or subsequent optimization. The sound was awful. Very colored, tubey, but of the worst kind - muffled, no transparency or detail, no emotion or dynamics. This kind of "hi-fi" gives tubes a bad name. Plus, the whole preamplifier was very microphonic. Just touching the control knobs would send a tingling sound through to loudspeakers.

The components inside looked very old and even used, with lots of scratches and wear & tear on them, especially the large electrolytic capacitors. It was as if somebody picked them up from a box of junk parts where they were rubbing against each other for years, if not decades.

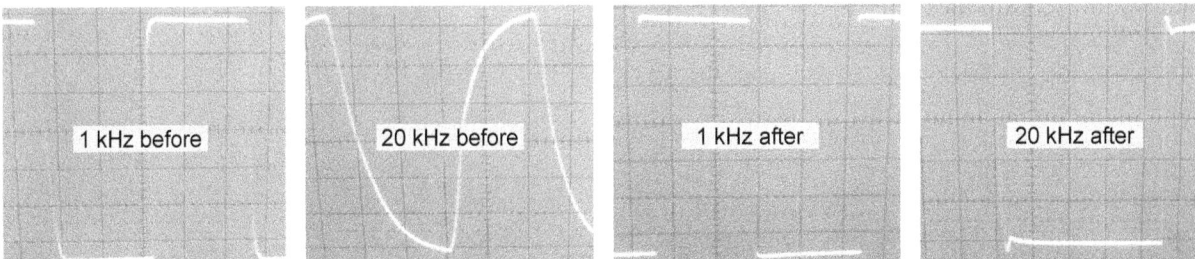

Fig. 7.38. (L-R): 1kHz and 20 kHz square wave reproduction before and after improvements
Fig. 7.39. The internal view of the Music Angel preamplifier

The preamp's steel chassis wasn't earthed at all (1), making it illegal and deadly dangerous! The power transformer (2) was enclosed (a good idea), and the mains switch (3) was far from the audio section. Fine.

A tube rectifier (4) was used for high voltage, and solid-state diodes rectify the heater voltage (5). The three inputs (6) are very close to the selector switch. Strantgely, the volume control pot (7) was at the output, close to the output RCA sockets (8).

To minimize the length of internal wire runs in the preamp, the selector switch and the volume control pots are operated through extension shafts (9), a good practice.

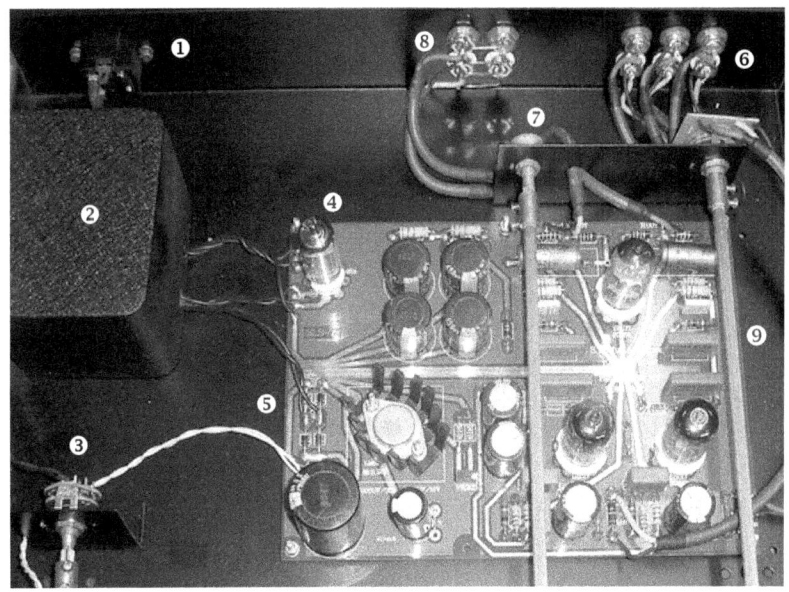

The frequency range was poor, 9 Hz - 28 kHz (-3 dB frequencies), and the square wave response confirmed that. The 20 kHz test signal degenerated at the output into a sawtooth waveform. The thick purple shielded cables were the culprit; their measured capacitance was almost 1nF per meter! All were replaced by a thin single-core silver plated wire (unshielded). The sound opened up, gained the missing top end, the previously mushy and dark-sounding preamp become infinitely more transparent and detailed.

The volume control pot was moved to the input but physically left in its location. There was far too much gain (12AX7 tubes used) despite very strong negative feedback. We reduced the anode resistors to lower the gain (4) while reducing NFB. The upper -3 dB frequency jumped to 185 kHz.

The microphony was reduced by placing the PC board on rubber mounts and replacing the Chinese tubes with NOS ones (Mullard, Ei, Amperex, Sylvania, RCA, ... all sounded way better than the stock tubes). However, some microphony remained, probably due to the printed circuit board and ceramic sockets.

PRINTED CIRCUIT BOARDS AND CERAMIC SOCKETS - A RECIPE FOR A SONIC DISASTER!
Ceramic tube sockets are among the most overrated of all audio components. In RF (Radio Frequencies) circuits, ceramic sockets may be advantageous, but at audio frequencies, they bring no benefits at all! Phenolic ("plastic") sockets are less microphonic and sound better.

CASE STUDY: *Dynaco (Dynakit) ST-70 power amplifier improved*

Dynaco (Dynakit) ST-70 is probably the most common and most upgraded push-pull tube amplifier of all time. Lots of cost-cutting measures: an undersized power transformer, a joke of a choke, an inadequate power supply, the list of its shortcomings is longer than my arm.

ST-70 designers suffered from severe delusions regarding both functionality and aesthetics. The controls that are used often (On-Off switch) are at the back of the amp, while the ones seldom used are at the front (stereo-mono switch, bias check terminals, and preamp power connections).

Originally, the front of the amp was not just ugly but impractical as well. The RCA sockets were too close together, so thick interconnects couldn't be used. The amp had no volume control, so a preamplifier was required. In short, a vintage amplifier that looked dated and sounded ordinary. Just because Dynaco sold a few hundred thousand of them over the decades does not mean the amp is worthy of its reputation.

The new front fascia was cut out of a large sheet of red Perspex® with silver backing. The new RCA sockets (1) are now further apart, the ON-OFF switch (2) is now a rotary-type, mounted at the front (left knob), while the right knob controls the volume (3). New silver feet match the knobs and the chrome chassis. The daggy brown cage was resprayed in metallic charcoal.

Small biasing sockets in front of the power transformer (4) replace large and ugly octal sockets that used to be at the front. After this photo was taken, the multi-cap (5) was replaced by a motor start capacitor, and a few smaller electrolytic capacitors were added under the chassis. All coupling capacitors were replaced with much better quality film & foil type.

Fig. 7.40. Dynaco ST-70 after a major cosmetic facelift and functional upgrade. The cage is still dated, but the front fascia looks elegant in red Perspex® with a reflective background.

Power supply improvements

To save page space, we will not publish the original circuit diagram; it is available online, anyway. Referring to the improved power supply's circuit diagram (next page):

1. The fuse and On-Off switch moved to the same line (Live).

2. The illegal and deadly dangerous 2-prong mains plug and cord was replaced with a 3-pin mains plug and cable. The amplifier is now grounded, safe and legal.

3. Larger electrolytic caps in the filtering circuit. The first 30 µF capacitor increased to the maximum allowed 47 µF, and three other filtering capacitors increased from 30, 20 & 20 µF to 220, 220 & 47 µF. You can calculate yourself the increase in stored energy! The last capacitor is now a 47 µF film & foil motor start unit, not electrolytic, for a significant improvement in sonics.

4. The ancient and unreliable selenium rectifier in the bias circuit was replaced with a modern silicon diode. The failure of this rectifier would result in a total loss of bias voltage, so all four power tubes would overheat (plates would start glowing cherry red) and ultimately self-destruct.

5. Zener diode and 0.1 µF film bypass capacitor were added to the bias circuit to make it more stable.

6. 6k8 resistor in the CRC filter replaced with a second (added) 15-17 H choke of a lower DC resistance. That will improve power supply regulation and also slightly raise the DC voltage in points A and B.

7. The amp already had a separate bias adjustment for each channel, so no improvements were needed there.

AUDIO AMPLIFIERS - HOW THEY WORK AND HOW TO IMPROVE THEIR SOUND

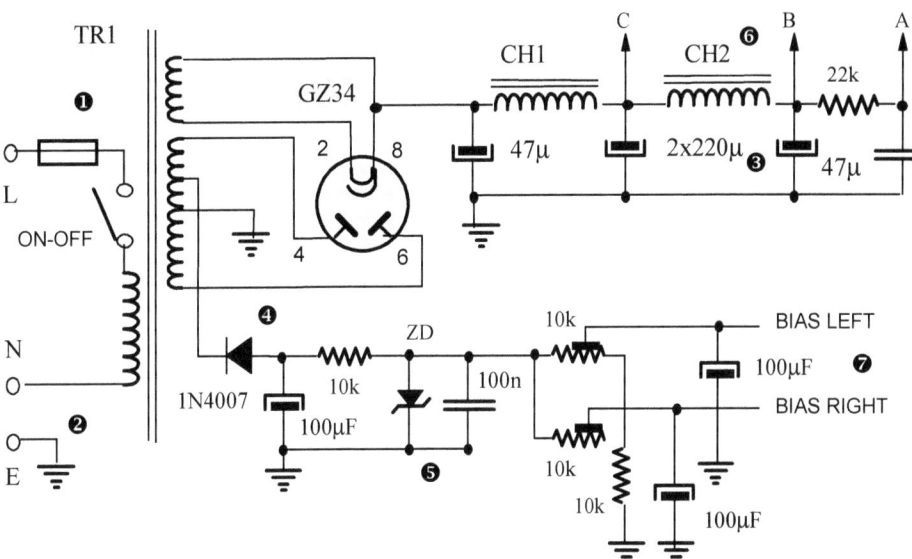

Fig. 7.41. The power supply section after a few easy modifications. Heater circuits remain unchanged and are not shown here.

Partial cathode coupling & other audio circuit improvements

The improvements #1 - #7 refer the the marked numbers on the audio-section's circuit diagram below.

1. Add a 10 kΩ trimmer potentiometer (1) in the cathode circuit of the phase inverter and change the 47 kΩ cathode resistor to 39 kΩ. This enables the AC (signal) balance between the two output tubes to be adjusted. This issue is critical at higher audio frequencies where the two grid driving signals tend to differ

2. Add a 5-10 kΩ trimmer potentiometer between the two 270 kΩ resistors that bring the negative bias voltage to the grids of output tubes. This enables the DC (bias) balance between the two output tubes to be adjusted. The two tubes are never perfectly matched and may age differently; adjust that balance after 400-500 hours of use.

3. Increase the value of 0.1 µF coupling capacitors to 0.22 µF or 0.33 µF.

4. Add 100Ω screen grid resistors to the output tubes.

5. EL34 suppressor grid (pin 1) does not have to be connected to the cathode (pin 8). You can connect it to the ground or to a source of low negative voltage; the bias voltage terminal is one such possibility. Experiment and see if you can notice any difference in sound and amplifier performance.

6. Implement partial cathode cross-coupling. Remove the 15.6Ω cathode resistor from the output stage. Add a 10Ω resistor between the 4Ω tap and ground, so you can measure cathode currents by measuring DC voltage drop on it. Disconnect the speaker COM terminal from the ground. Cross-couple the power tubes as per the diagram.

7. Finally, reduce the global negative feedback via the 1k resistor by increasing its value to 10k; you can even remove it completely since the partial cathode cross-coupling introduces NFB in the output stage instead. If you remove the global NFB, to reduce gain, change the 270 kΩ anode resistor (8) of the first pentode to 100 kΩ to get a triode-like sound.

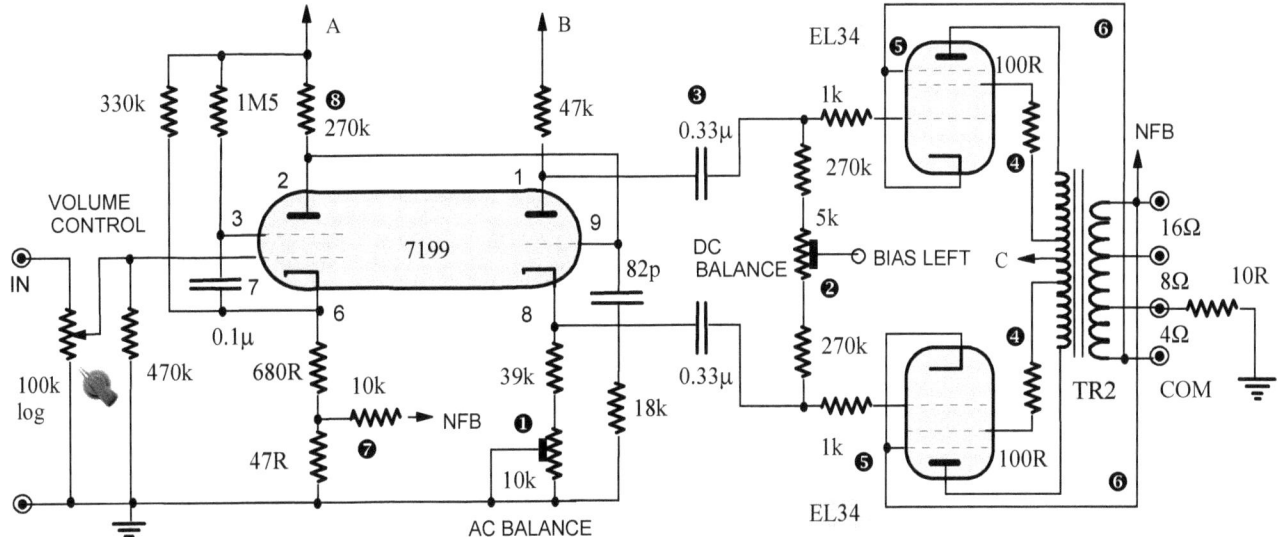

Fig. 7.42. The audio section of Dynaco ST-70 amplifier (one channel shown) after a few simple and easy modifications.

CASE STUDY: Cary SLP-90 line preamp improved

The first stage is a common cathode amplifier with both halves of the 6SN7 duo-triode strapped in parallel. However, paralleling triodes also doubles their inter-electrode capacitances and lowers the upper -3 dB frequency, rolling off the treble response. Cary specifies the upper -3 dB frequency as 163 kHz, while we measured only 43 kHz. The lower -3dB frequency was specified as 5 Hz, but we measured 19 Hz, not a great result either.

Initially, the preamp sounded sweet. No harshness, no shrillness, but also no top end at all! The cymbals, the triangles, and similar treble-dominant instruments were severely curtailed. The more we listened, the less we liked it. There was no detail, no transparency, the bass was weak, and the presentation veiled.

First, we removed the two Russian paper-in-oil (PIO) coupling capacitors between the two stages (1) and the output cap (2) and replaced them with film & foil polypropylene caps. PIO caps are highly regarded by audiophiles, but in this circuit, they sounded terrible. Into the bin! That immediately opened the sound up; the treble appeared, the level of detail increased. The preamp sounded faster; gone was the sluggish dynamics.

Notice how close the DC voltages on the anode (plate) of the first stage (pins 2 and 5) and the grid of the second stage (pins 1 and 4) are, 90 V_{DC} compared to 80 V_{DC}. To lower the anode voltage to 80 V_{DC}, the anode resistor would need to be increased to 51 kΩ. Not to be bothered with such a minor change, adopting a brute force approach, we connected the two points, which coupled the two stages directly, and everything worked like a charm. The sound improved even further.

The preamp's output impedance was around 400 Ω at 1 kHz but rose significantly at 20 Hz to around 7kΩ, due mostly to the limited size of the output capacitors. This was the cause of weak bass, especially with solid-state amps and tube amps whose input impedance was relatively low, less than 47 kΩ. After replacing the output capacitor (2) with a 6.8 µF film & foil unit, the improvement in bass was significant.

The highly-regarded PIO capacitors can be bad news. Cary's (physically) large capacitors had a significant parasitic capacitance between their metal body and chassis. This capacitance forms a low-pass filter that shunts high frequencies away to the ground and muddles the sound.

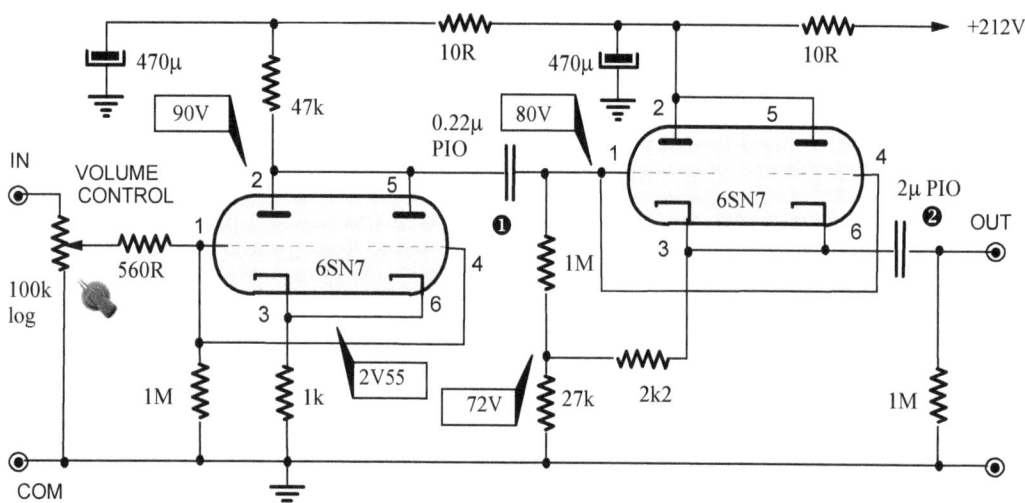

Fig. 7.43. Audio circuit before modifications (one channel only).
© Cary Audio

Fig. 7.44. Internal view of one channel, the original wiring with PIO coupling capacitors, (1) and (2).

No shielded cable in sight, the chassis is used as a ground plane, no star earthing point.

THE SONIC BEAUTY OF DC COUPLING

The elimination of the coupling capacitor between two amplification stages (directly coupling them) results in lower phase distortion. The benefits are most apparent at the frequency extremes, resulting in faster & tighter bass and higher treble precision & transparency.

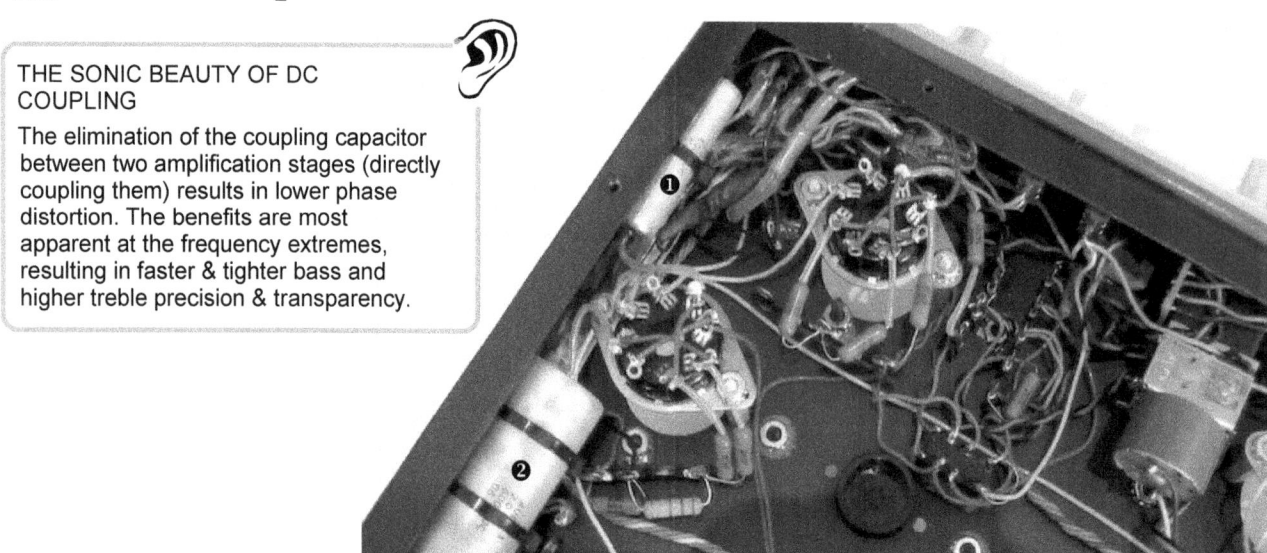

Modifications to allow different types of tubes to be used in the same amp or preamp

If possible, before purchasing an amp or preamp, get its circuit diagram, either from the manufacturer or the web. At least, identify the tubes it uses; find out their equivalents and similar tubes that could be used for tube rolling.

Reisong Boyuu A10 sells for US$243 for a PCB version, plus an optional $15 for a point-to-point wired version and $15 for a tube cage. Add US$105 for air delivery to Australia, slightly more to the USA.

This Made-in-China amplifier ticked all the boxes: it looked nice, it was cheap, had point-to-point wiring, and its circuit diagram was available online.

It sounded better than most similar budget amps that we had on our test bench and in our listening room; some were OK, others truly shocking (poor construction and horrible sound).

Two Chinese 6N2J duo-triodes provide voltage amplification (1). Russian 6N2P (Russian: 6H2P) (7) and Chinese 6N2 duo-triodes have their two filament elements connected in parallel (pins 4 and 5), with pin nine used as an internal shield connection (just like 6DJ8 or ECC88). The pinout of the other six triode pins is identical to 12AX7. So to use 12AX7 -12AT7- 12AU7 tubes (6), only pins 4, 5, and 9 have to be rewired.

The Chinese military NOS rectifier tube (5) was replaced with a better-sounding made-in USA 5V4 rectifier (2). 5U4 and GZ34 can also be used.

We added Nichicon Gold cathode bypass capacitors in parallel with the input stage cathode resistors (3).

Fig. 7.45. Reisong Boyuu A10 after the first wave of modifications

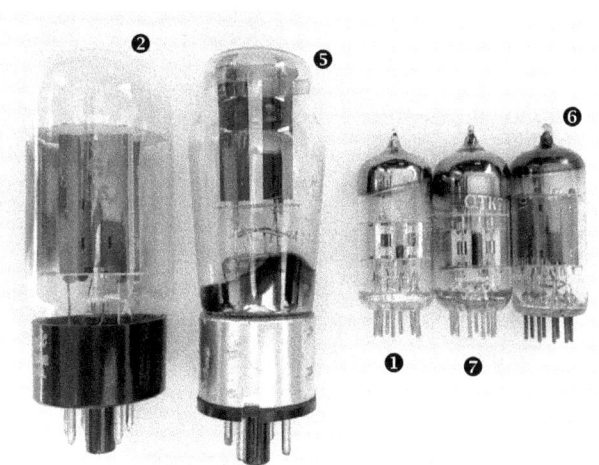

Fig. 7.46. The original and replacement rectifier and preamp tubes for Reisong Boyuu A10

After this photo was taken, we replaced the decent Philips metalized film coupling capacitors (4) with polypropylene film & foil caps. There's only one coupling cap per channel so buy the best ones you can afford.

There are only seven resistors in the audio circuit (per channel); replace all of them with your favorite audiophile brand.

There's so much space inside the otherwise compact chassis that all three electrolytic capacitors in the high voltage power supply section (8) can be upgraded to much larger values and physically larger units.

There was way too much gain and shrillness with 12AX7 (a tube whose sound I truly dislike), even with the medium gain 12AT7, so we settled on 12AU7. Now the amp sounded smooth, sexy, and seductive.

The hybrid rectification is provided by a tube rectifier and two ordinary 1N4007 silicon diodes in a full-wave bridge. Replace them with fast- or ultra-fast recovery diodes, which sound much better, most likely due to their soft recovery characteristics.

The EL34 power tubes are self-biased with 500 Ω resistance in their cathodes, but in that regime dissipate only 15 or so Watts out of the maximum 25 Watts rating, meaning they are cold-biased. Tubes sound much better when hot-biased, so we changed the 490 Ω cathode resistor to a 390 Ω wirewound resistor, which increased the cathode current from 48 mA to 59 mA and dissipation from 15 to 19 Watts. The best option would be to use 330-ohm resistors, which would run the tubes really hot, at around 24 Watts dissipation.

This diminutive amplifier, rated at 12 Watts per channel, drove Opera Terza floorstanders without any noticeable strain, with plenty of power in reserve. Since we had two of them, while writing the section on bridging and paralleling amps (a few pages ahead), we decided to parallel the two channels and use the amps as monoblocks.

I'd thought that nothing in audio would surprise me anymore and that I shouldn't be excited like a little boy with a box of chocolates, but the sonic improvement was immense. Everything improved, the bass, the slam, the speed, transients and microdynamics.

Distortion went down, too, resulting in cleaner, more transparent sonics. Each channel now had its own power supply (nothing shared between the two monoblocks). As a result, the crosstalk was eliminated and the soundstage improved.

Modifying 6L6 or KT88 tube amps to accept EL34 (6CA7) and 7027A as well

Every designer has their favorite power tube(s), and in the hi-fi realm, 7027 and 7027A are definitely on our list of desirables. While the distortion of the ubiquitous 6L6 in SE triode mode is up to 10%, triode connected 7027A in the SE output stage easily achieves 5 Watts at only 2.8% distortion.

Two 7027A in push-pull can produce 50 Watts at only 1.5% harmonic distortion and 76 Watts at only 2% THD! In other words, 6L6 is a high distortion Tube, more suitable for guitar amps, while 7027A is a cleaner hi-fi tube.

For the same heater power, 7027A has a 35 Watt anode power rating, higher than 30 Watts of 6L6 varieties, so it is also more efficient. After all, 7027 is a much more modern tube, developed 20 years after 6L6, which dates back to 1936 (same vintage as 300B)!

SIDE-BY-SIDE	EL34	6L6	7027A
Heater volts / amps	6.3/1.5	6.3/0.9	6.3/0.9
Max anode voltage [Volts]	800	500	600
Max screen voltage [Volts]	800	450	500
Anode dissipation [Watts]	25	30	35
Screen dissipation [Watts]	8	5	5

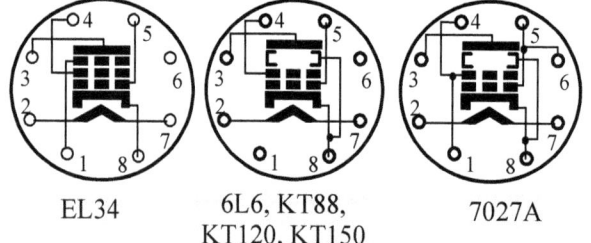

EL34   6L6, KT88, KT120, KT150   7027A

EL34 and 6L6 are pin-compatible, providing pin 1 on the tube socket is connected to pin 8, since EL34 does not have an internally connected suppressor grid but is taken out to pin #1 instead. So, modifying 6L6 amps to take EL34 involves simply soldering a wire between pins #1 and #8!

Luckily, 7027 and 6L6 are also pin-compatible beam power tubes, providing that in amplifiers designed for 6L6, nothing is connected to pins 1 and 6 (not used by 6L6 but internally connected in 7027A). Pin #1 is a screen grid in 7027A (also connected to pin 4, as in 6L6), and pin 6 is connected to pin 5, the control grid.

But what about pin #1 in EL34? Can we modify a commercial amp to take all three (6L6, EL34, and 7027A)? Yes, we can! A suppressor grid of a pentode doesn't have to be strapped to its cathode; it can be tied to the anode or the screen grid instead. Join socket pins #1 and #4 to strap EL34's G3 (suppressor grid) to its screen grid (as 7027A has done internally), and such connection will have no effect on 6L6 at all since its pin#1 isn't used.

Just make sure nothing is connected to pin #1 in your amp! Sometimes these unused pins on tube sockets serve as tie-in points for other components, and if that is the case with your amp, this mod is still doable, but you first have to transfer such parts to another tie-in point, perhaps by adding a small terminal strip or two.

If you want to use 7027A in a 6L6 amp, there are no heater issues, but notice a much higher heater draw of EL34, 1.5A versus 0.9A for 6L6 and 7027A, so the heater winding on the amp you are modifying must be capable of delivering additional 0.6A per tube. So for a stereo single-ended amp, that's 1.2A of extra load on the heater power supply. KT88 tubes pull 1.6A heater current each, so no issues there if you want to use EL34s instead.

All these mods are much easier to perform on "hardwired" ("point-to-point") amps; otherwise, PCB tracks need to be cut and rewired, a job best left to professionals.

Should you replace potentiometers with stepped attenuators?

Both sonically and from a reliability perspective (they become scratchy, lose contact and suffer from tracking errors), volume control potentiometers are usually the weakest link in the signal chain. Attenuators are considered superior because a switch and a resistive divider replace the carbon or conductive plastic film.

Most commercial attenuators use at least 23 or 24 position switches. Obviously, the higher the number of discrete positions, the finer the volume control resolution. These steps or jumps in resistance between the switched positions are usually expressed in dB. They may be uniform but need not be. At lower volumes, finer steps may be used, say 1.5 or 2 dB, and at higher volumes, bigger jumps can be tolerated, for instance, three or even 4 dB.

The simplest type of audio attenuator is the series type, where a string of resistors is connected between the input and common (ground). As the switch is turned from the lowest (zero) volume (the bottom position on the drawing where the output voltage is zero, $V_{OUT}=0$), more and more resistors are connected in series between the output and ground. All of the resistors are always in the signal path! The input impedance of the series attenuator is constant.

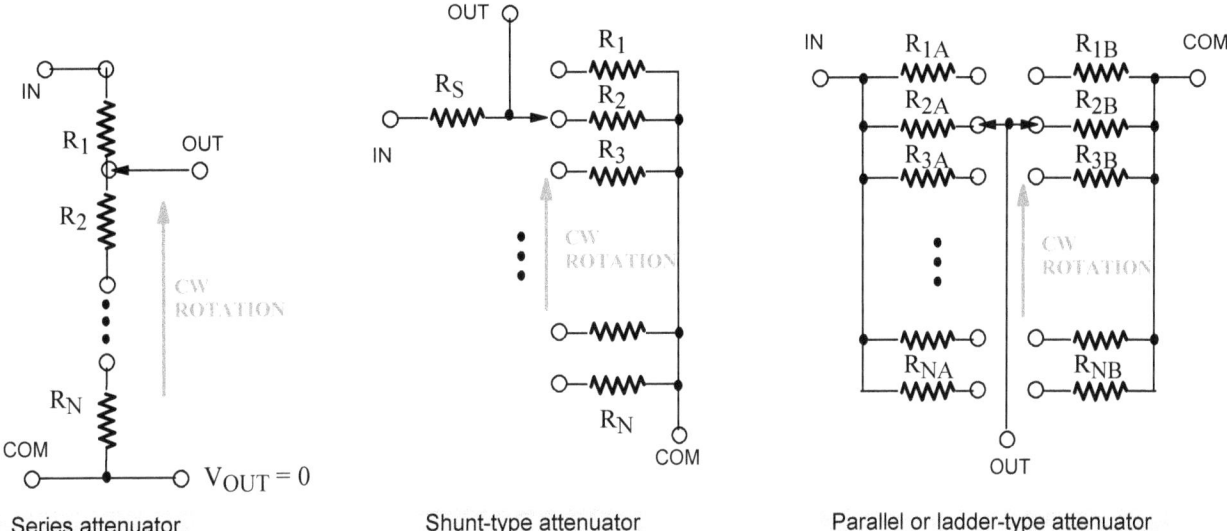

Fig. 7.47. The three types of audio attenuators

However, no matter in which of the 24 positions the switch contact is, the signal always goes through all 23 resistors (there are always N-1 resistors if N is the number of switch positions). That is the biggest weakness of the series type: there are just too many solder joints and too many noisy resistors in the signal path. One wonders if any improvement over a good quality potentiometer has been achieved, and my answer is a definite no.

The parallel or "ladder" type and the shunt attenuator both eliminate this problem by having only two resistors switched into the circuit at any time. The shunt type is simpler and cheaper since all positions share the same series resistor R_S, and only the shunt resistors R_1 to R_N are switched in and out.

Since the series resistor is fixed and the shunt resistor's value gets progressively smaller and smaller as the switch is turned CCW (counterclockwise), the input impedance of a shunt attenuator is not constant. This could become an issue with some sources, so the downside to the shunt type is that it presents a varying input resistance to the source driving the amp or preamp.

In the ladder-type, each shunt resistor (R_{1B} to R_{NB}) has its own corresponding series resistor (R_{1A} to R_{NA}). This enables the designer to choose the values, so a constant input impedance is achieved. The switches used in audio attenuators are the "shorting" type, called "make-before-break" switches. The wiper will make contact with the next lug before breaking the connection with the previous lug. That way, there are no "gaps" in the switching and no transients that may cause audible "bumps" or even damage sensitive speaker drivers.

Assembling attenuators requires manual dexterity and immense patience. If you value your time (hours can be spent sifting through resistors trying to find the needed values) and sanity, relatively cheap ready-made units are available online. Some are clunky and noisy due to poor quality switches used; others are quiet and refined, so buy one first, to ascertain its quality and "feel." Of course, there are also many brand-name (read "expensive") audio attenuators, but you'd need to get a second mortgage on your home to afford some of them.

All series-type and cheap, poor-quality shunt and ladder attenuators sound worse than good-quality potentiometers, whose performance is bettered only by expensive, top-quality parallel or ladder attenuators. Potentiometers are way down on my priority list of upgrades; there are better value-for-money upgrades to consider.

BRIDGING AND PARALLELING: TURNING STEREO POWER AMPS INTO MONOBLOCKS

Many audiophiles love the sound of their power amplifier but feel that their speakers need just a few more additional watts of power to "come alive" or truly start "singing." Or, for some reason, they've replaced their speakers with a less efficient model and now need a more powerful amplifier to drive them. They don't want to sell the amp they love so much and spend time and effort in auditioning others (not to mention losing lots of money in the process), so they'd rather buy another identical amp.

Often an opportunity presents itself for us to buy an identical used amplifier at a bargain price, so now we have two amps and need to figure out how to use both in the same system. Apart from bi-wiring, a topic covered on page 241, bridging or paralleling stereo amps into monoblocks are other exciting possibilities.

It may pay to outline the topology of a typical SET stereo amplifier first (above right), with signal polarities indicated by voltage pulses in the main points. Both channels share the common ground (-) as a signal return and the same HV (high voltage) power supply.

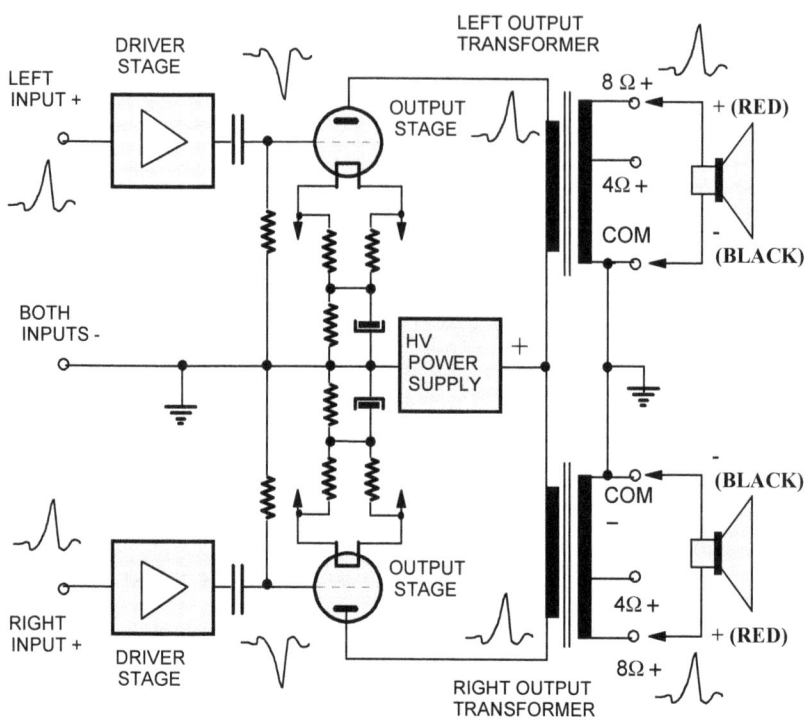

Fig. 7.48. (ABOVE) A topology of a typical stereo SET amplifier with a single driver stage

Fig. 7.49. (BELOW RIGHT) Bridging a stereo SET amplifier into a monoblock by reversing the polarity of one channel's input and disconnecting the COM output terminals from the GND (ground).

Fig. 7.50. (BELOW LEFT) Bridging a stereo tube amplifier into a monoblock by keeping the standard input polarity and connecting the output transformers' secondaries in series with the same orientation: +, -, +, -.

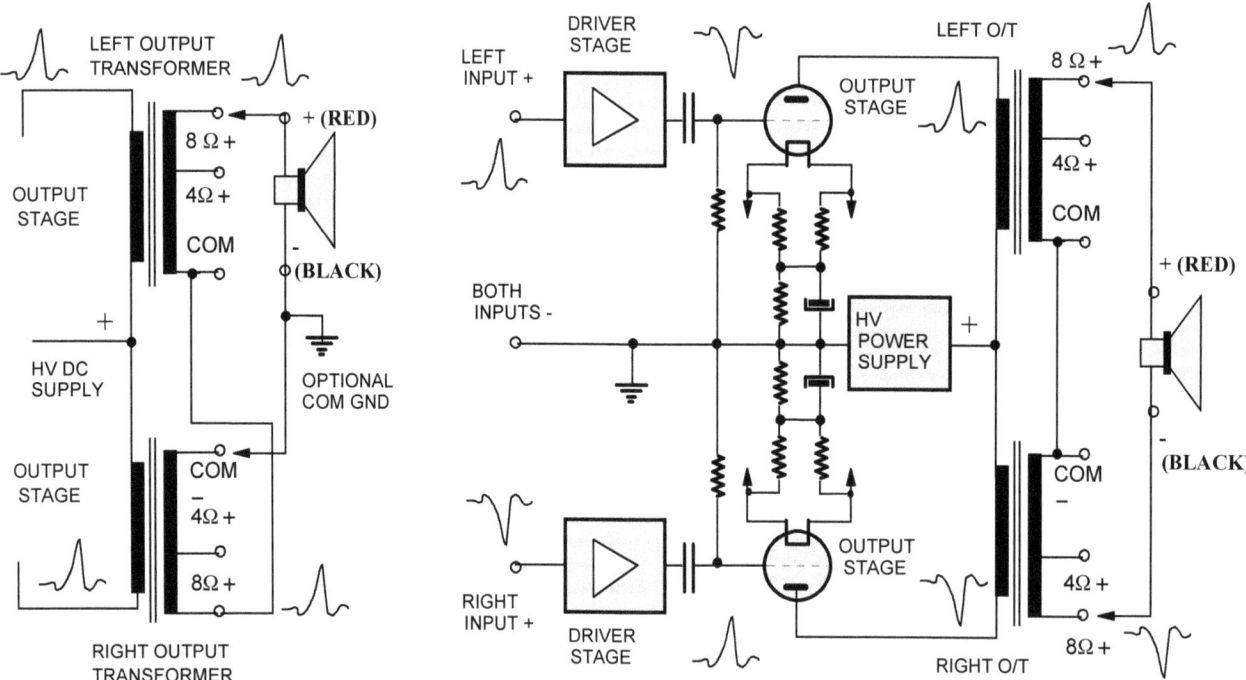

We will use a tube power amp as an example, but it is possible to bridge most solid-state (transistor) amplifiers as well. Please consult your amplifier's manufacturer or the dealer you bought it from for the exact information on how to do it. There is always a risk of improper connection and short-circuited output(s), so proceed with caution. Tube amps are much more resilient and forgiving than their transistor counterparts. If you notice tubes glowing red, a strange smell, or sudden buzzing, quickly turn the amp off, and all should be well again.

How to do it

The simplest way to parallel two channels is to connect the + inputs and parallel the output transformer taps (4Ω and 4Ω, 8Ω and 8Ω).

The bridging option requires the phase of one channel to be inverted so that one channel is propagating a positive signal, while the other amplifies the same signal but with an inverted phase (a "negative" signal). That is easy in a balanced system (using XLR connectors) but much harder in systems with unbalanced RCA interconnects. Many sources and preamps have the negative ends strapped (soldered) together internally, and it is impossible to reverse the polarity of one of them from the outside.

Normally the + (positive) on the unbalanced interconnects is the central conductor, connected to the center pin of the RCA plug, while the - (minus or negative) is connected to the outer ring of the RCA connector via the cable's screen (shield).

There is an option of bridging a stereo tube amplifier into a monoblock with tube amplifiers by keeping the standard input polarity and connecting the output transformers' secondaries in series with the same orientation: +, -, +, -, as we have seen on the previous page. Usually, the two negative ends of the output transformer windings are strapped together and then grounded, so in this case, that connection must be removed.

The benefits of bridging and paralleling

When the two channels of a stereo amp are bridged, the output voltage V_{OUT} is doubled, and since the output power is $P = V_{OUT}^2/R$, the power is theoretically four times higher.

Bridging works better with higher impedance speakers, which benefit more from higher voltage than from higher current. Each of the two channels "sees" only half of the speaker's impedance, so with speakers of four ohms "nominal impedance" that at some frequencies dips down to two or even one ohm, half of those figures would be almost like a short circuit across the amplifier's outputs! In this case, instead of 8Ω taps when in stereo mode, the common 8Ω speaker should now be connected between the two 4Ω taps! Likewise, low impedance, current-hungry speakers benefit more from paralleling since the output impedance of the amp is halved, and its current capacity doubled.

Another benefit of bridging and paralleling is the improvement in channel separation, better imaging and sound staging. Instead of the two channels sharing the power supply, now you have two separate monoblocks, each with its own power supply! The improvement in dynamics is also noticeable since the power supply energy capacity has now doubled for each channel.

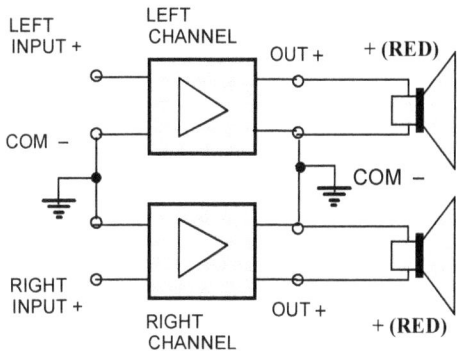

Fig. 7.51. Block diagram of a typical stereo amplifier with unbalanced (RCA) inputs and a common speaker GND or NEGATIVE connection

Fig. 7.52. If the negative terminals of the two channels' outputs are internally connected, paralleling the left and right channels' outputs requires only one jumper wire between the two (+) output terminals. That wire can be inserted externally, through the binding posts, or soldered permanently, internally.

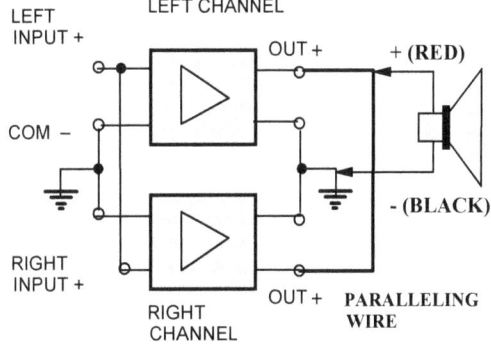

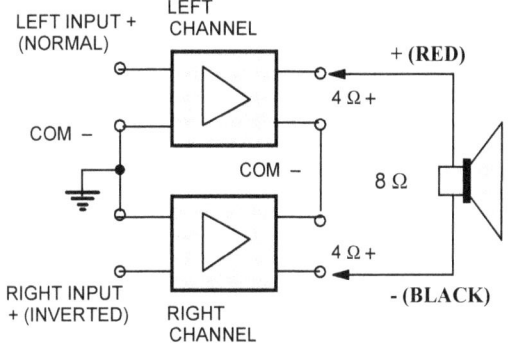

Fig. 7.53. Block diagram of a typical stereo solid-state and tube amplifier bridged into a monoblock. The input to one channel must be inverted at the output of the preamp or at the amplifier's input and the "COM" speaker terminals must be disconnected from GND or earth.

From my experience, paralleling the two channels results in faster, more "confident," and authoritative sound. The bass is tighter and more prominent. This can easily be explained by imagining two identical resistors in parallel, resulting in a halved overall resistance. The output impedance is halved, and the damping factor is doubled!

This is most noticeable in single-ended tube amplifiers whose damping factor is generally between two and five, so doubling that is a significant improvement. The doubling of a typical solid-state amp's damping factor of say 50 to 100 isn't noticeable at all.

While the output voltage stays the same when the two channels are paralleled, the amplifier can now provide twice the output current, so its speaker driving capabilities are significantly improved.

How to connect the bridged inputs & outputs using bridging adapters

While paralleling involves using jumpers on the outputs (speaker connections) and standard splitters on the inputs, the bridging option requires either internal modifications to the equipment or special bridging adapters as input interconnects.

One channel's balanced (XLR) input can easily be changed by swapping the wires on pins 2 and 3 (+ and -). Alternatively, the XLR to two XLR interconnect (male-male) or adapter (female-male) can be used (illustrated on the right).

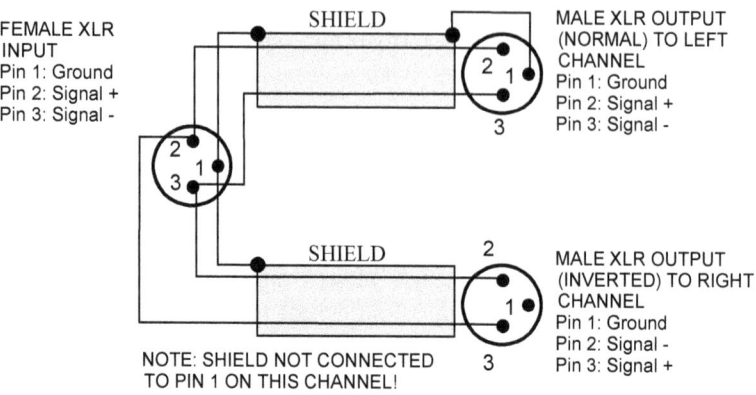

Fig. 7.54. The wiring of a balanced input bridging <u>adapter</u> to convert a stereo amp into a monoblock

The wiring of an RCA to two XLR adapters is simpler, as illustrated since either XLR pins 1 & 3 or 2 & 3 are connected together.

However, notice that using adapters introduces an additional cable and another RCA-RCA or XLR-to-XLR connection, which can only degrade the sonics. Thus, modifying one XLR input inside the amp is sonically a much better option, although it will void the warranty should your amp still be covered by it.

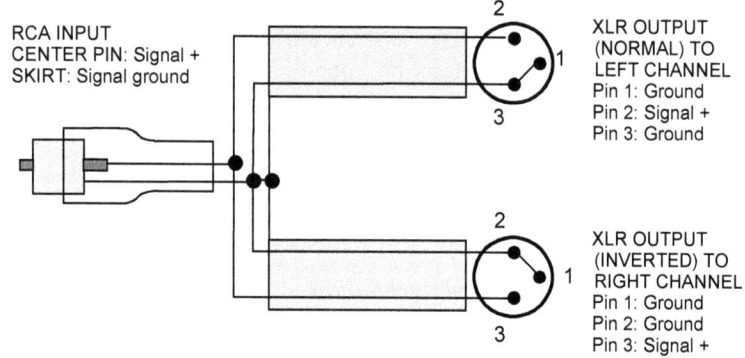

Fig. 7.55. The wiring of a unbalanced (RCA) input bridging <u>interconnect cable</u> (male RCA to two male XLR) to convert a stereo amp into a monoblock

In that case, use a custom-made interconnect, male XLR to two male XLRs, one of which must be with inverted polarity, or male RCA to two male XLRs, as illustrated.

As previously mentioned, most amps have the negative connections from both channels' RCA inputs strapped together at the input (1), so for bridging, these must be separated (2) and (3) in the diagram below. In that case, opening up and resoldering connections inside your amp is inevitable.

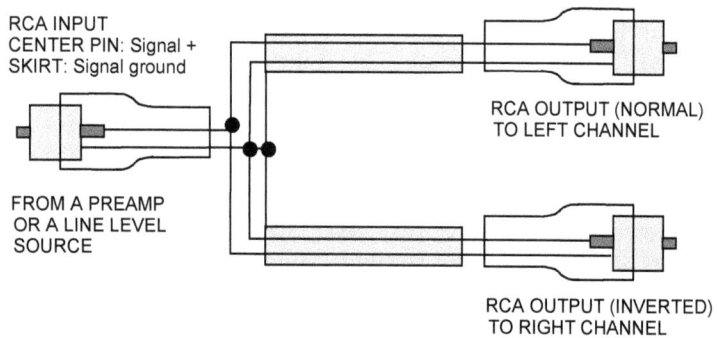

Fig. 7.56. The wiring of a unbalanced (RCA) input bridging <u>interconnect cable</u> into two RCA outputs to convert a stereo amp into a monoblock

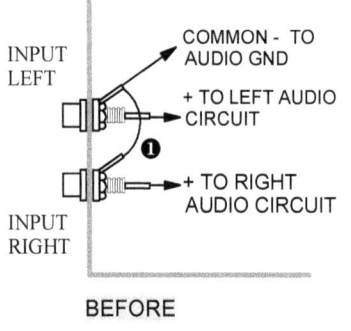
BEFORE

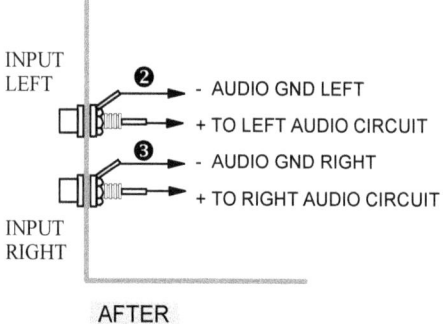
AFTER

Fig. 7.57. Most amplifiers with RCA inputs use common ground or "negative." Sometimes the two negatives from the RCA sockets are joined on the printed circuit board, but more often by bridging the negative tabs on the RCA input sockets together (1).

To use the bridging RCA-to-2 x RCA adapter illustrated above, such a link must be removed and separate "negatives" run to the audio circuit, (2) and (3).

8 | HEADPHONES AND HEADPHONE AMPLIFIERS

We've already seen how to make your own headphone cables, and headphone amplifiers aren't hard to construct either. In this chapter, we look at two tube designs and one solid-state option to whet your appetite.

A headphone amp can be connected in various ways to the rest of your audio components; each way of integrating it into your system will result in different sonics. So, as with other such options discussed in this book, it pays to experiment to find the sound you find most satisfying.

Many audiophiles find headphones an unnatural medium and don't even consider them a viable alternative to loudspeakers. Fair enough. However, even if you share these sentiments, I'd recommend you still read through this section because some valuable lessons are to be learned here. They can be applied to other aspects of the audio reproduction chain ...

- TYPES OF HEADPHONE TRANSDUCERS
- HEADPHONE AMPLIFIERS
- DIY PROJECT: SKIPPY, THE MINIMALIST SOLID STATE HEADPHONE AMPLIFIER
- INTEGRATING HEADPHONE AMPLIFIERS INTO AUDIO SYSTEMS
- DIY PROJECT: HEADPHONE INTERFACE BOX

> "God is in the details."
> Ludwig Mies van der Rohe (1886–1969), avant-garde architect, famous for his 'less-is-more' approach to design

The pros and cons of headphones

Top-quality headphones are generally much cheaper than hi-end loudspeakers. So one way of easing your way into audiophilia is to buy the best source and amplification components you can afford and start with a pair of headphones while you are saving for quality loudspeakers.

The headphones (closed back type) provide privacy and sound isolation, thus avoiding possible problems with cranky neighbors and housemates. However, open-back types do release some of their sounds into the room. Loudspeakers cannot compete with headphones in the level of detail, microdynamics, and transient reproduction.

Headphones eliminate the three main offenders in loudspeaker-based systems:

- The generally troublesome listening room acoustics (standing waves, reflections, and colorations).
- The detrimental sonic effect of speaker crossovers.
- The difficulties of integrating different speaker drivers into a coherent whole (different radiation patterns, beaming and lobbing effects).

Other problems such as amplifier-speaker interaction are also eliminated or greatly reduced. So what is then wrong with the headphones, and why don't all audiophiles use them?

While binaural-recorded material sounds incredible on headphones, such recordings are still relatively rare. Most recordings, mixed, compressed, and manipulated to death in various artificial ways, don't sound natural or even real. The soundstage is nonexistent; all the sound is inside the listener's head, between the ears. Lower piano keys could be in the left ear, with higher keys all the way to the right. I haven't tried sticking my head inside a grand piano, but that is neither how they are "sized" in real life nor how they sound in live performances.

Circumaural (over-the-ear or around-the-ear) headphones completely enclose or cup our ears, thus creating a sound pressure chamber around them, a feeling almost all audiophiles dislike, although some are obviously prepared to accept. No matter how much time & effort earphone designers and manufacturers dedicate to reducing their weight, earphones do feel like a foreign body around our heads, especially in hot and sweaty climates and to listeners with large or oddly shaped heads. You may be laughing now, but as a large-headed audiophile who is also a profuse sweater, I'm speaking from personal experience.

Open or closed back?

Headphone diaphragms radiate sound both forward, towards the listener's ear canal, and backward, towards the room. Closed-back headphones seal both the front and back chamber, thus eliminating the sound leakage into the room, while open-back designs allow the sound to escape and the sounds from the room to enter the listener's ear. Therefore, the two approaches result in a different overall feeling with respect to internal-external sound balance.

Closed headphones usually produce deeper and more prominent bass than the open type, although some open-back headphones overcome this problem by clever design and engineering. However, in the closed-back arrangement, the diaphragm size and mass must be increased to attain such low-frequency reproduction, which negatively affects (slows down) transient response. The open back means there is no air sealed behind the diaphragm that could raise the resonant frequency or slow down the transient response speed; there is no resonance of the sealed rear chamber that some listeners hear as artificially pumped-up bass.

Open-back headphones create a wider, more open soundstage and reduce that unnatural feeling of the sound being beamed into your head for an overall more realistic spatial presentation.

TYPES OF HEADPHONE TRANSDUCERS

Electrodynamic headphones

The electrodynamic drivers used in headphones work the same way as dynamic loudspeakers (they are, indeed, miniature speakers), so we will not dwell on their physical operating principles here. Just as with loudspeakers, the impedance of dynamic headphones varies with frequency but within a much narrower range, so headphones are considered (almost) constant impedance loads.

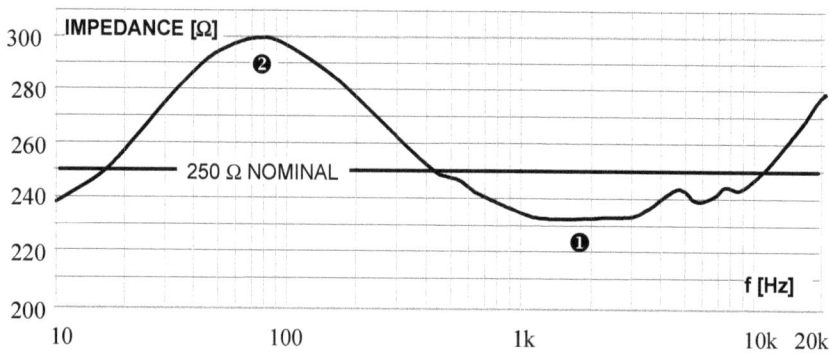

Fig. 8.1. Beyerdynamic DT880 250 ohms impedance version, based on measurements by *Inner Fidelity*, https://www.stereophile.com/images/BeyerdynamicDT880250ohm.pdf

Their impedance is usually in the 30-60 Ω region (although a few models go up to 300 and 600Ω), much higher than the nominal 4-8 Ω loudspeaker range. For example, the manufacturer declared Beyerdynamic DT880 headphones to have a nominal impedance of 250 Ω. The impedance deviation is minor, from the minimum of 230 Ω in the 1-3 kHz midrange band (1) to the wide resonant peak, 300 ohms at around 80Hz (2). That is a variation of (300-250)/250 = 0.2 (+20%), and (230-250)/250 = 0.08 or - 8%.

Electrostatic headphones

Like electrostatic speakers, electrostatic headphones require dedicated headphone amplifiers specifically designed to provide the diaphragm's high DC biasing voltage and drive it properly. Thus, you cannot use them with generic headphone amps; you are stuck with the amp the manufacturer supplied with the headphones.

Some manufacturers, for instance, AKG with their K340 model, have tried a hybrid approach, integrating a dynamic and an electrostatic driver in each headphone earcup, dynamic for bass, and electrostatic for treble frequency bands. The electrostatic panel is driven through a miniature voltage step-up transformer and sits in front of the dynamic cone, both axially aligned. For a detailed explanation of how electrostatic transducers operate, see page 206.

Planar magnetic headphones

While they may look as electrostatic drivers, planar magnetic drivers are also electrodynamic or "magnetic" drivers, which, instead of a cone of a standard dynamic driver, use a thin plastic membrane (1) or the diaphragm, sandwiched between two permanent magnets (2). The windings of a voice coil are evenly distributed across most of the diaphragm's surface (3).

The magnets must cover the whole membrane area on both sides, with numerous holes (4) through which the pressurized air can escape. Such large magnets also make planar magnetic headphones a little bit bigger and heavier than dynamic ones. Planar magnetic drivers can be designed as either monopoles (like traditional dynamic drivers) or dipoles. In that case, they radiate equally but with opposite phases in front of and behind the driver.

Since their diaphragm is thin and light with very low inertia (a quality they share with electrostatic transducers), planar magnetic drivers reproduce fast transients better than heavier and slower dynamic ones.

Usually, they also produce a tighter bass, and due to their distributed voice coil, they have a superior soundstage and lower distortion.

The distributed nature of the voice coil means planar magnetic drivers don't suffer from heat buildup issues as traditional voice coils do. Such effective heat dissipation results in superior power handling, making them harder to overdrive and damage than the more sensitive dynamic drivers.

The impedance of PM headphones remains constant over the entire audio frequency range and is predominately resistive in nature, although on the lowish side (15-20 ohms, for instance).

The parameters of PM drivers show more variance, making them more difficult to match. This could be one of the reasons why in general, dynamic drivers produce a better stereo image compared to planar magnetic ones.

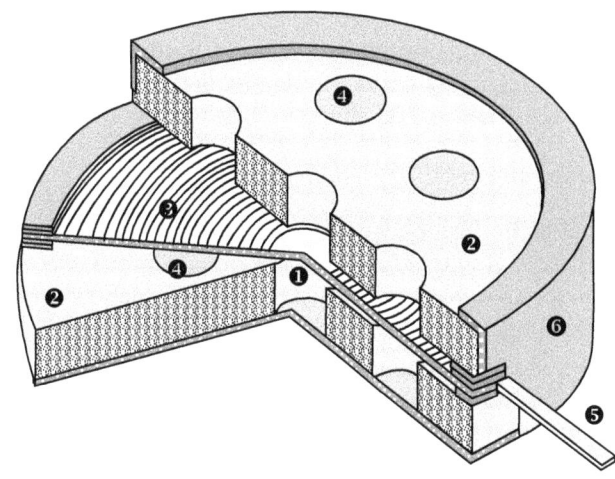

Fig. 8.2. Cross-sectional view of a typical headphone planar magnetic driver: 1) diaphragm 2) permanent magnet 3) voice coil 4) openings in the magnet 5) voice coil leads 6) casing

Ribbon drivers

In a true ribbon transducer, the diaphragm is both a moving element and a voice coil. As its very name suggests, it's a strip of metal foil attached at its ends, positioned in a permanent magnetic field produced by a row of magnets on its two sides.

True-ribbon or even quasi-ribbon transducers aren't often used in headphones, and for a very good reason. Due to their extremely short conductive path, their resistance is a dozen or two milliohms (0.01-0.02 ohms), which is equivalent to a short circuit to all amplifiers, tube or solid-state alike. That problem is solved by adding resistive interface boxes to increase the resistance the driving amplifier "sees" to standard 4-8 ohm levels. The main cause of future problems are past and present solutions. That means such headphones need to be driven by high-powered amplifiers (50-150 watts-per-channel). Only a tiny percentage of that power will be passed onto the ribbon transducer; almost all of it is dissipated as heat on such a box.

HEADPHONE AMPLIFIERS

How the line outputs and headphone outs are compromised on most CD players, DACs and amps

Looking at the typical block diagram of the output stages of CD players and DACs with a headphone output, the weakest links become immediately apparent. The headphone signal goes not just through the main output amplifier and usually a filter (1), but also through the volume control potentiometer (2) and its buffer amplification stage (3), usually . with cheap integrated circuits (op-amps).

Using the main (line) outputs to drive your super-duper headphone amp would at least bypass the buffer amps and cheap pot, but the best option would be to take the signal straight from the DAC chip's output (4) and bypass the whole solid-state amplification within the CD player or DAC.

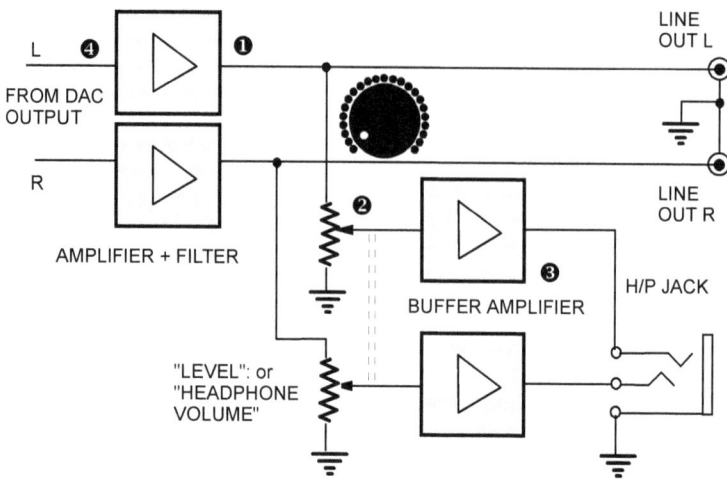

Fig. 8.3. A typical block diagram of the output stages of CD players and DACs with a headphone output

Headphones' impedance - the critical parameter

Headphones come in impedance ranging from 8 ohms to 600 ohms or more, with most in the 20-32 ohms region, for instance, Bowers & Wilkins P7 at 22Ω, Fostex TH-610 and Denon AH-D7200 at 25Ω, Grado PS500e, Meze Audio 99 Classic and V-MODA Crossfade M-100 at 32Ω. Higher impedance models include Sennheiser HD 800 at 300Ω and Beyerdynamic T1 at 600Ω, while others, like Focal Elear, sit somewhere in-between at 80Ω.

Higher impedance headphones usually sound better (more transparent, more spacious sound stage, tighter bass), primarily due to their lighter moving coils, resulting in faster transient response and lower distortion.

As an illustrative example, Audeze's headphones' impedance goes up with their price; the cheapest model in the LCD series, LCD-2, is a 70-ohm model, followed by the twice as expensive LCD-3 at 110 ohms and much dearer Audeze LCD-4 at 200 ohms.

Headphone amplifier topologies

Stand-alone headphone amplifiers of the solid-state variety aren't that different from those just mentioned and illustrated above (inbuilt into integrated amplifiers, CD players, or DACs). Transistors are low impedance devices and can drive very low resistance loads (such as 4-ohm loudspeakers), let alone headphones, generally of much higher impedance than loudspeakers.

Tubes are high impedance devices, and even high impedance headphones ("high" being the 300-600 ohms range) are a low impedance load for tube headphone amps. Thus, the first category of tube headphone amps uses an output transformer, just as with amps that drive loudspeakers. The other two topologies don't use output transformers but lower the output impedance through various other means.

The first is the use of a cathode follower output stage, an electronic impedance converter or "transformer" (very high input and low output impedance); this typically results in an output impedance of a few hundred ohms, a perfect match for high impedance headphones.

Cathode followers do not amplify voltage but can supply high current levels. Their low output impedance makes them ideal for driving low impedance headphones directly.

Since a cathode follower is an impedance transformer, no output transformers are needed. However, an output capacitor (an electrolytic type) is required at the output, the main sonic weakness of this class of headphone amps.

Some audiophiles don't like the sound of cathode followers, but we found that myth of bad-sounding cathode followers to be just that, a myth.

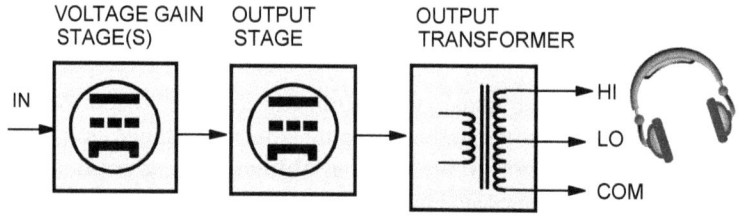

Fig. 8.4. Standard tube headphone amplifier topology with output transformers (the same as power amps that drive loudspeakers)

HEADPHONES AND HEADPHONE AMPLIFIERS

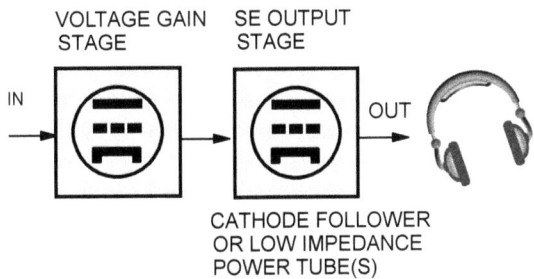

Fig. 8.5. OTL (output transformerless) topologies with either a cathode follower power output stage or using one or more (paralleled) low impedance power tubes

Some of the best-sounding amps and preamps use cathode followers (Audio Note's Ongaku, for instance).

Certain power tubes have much lower internal impedance than others and can be used instead of cathode follower stages. These tube types are used in OTL power amplifiers, 6080, PL519, PL36, 6C33C-B, and E130L.

Triode-strapped higher impedance tubes such as EL84, EL34, 6V6, and 6L6 work well, too, if two or three are connected in parallel to reduce the overall output impedance.

All three topologies discussed so far have a single-ended output stage, but push-pull arrangements can be used as well. However, with such low power levels needed to drive headphones (in milliWatts), why would you choose the sonically inferior and more complex push-pull amplifiers?

CASE STUDY: Triode-strapped EL84 headphone amplifier

A simple topology, one preamp stage driving the output stage using EL84 pentode in triode connection (pin 9, the screen grid, connected to pin 7, the anode), working as a cathode follower. The output is taken from the cathode, pin 3, and not from the anode.

The top half of the ECC82 triode is an "active resistor" for the common cathode stage of the lower triode. Without determining the exact parameters in the operating point from ECC82 (12AU7) graphs, in the first approximation, the mutual conductance is gm = 2.2 mA/V, voltage gain µ=17 and internal resistance r_I = 7.7 kΩ .

The voltage gain of the input stage can be calculated as 12.6 times. Still, the actual gain will be lower than estimated due to the local negative feedback (the un-bypassed 220 Ω cathode resistor). The measured value was A=10.

The power tube's cathode current is 30 mA (6.6 V/220 Ω). From the graph (EL84 data sheets), we find that the tube's internal resistance is r_I = 1.8 kΩ and the voltage gain is µ = 19.5, so we can calculate the voltage "gain" of the output stage: $A_K = µR_K/[(1+µ)R_K+r_I]$ = 19.5*220/(20.5*220+1,800) = 0.68! Since it is lower than 1 (unity gain), this means the output stage actually attenuates the voltage signal.

The voltage gain of the amplifier is thus A=10*0.68 = 6.8 times; for a 1 Volt input signal, the output will be 6.8 Volts.

The output impedance can be calculated as $Z_O = [r_I/(1+µ)] || R_K$ = (1,800/20.5) || 220 = 88 || 220 = 62.8 Ω, a very good (low) result.

The measured -3 dB frequency bandwidth without load was 5 Hz - 32 kHz, an excellent result.

Due to the headphones being a low-impedance load, a high-value coupling capacitor is needed. We bypassed it by a quality film & foil capacitor for better HF performance.

The amp sounded detailed and transparent, yet smooth and lush, a rare combination. Substituting 5963 duo triode for ECC82 made the amp sound even better.

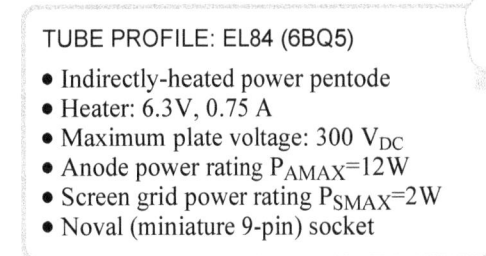

TUBE PROFILE: EL84 (6BQ5)
- Indirectly-heated power pentode
- Heater: 6.3V, 0.75 A
- Maximum plate voltage: 300 V_{DC}
- Anode power rating P_{AMAX}=12W
- Screen grid power rating P_{SMAX}=2W
- Noval (miniature 9-pin) socket

Fig. 8.6. OTL headphone amplifier with triode-strapped EL84 pentode in cathode follower topology

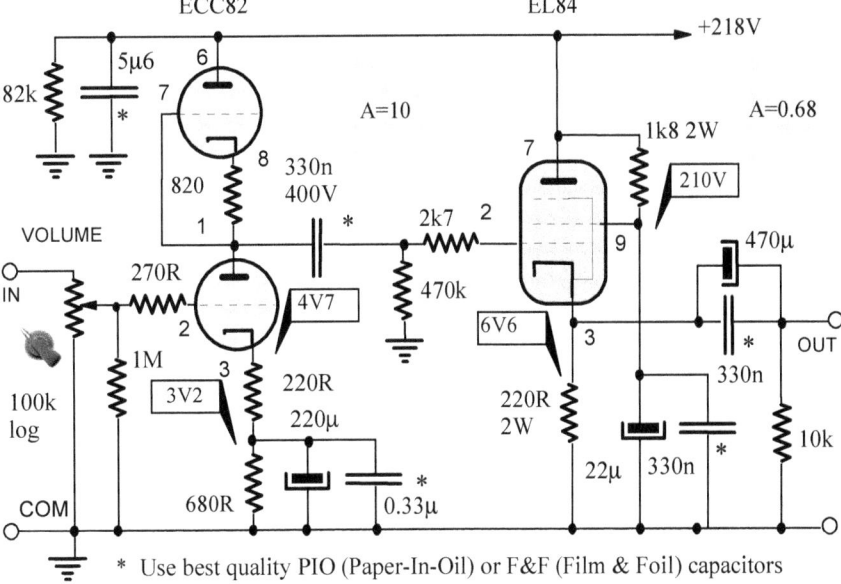

* Use best quality PIO (Paper-In-Oil) or F&F (Film & Foil) capacitors

CASE STUDY: Hagerman Tuba headphone amplifier

Hagerman Tuba is a single stage tube amp, EL84 pentode strapped as a triode (1). A pentode has a high internal impedance, which drops significantly once it's connected as a triode. A cheap commercially available audio transformer (Hammond 119DA) is used as the output transformer (O/T), with 600 ohms primary and 4 and 8 ohms secondary impedances to lower the output impedance even further.

Since that transformer cannot take any DC current through its primary (it would saturate), Hagerman had to use an LC-fed stage, or "parafeed" stage, where DC anode current flows through the anode choke (2), which acts as the anode load. The coupling capacitor (3) prevents this DC current from flowing through the transformer's primary.

From the 5V_{DC} voltage drop on the 330 Ω cathode resistor (4), we can calculate the cathode current as 5/330=15.16 mA. Tube datasheets give you the graph of tube parameters (internal resistance, voltage gain, and mutual conductance) versus anode current and also the anode characteristics of triode-strapped EL84. At that operating point its internal resistance is r_I =2,500 Ω and the voltage amplification factor µ=17.5!

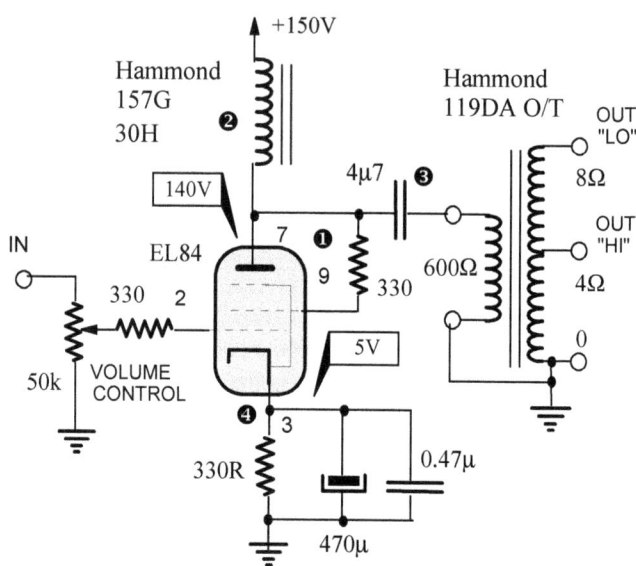

There are two outputs (marked "LO" and "HI"). The manufacturer specifies "5/35 ohm output impedance (LO/HI)". Notice that Hammond transformer's markings are 4 ohms (for "35 ohms" or "HI") and 8 ohms (for "5 ohms" or "LOW"). Although an average user would not study the circuit diagram, this wouldn't make sense until we realize that 5 and 35 ohms are arbitrary figures chosen by the manufacturer to signify low and high impedance headphones.

The output transformer tells us how the impedance of the load (headphones) will be reflected back to the tube. The impedance ratio of the 8-ohm output ("LO") is 600/8 = 75, so the EL84 tube will see 32 ohms headphones as 75*32 = 2,400 ohm load.

The impedance ratio of the 4-ohm output ("HI") is 600/4 = 150, so high impedance headphones (say, 300 Ω) connected there would be reflected back as 300*150 = 45,000 Ω. That doesn't seem right; it's too high (but the combination will work).

Fig. 8.7. The audio circuit of Hagerman Tuba H/P amplifier

Fig. 8.8. The operating point of the Hagerman Tuba headphone amplifier with two examples of load lines, 4.8kΩ and 22.5kΩ

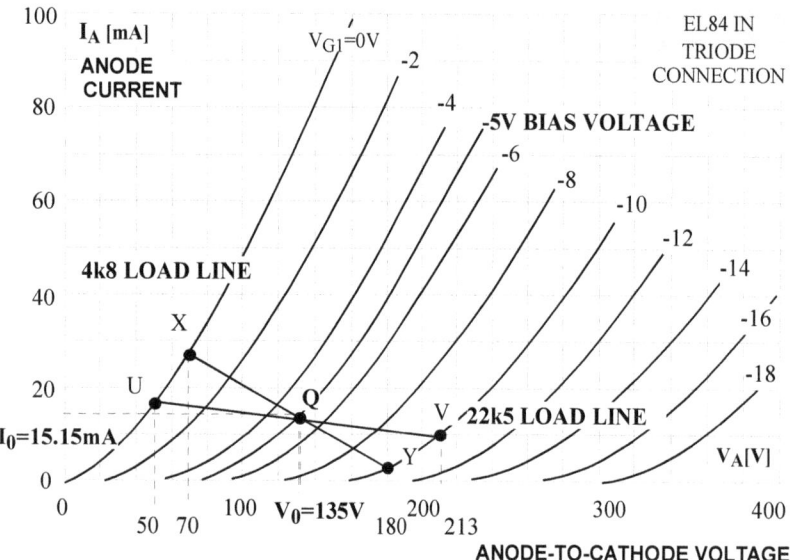

In my opinion, high Z headphones (300-600 Ω) should be connected to the 8-ohm transformer terminal (which Hagerman on their drawing marked as "LO" perhaps in error), and low impedance ones should use the 4-ohm (or "HI") terminal. Thus 300*75 = 22,500 ohms and 32*150 = 4,800 ohms.

Let's look at how this amp works with these two different loads on the primary side of the output transformer, 4,800 Ω and 22,500 Ω. We need to draw two load lines through the quiescent operating point Q (no signal).

How to graphically estimate the voltage gain and distortion of a single-ended output stage

Let's chose a grid signal of +/- 5V around the bias point of -5 V. With a 4k8 load, the input will swing between points X and Y, from 0 V to -10 V curves. We can read on the horizontal axis that the anode voltage will swing from 70 Volts (point X) to 180 Volts (point Y). The voltage amplification or gain is thus A=$\Delta V_A/\Delta V_G$ = (180-70)/-10 = -110/10 = -11! The negative sign means that the amplifier inverts the absolute phase.

Notice how the output isn't symmetrical for a symmetrical (undistorted) input; the positive swing is 135-70 = 65 Volts, while the negative swing is only 180-135 = 45 Volts. The stage is distorting severely. You can see that the distance X-Q is much larger than the distance Q-Y, indicating distortion even on the graph. For lower impedance headphones, the load line will be even steeper (more vertical) with an even higher distortion.

The 22k5 load line is more horizontal, so the anode voltage (V_A) swings are closer to each other, resulting in lower distortion (the positive swing is 135-50 = 85 Volts, while the negative swing is 213-135 = 78 Volts, only 7 V difference). Also, the voltage gain is higher than before, A = (213-50)/-10 = 163/-10 = -16.3!

This graphic analysis and the 3-point formula can be used to estimate the harmonic distortion (2nd harmonic D_2) of the output stage. For the load reflected as 4k8 to the primary side we have V_0=135V, V_{MAX}=180 V and V_{MIN}=70V, so ΔV_+=65V, ΔV_-=45V, D_2 =($\Delta V_+ - \Delta V_-$)/2ΔV= (65-45)/(2*110) = 0.0909 or approx. 9.1 %!

This high distortion is typical of this type of circuit. For instance, for their *Valhalla 2* phono stage Schiit specifies "Maximum Power, 300 ohms: 300mW RMS per channel at less than 10% THD".

For the load reflected as 22k5 to the primary side, we have D_2 = ($\Delta V_+ - \Delta V_-$)/2ΔV = (85-78)/(2*163) = 0.0215 or 2.15 %. The distortion is much lower now due to the higher load impedance and the more horizontal load line.

The impact of the anode choke and the output coupling capacitor

The analysis above would be correct for the standard, transformer-loaded stage, but in this case, we have an anode choke of 30 H inductance. The audio signal sees that choke in parallel with the reflected load impedance. However, the impedance of that choke will vary with frequency. At 20Hz its reactance (the reactive part of its impedance) will be X_{L20} = ωL = $2\pi f L$ = 3,770 ohms. It is the same order of magnitude as the reflected 4k8 load, meaning the load the output tube will "see" will be the two impedances in parallel, 3,770 || 4,800 = 2,112 Ω. The lowered load impedance means an even steeper load line and much higher distortion in the bass region.

At 20 kHz, the reactance is 3.77 MΩ. Such a high impedance in parallel with the reflected load impedance, either 4k8 or 22k5, will not change things much. Thus, the higher the signal frequency, the more accurate our previous simplified analysis becomes. Here you see a significant problem with LC-fed output stages: the frequency-dependent load on the power tube will make both the amplification factor and distortion levels vary a lot with frequency. The issue is more pronounced at lower (bass) frequencies.

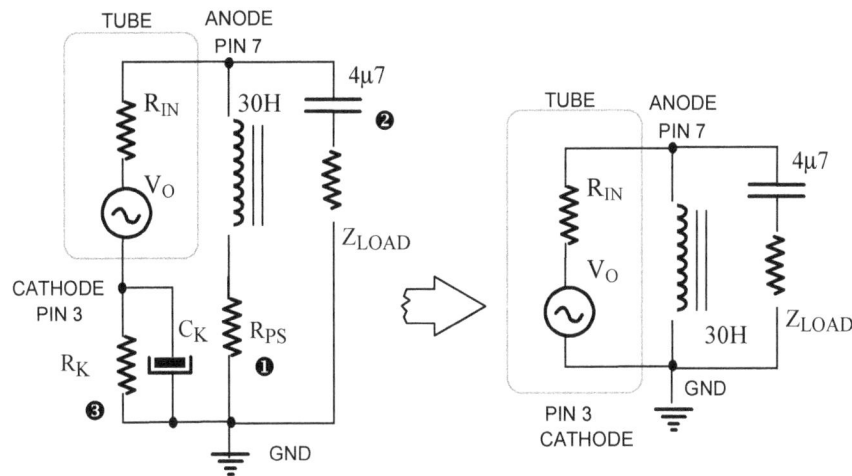

Fig. 8.9. The output circuit showing how
1. the internal impedance of the DC power supply is in series with the anode choke
2. the coupling capacitor is in series with the load (headphones' impedance reflected backward through the output transformer)
3. the cathode resistor and its bypass capacitor are in series with the tube

The simplified output circuit is based on two assumptions:
1. the impedance of the cathode bypass capacitor is zero, so it "shorts-out" the cathode resistor, thus grounding the cathode
2. the impedance of the power supply is zero, so it can be omitted from the circuit

The output circuit above (left) and its simplified version on the right show how various components impact the signal flow. The cathode resistor and its bypass capacitor are in the signal path, and so are the power supply, the coupling capacitor, the anode choke, and the output transformer.

At 20kHz, the reactance of the coupling capacitor is X_{C20} = $1/\omega C$ = $1/2\pi f C$ = 1.693 ohms, much lower than the load impedance; thus, it can be omitted. However, at 20 Hz, the other frequency extreme, the reactance increases to 1,693 ohms (1.7 kΩ). It is now of the same order of magnitude as the reflected 4k8 load, meaning the load the output tube will "see" is the two impedances in series. The capacitor will cause a significant phase shift of the output signal, a portion of which will be wasted on its reactance, thus further lowering or weakening the bass!

This particular amp was selected simply because its circuit diagram had been released into the public domain by the manufacturer; this analysis was not meant to be a criticism but a constructive discussion that illustrates performance issues of this class of headphone amplifiers.

DIY PROJECT: SKIPPY, THE MINIMALIST SOLID STATE HEADPHONE AMPLIFIER

How to save $500 in one afternoon

Skippy, the Bush Kangaroo was the cult 1968-1970 Australian TV series about a ranger's son and his best mate, a kangaroo named (for rather obvious reasons) Skippy. Due to its meager component count and a simple topology, I should have called this minimalist headphone amp Skimpy, but Skippy sounded better.

Most solid-state headphone amplifiers with a retail price under $800 are equally simple affairs, usually a single integrated circuit ("chip"), with one or two op-amps (operational amplifiers) inside. For less than $50 in parts and a few hours of your time, you can construct an equally good or even better-sounding headphone amp of your own. How do I know it would sound better? Because I've tried it and compared the sonics with headphone amps such as Grado RA1 and similar benchmarks.

You can use a self-contained (external) analog DC dual power supply (+/- 9V up to +/-24V) or power the amp from two 9V batteries. However, these will only last 10-15 hours, so if you do lots of headphones listening, the cost of replacement batteries will quickly add up. Larger, sealed lead-acid batteries (12 Volt 7.5 Ah) will last hundreds of hours and can be recharged using ordinary commercial battery chargers.

The naming of the parts

The best (sonically) IC to use is Burr-Brown OPA2604, a modern dual op-amp with FET input and optimized circuitry. Some of the best solid-state audio gear uses this chip, for instance, Mark Levinson No. 38S preamplifier.

Instead of wasting time making our own PCB, we bought one on eBay, $4 including postage. Populating the board with components took less than 10 minutes, but the mechanics (enclosure drilling, mounting the volume control pot, and input/output jacks) took a couple of hours.

Since there are only a dozen or so components, so use the best quality resistors and capacitors you can afford. The values are not critical; instead of 2.2 mF coupling capacitors, you can use any value between 1 and 5 mF. The same applies to the resistor values (instead of 100k, you can use 82, 120, or 150k).

The 390k and 100k resistors form a voltage divider in the negative feedback circuit; their ratio determines the strength of the feedback (and overall gain). The higher the R1 and the lower the R2, the weaker the NFB and the higher the amplifier's overall gain, so experiment with different ratios to find the sound you like.

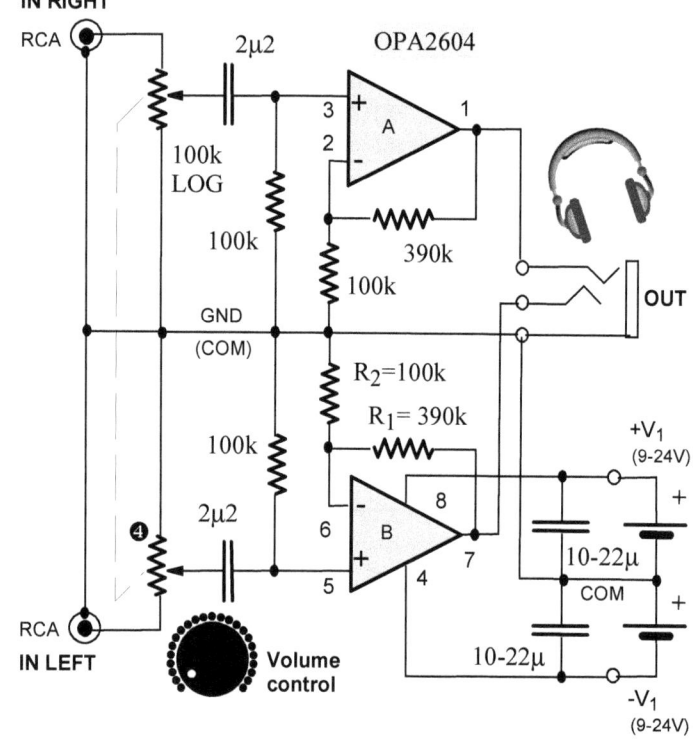

Fig. 8.10. The circuit diagram
Fig. 8.11. How to identify pin numbers of DIL (Dual-In-Line) integrated circuits

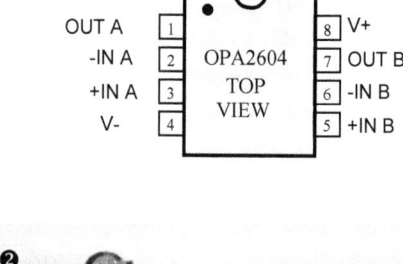

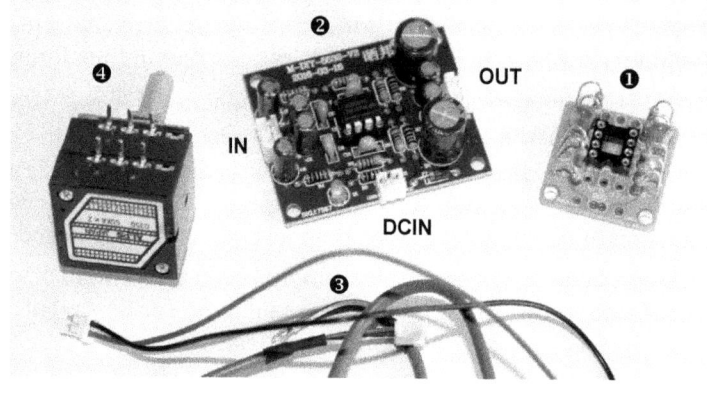

Fig. 8.12. Skippy's internal parts, ALPS volume control potentiometer 94) and two finished PCB options from eBay (only one is needed). While the smaller one (1) has only the essential components, the larger one (2) is a superior product. It came with snap-in connector cables (3), the IC (5532), eight additional electrolytic capacitors, a protection diode (against DC supply polarity reversal), and an "ON" LED indication.

Both PCBs were designed for the 5532 dual op-amp. Since OPA2604 features an identical pinout, both boards can be used for our project.

The enclosure, external power supply, and various connectors (input RCA sockets, output headphone jack, and DC input socket) are not shown.

INTEGRATING HEADPHONE AMPLIFIERS INTO AUDIO SYSTEMS

Headphone amplifiers with the input loop

Some headphone amps have two pairs of RCA sockets at the back, marked "Line IN" and "Line OUT," or something to that effect. This handy feature (a simple internal loop) doesn't add much cost but enables the H/P amp to be inserted between the output of any line-level source used (DAC, CD player, music streamer, tuner, tape deck, etc.) and the input into a preamp, integrated amp or power amp.

This is an elegant solution for a situation where a line-level preamp has only one pair of output sockets (connected to a power amp), with no provision for an H/P connection.

Headphone amplifiers without the input loop

In audio systems that use a preamplifier or an integrated amplifier with a spare line level (preamp) output (1), that output can be used to drive a headphone amp. However, some manufacturers advise against such a connection, warning that it may result in overdriving and damaging the headphones.

Instead, they advocate the use of "Tape out" or "Record out" output (2). Since this is typically a fixed output, the preamplifier volume control is out of the circuit, and the volume control of the headphone amplifier is used exclusively.

However, the tape or record output is usually taken before the final output stage of a preamp, so the two connections will sound different. Try both to see which works better with your headphones.

Driving headphone amplifiers from power amplifiers' speaker terminals

Audiophiles who subscribe to the minimalist school of thought claim that unnecessary audio components color the sound and degrade the sonics. How can you use headphones if you only have a single source (CD player, streamer, etc.) without the headphone output and a power amplifier (or an integrated amplifier) without the headphone, line-level, or tape/rec outputs?

One solution is to pay a knowledgeable technician to add a headphone output to either your source component or your amplifier and then use it to drive your headphone amp.

Alternatively, build a simple "headphone interface box" (3) which will use the signal from the speaker binding posts of your power or an integrated amplifier to drive your headphones.

Again, that option will sound different from the line-level output/headphone amp combination since the power amplifier is now in the audio chain. So are the passive components inside the interface box, introducing their own sonic coloration.

Let's look at a few ways to design & build such an interface.

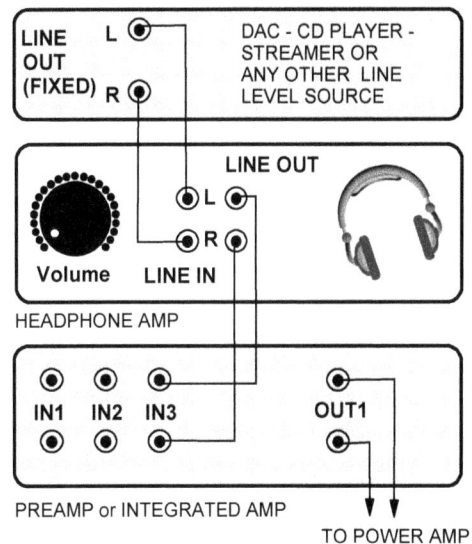

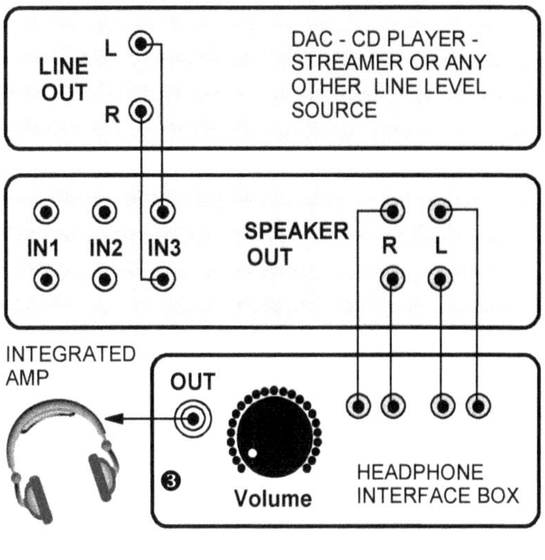

Fig. 8.13. (ABOVE) A headphone amplifier with the input loop is connected between the source and the preamp or integrated amp

Fig. 8.14. (BELOW) Either the second line level output or the "Tape out" ("Record out") signal can be used to drive a headphone amp

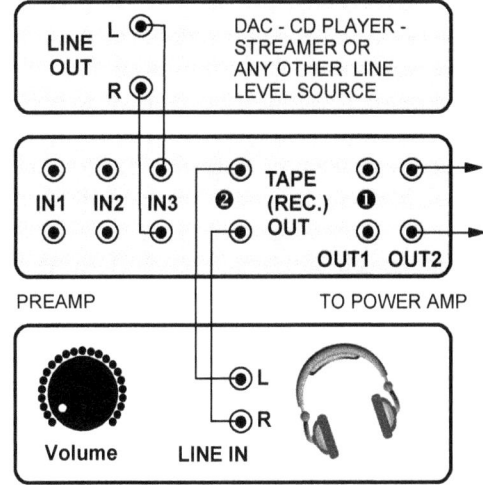

Fig. 8.15. (RIGHT) A "Headphone interface box" (3) enables headphones to be connected directly to power amp's speaker terminals

DIY PROJECT: HEADPHONE INTERFACE BOX

As of this writing, there are dozens of affordable solid state and even tube headphone amplifiers on the market in the $200-$500 budget category. However, each will introduce its own sonic coloration, and certain drive or compatibility issues may be encountered. Plus, why spend that money if you don't have to?

If your amp hasn't got a headphone output, as with most hi-end amps, it can be added for very little money. However, many audiophiles don't want to mess around with their expensive amps or modify them in any way. Plus, if you like the sonics of your power amplifier, why not get the same sonics with headphones? Why can't we just use the speaker outputs of your beloved amplifier and drive the headphones from there? We certainly can. Let's see a few ways that could be done.

A fixed resistive voltage divider drive

Most headphones require less than 500 mW (half-a-watt) of power to drive them to their full SPL level. We cannot connect them directly even to a low-powered amp, an 8 Watt 300B SET amplifier, for instance, let alone a 200 WPC solid-state monster. Plus, most amps expect low resistance loads of between 4 and 8 ohms, not high impedance (32-600 ohms) headphones. So, we need an interface circuit that will address these two issues, reduce the power fed to the cans, and at the same time still have the amp's speaker output loaded with a 4-8 ohms load.

Assuming an 8-ohm amplifier output, we want the amplifier to "see" an 8-ohm load even when the headphones are connected. So, let's choose the two resistors, so their sum is as close to 8 ohms as possible. 6.8 ohms and 2.2 ohms are standard resistor values and would give us 9 ohms in total, close enough.

Without the headphones connected, voltage V_1 going to the headphones is $V_1 = 2.2/(2.2+6.8)V_O = 0.24V_O$ or approx. one-quarter of the amplifier output voltage. Connecting low impedance (say 32 ohms) headphones in parallel with the 2R2 resistor results in the equivalent resistance of 2.06 ohms, so now $V_1 = 2.06/(2.06+6.8)V_O = 0.23V_O$!

Things haven't changed much, and connecting higher impedance headphones would result in even less change. We see that from the amplifier loading aspect, any load higher than 30 ohms can be neglected! For low-power amplifiers driving inefficient headphones, this 4:1 voltage reduction may be sufficient. In most cases, however, it won't - there'd still be too much power flowing into the headphones. We need to reduce the voltage further by adding a 100-180 ohms resistor (R_X) in series, forming another voltage divider with the headphone impedance.

For instance, assuming a 120R (120 ohms) series resistor and 32 ohms headphones, the voltage across the headphones will be $V_2 = 32/(120+32)V_1 = 0.21V_1$, and since $V_1 = 0.23V_O$ we get $V_2 = 0.21*0.23V_O = 0.048V_O$! The voltage signal on the headphones will now be only 4.8% of the amplifier's output voltage.

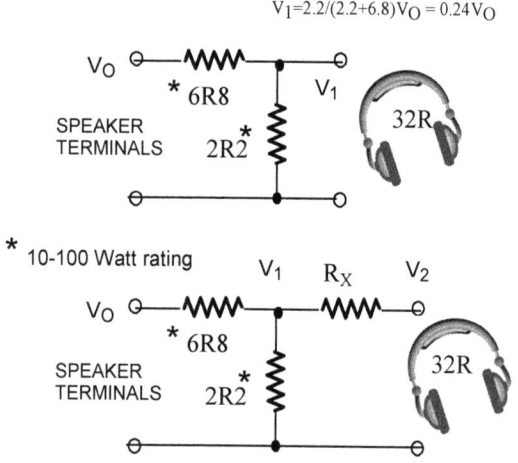

Fig. 8.16. (ABOVE) Voltage dividers to reduce the signal driving the headphones

Fig. 8.17. (BELOW) How to interface headphones with speaker terminals - the complete stereo circuit with volume control

Adding the output volume control

To use our HIB adapter with power amplifiers without any volume control, we need to include a dual 100 ohm logarithmic (marked "A") volume control pot. The only problem is that such pots aren't available; the lowest value available is 1,000 ohms. There are mono 100-ohm wire-wound pots made in China, so you may have to use two of those instead.

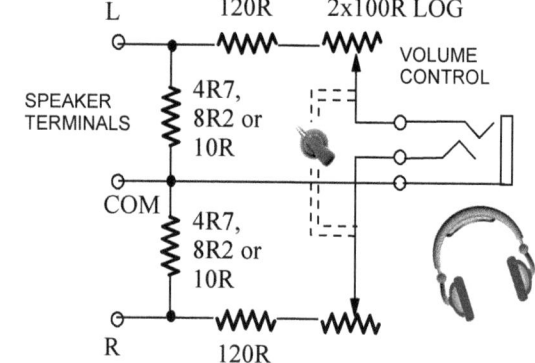

Practical implementation

A dual footswitch (photo on the next page) for a Marshall guitar amp gathering dust was just the right physical size for our HIB. The two switches and the cable were removed, and the top of the enclosure was resprayed to cover the screen-printed markings. The holes were the right size for the 6.3mm stereo socket (1) and the volume control pot (2). Since all of our power amps have a volume control, we didn't bother adding it here; the knob is just to cover the hole for picture-taking purposes. For the romantics amongst us, another option is to install a second jack, so two headphones can be used simultaneously.

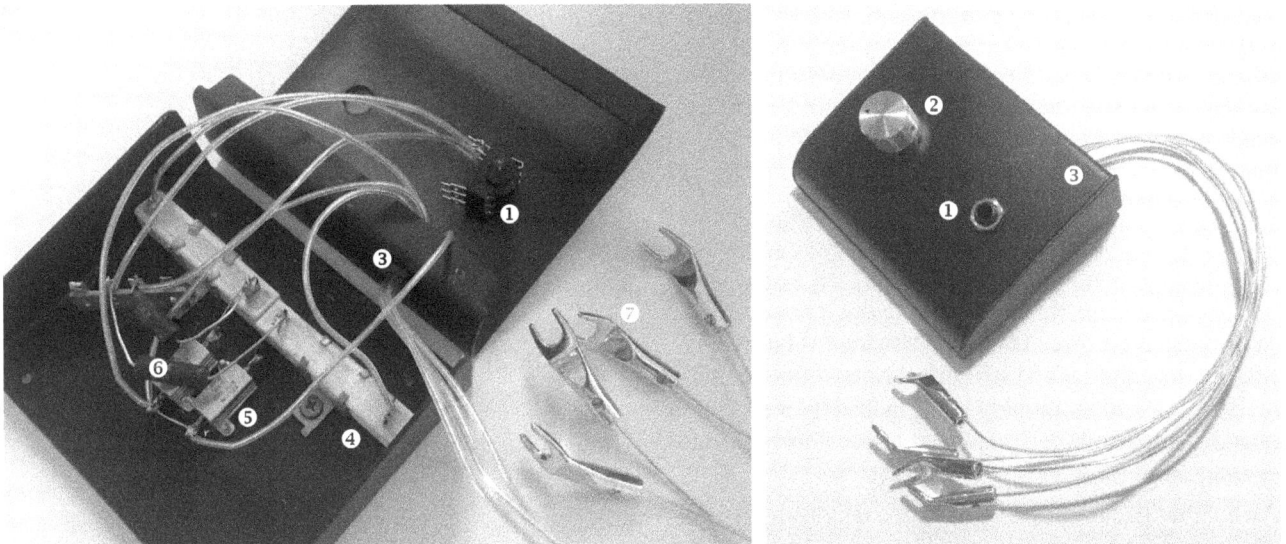

Fig. 8.18. The internals of the Headphone Interface Box (without the volume control potentiometer) ABOVE RIGHT: The finished Headphone Interface Box

20 Watt rated 6R8 (6.8 ohms) resistors were bolted to the bottom part of the Marshall enclosure (4), as were the 10 watt rated aluminium-cased 2R2 resistors (5). The 120-ohm series resistors were only rated at 3 watts, which is plenty (6); no dynamic headphone draws more than 3 watts of power. The existing hole at the rear (3) was reused, and the four single solid core cables fed through it, terminated with gold-plated speaker spades (7).

A permanently wired interface

The drawback of this design is that you need to unplug your speaker cables and plug this HIB into your amplifier every time you want to listen through headphones. A permanently wired interface would be a much more elegant option.

We could then use the second hole in the enclosure (2) for a DPDT (double pole double throw) changeover switch (8). The amplifier terminals are connected to terminal strip TS1, while speakers are permanently wired to the terminal strip TS2. In the "HP" position (as drawn), the switch will connect the HIB (terminals L and R) to the amplifier terminals at TS1. Flicking the changeover switch to the second ("SP") position would disconnect the headphones and connect speaker terminals at TS2 to amplifier terminals at TS1.

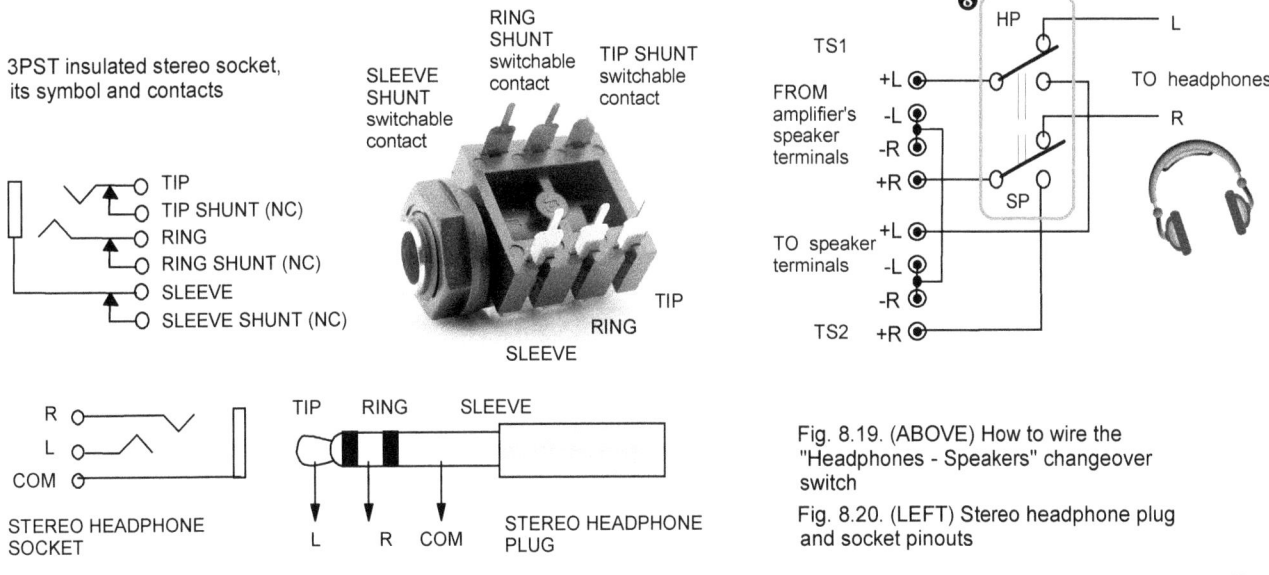

Fig. 8.19. (ABOVE) How to wire the "Headphones - Speakers" changeover switch

Fig. 8.20. (LEFT) Stereo headphone plug and socket pinouts

Due to the wide range of headphone impedances (20-600 ohms), the sonic character of a particular amplifier-headphone combination is highly unpredictable, even more so than with loudspeakers, since nominal loudspeaker impedances do not vary as much (1:30 ratio) between different models.
Generally, the higher the headphone impedance, the more horizontal the load line and the lower the distortion!

Driving low impedance headphones directly from a line stage outputs? Has Dr. Iggy lost his marbles?!?

#46 is a vintage dual grid directly heated triode with a 10 Watt anode (plate) power rating, meaning it can be used both as a driver and an output tube. This line-level preamplifier amplifies signal voltage up to the maximum of 3.9 times (14 dB).

Very few preamplifiers could drive low impedance (30 ohms) dynamic headphones at all, let alone with low levels of distortion, but this one did it beautifully. Removing the power amplifier between this preamp and the headphones resulted in a further refinement of the sonics. The stunningly vivid and detailed sound raised goosebumps many times during our evaluation.

An adapter had to be made, with two RCA plugs (6) at one end and a 6.3mm stereo socket (7) at the other, so a standard headphone cable (with a 6.3mm jack) could be used.

The Hagerman Tuba headphone amp used a small audio output transformer at its output, giving us an idea to check how one of our transformer-coupled line stages would work as a headphone amp.

Just like with the previous preamp, a single amplification stage per channel, this time with two NOS General Electric (USA) 12SN7 triodes in parallel (1), an output transformer (2) at the output, and not a single coupling capacitor in the signal path!

The power supply with a 6X4 rectifier tube (3) is in its own enclosure. A ladder-type attenuator as volume control (4) and a 3-position input selector switch (5) complete the spec sheet.

The "enclosures" are large cigar-cases made of cedar in burl finish, with gold hinges at the rear. The top and bottom parts are held together by the screw-on decorative front panels in gold mirror-finish, matching gold-plated solid machined knobs and gold-plated tube sockets.

A very elegant and classy-looking preamp, which sounded even better with low impedance headphones (32 ohms) than the #46 preamp above. Another example of the supremacy of transformer-coupling over the cheaper and simpler capacitive output.

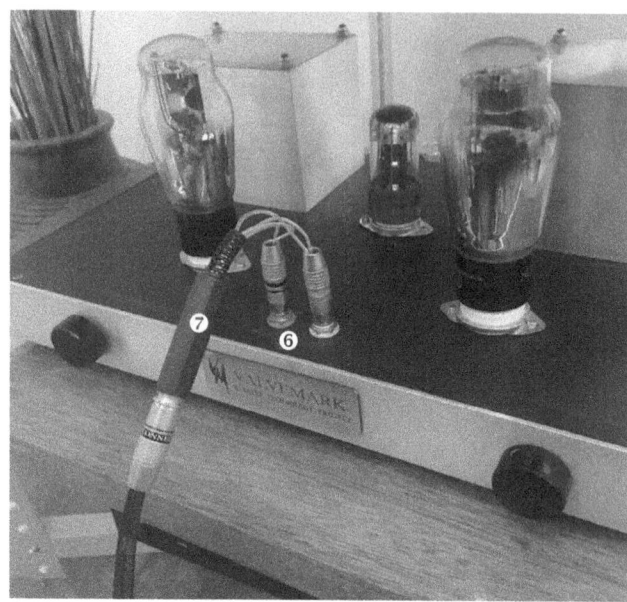

Fig. 8.21. (ABOVE) With this DIY adapter (2xRCA plugs to 6.3mm socket), a headphone cable with a 6.3mm stereo jack can be plugged into the output of a preamplifier, enabling headphones to be driven directly by the preamps capable of such a task.

Fig. 8.22. Although designed to drive higher impedance loads (power amps), with ten thousand ohms or higher input impedance, this tube line-level preamp with output transformers drove low impedance (30 ohms) headphones with no aurally noticeable stress or strain.

9 LOUDSPEAKER TYPES, TESTS, AND IMPROVEMENTS

Together with your listening room, loudspeakers are the most critical part of our audio setup. The first few sections of this chapter cover the operational principles and peculiarities of various speaker types (dynamic, electrostatic, planar magnetic, ribbon), the enclosures they are mounted into (sealed box, bass reflex, horns, and others) and the crossovers, the filters used to split audio signals between various drivers.

This fundamental knowledge will enable you to critically evaluate loudspeakers and choose the best speaker for your amplifier, your listening room, and the type of music you listen to.

Since this is also a DIY modification & improvement manual, a few setups are outlined to show you how to perform basic electrical and acoustic measurements on your speakers. The results of those tests will help you to decide on possible hardware tweaks and upgrades, such as internal rewiring and driver or crossover upgrades, which are also explained in detail.

- DYNAMIC LOUDSPEAKERS
- OTHER SPEAKER AND DRIVER TYPES
- SPEAKER ENCLOSURES AND CROSSOVERS
- CHOOSING A LOUDSPEAKER
- CRUCIAL QUESTIONS AND QUICK SPEAKER TIPS
- HOW TO MEASURE SPEAKERS' SENSITIVITY, IMPEDANCE, AND FREQUENCY RESPONSE CURVE
- SPEAKER MODIFICATIONS AND UPGRADES

> "If it sounds good, it is good."
> Duke Ellington

DYNAMIC LOUDSPEAKERS

Permanant magnet dynamic speakers

A dynamic loudspeaker is an electromechanical transducer, transforming the electrical energy of the signal current flowing through its voice coil ("spool") into a mechanical sound wave (air pressure) produced by its cone. The voice coil is wound on a cylindrical voice coil "former," attached to the cone and placed in a uniform magnetic field. It moves linearly between the poles of a magnet, surrounded by a narrow gap.

With no audio signal, the voice coil is at rest. An AC signal current flowing in the voice coil (from the output of an amp) generates a pulsating magnetic field that interacts with the magnet's stationary magnetic field and produces a mechanical force that is transferred onto the speaker's cone. The cone's movement "follows" the audio signal and produces sound waves.

Most modern loudspeakers use ferrite magnets to create this stationary magnetic field. Such permanent magnets are made by pressing ferrite powder into a ring-shaped structure (1).

The first permanent magnet speakers used Alnico (an acronym for Aluminium- Nickel- Cobalt mix), an alloy whose chemical composition is 8–12% aluminium, 15–26% nickel, 5–24% cobalt, up to 6% copper, up to 1% titanium, and the rest is iron.

In the 1970s, "ceramic" or "ferrite" magnets were developed, which replaced Alnico almost entirely because they were cheaper and easier to manufacture.

However, Alnico speaker drivers are back in fashion in guitar speaker cabinets and combo amps. Although much more expensive than ceramic magnet drivers, they are considered better sounding (more natural, more "musical" and engaging").

Alnico speaker drivers have also started their inroads into expensive hi-end audiophilia. Serbian firm Teresonic uses expensive A55 Lowther alnico drivers (US$1,750 a pair) in their Magus loudspeakers, but Lowther DX55 or DX65 drivers can be substituted.

Both of these drivers have modern (developed in the mid-1980s) neodymium magnets (an alloy of neodymium, iron, and boron), the strongest commercially available permanent magnet type.

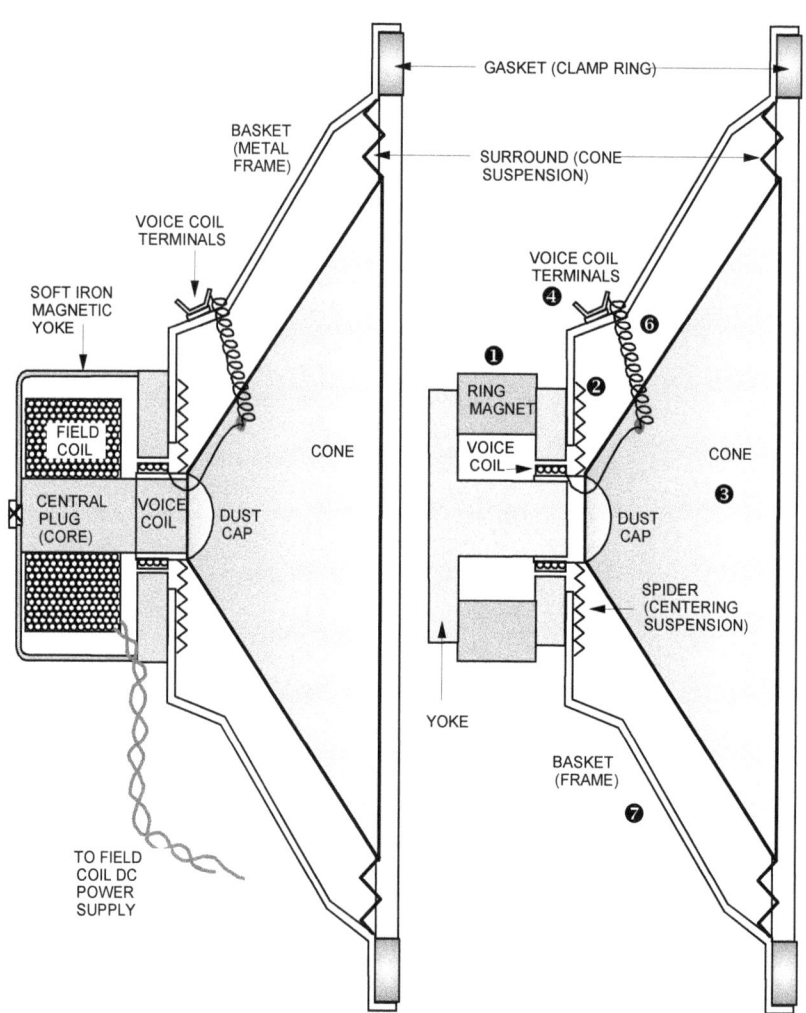

Fig. 9.1. A cross-sectional view of a typical permanent magnet (RIGHT) and electrodynamic loudspeaker (LEFT).

Fig. 9.2. Close-up view of a woofer, ring magnet (1), spider (2), paper cone (3), terminal strip (4) rivetted to the metal frame (7), incoming wires (5) and voice coil connection wires (6).

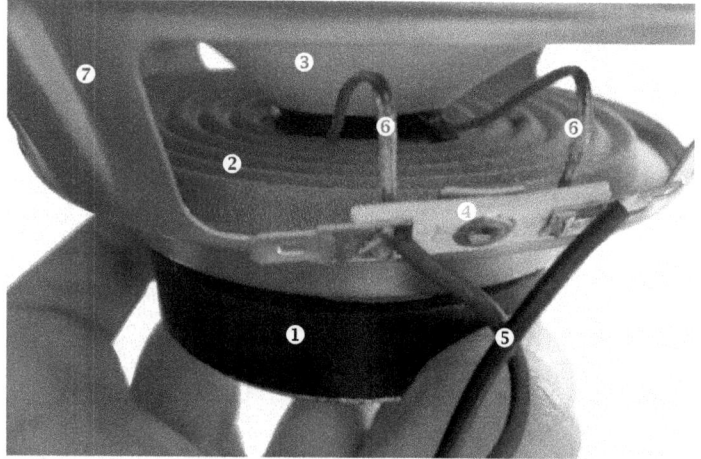

Electrodynamic (field-coil) speakers

Alnico and ceramic magnet permanent speakers are relatively recent invention. In the first half of the 20th-century, speakers did not have permanent magnets at all. Instead, a magnetic yoke with a center insert was used, and the magnetic field was provided by the flow of DC current through the "field coil." The field coil would have a few thousand turns of a suitably sized lacquer-insulated copper wire, wrapped around a spool or cylindrical bobbin, which would then snugly fit over the center plug (core).

With reference to the simplified diagram below, this discussion applies to permanent magnet speakers as well. When signal current I_S flows through the voice coil placed in the uniform magnetic field with magnetic flux Φ, the mechanical force acting on the voice coil is $F=B*I_S*l_C$. B is the magnetic flux density, I_S audio signal current, and l_C is the total length of the coil wire. Since l_C is constant for each speaker's voice coil, the mechanical force is directly proportional to B! If B is reduced, the force on the voice coil is also reduced; this results in smaller movement (excursions) of the cone, reduced air pressure created by the cone and lower SPL (loudness).

With permanent magnet speakers B depends on the type and size of the magnet ring and is thus fixed, and so is their efficiency and "loudness". The force is $F=k_1*I_S$ (k_1 is a different constant).

The magnetic flux density of electrodynamic speakers is proportional to the number of turns in the field coil N_F and its current I_F so the force on the voice coil and the cone is $F = k_2*I_S*I_F$!

This means that with electrodynamic speakers, we can vary the loudness not just by changing the amplitude of the current through the voice coil (by changing the volume control on an amplifier, for instance) but also by varying the field current I_F of the electromagnet. The higher the I_F, the higher the speaker's SPL (sound pressure level).

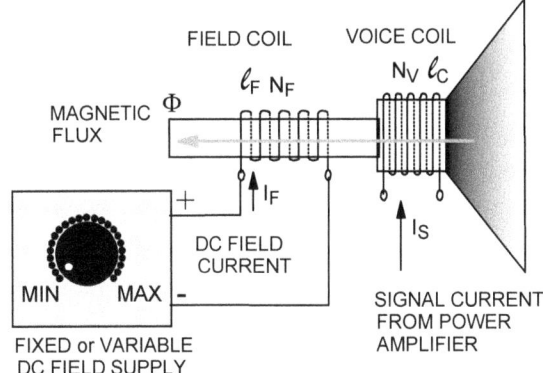

Fig. 9.3. The operational principle of electrodynamic speakers

Commercial audiophile field coil speakers

Two notable examples of currently produced audiophile field coil speakers are LM 755i by Line Magnetic Audio (made in China) and the Hommage Cinema by Auditorium 23, a German company. While LM755i are reasonably priced at $8,995 per pair, a pair of the Hommage Cinema costs a whopping $49,995 (in 2020), plus $5,495.- for the required AcousticPlan NT-1 field-coil power supply. The prices are absurd, more than five grand for a simple DC power supply, and in addition to $50 grand for the speakers that wouldn't work without it!

While the strength and density of a magnetic field provided by permanent (AlNiCo and ceramic) magnets are fixed, the magnetic strength of a field coil speaker will vary according to its power supply. Lowering the voltage lowers the strength of the electromagnet's field and thus the loudness (SPL) of the speaker.

Auditorium 23 Hommage Cinema speakers' power supply is nominally $7V_{DC}$, but its output can be manually adjusted (internal rheostats with 0.1 V jumps) from 4.0 V_{DC} to at least 8.0 V_{DC}! Such a wide control range will result in significant changes in the speakers' sensitivity (and thus loudness or SPL levels produced) but also in tonal balance variations.

The output voltage of LM755i speaker's power supply is variable by a potentiometer and can be monitored by an inbuilt meter. Each power supply (one for each speaker, left and right) is rated at 200 V, and 50 mA of direct current (power consumption 200*0.05 = 10 Watts) and uses one 300B triode as a series pass regulator tube.

There is also a cheaper option, a solid-state power supply, which, in the reviewers' opinion, is not as good sounding as the tubed variant. Again, we see the technically inexplicable situation of a DC power supply affecting the sound of the electrodynamic speaker whose field coil it energizes.

Why speaker drivers with ceramic magnets, Alnico magnets and electromagnets sound different

Despite their fundamental difference when it comes to distortion, audiophiles and guitar players have quite a few things in common. While certain types of distortion are valuable in guitar amplifiers, other types are not. Tube distortion is held in high regard, while distortion of solid-state transistors is not. That is why most guitar players prefer tube guitar amplifiers.

While audiophiles generally profess their desire to eliminate or at least minimize distortion, many prefer the sonic signatures of triodes and tubes in general, with relatively high 2nd harmonic distortion. As we have seen, such distortion makes music euphonic and the whole audio system more "engaging."

It is always interesting to look at the extreme ends of various audiophile beliefs and attitudes. At one end, there are those who claim that everything makes a difference to the sound, and by everything, I mean even things such as the color of the wire used inside an amplifier, bi-wiring of the speakers, and the cable risers that lift the speaker cables an inch or so of the floor. All of these issues and many more will be discussed in this book.

At the other end are skeptics (called "objectivists", which sounds more respectable), who discard subjective evaluations and impressions by claiming that only measurements count. For them, even relatively significant issues such as cables don't make any sonic difference. Obviously, they live in denial. Everything makes a difference, but not to the same extent. Some changes are obvious and striking, others subtle or hardly perceptible.

This particular issue is an interesting example of this clash of beliefs. While guitar amplifiers and speaker cabinets with electrodynamic speakers are still rare, most guitar players would agree that speaker boxes with drivers using Alnico magnets sound better than those with ceramic magnets. Likewise, most of those who have heard expensive audiophile electrodynamic speakers swear that they sound even better than the Alnico ones.

All three types of dynamic speakers work on the same principle. A voice coil placed in a stationary and uniform magnetic field, which, when the audio signal (current) flows through it, moves due to the interaction between the variable magnetic field created by the flow of the signal current and the stationary magnetic fields created by the permanent or electrodynamic magnet. The skeptics would claim that there should be and that there isn't any difference between the three types, that, assuming the magnetic fields are of the same strength, how the field was created should make no difference.

All other things being equal (the driver's construction, size, power rating, resonant frequency, etc.), the sonic difference comes from the source of the permanent magnetic field, or, more precisely, how such field was created. In other words, the supposedly "permanent" or steady (DC) magnetic field is not permanent at all. It varies (gets modulated) by its interaction with the AC magnetic field created by the audio signal current flowing in the voice coil. This changes the response and the dynamics of the speaker driver.

As the voice coil moves through a stationary magnetic field, its variable magnetic field (created by the audio signal flowing through it) distorts the stationary magnetic field and initiates a complex dynamic interaction. A dynamic signal compression results, with consequences such as reduced dynamics, increased driver distortion, and aberrations in its frequency response. This is the case with ceramic magnets, slightly less so with Alnico drivers (which is most likely why they sound better!) and much less so with electrodynamic drivers.

The magnetic field created by a permanent magnet can be weakened by the varying magnetic field of the voice coil. Since the strength of its magnetic field drops with the increase in the voice coil's field (louder music passages), this results in a lower SPL produced by the driver and a significant dynamic compression. Alnico drivers produce a stiffer magnetic field with minimal compression and improved dynamics. The electromagnet's flux field density also remains constant when a voice coil moves in the gap of the motor, and such compression and distortion effects are not present!

Dome tweeters

Dome tweeters operate on the same electromagnetic principle as woofers and midrange drivers, but instead of a cone, a much smaller (typically 1" diameter) dome diaphragm (4) is attached to the voice coil former (5).

Soft domes are made by thermoforming polyester film or impregnating silk or polyester fabric with a polymer resin. The hard dome varieties are made of lightweight metals such as titanium, aluminium, or aluminium-magnesium alloys.

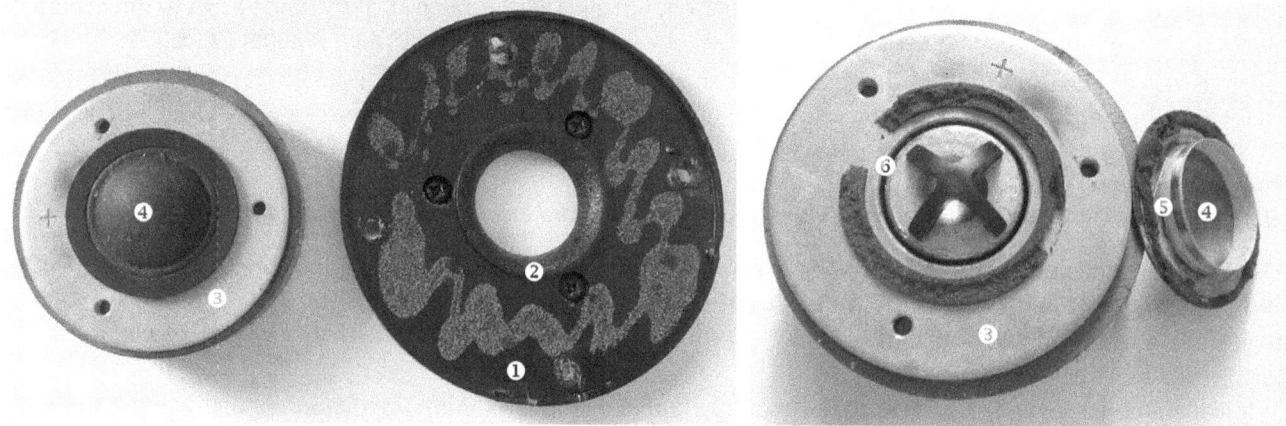

Fig. 9.4. The anatomy of a soft dome tweeter: Plastic faceplate/mounting flange, after the removal of the glued foam layer (1) with waveguide (2), Magnet assembly (3), Fabric dome (4), Kapton former with voice coil (5), Air gap (6)

Dynamic speaker as a complex impedance

A typical dynamic (moving coil) loudspeaker, both drivers by themselves, and especially a combination of drivers with crossover networks in a speaker box, are complex RCL networks whose impedance modulus and phase angle vary widely with frequency. This typical impedance (modulus) vs. frequency curve of a dynamic loudspeaker (driver itself, without an enclosure) shows a resonant peak at the resonant frequency (in this example, $f_R=100Hz$, the peak is around 20Ω). The rise of impedance at higher frequencies is due to the dominant inductance of the voice coil.

The "nominal" speaker impedance, usually 4, 6, or 8 ohms, "proclaimed" by manufacturers is almost meaningless, but, paradoxically, we still assume such a fixed load when buying power amplifiers.

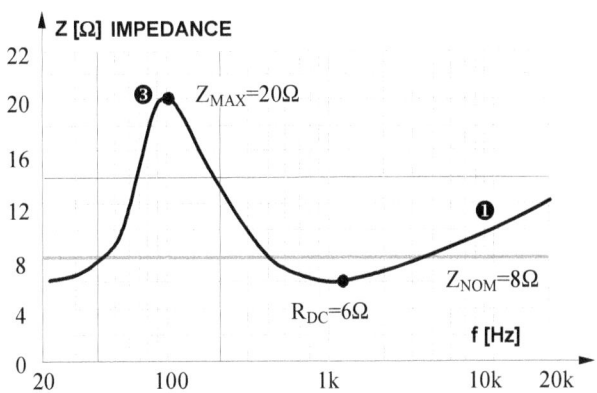

Fig. 9.5. A typical speaker (driver only) impedance versus frequency curve bears no resemblance to its nominal (constant) impedance, a figure declared by its manufacturer!

Zobel networks across speakers' terminals?

One commercial tube amp had a series RC combination (a 22 kΩ resistor and a 47nF capacitor) across its 8Ω output. What would be the purpose of such a circuit?

The impedance of a dynamic speaker increases (1) in a linear fashion with rising frequency f ($Z=2\pi fL$), so this simple type of RC circuit aims to compensate for this rising impedance and make the load (speaker's impedance) independent (or less dependent) on frequency. Remember, to put it somewhat crudely, the inductance and capacitance are electrically opposite and are often used to compensate for or cancel/neutralize each other!

The C_1R_1 product (2) is made equal to L/R_{DC} (speaker inductance and DC resistance), and if R_1 is made equal to R_{DC}, then $R_1=\sqrt{(L/C)}$! If the loudspeaker is of predominantly capacitive nature (such as electrostatic speakers), then this RC network (usually referred to as a "Zobel" network) is not needed and, if installed, should be removed from the amplifier's output terminals.

If you own such an amp, you can experiment by removing this RC pair from both channels' outputs and comparing the sound before and after. Should the sound worsens, the change is reversible. Likewise, you could try adding a Zobel network to your speakers, especially if you think their treble ("top end") is too harsh or prominent.

The resonant peak (3) can also be compensated for by installing a series RLC resonant circuit (4) in parallel with the speaker. The values of the components in the series circuit (a "notch" filter tuned to one specific frequency) should be calculated so that its attenuation curve (5) is a mirror image of the speaker's impedance curve, so the two effectively linearize each other.

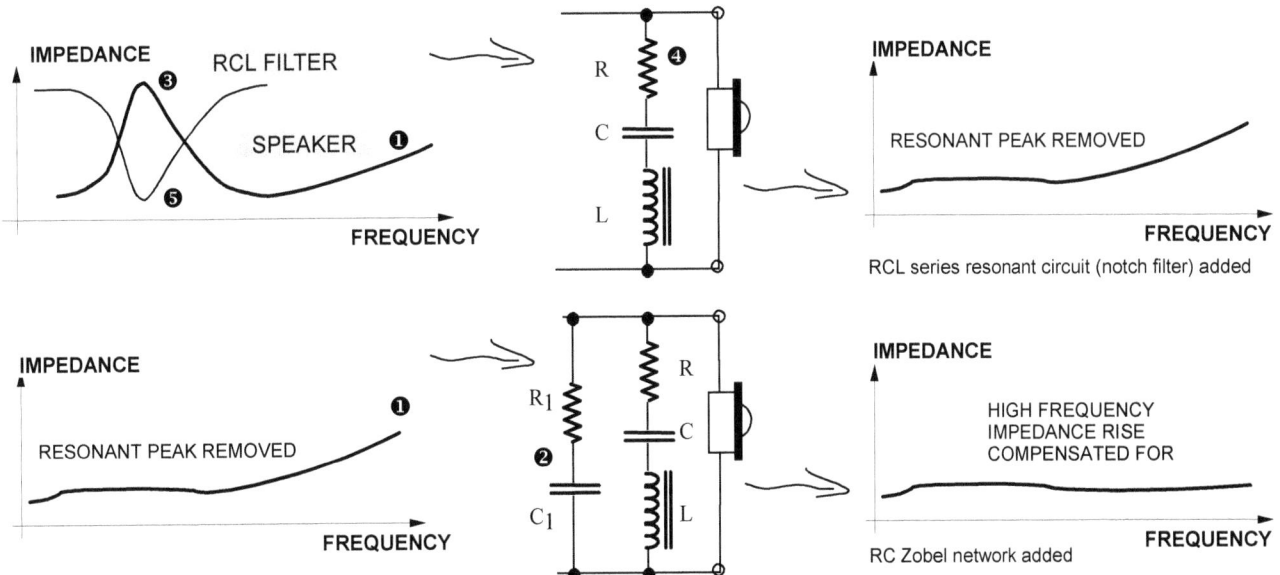

Fig. 9.6. Two simple passive networks, a Zobel RC circuit to compensate for the constant rise of speaker's impedance at higher frequencies (1), and a series resonant RCL filter (4) to flatten or even completely remove the speaker's resonant peak (3).

OTHER SPEAKER AND DRIVER TYPES

Electrostatic loudspeakers

The name "electrostatic" comes from the electrostatic field generated by two stationary grids or stators, which are driven in anti-phase (push-pull) by the audio signal from the secondary of the input transformer (diagram below). A thin flat plastic sheet (diaphragm), usually made of polyester, is coated with graphite or a similar moderately conductive layer, whose task is to accumulate charge but not be able to conduct current, which would allow the charge to flow away. There is a small air gap "d" between the diaphragm and each stator.

The charge Q is generated by V_P, a DC source (power supply) of extremely high polarizing voltage (1-6,000 Volts) and applied through a very high resistance R. The mechanical force F, proportional to the audio signal, moves the diaphragm, which in turn, due to its large area, creates air pressure changes on both sides of the speaker panel.

The conductive electrodes must be as acoustically transparent as possible to allow the air pressure created by the movement of the diaphragm to pass through. They are usually made from a perforated metal sheet or formed from a wire mesh or an array of parallel wires.

To generate a sufficient field strength, the audio signal on the grids must be of high voltage. The electrostatic construction is a capacitor, and current is only needed to charge the capacitance created by the diaphragm and the stator plates. While dynamic loudspeakers are low impedance transducers of predominately inductive nature and require high currents through their coil, this transducer is, in its essence, a high impedance, voltage-driven capacitor, which requires a very low or no current at all but very high voltage swings.

An audio amplifier provides a low voltage signal, which thus must be transformed by the input step-up transformer in each panel to a low current high voltage push-pull pair of signals.

The force on the diaphragm is directly proportional to the audio signal (voltage V_S) and the static charge Q on the diaphragm, and inversely proportional to the total distance between the two stator electrodes (2d). "k" is a constant, a factor dependent on the design of each speaker.

$$F \approx k * \frac{V_S * Q}{2d}$$

Low distortion and a flat frequency response curve are two main strengths of this type of speaker, making them sound detailed, transparent and "airy". However, just as dynamic and other speaker types (ribbon, air-motion, etc.) each face their own significant design and construction problems, there are many such obstacles here. The transformer is obviously in the signal path and its quality is critical for the performance of the whole speaker.

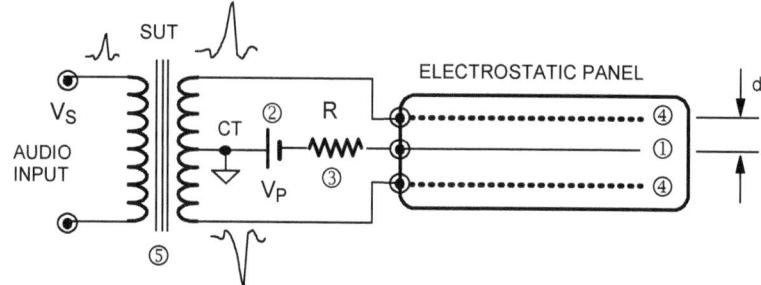

Fig. 9.7. The operating principle of electrostatic loudspeakers
1) Diaphragm (membrane)
2) High voltage DC power supply
3) High value resistor
4) Grids or stators (stationary electrodes)
5) Step-up push-pull audio transformer

Difficult to drive, fiddly to setup, unreliable and weak in bass

The subtitle above was not meant to denigrate electrostatic speakers but to summarize their critics' most vocal objections. Although all four claims are valid, everything in the audio game is a matter of "degrees" or "shades of gray." There are many sealed box speakers, horns, omnidirectional, and bass reflex speakers using dynamic drivers that are also hard to drive, weak in bass, a nightmare to optimally position in a room, or, in some unfortunate cases, "all of the above".

The dipole radiation pattern of electrostatic speakers means that the front and rear sound waves are in phase. If placed close to the front wall, after their reflection, the back waves will be of the opposite phase to the front ones, resulting in noticeable bass cancellation. This speaker placement difficulty is further exacerbated by the fact that electrostatic speakers are very directional. The beaming or narrowing of the forward response angle is especially pronounced at high frequencies. After a long and tedious setup, a slight movement away from the sweet spot (after a long and tedious setup) affects sound staging and spectral balance.

Interestingly, the original Quad ESL (sometimes referred to as ESL 57) had a layer of sound-absorbing material behind the back stator panel to attenuate the back sound radiation. Later models were true dipoles, with free radiation from the back.

The best known and most serious weakness of electrostatic speakers is their feeble bass. Despite its large area, a diaphragm only moves a very short distance and thus cannot push large volumes of air.

Some manufacturers, such as Martin Logan, combine them with a dynamic bass unit, a powered woofer. However, having an integral bass amplifier and active crossover introduces new problems. Also, the radiation patterns of cones (dynamic speakers) and large membranes (electrostatic panels) are very different and thus impossible to blend seamlessly, causing tonal imbalance and making speaker positioning even more difficult.

From the system integration point of view, electrostatics are hard to drive, making them unsuitable for many amplifiers, especially thermionic ones. The first reason is their below-average sensitivity, but the main factor is the impedance, which dips to very low values (almost a short circuit) at bass frequencies, and changes with frequency, from being initially highly inductive to almost purely capacitive.

Both mechanical (grid-diaphragm spacings and tolerances) and electrical (high voltage) reliability are also questionable; the failure rate of electrostatic speakers is much higher than that of the dynamic speakers.

I have never warmed up to electrostatic speakers, finding them too frigid and uninvolving. Sure, they are detailed and transparent, but a superior amplifier and a well-designed dynamic speaker combination can get you 95% there, with more warmth, musicality, and emotional expression.

Even when properly and optimally placed, most ESPs suffer from a narrow "sweet spot"; move your head a bit in any direction, and the sound changes significantly. Some designers use curved panels to broaden the sweet spot, but apart from technical and manufacturing difficulties, that solution usually negatively affects the imaging.

On the plus side, due to their more directional radiating patterns, ESPs can be placed closer to the sidewalls since early side wall reflections are not such a big issue in their case.

Planar magnetic speakers

The planar magnetic speaker looks like a love child of the electrostatic and dynamic speaker. It works on the electromagnetic principle - a current flowing through its voice coil placed in the uniform and steady magnetic field produces a mechanical force on the moving element. That moving (vibrating) structure is not a stiff cone but a plastic membrane, like in electrostatic speakers. Just to confuse the uninitiated, some makers of speakers and headphones using this technology call their wares "isodynamic" or "orthodynamic", but it's all the same stuff.

Their radiation pattern is similar to that of an electrostatic loudspeaker, meaning they are true dipoles; the sound waves from the front and back are identical but of the opposite phase. They can also be made to work as monopole sources, with forward radiation only, just like conventional dynamic speakers.

The "voice coil" is not a tiny cylinder squeezed inside the tiny gap in the magnetic structure anymore, as in classic dynamic speakers, but a 2-dimensional "wiggle" (aluminium or copper wires or foil glued to the film) across the whole surface of the membrane. Such a large surface area partially open on both sides makes heat dissipation a breeze (no pun intended), requiring no special designs and cooling solutions as with traditional voice coils.

Instead of perforated metal stators of electrostatic loudspeakers, thin, flexible magnetic strips (vertical rows of alternating north and south poles permanent magnets) are placed on both sides of the diaphragm. There is no need for high voltage to polarize the membrane or the large and expensive step-up push-pull transformer as in electrostatic speakers.

Unfortunately, just like you or me, the PM drivers could not escape their genetics either. They've inherited the weaknesses of their electrostatic parent, namely their large physical size, low sensitivity (they are power-hungry), weak bass, and a fussy personality (their placement is critical). Once their low sensitivity is combined with low impedance, such high current demands make them a difficult load to drive, so don't even think about pairing them with the low-powered triode lovelies.

Sure, you get the speed, transients to die for (or of?), and transparency, but at what cost? Do I have to use 200 WPC transistor amps?!? No thanks!

However, modern neodymium magnets can provide much stronger magnetic fields and improve planar magnetic speakers' efficiency. The smaller size of such magnets results in more open space on both sides of the diaphragm (the magnets block a much smaller percentage of the total radiating area), which means even more open sonics and reduced colorations. So, we could be poised for a revival of this promising technology.

AMT tweeters

In the "Air Motion Transformer" (AMT) tweeter, a polyimide diaphragm is coated with a thin conductive layer which functions as a voice coil, and folded in a curtain-like manner (or like an accordion folds). The diaphragm is suspended in a strong magnetic field, and, due to its very low mass, is capable of fast response time and superior transient reproduction.

Also referred to as "Heil transducers," after their inventor, they come under various trade names: high-velocity folded ribbon, folded motion tweeter, unique accelerated ribbon tweeter, and jet emission tweeter.

The "air motion," "jet," and "accelerated" descriptors come from the fact that the air velocity produced by such transducers is much higher than the velocity of the folded diaphragm that produced it.

Compared to dome designs, AMT tweeters have a wider horizontal dispersion, mostly due to their order-of-magnitude larger projection area. However, their vertical dispersion is narrower; this may not be bad since the reflections from the ceiling are reduced.

Ribbon speakers

We covered the basics of ribbon transducers in the chapter on headphones and headphone amplifiers (see page 191). True ribbon speakers could be considered a subclass of planar magnetic transducers. The "quasi-ribbon" arrangement means that the membrane that produces the sound has a separate "voice coil" etched or attached to it. Since the signal current flows through it, the ribbon of a true ribbon speaker performs both functions; it is the sound pressure producing element and the voice coil. Aluminum foil, being thin, light, and conductive, is typically used in this application, either straight or, more commonly, pleated.

Ribbon drivers have a spherical radiation pattern at lower frequencies, where the ribbon's dimension is small compared to very long wavelengths of the bass signals. Like dome tweeters, at higher frequencies and shorter wavelengths, the ribbon's radiation pattern narrows and resembles a line rather than a point (as in point-source drivers or ribbon at low frequencies).

For that reason, ribbons are called and classified as line-source transducers. Ribbons are dipole radiators, just like their electrostatic cousins, so all the placement problems and necessary precautions already mentioned a few pages back also apply to ribbon panel speakers such as Magnepan and Apogee.

A step-down transformer reduces the primary impedance from 8Ω needed by the driving amplifier to 10 mΩ ribbon's resistance, an impedance ratio of 8/0.01 = 800! The turns and voltage ratio is a square root of that, or 28.3; the amplifier's output voltage will be reduced 28.3 times. This means that ribbons require very high driving voltages from audio amplifiers.

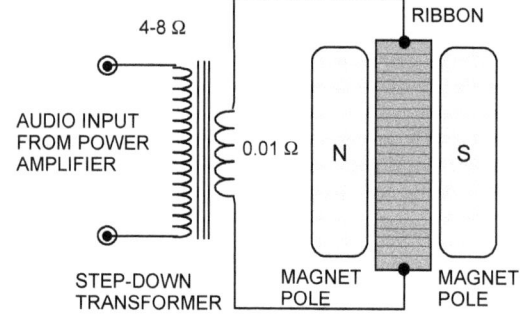

Fig. 9.8. The operating principle of ribbon drivers and loudspeakers

SPEAKER ENCLOSURES AND CROSSOVERS

The ultimate loudspeaker problem: How to faithfully reproduce bass frequencies

Due to the high frequencies (short wavelengths) of the sound waves they have to produce, tweeters and midrange drivers don't need large or any enclosures at all. However, an enclosure of some sort is required in order to reproduce low midrange and bass frequencies, whose wavelengths are much larger than the drivers' dimensions. This is to "separate" the front and rear sound waves produced by woofers, so they don't cancel each other (next page).

Most loudspeaker problems fall into the bass and low midrange frequency bands, typically below the frequency at which the speaker's cone's circumference equals the wavelength of the sound wave.

For a 6.5" woofer with a 6" cone diameter, the circumference is $C = D*\pi = 48$ cm. Since wavelength $\lambda = v/f$ (v is the speed of sound in air, 334 m/s and f is the frequency of the sound), we can calculate that critical frequency as $f = v/\lambda = 334/0.48 = 696$ Hz.

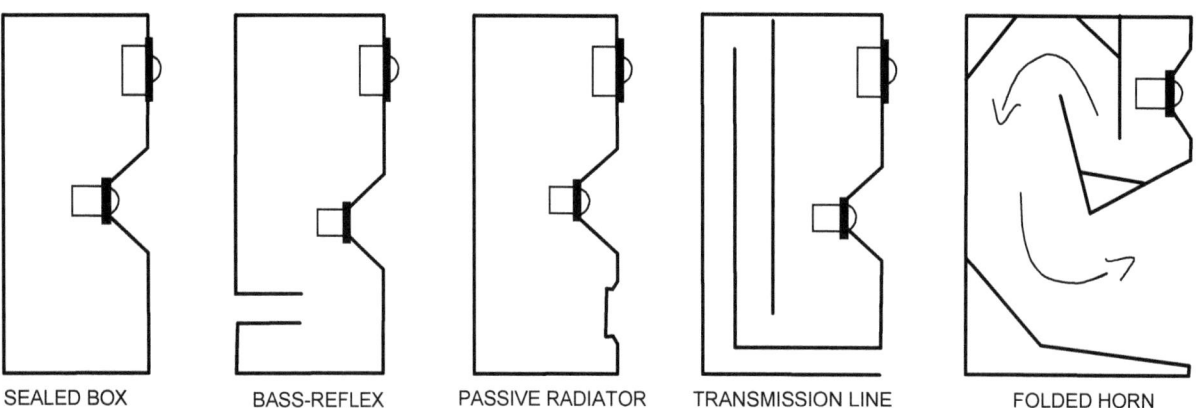

Fig. 9.9. A few of the most commonly used enclosure types using dynamic speaker drivers

Acoustic suspension (sealed box)

Mounting a driver onto a flat baffle reduces the cancellation of front and rear sound waves, but it's still far from a feasible arrangement. For instance, to reproduce a modestly low frequency of 50 Hz (generally considered insufficiently low for top-quality audio systems), the baffle would need to be at least 4 x 4 meters in size!

Folding the baffle a few times into a cabinet open at the back gets us closer to the ultimate goal of a reasonably sized loudspeaker, an approach used in many guitar amplifiers. The rear sound wave's path is much longer, yet the overall cabinet dimensions are much smaller.

However, an electric guitar amp doesn't have to go below, say 80 Hz, so open-back cabinets are fine. Even then, the open box acts as a resonant tube or cavity, boosting frequencies around its audible resonance, resulting in a dominant "one-note bass" or "boombox" character.

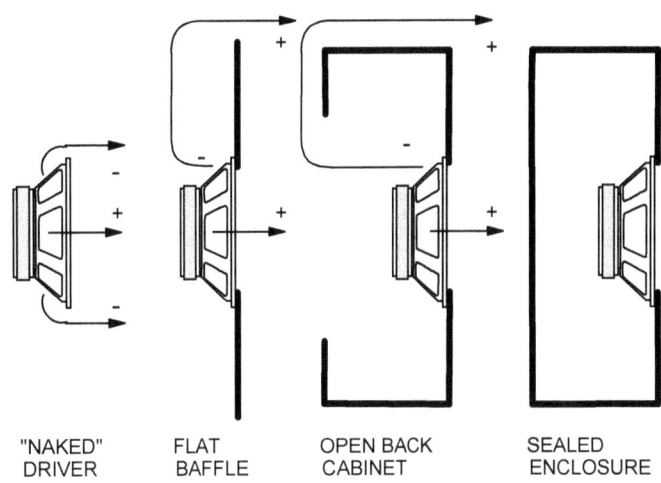

Fig. 9.10. The progression from "naked" driver, through open baffle/open back cabinet, to the closed (sealed) box

Bass guitar amps, on the other hand, require closed cabinets for proper bass extension, as do their hi-fi brothers. Closed cabinets eliminate all the radiation from the driver's rear, but solving one problem causes a few new ones. The trapped air behaves like a spring, so its "springiness" or "compliance" (a fancy word for elasticity, the opposite of stiffness, just as with phono cartridges) increases the resonant frequency of the sealed speaker box above the resonance of the driver itself.

To make matters worse, as the frequency of reproduced sound increases, the speaker box exhibits a series of minor resonances evident as peaks and dips on the speaker's frequency response. These can be minimized (but never completely eliminated) by padding or stuffing the enclosure internally with fiberglass, long fiber wool, synthetic products such as polyester filling (Foam Fill, Poly Fill, Fairy Floss Fill, Acousta-Stuf, etc.), or bonded acetate fiber (BAF) wadding. The effect of such stuffing is also an increased apparent size (volume) of the enclosure.

Bass-reflex cabinets

Instead of trapping the woofer's rear sound waves and dissipating their energy into heat, a bass reflex cabinet utilizes a long duct to delay the back waves enough to invert their phase. They exit the enclosure in phase with the main, front-radiated sound waves, thus reinforcing them and producing higher SPL levels than an equivalent sealed box speaker.

As an oscillating system, the bass-reflex speaker has two components coupled together, the driver (woofer) itself and the air from the port, making it a tuned system. The tube's size and length determine the system's Q-factor (quality factor) and its resonant frequency. The tube's cross-sectional area and its length are chosen to extend the lower bass so that the resonant frequency of the enclosure (a Helmholtz resonator) equals the resonant frequency of the woofer (f_0), hence "tuning."

A wider port (cross-sectional area) raises the resonant frequency of the enclosure, while a longer neck (tube) lowers it. When properly tuned, the two oscillating components become so strongly coupled (in the acoustic sense) that the resultant amplitude-vs-frequency characteristic does not feature the peaks at the original resonant frequency.

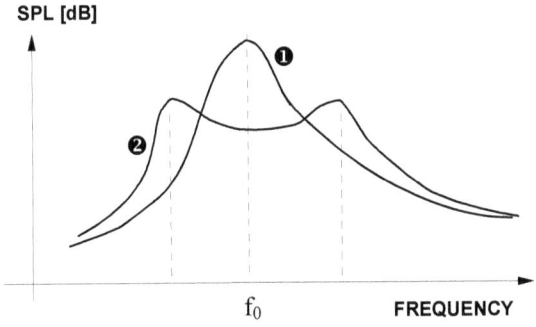

Fig. 9.11. Typical sound pressure level versus frequency response of tuned port ("bass reflex") enclosure (2) and the woofer mounted on an infinite baffle (1)

Instead, two lower amplitude peaks appear, one corresponding to the in-phase motion and the other resulting when the two oscillatory components move out of phase. Compared to the sealed box, the overall sensitivity is also increased. The port can be facing forward ("front-firing") or rear ("backfiring").

However, while such a peak boosts specific bass frequencies, those below the peak are attenuated faster since the speaker's frequency response curve drops at a higher rate (steeper curve) below the peak. Thus, compared to the more forgiving sealed-box type, the room positioning of bass-reflex speakers is more critical, especially of the rear-firing ones.

Passive radiators

A passive or auxiliary bass radiator enclosure replaces the conventional bass reflex port with a moving diaphragm, usually of a standard woofer or large midrange driver. The frame with the cone (often called a "drone" cone) and its suspension (spider) is retained, but the magnet assembly and voice coil are removed.

A passive radiator with a decent cone area and excursion capability helps achieve a tout, low distortion bass. The "chuffing" (wind noise) effect of poorly designed ports is also eliminated, especially at higher volumes. Passive radiators can be used in small or shallow enclosures where the required length of the vent would exceed the enclosure's dimension(s). While they share vented enclosures' qualities such as lower distortion and increased power capability, on the debit side, passive radiators have a higher cutoff frequency and steeper drop rate than equivalent vented designs. They are also less efficient.

Instead of using just its cone for acoustic damping, some designers retain the entire auxiliary driver and load it with a resistor across its voice coil terminals. As the "drone" cone moves its voice coil in the magnetic gap, the induced voltage in the coil pushes current through the load resistor. Instead of working as a linear motor in regular speaker duty, it acts as an AC generator. The current flowing in this simple circuit (voice-coil + resistor) then produces its own magnetic field that tries to oppose the motion of the coil & cone, resulting in electromagnetic damping. The energy is dissipated in the terminating resistor as heat.

Transmission line speakers

A transmission line in electrical engineering transmits power with as little loss as possible; thus, its use in this context is somewhat misleading and inappropriate, although the transmission line does act as an acoustic waveguide, akin to electric power lines.

The folded labyrinth at the back of the woofer in transmission line speakers is designed to absorb as much of the driver's rear-propagating mid- and high-frequency energy as possible. It allows only the lowest bass signals to pass through and exit the enclosure, making it an acoustic low pass filter.

The total length of the folded transmission line should be equal to $\lambda/4$, a quarter wavelength at the lowest bass frequency the speaker is designed to reproduce. It must be lined with damping material to avoid a series of minor resonant peaks and dips in the frequency response.

Well-designed transmission line speakers have smooth impedance curves and roll of the low bass frequencies more gently than bass reflex designs, thus providing better woofer control and cleaner bass.

Proponents of this approach claim that it avoids the problems that sealed boxes suffer from without introducing any new ones, thus avoiding the drawbacks of bass reflex designs. However, as we have learned by now, there's always a price to be paid. In this case, it's the large size of transmission line cabinets that makes them unpopular, just as with another class of loudspeakers that must be large to work correctly, namely horns.

Isobarik loading

Two identical woofers are connected in parallel (driven in phase by the same signal from the amplifier), thus halving the impedance and doubling the current the amplifier has to supply. Mechanically and acoustically, however, they are arranged "in series," the front driver fires into the listening room while the back driver pressurizes the small sealed chamber behind the front one. This tandem operation is equivalent to that of a single driver with half the compliance and twice the moving mass (the same resonant frequency) in a cabinet twice as large.

Compared to the single driver situation, nonlinear distortion is slightly reduced, but the system's resonant frequency is lowered by 30%, thus achieving a lower bass extension. However, the sensitivity is also lower by 3dB, which can be intuitively figured out; an amplifier provides power to two drivers, but only one generates acoustic energy into the room. Linn Products gave this approach their trade name "Isobarik."

Another approach used by Linn in their Majik model and other speaker makers (for instance, Jadis in their Eurythmie II horn speakers) is to position the two drivers in a face-to-face arrangement (next page). Of course, to ensure that both cones move in unison (in the same direction), the speakers must be driven out-of-phase in a push-pull arrangement. No signal manipulation or inversion is necessary. That can be achieved by simply swapping around the connections on one driver: signal + to the positive terminal of one driver and the negative terminal of the other, and vice versa.

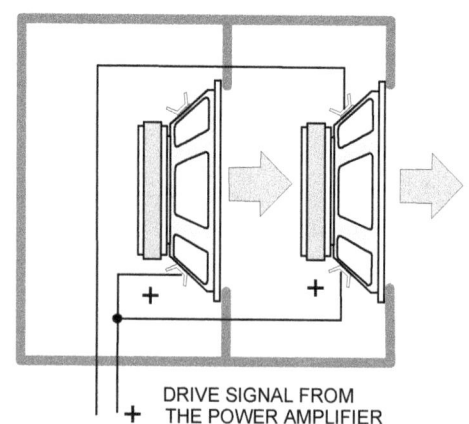

Fig. 9.12. The front-to-back or cone-to-magnet isobarik arrangement

Apart from the benefits in the lower-to-mid bass region, some experimental reports claim a significantly smoother response of isobarik loading in the upper bass and lower midrange bands. Also, the coupling of non-identical drivers would give designers much more flexibility than identical ones. Hence, it is perhaps surprising that more investigation hasn't been carried out in this regard.

D'Appolito or MTM arrangement

The MTM (midrange - tweeter - midrange) vertical arrangement is better known as the D'Appolito configuration, named after Joseph (Joe) D'Appolito, who improved the 1960s idea enough to be successfully used in commercial loudspeakers.

The obvious benefit of this symmetrical arrangement (two drivers instead of one, connected in parallel, with a tweeter in between) is an increase in sensitivity by three or so dB.

However, the main goal was to achieve more uniform dispersion than the imperfect dispersion of the tweeter-midrange - woofer designs, whose lobbing pattern tilts downwards towards the woofer.

However, this arrangement requires the listening height to be precisely at the tweeter level; otherwise, the two different arrival times from the two drivers could be audible, affecting clarity and sound staging (imaging).

Also, while its on-axis response is exemplary, D'Appolito's configuration's radiation pattern is narrower compared to the standard tweeter - woofer - midrange arrangement.

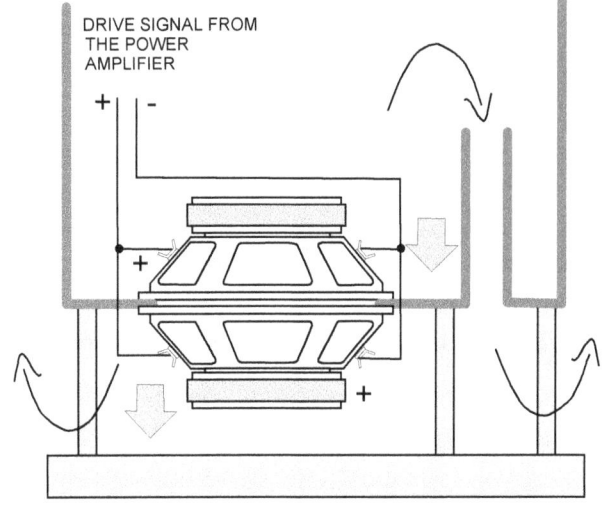

Fig. 9.13. (ABOVE) The cone-to-cone or front-to-front arrangement of the two woofers in Linn Majik loudspeakers

Fig. 9.14. (RIGHT) D'Appolito configuration

Fig. 9.15. (BELOW) The saxophone - horn speaker analogy

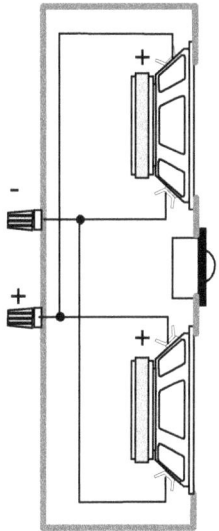

Le coup de foudre: The most beautiful speakers I have ever seen

In late 2001 I was contacted by a loudspeaker designer from Sydney, Australia. He needed a couple of our low-powered triode amplifiers to demonstrate his horn loudspeaker at the upcoming audiovisual fair. I traveled with two of our best amps to join him at the show.

The speaker was a very tall and relatively narrow horn in a shape of a giant saxophone. A single Fostex full-range driver was at its throat, facing forward, with its back horn-loaded. The mid and high frequencies had nothing to do with the horn - they were radiated straight ahead.

The speaker's body was painstakingly hand-crafted by joining dozens of mitered solid timber slats and mounting (interlocking) them onto a wooden frame.

I had never seen a speaker more beautiful; it was a sculpture of stunning beauty, a true work of art. He was selling it manufacturer-direct and priced it at AU$18,500. At the time, that seemed like a fortune to me; now, I'd say that was indeed the bargain of the century. To give you an idea, it took him two months to make a pair!

I should have asked for a photo, but at that stage of my audio journey, I hadn't even contemplated becoming a writer of books on DIY audio. I did not envisage that I would be writing about his incredible creation for a worldwide audience one day. We became good friends during those four days, but sadly, Zeka (his nickname, meaning "bunny" in Serbian) died a year or so later, and his speaker designs died with him.

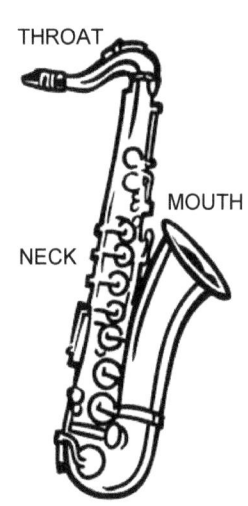

Horn enclosures

Apart from honoring my friend and his creation, I mention this story for another reason. A saxophone is an acoustic horn, a small cross-sectional area (the "throat") opening up through its "neck" (a flared body) into a much larger "mouth." It works on the same acoustic principle as horn-loaded speaker enclosures. However, unlike Zeka's stunners, very few such fridge-sized speakers (Klipschorn anyone?) could be called pretty.

There are so many types of horn enclosures and variations on that theme that it's impossible to proclaim one design to be a "typical" horn. In the 1950s and 60s, the golden era of horns, apart from Paul Klipsch and his company, many types of horns were marketed by various well-known brands, Altec Lansing being one of the most prominent.

Zeka's horn was of the purist type - a single driver with no crossovers. Apart from reducing speaker sensitivity, crossovers introduce sonic degradation due to the undesirable phase shifts they create.

Many other "horns" are actually hybrids, using a standard enclosure for the woofer, while only the midrange driver and perhaps the tweeter are horn-loaded.

Also, most folded horn cabinets feature straight sections, which are only a crude approximation of a true horn's smooth, exponentially flaring curvature. A smoother flare means less chance of standing waves developing and lower acoustic loss (sound attenuation) along the way.

A horn is an acoustic transformer; it couples the small area of the throat to a large area of the mouth. The high pressure and low volume velocity at the horn's throat are gradually transformed to low air pressure and high velocity at its mouth, thus achieving better coupling to the surrounding air and much high power transfer efficiencies.

Again, we see the mechanical analogy with cars (coupling high-RPM engine to low-RPM wheels through a gearbox) and an electrical analogy with tube amps, whose output stages must be coupled to low impedance speakers by output (impedance) transformers.

Various tapers are used, hyperbolic, exponential, and tractrix being the most common. The tractrix contour expands faster than the exponential, resulting in shorter horn lengths.

The diameter of the mouth (or, more commonly, the width of a rectangular cross-section) must be at least $D = \lambda/4$, 1/4 the wavelength λ of the lowest frequency f_C to be reproduced, and since $f_C = v/\lambda$, we get $f_C = v/4D$. A designer can decide on the overall size of the mouth and then calculate the cutoff frequency or decide on the cutoff frequency and find out how large the enclosure will be.

A designer can decide on the overall size of the mouth and then calculate the cutoff frequency or decide on the cutoff frequency and find out how large the enclosure will be.

Since $A_2 = A_1 e^{kX}$, we have to chose the throat area A_0 and the mouth area A_M. For $A_2/A_1 = 2$, we can calculate the distance resulting in doubling of the cross-sectional area, $e^{kX} = 2$, so $kX_D = \ln 2 = 0.7$.

Since $k = 4\pi f_C/v$, we get $k = 4\pi 40/344 = 1.461$, and now we know kX_D and k, so $X_D = kX_D/k = 0.7/1.461 = 0.479$ m, which is our doubling distance for 40 Hz cutoff.

Based on the size of the driver, the designer has to choose the size of the throat A_0, which will depend on the effective area of the driver's cone.

To maximize the bandwidth of the horn's operation, throat-to-driver ratios should be in the 0.3-0.5 range and 0.5-0.7 to maximize efficiency. For example, for a 220 cm² driver, A_0 should thus be in the 66-88 cm² range.

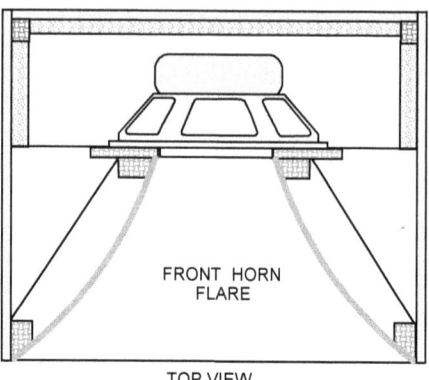

TOP VIEW

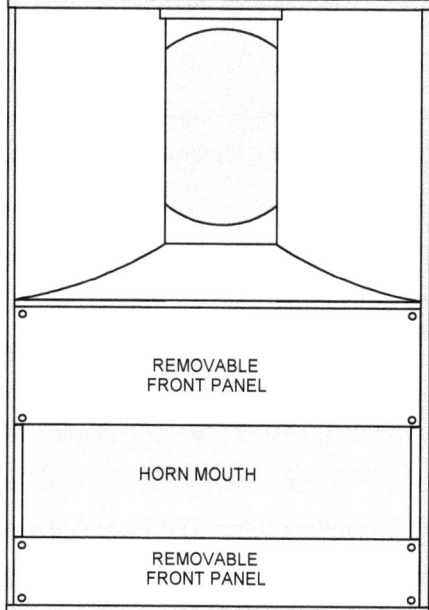

FRONT VIEW

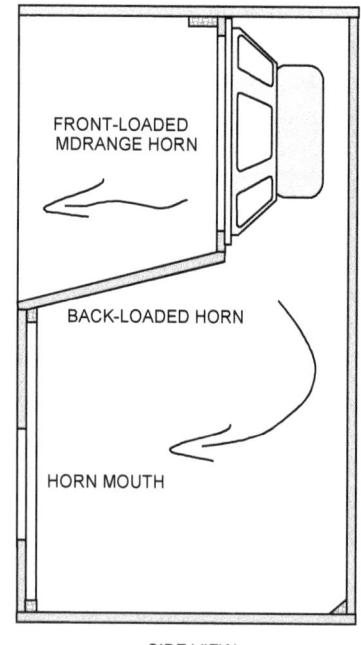

SIDE VIEW

Fig. 9.16. The drawing of Altec Lansing A7-500 main enclosure, one of many versions produced (1962). The externally mounted HF (tweeter) horn is not shown. The 15-inch woofer-midrange driver is both front- and back-loaded with simple short horns.
Dimensions (HxWxD): 107 x 76 x 61cm (cabinet only, without the top horn), 80 kg, 102.5 dB/W/m sensitivity

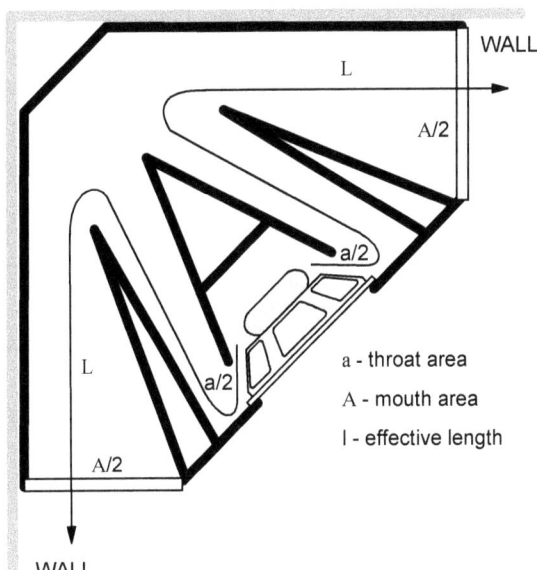
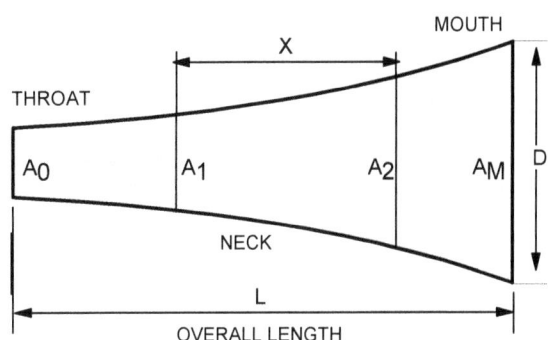

Fig. 9.17. (ABOVE) The exponential flare and its rate of expansion, $A_2 = A_1 e^{kX}$, where k = flare constant, e = base of the natural logarithm (e=2.718)

Fig. 9.18. (LEFT) The horizontal section of a back-loaded horn (such as Klipshorn) designed to be placed in a corner so the walls act as the horn extension, increasing the effective length "L"

As you can see, there are many parameter choices to play with. Now that you have the doubling distance and the mouth and throat areas, you can calculate the required length of this horn (the number of doubling distances between the mouth and throat, multiplied by the doubling distance).

Detailed horn design is best left to the specialists, but I hope this short introductory overview has given you a better understanding of the issues and compromises involved.

How do horns really sound?

Although a horn loudspeaker designer's first and foremost goal is to achieve clean, tout bass reproduction, the first impression that listeners usually get is that horns are bass-shy. This seeming contradiction can be explained by the fact that most of us are used to the pumped-up nature of most bass-reflex speakers and have thus been conditioned to accept their sonic character as the "bass standard"; it isn't.

Also, to reduce their size and cost, the design of some horns has been compromised (foreshortened too much), thus increasing the cutoff frequency, making their bass sound thin or "empty."Well-designed, properly built horns produce clean, clear, and crisp bass. This has a positive flow-on effect on the midrange as increased transparency and airiness of presentation; again, for want of a better word, "clean" comes to mind.

This is understandable. Amplifiers driving horns work below 1 watt of output power at most times. At such levels, the distortion is minimal since both amplifier and horn distortion are way lower than that of a standard setup, where a 100 watt per channel amplifier is likely to be struggling to drive 82 dB (in)efficient speakers.

Apart from this difference in tone, horns also produce a different sound stage. Conventional boxed speakers in a stereo arrangement produce a 3-dimensional soundstage behind the speakers, with only some extending the soundstage sideways around the speakers. The soundstage of horns projects forward, enveloping the listener like giant headphones to use a crude analogy. Some audiophiles like it; others don't.

It would be wrong to proclaim that horns generally sound more dynamic than bass-reflex and other designs, many of which are very lively and engaging. I would say that to me, they sound more effortless, more lucid, or liquid, if you will. Perhaps that is also due to the effortless operation of the amplifiers driving them, as already mentioned, or, in single driver horns, to the absence of complex crossovers.

They create aural havoc in various ways, none conducive to natural sound. The lobbing effects and acoustic mismatching (especially around the crossover frequencies) between drivers of multiple sizes and radiation patterns are also on the suspect list.

The second major loudspeaker problem: The limited frequency range of dynamic drivers

A speaker cone should be light (so it can move fast and reproduce fast signal transients better) yet rigid (so it doesn't flex and distort), meaning it should have a high degree of internal damping. As it happens almost always in electronics and acoustics, these are contradictory requirements.

Speaker designers and builders have been trying various materials, almost all of which are still in use - paper (cut and seamed), molded paper pulp, metal, plastic (polypropylene, for instance), and various types of composite materials and layered (sandwiched) hybrids.

No matter how large a cone of a dynamic driver may be or what material it's made of, all cones are driven by a relatively small area of the voice coil and not through their whole surface, as with planar magnetic or electrostatic speakers. At low frequencies (wavelengths much larger than the cone size), that is not an issue; a cone moves as a rigid whole (a "piston").

As the frequency increases and wavelengths become smaller than the cone, the cone reaches the limits of its rigidity. It starts to act as a mechanical transmission line with distributed parameters (mass, resistance, and compliance).

Such flexing is comprised of a multiplicity of mechanical waves traveling outward as pictured (highly exaggerated for illustration purposes) and then reflecting back onto themselves with little attenuation. This unpleasantly sounding effect is known as the "cone breakup."

Cone breakup results in standing waves in various directions, for instance, around the circumference and along the flared edges. It causes minor resonant peaks of the frequency response curve.

This means that a single dynamic speaker driver can't reproduce the whole audible frequency range with a relatively flat amplitude characteristic. Avoiding the cone breakup frequencies means that the usable (where they behave in a fairly linear and reasonably efficient way) bandwidth of all drivers (woofers, midrange drivers, and tweeters) is limited. At least two, and often three or more different size drivers, must be combined in a single speaker box.

However, simply connecting them in parallel and feeding them all with the same signal would not achieve anything. Each driver must be fed only the frequencies that fall into its usable frequency band.

Electronic circuits that perform such a frequency-selective attenuation or filtering are called filters. The filters used in loudspeakers are pejoratively called crossovers since they have to cross over to each other in a smooth and seamless way.

Electronics students study filters in great detail, but all we can do in a book of this kind is to briefly look at some simple yet often used and thus important filter types.

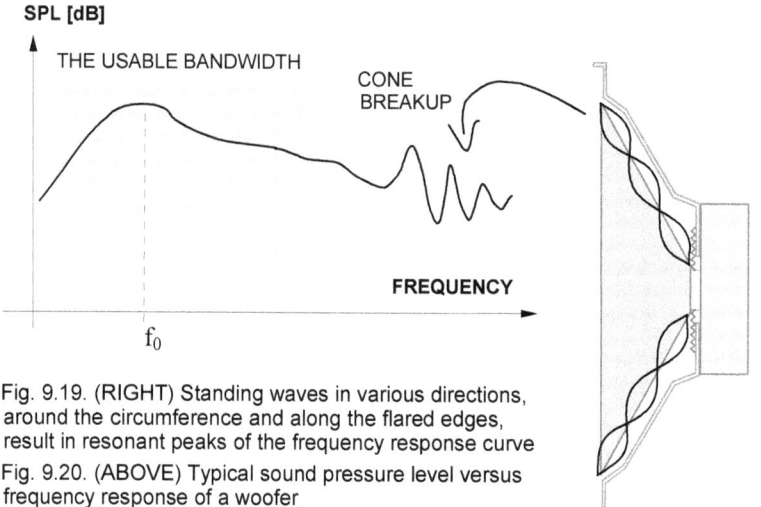

Fig. 9.19. (RIGHT) Standing waves in various directions, around the circumference and along the flared edges, result in resonant peaks of the frequency response curve

Fig. 9.20. (ABOVE) Typical sound pressure level versus frequency response of a woofer

Crossover filters

Speaker crossovers filter out the range of frequencies the particular driver cannot or should not reproduce (due to high distortion or irregular frequency response, with lots of peaks and dips). Crossovers can be passive (usually placed inside speaker enclosures) and active.

Active crossovers are electronic filters using active components such as vacuum tubes, transistors, and integrated circuits, placed between the preamp and power amps. For a two-way active crossover system, two power amps are required (per channel!), three amps for a 3-way system, and so on. This adds complexity and cost, not to mention the introduction of new, system-specific problems.

Passive filters in speakers crossovers use passive components, resistors (R), capacitors (C), and inductors or chokes (L). They are called passive because they do not amplify the signal and thus require no external power to operate.

However, they filter higher voltage and current signals, so the L, C, and R components used in passive filters must be rated for such high power levels. They are physically larger and more expensive than those used in active crossovers.

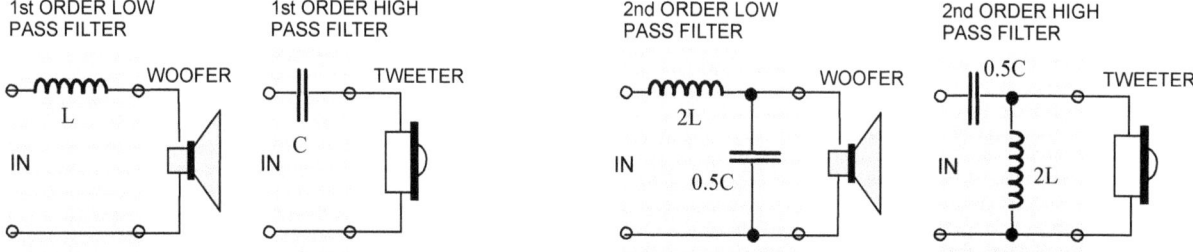

Fig. 9.21. The first and second order Butterworth filters (low-pass and high pass) for 2-way loudspeakers

LOUDSPEAKER TYPES, TESTS, AND IMPROVEMENTS

Butterworth-type filters are most common since they have the smoothest response and the sharpest possible roll-off without peaking. Linkwitz-Riley, Chebyshev, and Bessel filters are occasionally seen in some speaker designs. The filter types were named by the mathematicians or engineers who first described them.

In crossovers below, $C = 1/(2\pi f_0 R)$ and $L = R/2\pi f_0$, $\pi = 3.14$, f_0 is the crossover frequency (-3dB), and R is the coil resistance of the driver.

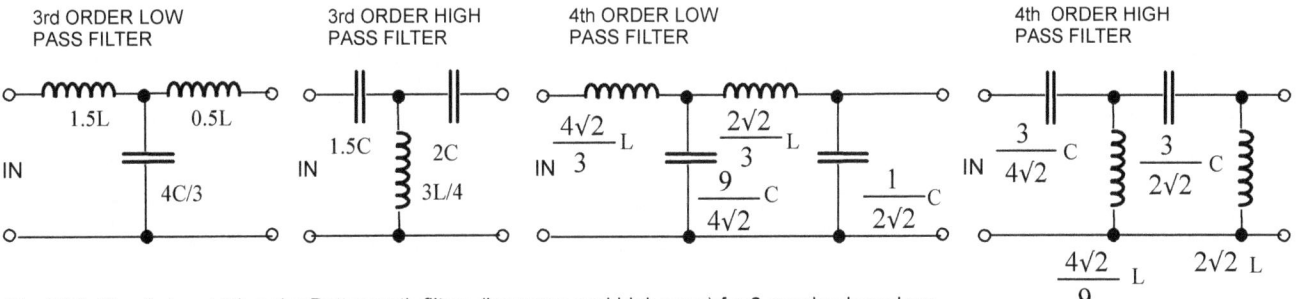

Fig. 9.22. The 3rd and 4th order Butterworth filters (low-pass and high pass) for 2-way loudspeakers

Crossovers range from the very simple 1st order types, with only one reactive element (capacitor or inductor) per driver, through 2nd order designs, with two LC elements per driver, to third, fourth, and higher-order filters.

Notice that filters start attenuating signals way before the crossover frequency f_0 (1). 1st order filters attenuate signals at a rate (slope) of 6 dB per octave (2:1 ratio of frequencies), so at a frequency twice f_0 (2), the curve is down -6 dB, and another octave up in frequency ($f/f_0 = 4$) the curve is down -12 dB (3).

Expressed in a different way, this attenuation is 20 dB per decade (a decade is a 10:1 ratio of frequencies), so in point (4), where the frequency is $10f_0$, the attenuation is -20 dB!

2nd order filters double that attenuation to 12 dB per octave, and 3rd order crossovers have attenuation curves with 18 dB/octave slopes.

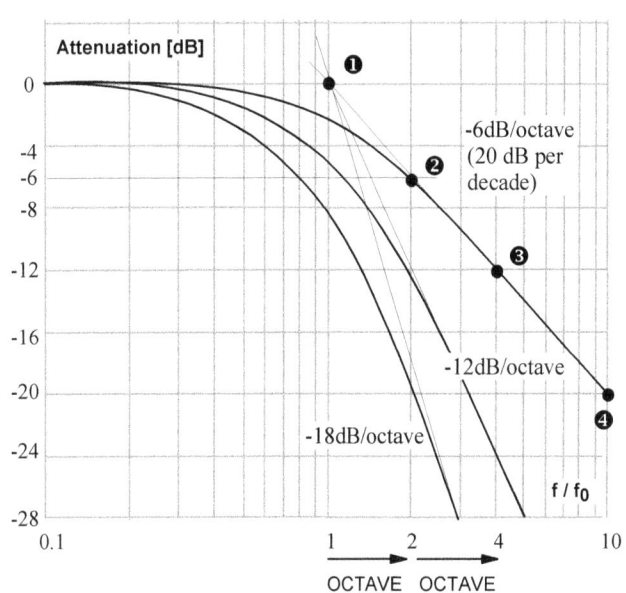

Fig. 9.23. Attenuation curves and asymptote slopes for the 1st, 2nd and 3rd order low pass filters

CHOOSING A LOUDSPEAKER

Speaker size

There are more factors in choosing loudspeakers than in selecting any other audio component. The speakers are the final transducer, which converts the electrical signals into sound. In such complex electromechanical conversion (the listening room plays a huge part), many things can go wrong.

Price aside, as with (to a lesser extent) amplifiers, the kind of music you predominantly listen to is arguably the most critical factor. Is it bass-dominant, or is the midrange the most prominent aspect of it? How loud do speakers need to be in your room to create the right audio illusion and play music the way it should be played? Simple, acoustic, and vocal pieces or small jazz bands are far less demanding than reproducing the full force of a symphonic orchestra during crescendo peaks.

Your room size and the distance between the speakers and your listening spot will immediately eliminate the whole category of speakers as unsuitable. Placing large speakers in small rooms is never a good idea. They would dominate the space and look overbearing. More importantly, their large woofers will energize the room and make the bass dominant and boomy. The soundstage will not be full-sized since such speakers will be too close together relative to their size and too close to the front and side walls.

On the other hand, small or close-field monitors designed to be listened to "up close and personal" would sound super focused in a smallish room but would be lost in a larger one. They would not just look silly, out-of-proportion, but would sound tiny and tinny, lacking the required energy to move the large volumes of air that bass frequencies demand.

Generally, smaller speakers (but also the tall & narrow ones) have a better soundstage and microdynamics; they are more subtle and nuanced. They cannot compete with larger ones in macrodynamics; they cannot energize a larger room as larger ones can or scale the dynamic heights required by large-scale or loud orchestral music.

The room size will also be a factor as far as dipole-type speakers (such as electrostatics) are concerned since their rear radiation pattern mandates a considerable distance from the front wall, something a small room cannot provide.

Finally, matching the speakers and the amplifier(s) that drive them is as important (and perhaps even more so!) as matching the speakers to your listening room. More on that crucial issue in the next chapter.

The size of drivers and their arrangement

If you flip through audio magazines from the 1970s, you'll notice a common trend: a (very) large-diameter woofer, a small midrange driver on top of it, and an even smaller tweeter. Soon after, Small and Thiele's seminal work systematized the parameters affecting speakers' performance, and bass reflex designs allowed the use of smaller diameter woofers.

Speaker designers realized that the crossover from the woofer to the midrange driver is critical for tonal coherency; it falls into the upper bass/lower midrange band, where the human ear is most sensitive.

In terms of coherence, the smaller the size difference between the woofer and midrange driver, the better. The best recordings for the test of such seamless coherence (or its absence!) are those featuring a male voice (especially bass or baritone) and acoustic instruments such as double bass and cello.

On the other hand, the midrange driver-to-tweeter crossover is critical in terms of beaming since the two drivers' diameters are usually very different. Using the same (smallish) driver as a woofer and midrange solves the first problem but limits the choice to 5" and 6.5" drivers, whose bass is feeble

Using drivers of a relatively small diameter for those lower- and mid-frequency duties keeps their physical size difference (in comparison to the tweeter) and thus the associated variation in their dispersion and beaming patterns to a minimum. Identical bass and midrange drivers also allow the use of lower-order crossover filters with gentler slopes and better phase and transient response.

Impedance variations with frequency as an indicator of speakers' character

Dynamic speakers' impedance curves typically exhibit at least two peaks, as illustrated below. The impedance of this nominally 8-ohm speaker shoots up to 40+ ohms (1) and dips to 4 ohms between 30 and 40 Hz (2). These bass frequencies are of high energy, so any amplifier driving this speaker will have to push high currents through the woofer's voice coil in the frequency range.

Even in the upper bass region (3) between 100 and 300 Hz, the impedance dips to around five ohms; it could be argued that this isn't an 8-ohm but a 4-ohm or 6-ohm speaker.

As for the phase, the good news is that in the upper treble region of 10 - 20 kHz (3), this speaker is an almost purely resistive load (the phase shift around zero degrees). The phase then becomes negative, dipping to 65° at 1.7 kHz (4), dead smack in the midrange, where human hearing is the most sensitive. The speaker exhibits a strongly capacitive character there, which is bad news for any amplifier driving it (increased distortion and reactive power swings).

The phase then swings to 30 degrees positive (at 500 - 600 Hz), the load takes on a slightly inductive character (5), and, as we move down into the bass region, such a wild swing happens twice more.

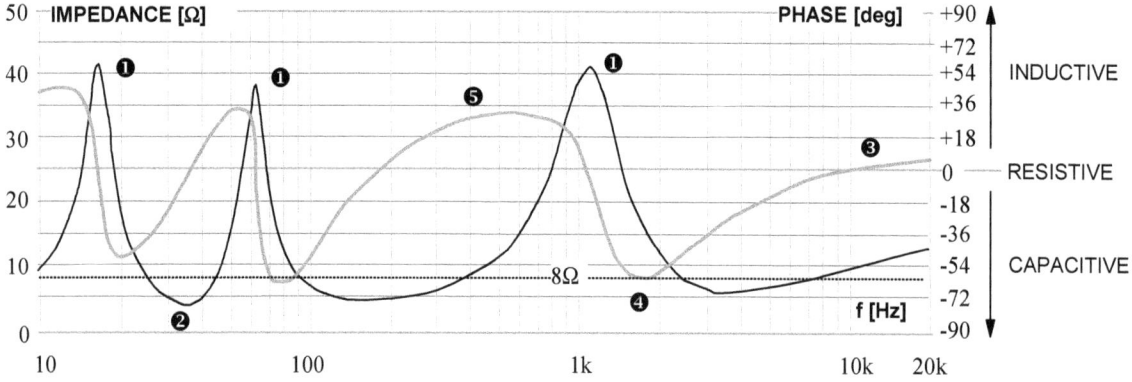

Fig. 9.24 The impedance magnitude versus frequency (black curve) and impedance phase- versus-frequency characteristics (thicker gray curve) of a certain 2-way bass reflex loudspeaker

Although this speaker presents a relatively high impedance without dipping too low, the three wild impedance swings, two of which are in the bass region (below 100 Hz), both in phase and magnitude, indicate a possible high degree of coloration, with some frequency bands emphasized and others highly attenuated (suppressed).

> **THE SPEAKER IMPEDANCE RULES**
>
> Before shortlisting speakers for a detailed audition, use these rules to narrow down the field:
>
> Rule #1: The magnitude (modulus) and the phase of the speaker's impedance should be as flat as possible (not too many sharp peaks and dips)
>
> Rule #2: The impedance magnitude should not drop to very low levels, especially not at low frequencies (which would make speakers current-hungry and difficult for amplifiers to drive)
>
> Rule #3: The phase variations should be as small as possible, close to a purely resistive load (phase angle of zero degrees).

Matching speaker's personality with yours - your ideal speaker is out there somewhere ...

Of all audio components, speakers vary the most in terms of their technologies, designs, parameters, and, ultimately, sonic "personalities." No other audio component's design is wrought with so many issues, contradictory requirements, and physical (electroacoustic) limitations. Each speaker designer deals with such restrictions in his own way, thus imparting the particular sonic "personality" to his speaker. Like audiophiles, whose needs and tastes in loudspeakers' sound vary enormously, speaker designers interpret sonics in various ways.

Consequently, when it comes to music material, no speaker reproduces all types of music equally well, or, in terms of loudness, sounds equally good at widely varying volumes (sound pressure levels). That makes it necessary to audition speakers in the actual listening room you would be using, playing the music you usually listen to. The music material chosen by the high-end dealers is always carefully selected to showcase the speakers' best aspects while concealing their weaknesses.

One of the problems we'd encountered in matching low-to-medium power tube amps with loudspeakers were the speakers that needed lots of power to "come alive." Some would sound flat and boring at lower SPL levels, while others would do a decent job but still fall very short of their performance at higher SPL levels.

Harmonic imbalance is also an issue. Some speakers have an excellent midrange but lack the top and bottom extension; others produce a fast, tight, and prominent bass but have a recessed midrange and a metallic, harsh, almost shrill treble. Great for parties but very fatiguing and irritating in an audiophile setting

The smooth on-axis response (a flat frequency response), as measured by manufacturers and audio magazine reviewers, isn't that important on its own. Since we seldom listen on-axis, what matters much more is the speaker's off-axis character. The off-axis sound levels are always attenuated somewhat, down a few dBs. Still, the off-axis performance should be as smooth as possible, without significant or numerous dips and peaks. Very few speakers are equally good in all these aspects.

CRUCIAL QUESTIONS AND QUICK SPEAKER TIPS

Breaking in loudspeakers

Electromechanical audio transducers, namely microphones and phono cartridges at the audio chain's recording and signal source end, and loudspeakers at its other end seem to benefit the most from a break-in period.

The fastest way to break a pair of speakers in is to wire them out of phase and turn them to face each other in a closed room. You may even throw a thick blanket or a curtain over them to minimize the sound pressure level. Play a continuous track containing pink noise (from a CD or music streamer) for 50-100 hours.

Speaker grilles - on or off?

Audiophiles automatically assume that all speakers should be listened to with their protective grilles removed. However, that may or may not be the case. It all depends on the acoustic transparency of the cloth the speaker grilles are made of.

Some speaker makers deliberately voice their speakers with grilles on, so removing them may result in slightly harsher and, since the cloth affects high frequencies more than the bass, overemphasized treble.

This was certainly the case with our Opera Terza speakers. With their grilles on, they sounded smoother, with a more refined and precise treble. Since you'll never know what their designer had in mind, listen to speakers with grilles on and off and make your own conclusions.

Should I use spikes or acoustic isolators under my speakers?

Even speaker designers and manufacturers disagree on the best way to interface speakers with floors. Some believe that rigid coupling between them is best (using spikes) because isolating speaker enclosures from the floor prevents the unwanted vibrational energy from being transferred to the floor, thus trapping it within the enclosure and making it vibrate. Such parasitic vibrations smear transients and affect microdynamics, resulting in a loss of detail and a haze or veil over the sonics.

However, passing on vibrations onto the floor is a problem since a floor can be considered a large panel resonator that will absorb, delay, and re-radiate such vibrations as unwanted sound back into the room. As always, there is no hard-and-fast rule. If you are concerned, try both and decide for yourself.

In general, concrete floors will vibrate less, making rigid coupling a better option. In contrast, suspended wooden floors, due to their propensity for significant vibration, are prime candidates for the use of speaker isolation feet.

Port tuning on bass-reflex speakers

The size of the bass reflex port or ports (two or more may be used) should have been carefully sized by the speakers' designer. However, the designer had no control over your room or your choice of the speakers' position, and those factors may result in the different optimal size of the vent. This applies to back-firing ports even more so.

Depending on how close the back-firing ports are to the front wall, to change the level and the character of the bass, you can experiment with partially or even wholly blocking one or both ports.

HOW TO MEASURE SPEAKERS' SENSITIVITY, IMPEDANCE, AND FREQUENCY RESPONSE CURVE

Most audiophiles have no idea about their amplifier's or speakers' technical specifications and performance parameters or even an inclination to find out. If you want to know more, performing some basic tests isn't difficult, although it requires access to a few basic instruments. The good news is that quality multimeters and LCR meters have become very affordable, costing less than a budget-priced audiophile cable. Even function generators and handheld LCD (liquid crystal display) oscilloscopes are getting cheaper by the day.

Just as we have done with amplifiers in chapter 7, in this section, you will learn how to perform a few basic tests that will give you valuable insight into the speakers you have or are auditioning and planning to buy.

The constant current source method of plotting loudspeaker impedance curve

Since the 220 Ω series resistor has a much higher DC resistance than the impedance of the measured loudspeaker, this simple circuit approximates a CCS (Constant Current Source). That same current flows through both the series resistor and the speaker's voice coil.

You are interested in the difference between the readings of two AC voltmeters: V_1 indicates the output voltage of a function generator (sinewave), and V_2 displays the AC voltage drop across the loudspeaker's impedance at various frequencies.

Had the speaker's impedance been constant with changing signal frequency, the ratio of the two measured voltages would remain constant.

However, since the speaker's impedance is highly frequency-dependent, the ratio of the measured voltages will also change with frequency.

The voltmeters must be capable of operation up to at least 20 kHz, so cheap multimeters are out of the question.

In the experiment described here, from 10 to 150 Hz, we took measurements every 10 Hz, then from 200 to 1,000 Hz every 100 Hz, and then every 1 kHz. After punching in all the figures into a spreadsheet, we got a recognizable dynamic speaker impedance curve, shown on the next page.

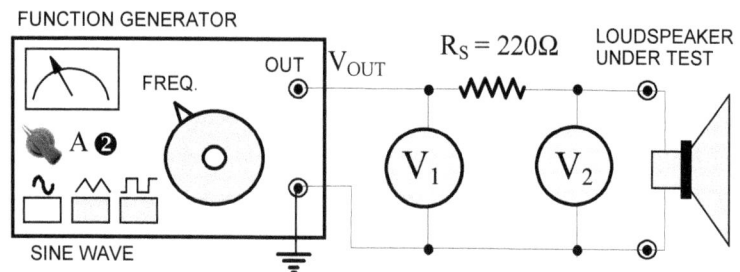

Fig. 9.25. Test setup for manual (point-by-point) plotting of a loudspeaker impedance versus frequency curve

SPEAKER IMPEDANCE BY TWO VOLTMETER METHOD

$Z(f) = R_S V_2(f)/[(V_1(f)-V_2(f)]$

$V_1(f)$ and $V_2(f)$ are measured results at various frequencies, $Z(f)$ is the calculated impedance modulus at each test frequency "f".

EXPERIMENT: *Loudspeaker impedance curve for Opera III (Terza) loudspeaker*

A pair of these speakers has served us faithfully for more than 25 years as one of our test speakers. Technically, due to their complex crossovers (a 3-way design with high-order filters) and relatively low sensitivity (86-87 dB/W/m), these should be totally unsuitable for low-power SET amplifiers.

Yet, for some reason, even a lowly 3 Watt amp can drive them relatively easily. The bass drivers are small, so don't expect ceiling-collapsing bass levels. They are not the most accurate or transparent speakers around. But, they are musical, engaging, and, above all, great looking - solid mahogany, slim and elegant.

Our low distortion function generator kept V_1 constant so we only had to measure V_2. Once the impedance curve was plotted, it became clear that these speakers are a very easy load to drive. The impedance never dropped below the DC resistance of the voice coil (6.2 Ω), while in the bandwidth between 3 and 20 kHz (upper midrange and treble), it stayed in the 12-16 Ω range.

The three resonant peaks are evident at around 30 Hz, 80 Hz, and 2,500 Hz. The most significant impedance drop is in the upper bass and lower midrange region, from about 110 Hz to almost 400 Hz, with a virtually constant 6-7 ohm impedance (4). The manufacturer declared these to be 6-ohm speakers. Since their impedance remains between 10 and 17 ohms at all frequencies above 600 Hz, as far as I'm concerned, they could have slapped an 8 ohm or even a 16-ohm rating on them.

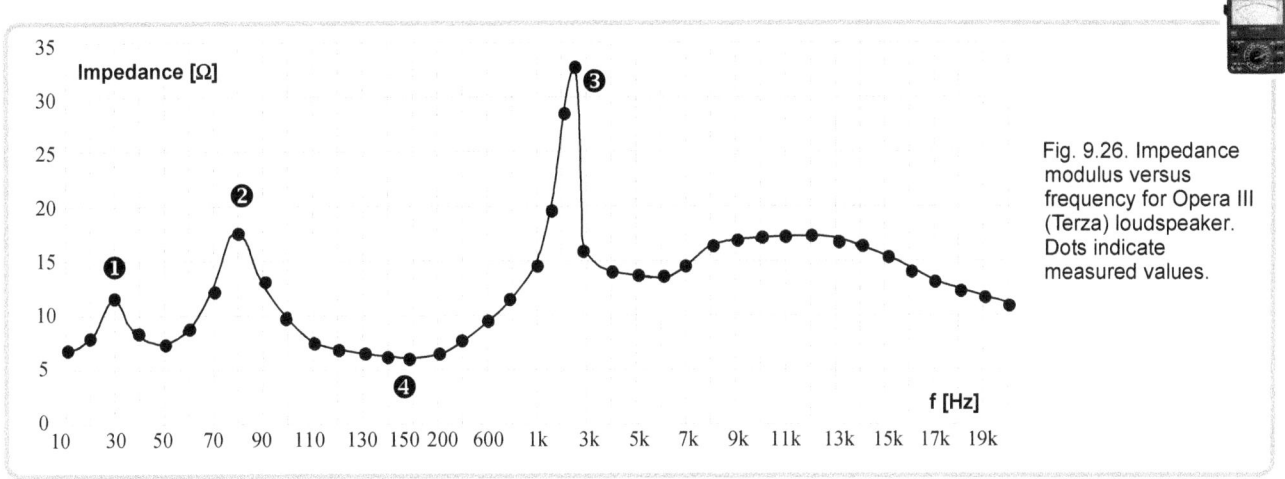

Fig. 9.26. Impedance modulus versus frequency for Opera III (Terza) loudspeaker. Dots indicate measured values.

EXPERIMENT: *The frequency response of an amplifier with a dummy load and dynamic speakers*

With a dummy (purely resistive) 8Ω load, the amplitude-versus-frequency characteristic of this single-ended triode amplifier with EL156 output tubes was flat over its entire bandwidth, dropping to -3dB at 15 Hz and 41 kHz. The curve is shown below and marked (5).

We chose 1 Watt at 1 kHz to be the 0 dB reference level. The speakers' impedance changes heavily influenced the amplifier's amplitude-vs-frequency characteristics when loaded by Opera II speakers. This is due to its relatively low damping factor of 3.6 in the triode mode and only 2.5 in the ultra-linear mode.

The three resonant peaks of the speaker result in three amplification peaks (marked 6, 7, and 8), but the overshoots are not severe in any of the cases. The peak (7) is only 0.7 dB over, and the peak (8) is just 1.2 dB above the referent 0 dB level.

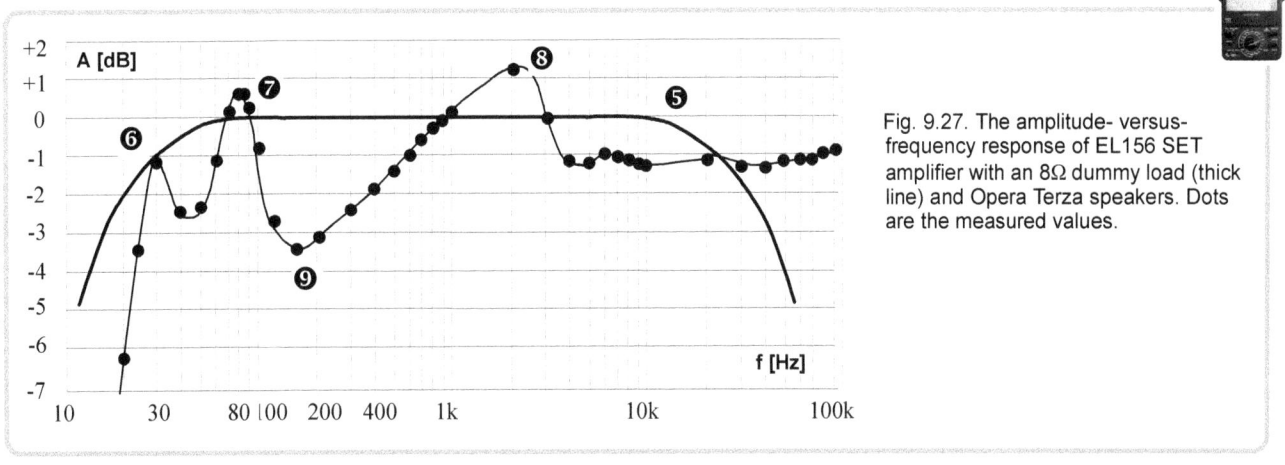

Fig. 9.27. The amplitude- versus- frequency response of EL156 SET amplifier with an 8Ω dummy load (thick line) and Opera Terza speakers. Dots are the measured values.

The biggest drawback is two dips, one in the lower bass region, around 45 Hz, the other in the upper bass, with a minimum of -3.5 dB at about 160 Hz (9). These two "suck out" regions are the main reason for the relatively weak bass of these speakers when driven by most SET amplifiers. The midrange frequencies between 1 and 3 kHz will be slightly emphasized (8), up to +1.2 dB, but that is such a modest rise that it is sonically hardly perceptible.

How to measure the sensitivity of speaker drivers and speaker boxes

You can use this simple setup to measure the sensitivity (dB/2.83V) of any other loudspeaker type. The less power a speaker needs to produce the same sound pressure level, or, in other words, the higher its produced SPL level for the same input power level, the higher the speaker's sensitivity.

You can measure speaker drivers by themselves (tweeters, midrange drivers, woofers) or whole speaker boxes. Strictly speaking, since this test should be done in a special anechoic test chamber, just as speaker manufacturers do it, your results will be different from theirs. The test room will include sound pressure from both direct sound waves and the reflected ones; it will also include the background noise level, which can be significant.

Thus, it pays to do this in a room well-insulated from outside noises and make that room as acoustically "dead" as possible. Minimize hard, reflective surfaces and increase the surface area of soft furnishings, carpets, blinds, curtains, and absorption acoustic panels (if any). Also, make sure the speaker measured is as far as possible from walls; the best position is often in the middle of a large room.

You have to stand behind the SPL meter to take readings, which may impact the sound field around the meter and the speakers and slightly alter the measured results, so make sure you always stand in the same spot and don't move around while taking the reading. You can measure the sensitivity at one particular frequency or repeat this test at various frequencies and get the speaker's whole SPL/efficiency versus frequency curve.

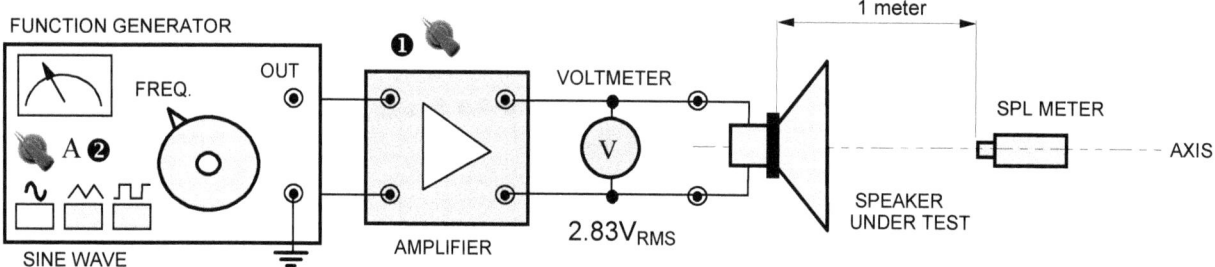

Fig. 9.28. The test setup for measuring the sensitivity (dB/Watt) of loudspeakers. Test tone(s) from a digital or analog source can be used instead of the function generator.

Apart from the variable frequency of the test signal (produced by a function generator), all other variables must remain constant. Both the function generator and the amplifier must have a flat gain-vs.-frequency characteristic over the entire measurement frequency range. You must ensure a constant 2.83 V_{RMS} reading on the voltmeter. If the voltage drops or rises, adjust either the volume control on the amp (1) or the amplitude control on the function generator (2) to bring it back to the required constant level.

There is no need to perform this test at hundreds or even dozens of frequencies. Five or six measurements at suitably chosen frequencies (for instance, at 100 Hz, 400 Hz, 1 kHz, and 3 kHz) are sufficient to give you an idea of what kind of speaker you are dealing with.

EXPERIMENT: Determining the crossover frequencies of your speakers

Once you have the circuit diagram of your speakers' crossover, you can determine the crossover frequencies in three different ways. You could calculate it, which is easy for simple crossovers, but finding such formulas is impossible for complex circuits. You could draw and simulate the circuit in one of the circuit simulation software packages and read the -3dB frequencies for the low-, mid-, and high-pass circuits from the frequency response curves.

Or, you could hook up a couple of test instruments and measure the actual crossover behavior of your speakers, the only exact method of all three.

Most function generators can produce output signals of at least 7-10 V_{RMS}. Still, not many can supply the current needed to drive the speakers directly, so you may have to use a power amplifier in between the generator and the speaker to provide such a current drive.

You must make sure that the amplitude of the signal going into the speaker's input terminals ("IN1" and "IN2") remains constant as you change the frequency of the sinewave test signal on the function generator (1).

Two sets of measurements are needed for a two-way system by connecting a true-RMS multimeter on the "AC Volts" range to woofer terminals W1 and W2 and then to tweeter terminals T1 and T2.

A third test is also required for a 3-way system at midrange terminals M1 and M2.

You don't have to plot the graphs as we did here; it is enough to identify the flat part of the curve or the reference level. For instance, at 30 kHz, 20 kHz, and 15 kHz, the measured level was 10.5 mV (2), so we calculate the -3 dB attenuation level (multiply the figure by 0.71), as 10.5*0.71 = 7.5 mV.

Keep lowering the signal frequency and observing the voltmeter readings (which will start dropping) until the voltmeter shows 7.5 mV. Then we read the frequency at the calibrated dial of the function generator (1). In this case, the -3 dB frequency was at 2,300 Hz (3).

For the woofer, the procedure is identical, but we start at a very low frequency, say 20 Hz, and keep increasing it.

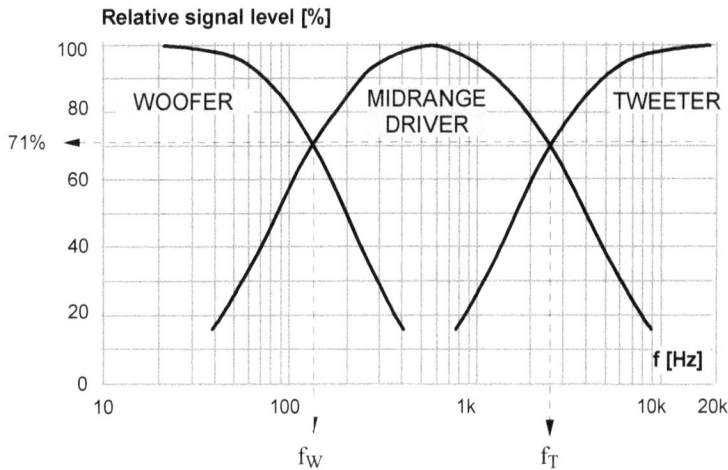

Fig. 9.29. How the outputs of various crossover filters vary with frequency for a 3-way loudspeaker: low pass filter for the woofer with the upper -3dB frequency f_W, the bandpass filter for the midrange driver with its low and high frequency limits, and the high pass filter for the tweeter with its f_T lower limit

The voltage level should be steady for a while until it starts to drop. Once it reduces to 71% of the reference voltage, we read the -3 dB frequency, in this case, we measured 140 Hz.

The midrange crossover filter will have two -3 dB frequencies, the lower and the upper. Usually, the lower f_L equals the woofer's cutoff frequency $f_L = f_W$, and the upper f_U equals the tweeter's $f_U = f_T$, but not always. In any case, the midrange frequency limits should be near the tweeter's and woofer's figures.

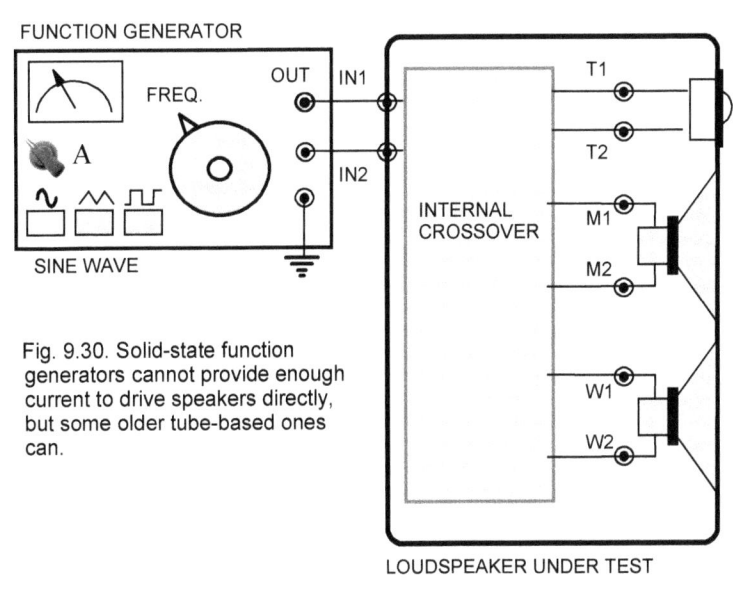

Fig. 9.30. Solid-state function generators cannot provide enough current to drive speakers directly, but some older tube-based ones can.

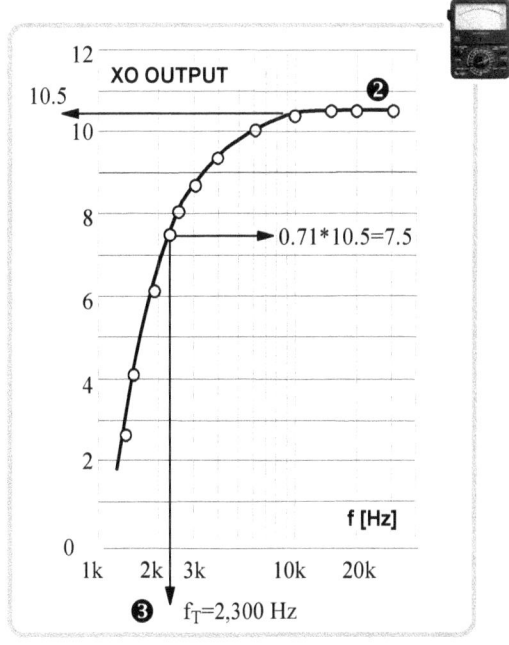

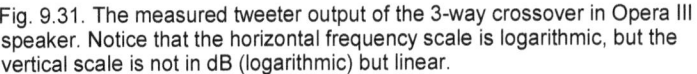

Fig. 9.31. The measured tweeter output of the 3-way crossover in Opera III speaker. Notice that the horizontal frequency scale is logarithmic, but the vertical scale is not in dB (logarithmic) but linear.

SPEAKER MODIFICATIONS AND UPGRADES

TWEETER ROLLING

If your tweeters are of a standard shape and size, trying alternative tweeters is easy. Just make sure the replacements are of the same impedance and have a similar power rating as the originals. The sensitivity and frequency response will be different, which will change your speakers' sound for better or for worse, making tweeter rolling lots of fun.

SPEAKER STANDS

Most speaker stands are of a fixed height, but if the height is adjustable on yours (a big plus), experiment with the optimal tweeter height and tune by ear. Fill them up with heavy granular material such as dry sand. That will lower the stands' resonant frequency and dampen any likely vibrations, both from the speaker cabinets themselves and those propagating from the floor upwards through the stands.

CROSSOVERS (XOs)

Capacitors, resistors, and inductors in speaker crossovers affect the overall sound. Start with electrolytic (bipolar) capacitors, then move onto iron-cored inductors, replacing them with film & foil caps and air-cored inductors. If you are fussy (or fanatical), replace the ordinary wire-wound resistors (most commonly used) with audiophile-grade, non-inductive exotic resistors (read: "expensive ones").

INTERNAL REWIRING

Most speakers are wired using thick & stranded hookup wire. Replace it with a thinner solid core copper alternative, or, if you like their sonics (some audiophiles aren't too crazy about them), Teflon® insulated silver-plated copper wire.

BASS-REFLEX PORT TUNING

Try partially or even completely blocking the port(s) with sound-absorbing foam, wool padding, or synthetic fiberglass. Even cotton balls and rolled-up socks can do the job! That will change the bass reproduction of your speakers and move their enclosures towards the sealed box design.

Since altering the size (diameter) of the bass-reflex port isn't feasible (but it is possible), you can try changing its length, which will also change the bass tuning.

SPIKES

Unless done as a part of crossover design to achieve phase and time coherence, a tweeter should be set back from the woofer/midrange driver. Adjusting the length (height) of the front spikes makes that a breeze.

EXPERIMENT: Drawing a circuit diagram of your speakers' crossovers

To identify possible improvements to your speakers' crossovers, the first step is to locate them and get them out of the speaker enclosure. Assuming they are identical, only one speaker needs to be opened. In our case, we had to remove the midrange driver since the crossover board was glued to the back right behind them.

Opera III speakers, used as an example here, are bi-wirable, so there are two pairs of binding posts; we named them + IN LOWER & - IN LOWER, and + IN UPPER & - IN UPPER. The white-colored rectangular components in the photo are wire-wound resistors (1). The values and power ratings are stamped; for instance, 4R7 means 4.7 Ohms, R is often used instead of the Greek letter Ω (omega), and 9WK means the resistor's power rating is 9 Watts.

The white, yellow, blue, black, or red-colored components are film capacitors; they can be box-shaped (2) or tubular (3).

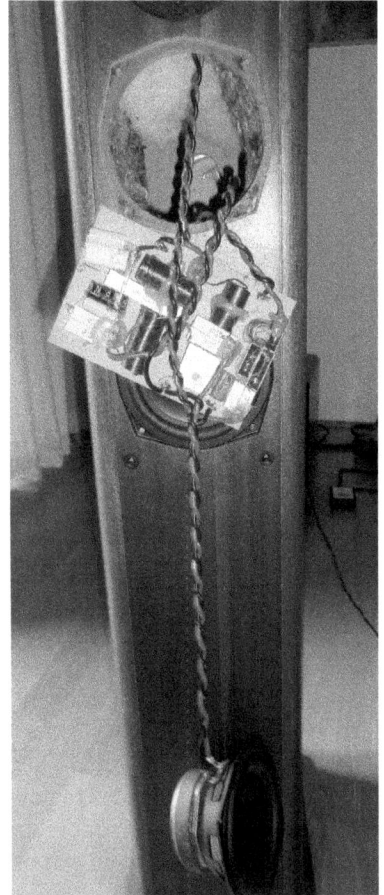

Fig. 9.32. (LEFT) The crossover is wired point-to-point on a piece of craft board, which is then glued to the back of the cabinet, just behind the midrange driver (the opening above it, through which we took it out of the speaker enclosure). Notice that all components are also simply glued to the craft board.

Fig. 9.33. (ABOVE): The close-up of the component side of the crossover board. On the back are two backlinks, linking the (-) or negative inputs and the (-) or negative outputs (to individual speakers or drivers). The light gray wires (+) are actually RED in color, the dark wires (-) are BLACK.

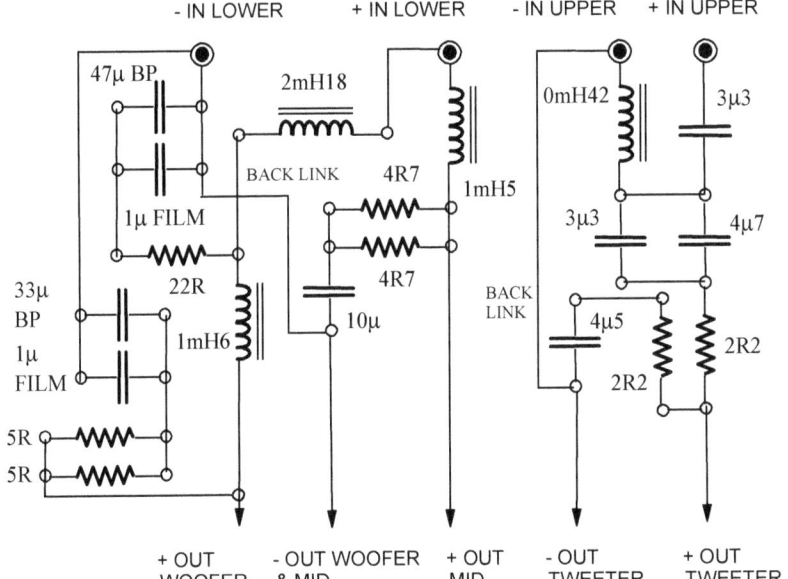

Bipolar electrolytic capacitors (4) are also tubular but have indentations or "pinched" bodies (outer casings) at both ends. Their values are generally much higher than those of the film capacitors, for instance, 33 or 47 µF (microFarads) compared to the film caps' 0.22 µF or 1 µF capacitance values.

The film capacitors are added in parallel to the electrolytic ones to improve their high frequency and transient response.

Fig. 9.34. (LEFT) Step 2 - drawing the wiring (or interconnection) diagram

Fig. 9.35. (BELOW) Step 3, drawing the circuit diagram of the speakers' crossover

Inductors, both air-cored (usually larger) and iron-cored (smaller in size), are simply many wound turns of a magnet (transformer) wire and can easily be identified. There are four of them here, three larger ones mounted horizontally (5) and a smaller one placed vertically (6).

The easiest way to draw the wiring diagram is to draw components exactly as they are placed on the board, using rectangular boxes for all of them. Alternatively, as a step closer to the wiring diagram, use the standard electrical symbols for resistors, capacitors, and inductors instead of the rectangular boxes. We've opted for the latter option here.

Step #3 is to convert the wiring diagram into a circuit diagram. It should be drawn in a conventional manner, the inputs on the left side, and the outputs to the right, so the signal flow is generally from left to right. Both of these steps are critical, so make sure you double-check everything to minimize the possibility of error(s).

The 3rd order tweeter high pass filter is followed by the Zobel network across its terminals to compensate for the impedance rise at high frequencies. The low pass woofer filter is also a 3rd order unit (LCL), also followed by the RC Zobel combination in parallel with the voice coil.

Instead of the usual band-pass filter (high pass followed by a low pass filter), only a 2nd order (LC) low pass filter is used for the midrange/woofer driver.

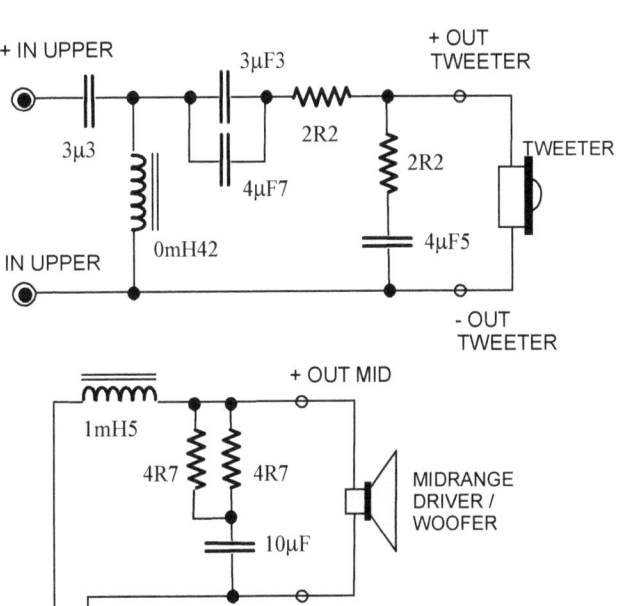

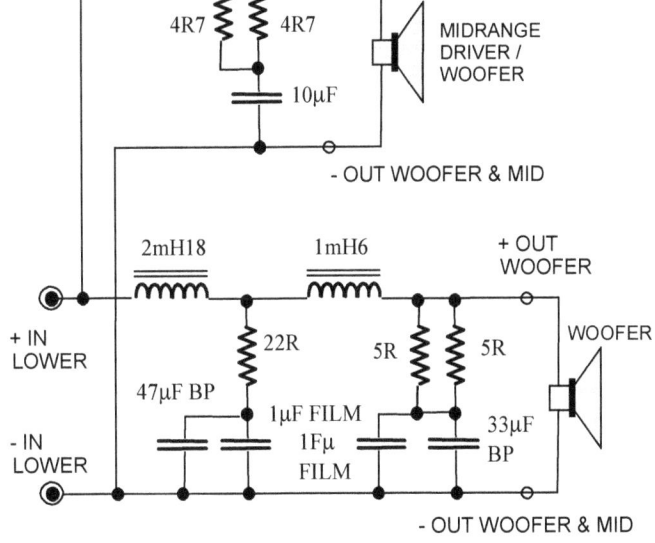

That means the lowest bass frequencies will be reproduced by both woofers, with the upper woofer also continuing to work as a midrange driver once the bottom woofer's contribution is filtered out by its LCL low pass filter.

Resistors are used to adjust voltage levels and don't determine the order of the filter. Notice separate terminals for the tweeter crossover and the combined woofer-midrange part. Once the external shorting links are installed, the two + terminals are connected, as are the two negative or (-) terminals.

Going "The Full Monty" - replacing the whole crossover with an external one

Once you have the crossover circuit diagram and understand its design, you need to choose the components to upgrade. Should you wish to keep the existing assembly (or the "board"), you may not be able to upgrade iron-cored inductors with air-cored ones since the air-cored ones are physically larger and may or may not fit onto the board. The only option, in that case, is to replace the bipolar electrolytic caps with film ones, but again, these are larger and may not fit either. It all depends on how closely the original components are placed on the crossover board.

As you can see from the photo on the previous page, they are pretty darned close, without any space between them. If this seems like a low-cost crossover to you, you are right, but keep in mind these solid mahogany speakers had a retail price of only £999.- in the late 1990s, so something had to give to keep the price so low.

If you are serious, you will have to upgrade the whole crossover, including the board. Some pedantic audiophiles with superior DIY skills even go one step further. They build new (still passive) crossovers in separate enclosures, which can be attached to the backs of the speaker enclosures (where the binding posts were originally located) or even placed on the floor behind each speaker, so there's no physical contact between the two.

Just as with turntables and vacuum tubes, the rationale behind this relatively drastic step has to do with vibrations and microphony. The vibrations of both the speaker cabinet (to which the boards are attached) and the air inside it are transferred to the internally mounted crossover boards, resulting in possible sonic degradation. Audiophiles who performed such an upgrade often reported reduced "glare" and the lifting of a slight "veil" from the notes, improved resolution and microdynamics, and generally more "relaxed" sound.

Iron-core inductors out, air-core inductors in

Assuming equal inductance, air-cored inductors require many times the number of turns of the winding wire than the iron-cored ones. Thus, the iron-cored ones are smaller, cheaper to produce, and have a lower resistance (shorter total length of the wire).

However, iron cored inductors suffer from the hysteresis effect and retentive magnetism, making them nonlinear, introducing various types of distortion. The upgrade from iron-core to air-cored inductors results in cleaner, more detailed, and transparent sound.

The standard 0.47 mH value is close to the required value of 0.42 mH, and 1.5 mH is close to the required 1.6 mH, so all is well here.

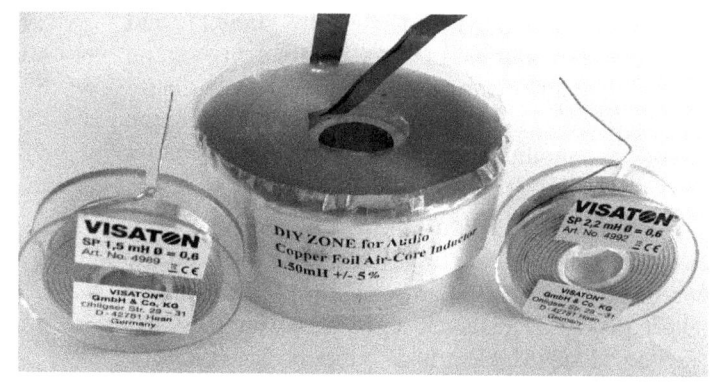

Fig. 9.36. Instead of ordinary copper wire, air-core inductors were wound with oxygen-free copper foil. That made them much bigger and heavier than the standard air-core inductors. The inductor at the far left has the same 1.5 mH inductance as the copper foil one in the middle!

If it bothers you, you can easily make a 0.42 mH inductor from 0.47 mH by unwinding and removing a few turns of wire. Making 1.6 mH out of a 1.5 mH spool is a bit harder; you have to solder a piece of wire onto the existing coil and wind it onto the spool. This assumes that there is physically enough space for additional turns on the plastic spool; usually, there is plenty (1).

Electrolytic capacitors out, film capacitors in

The reasons for the use of electrolytic capacitors in amplifiers and speaker crossovers are analogous to the use of iron-core inductors instead of air-cored ones. Still, the contrast between the two alternatives here is even starker. Electrolytic capacitors are cheaper and physically smaller than film capacitors but introduce significant sonic degradation.

Fig. 9.37. The coding system for film capacitors.

NUMBER	Multiplier	LETTER	Tolerance
0	1	F	+/- 1%
1	10	G	+/- 2%
2	100	H	+/- 3%
3	1,000	J	+/- 5%
4	10,000	K	+/- 10%
5	100,000	M	+/- 20%
6	NOT USED	P	+100% -0%
7	NOT USED	Y	+50% -20%
8	0.01	Z	+80% -20%
9	0.1		

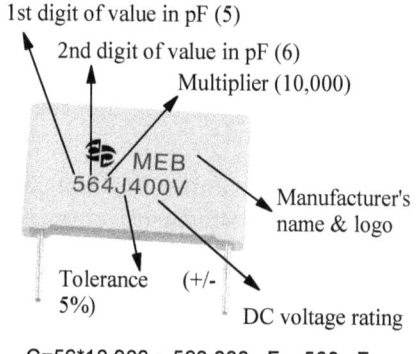

1st digit of value in pF (5)
2nd digit of value in pF (6)
Multiplier (10,000)
Manufacturer's name & logo
Tolerance (+/- 5%)
DC voltage rating
C=56*10,000 = 560,000 pF = 560 nF

LOUDSPEAKER TYPES, TESTS, AND IMPROVEMENTS

So, if you aren't keen on replacing inductors, at least replace the bipolar electrolytic caps. The coding system for film capacitors is outlined in the table below. Crossover capacitors don't need to be rated at 630 or even 400 V_{DC}, a 100 V_{DC} rating is usually sufficient.

The lower the capacitor's voltage rating, the smaller (physically) the capacitor, since thinner and less bulky insulation film can be used in its construction. Likewise, the thicker the insulation film between the capacitor's metal foils (electrodes), the higher its voltage rating. Lower voltage-rated caps are also cheaper.

The last upgrade option (or should that be the first?): speaker's internal wiring

Finally, we should mention a "component" that is often overlooked in upgrade discussions: the internal speaker wiring. Some more expensive speakers use OFC, silver-plated, or similar exotic and even brand-name wires (Monster Cable, Kimber Cable, etc.); however, most were wired with thick multistrand hookup wire of unknown origins and thus of dubious sonic quality.

If you subscribe to the "thin solid core wires sound best" school of audio tweaks, replacing all of the wiring with thinner, single solid conductor copper or silver wires may result in the most significant sonic improvement of all. So, if they seem to be of decent quality, leave the crossovers alone and upgrade the internal speaker wiring first.

DIY PROJECT: Redesigning (simplifying) and rebuilding speaker crossovers

Instead of a 3-way or 2.5-way crossover design, what would happen if the complex crossovers (such as those of Opera III speakers) were replaced by a simpler, 2-way design? Only two LC components would be needed for each driver, so we decided to find out.

Since the woofer and midrange driver/woofer are identical, let's connect them in parallel so the two will behave effectively as one larger 4-ohm driver. The resisistive 2R2-2R2 voltage divider will be retained, to bring the tweeter SPL level down to the woofer, providing a 50% reduction in the tweeter voltage, or, expresseed in dB, the attenuation is $A = 20\log V_{OUT}/V_{IN} = 20\log 0.5 = -6$ dB.

As for the complexity level, instead of 20 RLC components in each crossover we will now have only six!

The classical 2nd-order filter formulas are $C = 1/(2\pi f_0 R)$ and $L = R/2\pi f_0$, where $\pi = 3.14$, f_0 is the chosen crossover frequency (-3 dB), and R is the coil resistance of the driver. Let's choose the crossover frequency slightly lower than the existing 2,300 Hz, say $f_0 = 2,000$ Hz.

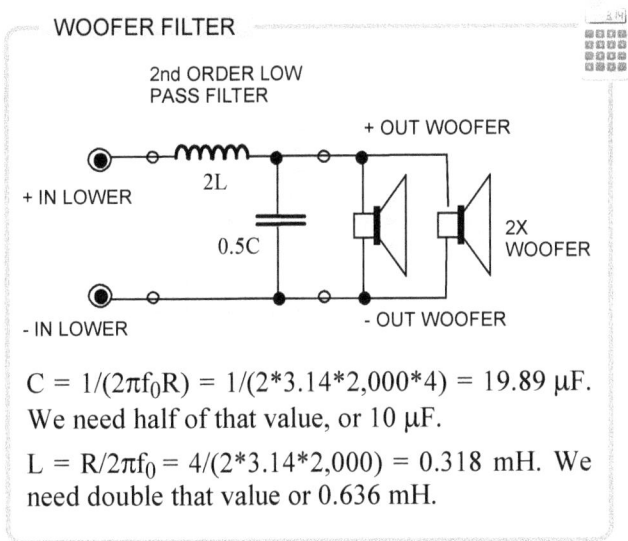

$C = 1/(2\pi f_0 R) = 1/(2*3.14*2,000*4) = 19.89$ μF. We need half of that value, or 10 μF.

$L = R/2\pi f_0 = 4/(2*3.14*2,000) = 0.318$ mH. We need double that value or 0.636 mH.

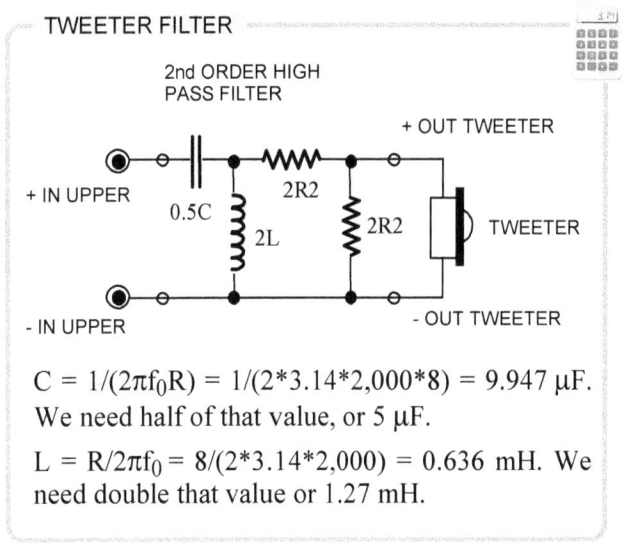

$C = 1/(2\pi f_0 R) = 1/(2*3.14*2,000*8) = 9.947$ μF. We need half of that value, or 5 μF.

$L = R/2\pi f_0 = 8/(2*3.14*2,000) = 0.636$ mH. We need double that value or 1.27 mH.

Five μF is not a standard capacitance value (although there are some capacitors with such capacitance, primarily for filtering duties), but 4.7 μF is. With the usual +/-5% tolerances of film caps, the capacitance can vary between $0.95*4.7 = 4.465$ μF, which would be too low, and $1.05*4.7 = 4.935$ μF, which could be OK. Alternatively, we can use four μF in parallel with 0.33 μF, another standard value.

Each crossover was built on a double-sided terminal strip (1) with ten pairs of lugs on each side. The high-pass filter was built on one (2) and the low pass on the other side (3) of the terminal strip.

Instead of buying lower quality 2R2 wire-wound resistors sold in electronic parts shops, we had a bagful of Ohmite (USA-made) 3R9 resistors (4), so we strapped two in parallel (for an effective resistance of 3.9/2= 1.95 R or approx. 2 ohms (5). The 600V-rated MKP capacitors were branded Audyn by Intertechnik Germany; 10 μF were red (6), while 4.7 μF were black (7). A code imprinted on the body of a film capacitor usually indicates the dielectric.

The codes are: KP-polypropylene film & foil (Kunststoff Polypropylen in German), the best sounding type, KS-Polystyrene film&foil, KT-Polyester film & foil, MKC-metallized polycarbonate, MKP-metallized polypropylene (Metallisierter Kunststoff Polypropylen in German), MKT - metalized polyester, MKY-metallized low-loss polypropylene, MKL or MKU- metalized lacquer (cellulose acetate).

Fig. 9.38. Crossover components before the final assembly and the finished crossovers

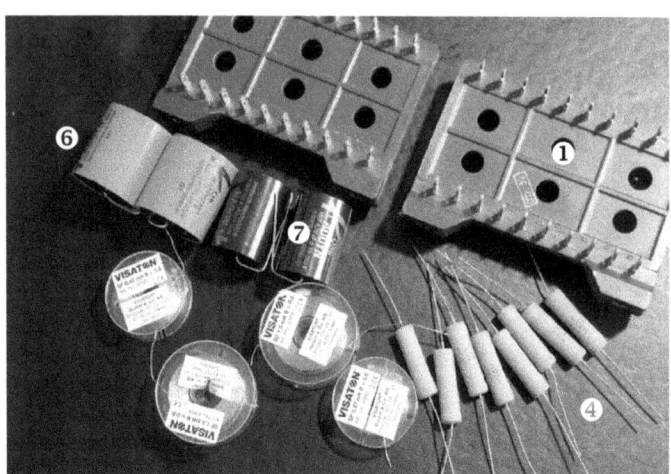
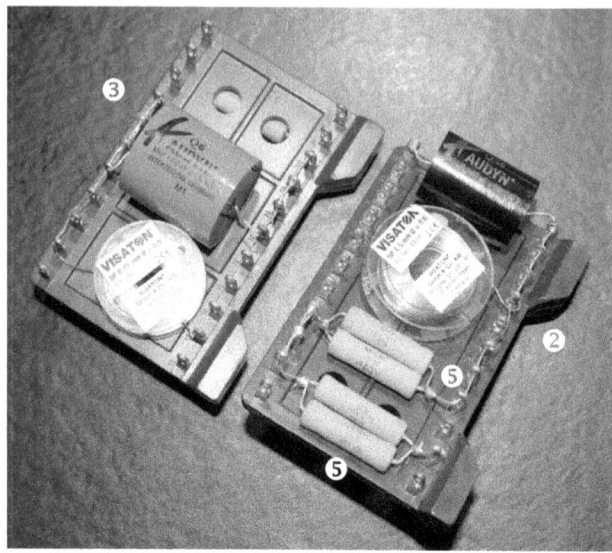

TW versus TMW and TWW arrangements

The original driver arrangement in Opera II is T-MW-W (tweeter-midrange driver/woofer-woofer) or "2.5 way", where both woofers start reproducing the lowest frequencies. The bottom woofer is rolled off, and the middle woofer continues working as a midrange driver, finally crossing over to the tweeter. This arrangement is almost always used with slim (narrow) cabinet speakers.

With 2-way speakers, the physical distance "d" between the tweeter's and woofer's acoustic centers (axis) is always minimized, something achieved in the present 2.5-way arrangement (d1 and d2 are both minimized).

However, as in this experiment, a change to the 2-way, dual woofer topology (T-W-W) will position the acoustic center of the woofer pair in between them, increasing the distance d3 to the tweeter's axis by 85 mm (from 125 to 210 mm). I'm not a professional speaker designer and cannot predict if that was going to become an audible issue or not, but such a slight increase does not seem significant enough to cause concern. Lowering the crossover frequency, as we have done here, should help address that issue.

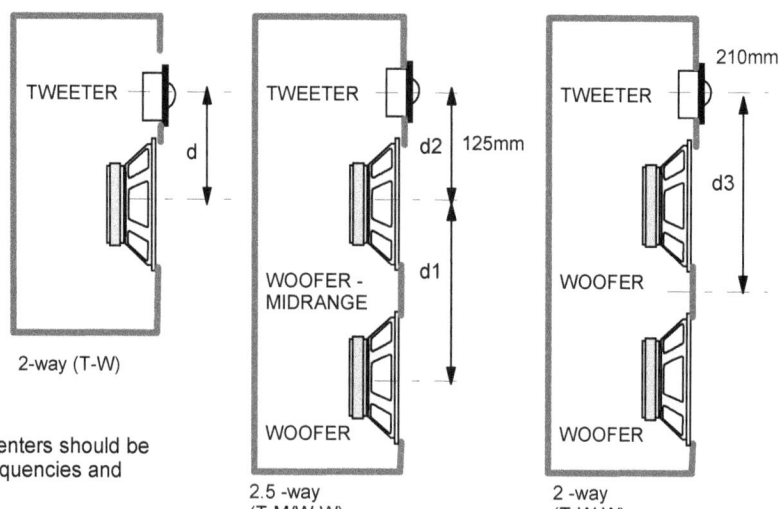

Fig. 9.39. The distance between the drivers' acoustic centers should be minimized for a smooth transition around crossover frequencies and improved tonal balance.

Adding an L-pad attenuator for tweeter volume control

If you feel that your tweeter level is a bit high, use an L-pad audio attenuator instead of trying out various fixed resistive dividers (to attenuate the tweeter signal). That way, you can precisely control the balance between the tweeter and the woofer/midrange driver.

Rated at 50 Watts, this Dayton Audio speaker L-pad attenuator was designed to be used with 8-ohm drivers. The faceplate, volume control knob, and all mounting hardware are included. It can be found at Parts Express (part number 260-255, $10.90). There is also a 100 Watt version priced at $12.95.

At its maximum (fully CW), the wiper of the 8Ω section is at the input terminal (3), and the wiper of the 30Ω section is at the output terminal (2). The 8Ω resistor is out of the circuit, and the 30Ω resistor is connected across the load.

As the control knob is moved counterclockwise towards the MIN position, the larger and larger upper section of the 8 Ω resistor is connected between the IN and OUT terminals, thus acting as a voltage divider and reducing the voltage across the tweeter.

At the same time, a smaller and smaller section of the 30 Ω resistor is connected between the OUT and COM terminals, in parallel with the tweeter. Due to such reducing resistance, more and more of the load current is diverted away from the load and shunted to COM.

At the minimum position (fully CCW), the 30 Ω resistor is out of the circuit, and the OUT terminal is shorted to GND, completely bypassing the load. The 8 Ω resistor is now connected between the IN and OUT terminals, the speaker terminals "OUT" and "COM" are short-circuited, so there'll be no sound from the tweeter at all.

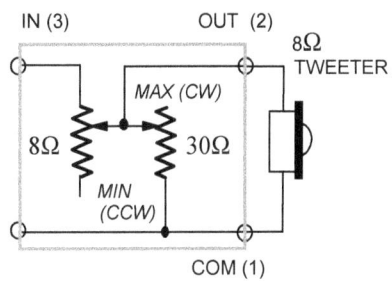

Fig. 9.40. The 8Ω attenuator consists of two sections of different resistance, 8Ω, and 30Ω.

Tweeter rolling

If your speakers' tweeters are of a standard size and shape, as ours are on Opera III speakers (4" diameter), and if you are happy with your speakers' bass and midrange, instead of buying a new pair of speakers, try various tweeters; you may be pleasantly surprised.

Made in Poland by Tonsil, GDWK 10/80 is a 4" diameter ferrofluid-cooled tweeter with 90 dB/W@1m sensitivity and 8-ohm impedance, rated at 40 Watts (80 Watts maximum) and 4-20 kHz frequency range.

AL-100, the first replacement tweeters we tried, are AMT (Air Motion Transformer) type and rated at only 15 Watts (maximum 30 Watts). Still, they only cost around US$30 a pair (AliExpress), so we thought it was worth the risk. At our modest listening levels, they worked fine.

The 1.5-40 kHz range of AL-100 tweeter is much wider then Tonsil's, but it's resonant frequency (1.33 kHz) is much lower, so no issues there. With 7Ω DC resistance and 8Ω nominal AC impedance their 91 dB/Watt sensitivity was almost identical to that of the Tonsil original, so on paper it all looked good enough to proceed.

The second pair we tried were Kasun QA-2100 dome tweeters, which sell for around US40 a pair. A 4" (104mm) diameter tweeter with a 1.6 kHz resonant frequency, 88 dB/W @1m sensitivity and 8 ohm impedance, it's rated at 60 Watts maximum.

Kasun dome tweeters sounded very similar to Tonsil's dome tweeters, but cleaner and more refined, especially after 20 or so hours of break-in time. The AL-100 tweeters sounded different, probably because of their inherently different operating principle. The treble wasn't harsher than Tonsil's but definitely more prominent.

Since we aren't interested in the frequencies below 3kHz, the significant differences between the three tweeters in that range aren't relevant (below the crossover frequency). Between 3 and 7 kHz, the AMT tweeter is the most efficient, 92-94 dB/W, followed by Tonsil (around 90) with Kasun between 87 and 88 dB/W.

According to the supplied frequency response curves, the upper midrange difference of 5-6 dB between AMT and Kasun tweeters should be noticeable, and it certainly was.

Fig. 9.41. AL-100 tweeters installed in Opera III speakers.

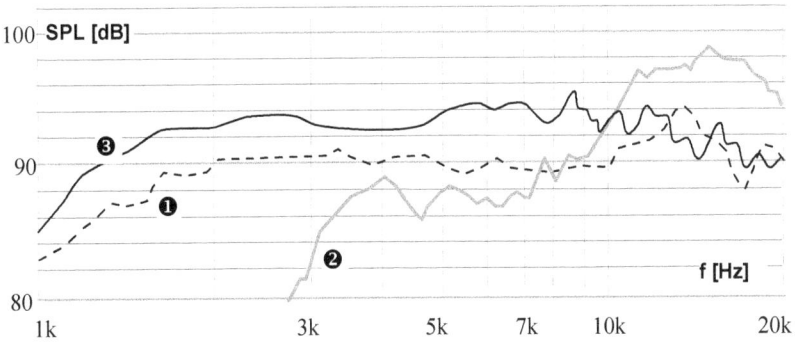

Fig. 9.42. The SPL versus frequency curves for the three tweeters tried, Tonsil GDWK 10/80 (1 - dotted line), AL-100 AMT tweeters (2), and Kasun QA-2100 dome tweeters (3 - thicker gray line)

From 7-10 kHz, there isn't much of a difference between Tonsil and AMT tweeter. Above 10 kHz, the AMT tweeter is the most efficient (94-98 dB/W), so it was the loudest in listening tests.

In conclusion, a quick glance at a particular tweeter's frequency response curve should give you a pretty good idea of its sonic nature and how it is likely to tilt the sonic balance compared to your speakers' existing tweeters.

Even a small change in the treble levels as the result of tweeter rolling could have a noticeable impact on the overall character of the whole system. With a treble set too low, the system will sound too warm, lacking sparkle and transparency, with compromised microdynamics and transient response. Installing a tweeter such as AL-100 could improve the sound of percussion instruments and the speed of transients but tilt the overall tonal balance towards the "too harsh" extreme, with sibilants and hissy/steely edges around voices, strings, and brass instruments.

Mismatched drivers

One of the paradoxical and often futile audio endeavors is aiming for ideally matched gains throughout the electronics chain (cartridge-phono stage-preamp-power amp). This includes trying to find perfectly tracking and matched potentiometers or attenuators. The paradox lies in feeding such perfectly balanced signal to a pair of loudspeakers whose left and right-channel drivers could often differ by as much as 3-4 dB!

Like vacuum tubes, loudspeaker drivers are hand-built from individual parts, and such mechanical tolerances result in significant distribution of their parameters. Very few speaker manufacturers publish their speaker tolerances or disclose them in interviews. Some don't care much about driver matching, while others invest considerable time in selecting and matching the drivers they make in-house or those supplied by others (OEM).

Matching two tweeters to 0.5 dB or better requires quite a few hours of someone's time, not to mention the need to buy hundreds or even thousands of damned things, significantly increasing the manufacturing cost. And what would they do with drivers that cannot be matched to such stringent standards? Return them to the supplier or sell them to other speaker makers or DIY speaker builders?

DIY tweeter taming

Sound diffraction of the speaker's edges affects the phase and frequency response at higher frequencies, which are primarily responsible for stereo imaging, sound staging, and focus.

The efforts towards smoothing out and absorbing diffraction ripple led to speakers with mitered or rounded edges, like our Opera III speakers here. However, due to additional machining, such speaker cabinets are much more costly, so speaker manufacturers often resort to cheaper measures, such as foam or felt rings around tweeters.

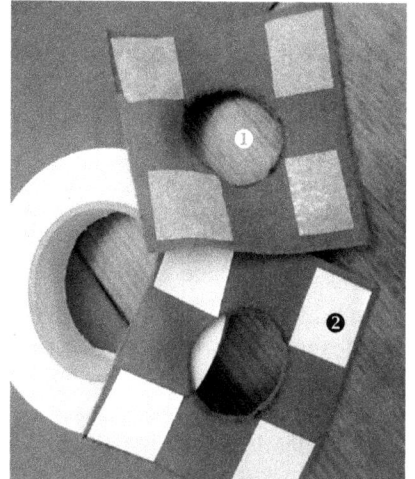

The original tweeters in our speakers had their whole plastic surround lined with a thin sponge layer, while the replacement tweeters used in our tweeter-rolling experiments had none. So, we wanted to hear if adding simple absorption pads would result in any audible improvement.

The felt in our local haberdashery shop was thin synthetic material; he felt we chose was slightly thicker, a wool/polyester blend selling for AU$17/m. It was easy to cut with scissors, and the round opening for a 25 mm tweeter dome was chosen to be 50mm in diameter.

A square of double-sided adhesive tape was attached to each of the four corners of the rectangular pad (2). The finished treatment didn't look too bad, but once speaker grilles were attached, it could not be seen at all.

The sonic improvement was subtle but unmistakable - the treble became softer and smoother. On superior recordings, the felt pad removed the last traces of harshness while the hard, sibilant inferior recordings became much more listenable and less fatiguing. The upper midrange improved as well, with better and more precise imaging. The improvement reminded me of replacing standard metallized polypropylene coupling capacitors in a tube amp with expensive tin foil or silver foil paper-in-oil caps.

Fig. 9.43. A single layer of felt around the tweeter, held in place by double-sided tape, makes an effective diffraction absorption pad. Two, three, or even more layers can easily be piled up.

Adding a second and a third identical layer to make the pad thicker did not result in any significant changes; we thought we heard a marginal improvement, but it is possible that the dreaded placebo effect and autosuggestion were at work.

10 | COMPONENT MATCHING AND AUDIO SYSTEM INTEGRATION ISSUES

In the introduction to this book, we listed the top five audiophile system issues. Suboptimal system integration (poor component matching) was one of them. This key chapter looks at a few critical aspects of audio synergy, such as the matching of amps and speakers, gain optimization in the signal chain, and various ways of integrating other audio components into it.

Apart from active crossovers and powered subwoofers, bi-wiring and bi-amping are also discussed.

- GAIN OPTIMIZATION IN THE SIGNAL CHAIN
- MATCHING SPEAKERS AND AMPLIFIERS
- BI-AMPING
- ACTIVE CROSSOVERS
- BI-WIRING
- INTEGRATING SUBWOOFERS INTO AN EXISTING STEREO SYSTEM
- AUDIO SYNERGY
- TUBE PREAMPS DRIVING SOLID-STATE POWER AMPS

"Perfection is not when there is no more to add, but no more to take away."
Antoine de Saint-Exupery, aviator and author of *The Little Prince*

GAIN OPTIMIZATION IN THE SIGNAL CHAIN

What is gain matching and why it matters

Apart from impedance matching between various components in an audio system, gain matching and optimization is arguably the most critical system integration issue. Over the years, we have designed and made dozens of different line-stage preamplifiers. Most were of a low gain variety, typically amplifying the input voltage 2-7 times (with the volume control turned up to the maximum).

However, there are many (very) high gain commercial designs with voltage gains of 15, 20, even 30 times, meaning that for a typical 2 Volt signal level at their inputs (from a DAC or another source), we get up to 30, 40 and even 60 Volts at their outputs. Since power amps usually require 1-2 V inputs for maximum power output (which could be 100, 200, or more Watts), why would someone want to overdrive them with such incredibly high input signals is beyond comprehension.

Most of these preamps are more or less copies of vintage 1950s and 60s designs, usually with 12AX7 (ECC83) tubes, which have the highest voltage gain of any audio tube ($\mu=100$). 12AX7 is great in guitar amps, which require high gain and "overdrive" stages, but, in my opinion, has no place in an audiophile line-level preamp.

We first need to understand the basics, such as Ohm's Law, and get a feel for voltages, currents, and power levels in a typical audio chain. All voltages discussed are RMS or Root Mean Square, also called the effective value of an AC signal, in contrast to the peak, peak-to-peak, and average values, which are less commonly used but are helpful in some cases.

Voltage, current and power amplification through the audio chain

Let's use a low-powered tube amp and a higher-powered solid-state amp as examples. For a 2A3 SET amplifier supplying 5 V_{RMS} to an 8 Ω load, the output power is voltage squared divided by resistance (impedance): $P_{OUT} = V^2/R = 25/8 = 3.125$ Watts. The AC (signal) current through 8Ω load is $I_L = V_L/R = 5/8 = 0.625$ A (Amperes).

For a high-power tube amp supplying 40 V_{RMS} to an 8 Ω load, the output power is $P_{OUT} = V^2/R = 40^2/8 = 200$ W (Watts), while the current through the load is $I_L = V_L/R = 40/8 = 5$ A.

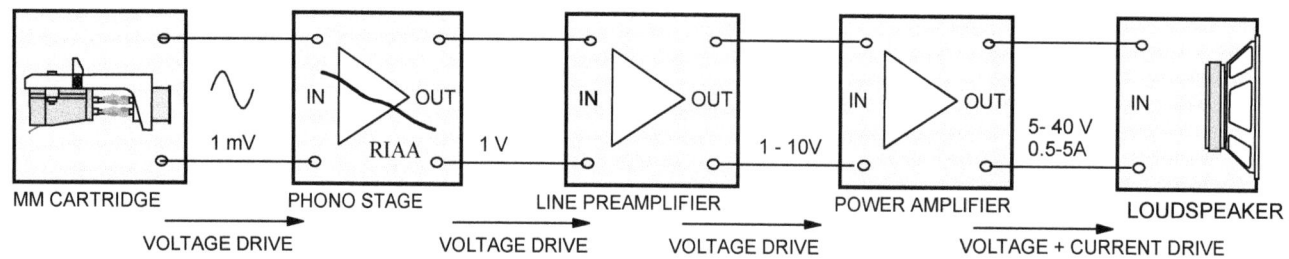

Fig. 10.1. Except for power amplifiers, which provide both voltage and current gain (amplification), all other components, such as phono stages and preamplifiers, provide voltage amplification only.

All amplification devices upstream of the power amplifier provide only voltage amplification, while the power amplifier amplifies both the voltage and current.

We can now proceed with the analysis of a multistage system or amplifier. We have an audiophile system comprising a signal source, a turntable with a moving coil cartridge, an MC step-up transformer, a MM (moving magnet) phono preamplifier, and a power or integrated amplifier. The amplification factors are given for each component, A_1 to A_3. What is the overall amplification factor in dB?

There are two ways to do this. You can calculate the gain of each stage or device in dB and then simply add them up A [dB] = $A_1+A_2+A_3$ = 103.5 dB. Alternatively, calculate the overall gain of the system A=$A_1*A_2*A_3$ = 150,000 and then convert it into dB: A [dB] = 20*log A = 10*log150,000 = 20*5.176 = 103.5 dB.

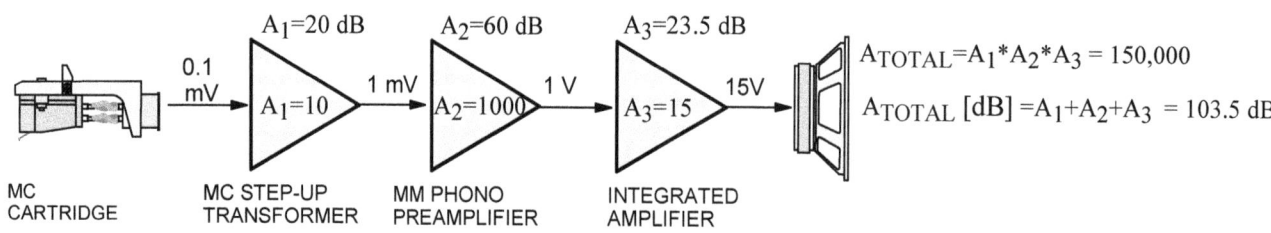

Fig. 10.2. An audio system with one source (a turntable with MC cartridge) and three amplification components

CASE STUDY: How to decipher manufacturers' claims such as dB gains and step-up voltage ratios

Zesto Audio Andros 1.2 phono stage has MC step-up transformers at its inputs, with an overall MC gain of 67 dB and moving magnet gain of 47 dB. That means the step-up transformers' voltage gain is 67-47 = 20 dB. What is their voltage gain in absolute terms and the output voltage if a cartridge used produces 0.1 mV at 1 kHz?

Since A [dB] = 20*log A, we have 20 = 20logA so log A=1 and A=10! With 0.1 mV input, the output of the MC transformer will be 1 mV. The gain of the MM stage is 47 dB, so 47/20 = 2.35 = $logA_{MM}$, so the MM phono stage amplifies A_{MM} = $10^{2.35}$ = 223.87 times. Thus the output voltage of the whole phono stage will be V_{OUT} = 1 mV*223.87= 224 mV or 0.224 volts.

Zesto Audio Andros Allasso MC step-up transformer has four gain settings (12, 16, 18, and 22 dB) and four step-up ratios, 1:4, 1:6, 1:8, and 1:12. Each step-up ratio determines the voltage gain of the transformer, so let's check if these claims match.

A_1 = V_{OUT}/V_{IN} = 4, so the gain in dB is A_1 = $20logA_1$ = 20log4 = 20*0.602 = 12dB. Correct. For 1:12 voltage ratio A_4 = 12 so A_4 = 20log12 = 20*1.08 = 21.6 dB. I guess 22 dB is close enough. For practice purposes, check the other two step-up ratios and the corresponding gains yourself.

Excess gain in the audio system

Apart from matching input and output impedances of the audio components in the signal chain, it is of paramount importance to consider their gain and input sensitivity figures.

This example is for an audio system (illustrated below) consisting of a MM phono stage with a gain of 1,000 (times), or 60 dB, a line-level preamp amplifying ten times (20 dB) and a power amp with a nominal input sensitivity of 2 V (for full rated power at the output).

With the nominal 1 Volt at its input (upper values in the block diagram below), a typical line-level preamp with a gain of 10 times (or 20 dB) would produce 10 Volts at the input of a power amp. However, most of those have around 2 Volts input sensitivity, the standard output level of DACs, both stand-alone and in CD players.

Clearly, all that excess gain that the line-level preamp brings into the signal chain isn't needed at all, and a passive (up to unity gain, A_2 = 1) preamp would do just fine. The line preamp will never even work as a voltage amplifier but rather as an elaborate and expensive attenuator, attenuating (reducing) the input signal from the phono stage.

This is even more evident in the bottom case (2), with a cartridge producing 5 mV, resulting in 5 Volts at the preamp's input. Only 2 V is needed for the power amp to go full blast, so the preamp needs to attenuate the signal at least 2.5 times (5 V to 2 V), and, since nobody listens to their systems at the full volume level, more likely ten or more times.

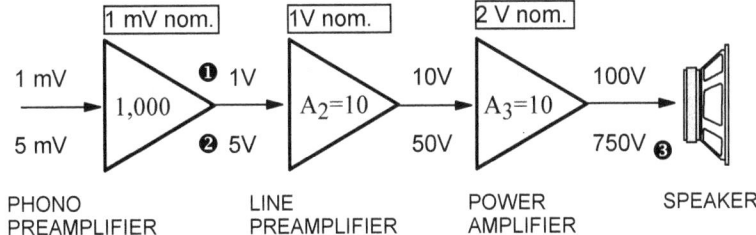

Fig. 10.3. Audio system with three amplification components. The input sensitivity figure for each component is framed, and two examples (scenarios) are given, one above and one below the signal arrow.

Some tube preamplifiers can produce up to 50 V_{RMS} at their outputs. Still, no power amplifier can take such a large signal at its input; their input stages would be fully saturated. Hence, the 750 Volt output to the speaker (3) is purely a theoretical value, for our discussion purposes, beyond the realm of possibility.

Even with only 10 Volts at its input (as in the upper case), the output voltage would be 100 Volts. This is way beyond the clipping levels of this power amp, which was designed to produce 20 Volts at its full output, so its power rating is P = V^2/Z = $20^2/8$ = 400/8 = 50 Watts on an 8-ohm nominal speaker load.

Thus, even 100V_{RMS} at this particular power amp's output is a laughable suggestion; the amp would go into severe clipping way before that level is reached. 100 V_{RMS} on an 8-ohm load would be 1,250 Watts of power!

Sonic consequences of excess gain

Excess gain may not seem a serious issue but can cause noticeable sonic degradation. Using volume control in a preamp to significantly attenuate signals and then amplify them again (in a power amp) increases noise and distortion while reducing resolution (microdynamics). The dynamics seem to suffer as well.

The frequency response and bandwidth of many preamps depend on the setting of their volume controls. The bandwidth is usually widest at around unity gain and drops both at the low and high volume settings.

The tracking of dual potentiometers in a stereo amp is often the poorest at the beginning of the track (fully counterclockwise or at the lowest volumes), which can cause a severe volume imbalance between the two channels. Also, with so much gain, it is hard to adjust the volume precisely; turning the volume control knob just a bit clockwise causes a considerable jump in loudness.

All of these are profoundly important reasons to pay attention to the matching of audio component's voltage and sensitivity levels.

The power amp in the discussion above had an assumed input sensitivity of 2 V_{RMS} (the input signal that will produce the full output power at its output). However, many power amplifiers have a much higher input sensitivity, even down to 0.25 V (250 mV), which would make the excess gain problem even worse!

MATCHING SPEAKERS AND AMPLIFIERS

The amplifier-speaker interface

One of the greatest paradoxes in audio is that we focus so much attention, spend so much time, effort and money to make audio amplifiers as close to perfection as possible (low output impedance, extremely low distortion, as little phase shift as possible, super linear amplitude-versus-frequency characteristic, etc.), and then connect them to (apart from a turntable) the worst link in any audio system - the imperfect loudspeaker.

A loudspeaker is a frequency-dependent complex-impedance load of extremely low efficiency and (compared to other audio components) very high distortion levels. It is impossible to predict how a particular amplifier-speaker combination will sound. The only way to ascertain that is to try the two together.

What matters most isn't how an amplifier amplifies the signals at its input but how it drives and controls the speakers it is connected to. This interaction is akin to the link between the engine and transmission/drive train in a car to use the automotive analogy again.

Of course, car makers can design the two systems in unison and optimize their overall performance (although it doesn't always happen). In contrast, audio amplifier makers seldom design their amplifier with a specific model or brand of speakers in mind, or even a particular type of speakers (dynamic, horns, electrostatic, etc.). The same applies to speaker designers and manufacturers.

So, when we audiophiles have an amp and are searching for and auditioning various speakers (or the other way around), the situation is akin to buying a car body from Ford, an engine from Lexus, and transmission from BMW. No sane car buyer would even contemplate such an approach, yet that is exactly how most audiophiles put together their systems. And then we wonder why the overall result is less than stellar, in many cases downright awful, despite all those high amounts of money spent!

The importance of speakers' sensitivity

Consider two audiophile setups, a low power (10 Watts) triode amplifier driving efficient dynamic speakers of 93 dB/W sensitivity and high power (100 Watts) push-pull tube amplifier driving hybrid electrostatic speakers (with a dynamic woofer) of a low 83 dB/W sensitivity. Which one will sound louder in the same room?

If we express the ratio of power outputs of the two amps as a ratio we get $P_2/P_1 = 100/10 = 10$ or in dB terms $10\log P_2/P_1 = 10\log 10 = 10$ dB.

I deliberately chose these friendly & easy (but very realistic) numbers. Those who understand decibels could have quickly figured that result out since the ten times higher power of the 100 W amp is +10 dB above the output level of the low-powered amp.

However, its speakers have ten dB/W lower sensitivity, so to get the same SPL (sound pressure level) as the low-powered amp with high-efficiency speakers, the amp must have 10 dB higher amplification, which it has.

Therefore, both setups will sound equally loud. In the second case, what we gained in amplifying power we lost by using very inefficient speakers!

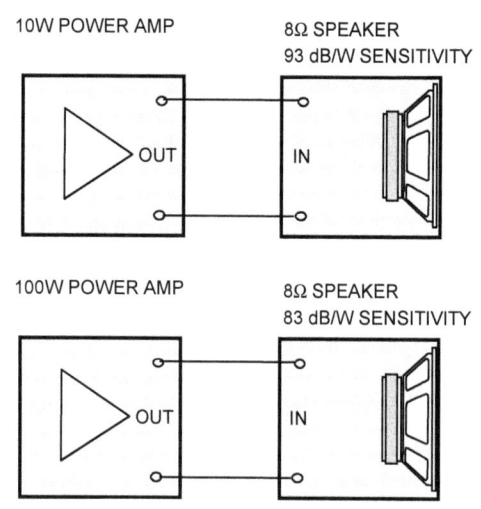

Fig. 10.4. Since we don't listen to amplifiers alone, but through loudspeakers, the maximum power of an amplifier does not matter on its own; the loudspeaker sensitivity also has to be taken into account. The two systems illustrated produce the same SPL (sound equally loud).

The importance of amplifiers' low output impedance

The following discussion is somewhat simplified. It assumes that the amplifier's output impedance is constant throughout the whole frequency range, which isn't strictly true (as we will see very soon). It also assumes that in all other respects, the amplifier is ideal, that it can provide any current demanded by the load, and that is only true for a limited range of load impedances. Nevertheless, for this purpose, it's close enough to reality.

This scenario illustrates issues arising from the high output impedance of an amplifier (especially single-ended amplifiers without negative feedback) and its interaction with the varying load impedance (the voltage divider effect). At frequencies where speaker's impedance is at its minimum (say 2 Ω), $V_L/V_0 = R_L/(R_L+Z_{OUT}) = 2/(2+3) = 0.4$ or 40%, but at resonant frequencies where speaker's impedance is at its peak (say 22 Ω), $V_L/V_0 = R_L/(R_L+Z_{OUT}) = 22/(22+3) = 22/25 = 0.88$ or 88%! The output voltage will vary $20\log(0.88/0.40) = 20\log 2.2 = 6.85$ dB, which will be very noticeable!

Assuming an ideal amplifier with zero output impedance ($Z_{OUT} = 0$), the output voltage of a 300B amplifier capable of producing 8 Watts on an 8Ω load would be $V = \sqrt{(P*R_L)} = \sqrt{(8*8)} = 8$ V.

A real amplifier with $Z_{OUT} = 3Ω$, driving a speaker whose impedance at some frequency dropped to, say, 2 Ω, would have its output voltage divided between its own output impedance and the speaker. The voltage on the speaker is now $V_L = V_0 R_L/(R_L+Z_{OUT}) = 8*2/(2+3) = 8*2/5 = 3.2$ V, and the power fed to the load is only $P = V^2/R_L = 3.2^2/2 = 5.12$ Watts, instead of 8 Watts!

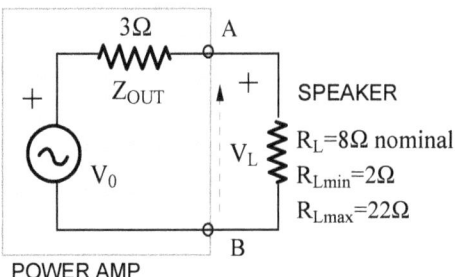

Fig. 10.5. The output of a real amplifier with output impedance Z_{OUT} behaves as a frequency-dependent voltage divider.

Output impedance versus frequency

Damping factor (DF) is the ratio of load impedance and amplifier's internal or output impedance: $DF = R_L/Z_{OUT}$. It helps us predict how well a particular amp would control the speaker cone. An ideal amplifier would have an infinitely high DF since its internal impedance would be zero.

Due to the output transformer's leakage inductance that increases with frequency, the output impedance of a tube amplifier rises with frequency, especially in the absence of negative feedback.

Even with relatively mild feedback, the same amplifier exhibits a much lower rise in the output impedance than the rise when NFB is removed. For that reason low levels of feedback are sometimes beneficial and sonically preferable to no NFB at all.

This significant rise of Z_{OUT} at high frequencies may seem like a serious problem, but luckily, there are mitigating factors. The most significant rise is in the 10-20 kHz band, and very little power is needed in that frequency band to drive a typical tweeter in a dynamic loudspeaker system.

Secondly, the impedance of a typical dynamic loudspeaker also rises at high frequencies, so while the output impedance of a typical amplifier Z_{OUT} increases in the absolute sense, the rise in the load impedance tends to offset it to a large degree.

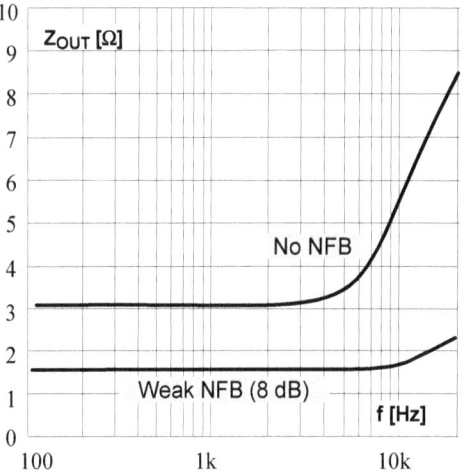

Fig. 10.6. How the output impedance of a certain 300B single-ended triode amplifier (8 Ω output) varies across the audio frequency range (with and without negative feedback).

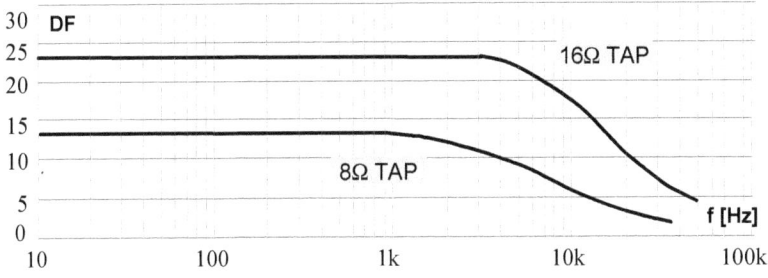

Fig. 10.7. Damping factor versus frequency of a typical push-pull tube amplifier

Finally, very few older audiophiles can hear frequencies above 15 kHz, so even a significant drop in the upper-frequency power levels may go unnoticed and actually "mellow" the amplifier's sound.

Damping factor is of most interest and importance at low or bass frequencies, where an amplifier must supply lots of current to the woofer coil and where damping of the woofer's cone is of paramount importance for the speed, definition, and prominence of the bass.

Loudspeaker sensitivity figures - a deliberate obfuscation by manufacturers?

In June 1990, *Stereophile* magazine published a review of Thiel Cs5 loudspeaker. You can read it online at https://www.stereophile.com/content/thiel-cs5-loudspeaker-specifications-0. One line in its "Sidebar 3: Specifications" raises an important point, one that most audiophiles don't fully understand. It says: "Sensitivity: 87dB/2.83V/1m (equivalent to around 82dB/W/m, given the low impedance). Impedance: 3 ohms (2 ohms minimum)." What does that really mean?

Typically, loudspeaker sensitivity is expressed in the number of dB (decibels) of sound pressure the speaker will produce for one Watt of input electrical power at a distance of one meter. So, a speaker may have a sensitivity of 87 dB/W/m, although it should really be written as 87 dB/W@1m, but we are nit-picking now.

Some speaker makers, such as Thiel in the example here, use 2.83 Volts instead of 1 Watt. Why would they do that? Audiophiles can relate to "1 Watt", but why the silly-looking "2.83 V" part?

Since power is $P = V*I = V^2/Z$, assuming speakers' impedance of $Z = 8\Omega$, the voltage of 2.83 V is equivalent to the power of $P = 2.83^2/8 = 8/8 = 1$ Watt. The two kinds of specs are identical *if* the speaker's impedance at that test frequency is 8Ω. However, the impedance of all speakers goes up and down with frequency, for example, from 2Ω to 32Ω. Does that mean that their sensitivity will also vary, that it's frequency dependent? It sure does!

Going back to the same online review and looking at the measured impedance of the CS5 speaker (Figure 1), we see a few minor dips, but generally a steadily rising impedance, from around 1.5Ω at 10 Hz, through 2.8Ω at 100 Hz and just under 4Ω at 1 kHz.

Stereophile's technical guru and long-time editor John Atkinson declared this to be a 3Ω speaker, probably because between 80 and 3,000 Hz, its impedance hovers around that figure. Fine. Proclaiming a speaker to have a constant impedance is a guess at best but could be wildly misleading at worst. You may declare this to be a 4Ω speaker, while I could call it an 8Ω speaker, looking at its treble range. We would both be right (up to a point) and, as with all arbitrary proclamations, very wrong.

Finally, how did they end up estimating this speaker's sensitivity to be "equivalent to around 82 dB/W/m, given the low impedance"? At 1 kHz, the impedance is approx. 4Ω, so comparing it with 1 Watt reference level at 8Ω the impedance is 4/8=0.5 (half), so we get a change in power of $\Delta P = 10*\log P/P_0 = 10\log 0.5 = 10*(-0.301) = -3$ dB. At that frequency, the sensitivity will be 3 dB lower, not 87 but 84 dB/W/1m!

At 3Ω impedance, the signal of 2.83 V_{RMS} means the power the amplifier needs to produce is $P = 2.83^2/3 = 8/3 = 2.67$ Watts, and the ratio is now $\Delta P = 10*\log P/P_0 = 10\log(1/2.67) = -4.26$ dB.

Thus, at all frequencies where the impedance modulus is 3Ω, the speaker's sensitivity will be 4.26 dB lower or 82.74 dB/W/1m, hence the *Stereophile* estuimated figure of 83 dB/W/1m.

Let's go back to our imaginary speaker with a simple single-peak impedance curve (right). Say the manufacturer specified its sensitivity as 91 dB/2.83V/1m at 1 kHz. How would we convert that figure into dB/Watt/1m? Is it possible to do that from its impedance graph? It sure is. This is a useful skill for those who like comparing speakers "on paper" before purchase.

At 1 kHz, the speaker's impedance is 4.0Ω, so the power is $P = V^2/R = 2.83^2/4 = 8/4 = 2$ Watt. That is the required amplifier power into that speaker to produce 91 dB of sound pressure at that frequency, measured at the distance of 1 meter using a 1 kHz test tone.

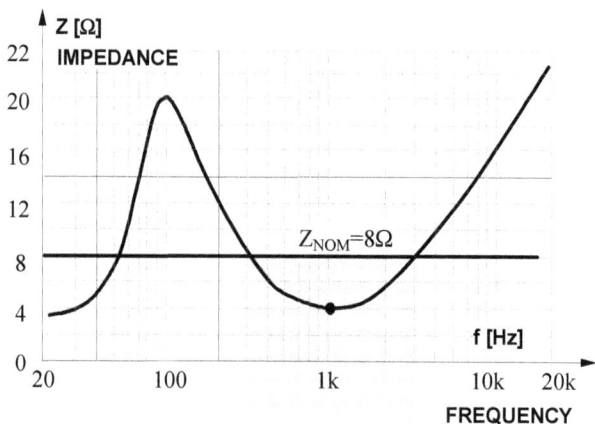

Fig. 10.8. The impedance-versus-frequency of a typical dynamic (moving coil) loudspeaker

We have seen that doubling the amplifier power is an increase of 3 dB and that halving the amplifier power is a reduction of -3 dB. So, with only 1 Watt into that speaker in the same test point, the speaker will produce a 3 dB lower SPL, meaning its sensitivity will be 91-3 = 88 dB/Watt/1m!

Try your speakers with all amplifier outputs

While solid-state amplifiers generally have only one pair of output posts for speaker connection, many tube amps offer multiple outputs, typically 4 and 8 ohms. Some (primarily vintage amps for the 1950s and 60s) even have a 16 ohms tap, but that is rare on currently-produced models.

If your tube amp has taps for more than one speaker output impedance (the one pictured has 4- and 8-ohm outputs), try both connections and decide which pairing sounds better to you. You cannot damage anything, so experiment as much as you want.

Your speakers' output sound pressure level will change when connecting your speakers to different output taps. This makes it difficult to distinguish if the tonal improvement on a particular tap is due to higher volume (louder usually sounds better) or other factors. As we will see in a moment, the distortion levels will also change with changing taps.

Of course, you can always run test tones or pink noise and use an SPL meter to adjust the amplifier volume to get identical sound pressure from different speaker connections, but that takes time.

What happens to the operating regime of a tube amplifier when we connect speakers to various output taps, in other words when the load impedance of output tube(s) changes? Let's consult a few graphs. We are primarily interested in output power and distortion levels, which can vary significantly with the changes in loading.

The optimal anode load and distortion of triodes

The choice of the anode load for output tubes is a compromise between maximum power and minimum distortion, especially for triodes.

For triodes, the load impedance should be approximately 3-4 times the internal impedance R_I of the tube. For 2A3 triode with internal impedance R_I = 800-900 ohms, R_L should be between 2k5 and 3k5, while 3k (3,000 ohms) is commonly used.

These are the maximum power rules. For lower distortion levels, higher load impedances should be chosen.

Notice that at the load impedance producing maximum power (2,500 ohms for 2A3 triode), the distortion is higher (5.7 %) compared to slightly larger load impedances, such as 3,500 ohms, at which the distortion drops to 3.2 %.

The output power and distortion drop off at approximately the same rate with increasing load impedance. So what really happens if we connect speakers to lower impedance speaker taps of triode amplifiers?

For instance, assuming a constant 8-ohm speaker impedance and output transformers that reflect an 8-ohm load back as 2,500 ohms (1), by connecting an 8-ohm speaker to amplifier's 4 ohm tap, the reflected load back onto power tubes is doubled to 5,000 ohms (2), resulting in much lower distortion (5.7% down to 1.1%), albeit at a reduced maximum available power, a drop from 3.4 to 2.2 Watts.

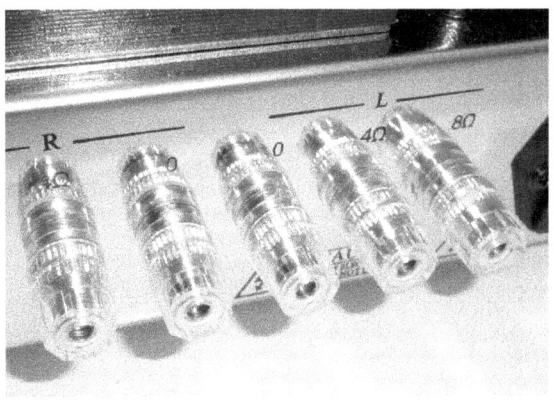

Fig. 10.9. (ABOVE) The "0" (Common), 4 ohms and 8 ohms output binding posts at the back of a tube amplifier

Fig. 10.10. (BELOW) Output power P_A and total harmonic distortion THD curves as a function of anode load impedance for 2A3 output triode in a sngle-ended amplifier.

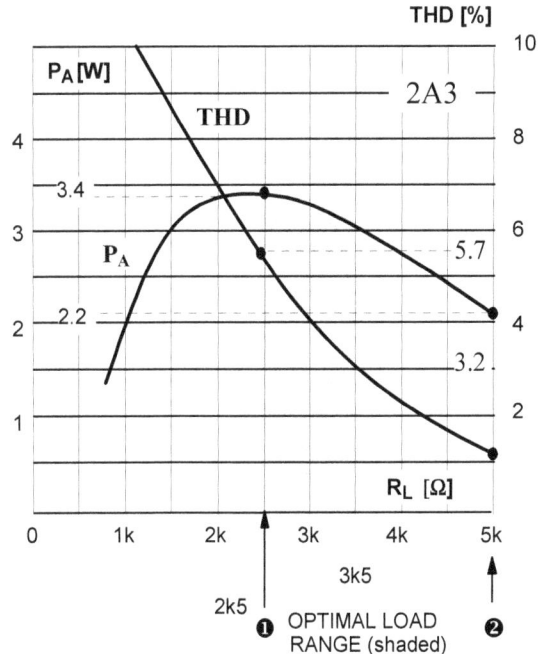

The loading and distortion of pentodes and beam power tubes

The graph on the next page shows that in comparison with triodes, the second harmonic curve D_2 for pentodes dips to zero at a certain load impedance R_X. The output power is now at its maximum, which is considered the optimal load (to get the full output power).

Unfortunately, the discordant (unpleasant sounding) 3rd harmonic has no such minimum and keeps rising with increasing load impedance. Also, at R_X, the 3rd harmonic is exposed since there is no 2nd harmonic to mask it, resulting in harsher, more stringent, and thus more fatiguing sound. Using a lower load impedance, such as R_Y, reduces the harsh-sounding 3rd harmonic, which is now also "masked" by the pleasant-sounding 2nd harmonic. The output power is slightly lower, but the sonic benefits are indisputable.

For instance, by connecting 8-ohm speakers to the amplifier's 16-ohm tap, the reflected load back onto power tubes is halved, reduced from, say, R_X to R_Y or even lower (R_Z), where the 2nd harmonic D_2 dominates the 3rd harmonic D_3 (2-3 times higher). This results in a smoother, more "musical" sound, and only a modest reduction in anode output power P_A.

However, all this assumes that the load (speaker) impedance is constant, which is only true for a resistive dummy load on a test bench. The impedance of real speakers varies with frequency, so different frequencies will be distorted to a different extent, depending on the speaker's impedance curve. That is another reason why some speakers sound fine with certain amplifiers but awful with others.

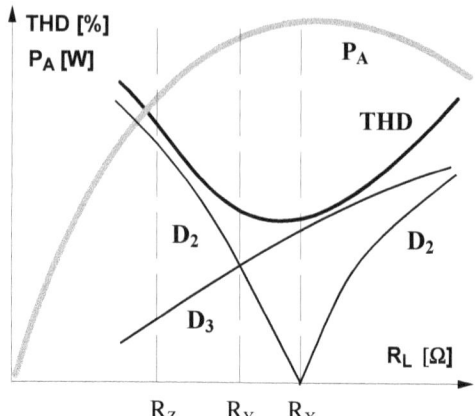

Fig. 10.11. Typical pentode's power and distortion curves as a function of anode load. D_2 and D_3 are the 2nd and 3rd harmonics. THD is the total harmonic distortion (the sum of all distortion harmonics).

> DISTORTION VARIES WITH THE CHANGING LOAD (LOUDSPEAKER) IMPEDANCE
> As the loudspeaker load impedance changes with the audio signals' frequency, it causes continuous changes in both the amplifier's output power and its distortion level.

Speaker crossovers - the main offenders

Speaker crossovers are usually designed to achieve the desired frequency response of a speaker as a whole, not to provide a constant impedance (as close to pure resistance as possible) loading on the power amplifier driving them. In other words, the issue of matching the particular speaker to the amps that would drive it seems to be of secondary importance to speaker designers. Judging by the proliferation of very hard-to-drive speakers, whose impedance jumps up and down like a yo-yo, many don't consider it at all.

Some amplifier manufacturers also make audio amplifiers (and the other way around), so you'd imagine they would pay more attention to this impedance matching issue. Still, I have not researched this aspect and cannot comment. Although, the old cynic in me would not be surprised to find out that in many, even most of those cases, the crossovers had also been designed for optimal driver equalization (making them "mismatched" from the amplifier's ease of drive point-of-view).

How much amplifier power do you really need?

I wish I had a dollar for each time a prospective buyer of our tube amplifiers asked, "Do you make any 100- or 200-watt amplifiers?" After my gentle and polite probing, it turned out they just pulled that figure out of their (thin air), that they had no idea whatsoever why they asked for such high power figures.

A few who lived locally would come over and listen to one or two (usually 10-15 watt) SET amplifiers in our demo room and were very happy with the sound levels produced. Then I had an idea and asked them, "What do you think, how many watts of amplifier power are we listening to right now? Inevitably, perhaps sensing a loaded question, they'd lower the number they originally had in their minds but still said, on average, around 30-40 watts. "How about less than one Watt?" I'd reply. To cut a long story short, they had no feel for the electrical power involved and could not relate it to the acoustic power (sound pressure levels) they were hearing.

The volume of our listening room is approximately 65 m³ (cubic meters). How much amplifier power (in Watts) would we need to achieve reasonable sound pressure levels in that room?

The answer depends on a few factors, which we'll identify as we go along. You'd need to perform your estimation of this kind, using the graph (next page), for your listening room. As an example, starting with 65 m³ volume on the X-axis (horizontal axis), we go up until we hit, say, 90 dB line (point X).

Notice that this chart is of the log-log kind; both X and Y scales are logarithmic, not linear, that is why the spacings reduce. For instance, the space between 10 and 20 is much larger than between 20 and 30, and so on. Why 90 dB? That is the first choice you need to make. 90 dB is damn loud, so to be on "the safe side," let's stick with that worst-case assumption.

Then we go horizontally from point X until we hit the vertical axis. We are just under 0.03, so let's say the acoustic power needed to produce 90 dB is $P_A = 0.28$ Watts. Precision is not required here; it's just a ballpark estimate, anyway, but more than sufficient for our intents and purposes. Now we face the loudspeaker sensitivity and efficiency issue.

COMPONENT MATCHING AND AUDIO SYSTEM INTEGRATION ISSUES

If you can obtain the efficiency figure for your speakers, great. If not, don't worry; most dynamic loudspeakers have an efficiency of between 0.5 and 1.5%. Only 0.5 -1.5% of the electrical power supplied by the amplifier is converted into sound pressure - the rest is wasted as heat. Depressing, isn't it?

So $P_A = \eta * P_E$, where η is efficiency (Greek letter "eta"), P_A is acoustic and P_E electrical power. Thus, $P_E = P_A/\eta = 0.028/0.005 = 5.6$ Watts.

Again, we have assumed the worst case, the lowest efficiency of 0.5% (0.005), which gives us a target of an amplifier with a minimum output power of 5.6 Watts.

Surprised? This is SET territory, where 8-10 Watts amps with 300B tubes rule the waters. And, indeed, our listening experience confirms this quick, back-of-the-napkin estimate. Even 3-Watters (2A3 or VT-25A triode-based amps) have enough power to "energize" our room, driving relatively inefficient speakers of 86-88 dB/Watt sensitivity.

To produce a noticeable 3dB increase in the SPL to 93 dB, we would need to double the amplifier power to around 11 Watts, still way down the low-powered end of the amplifier scale.

To make the sound appear twice as loud (a jump of 10 dB to the 100 dB line), we'd need ten times the amplifier power, $P_E = P_A/\eta = 0.28/0.005 = 56$ Watts. The only reason for aiming at such ear-piercing SPL levels is to faithfully reproduce the dynamic peaks in the recorded program material, those crescendos that are seldom present in pop and jazz albums but common in classical music recordings.

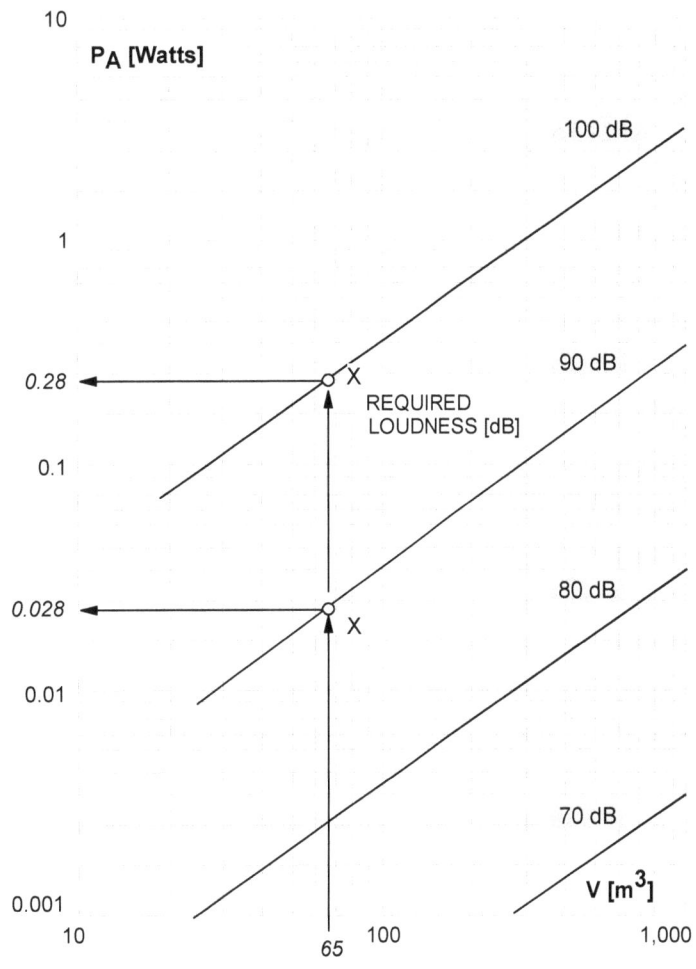

Fig. 10.12. The room size (volume) - acoustic power - SPL chart

Surround sound (home theater) speakers in hi-fi service

Together with triangular-shaped SCM8 rear speakers, B&W PCS8 subwoofer and FCM8 front and center speakers (pictured) were part of the (very expensive) mid-1990s TXH package. A friend asked us to check them out, fix anything that needed fixing and sell them on his behalf.

I've seen quite a few of the buyers of our tube amplifiers use surround sound (or "home theater") speakers in their audiophile systems and, in general, was not impressed with the sonic results of such pairings.

However, I was seldom impressed even with the sound of the setups using $10,0000 "audiophile" speakers either, but that is another story, the main reason for this book's existence.

Using just the FCM8 top speaker boxes (without the PCS8 subwoofers), the sound was dull and muffled. Removing the grilles did not improve things one bit.

Fig. 10.13. The visual aspect of audio equipment, particularly speakers, is essential for enjoying the whole audiophile experience. I could not live with such ugly and visually overpowering speakers even if they sounded incredible (which they didn't).

Adding the PCS8 subwoofers, driven by the same 300B monoblocks (35 watts in Class A2, transformer-coupled parallel single-ended output stage, as illustrated), the bass became more prominent. Still, the overall awful tonal balance (too much midrange and no treble sparkle to speak of) and the lack of dynamics remained.

Of course, some speakers sold as the main (front) surround sound or home theater speakers are of decent quality and sound quite good; you should not dismiss them outright. But, unless you find an absolute bargain, with so many great audiophile speakers around, why waste your precious time on trying them out?

BI-AMPING

Bi-amping, as its name suggests, involves using two stereo amplifiers to drive various woofers and tweeters. Using two monoblocks, each driving one speaker, is not bi-amping.

In vertical bi-amping, one channel of a stereo amplifier drives the tweeter while the other powers the woofer (of the same loudspeaker). In horizontal bi-amping, the two channels of one amplifier drive the two woofers, while the two channels of a second amp drive the two tweeters.

Bi-amping of both types can be done simply by removing the links between speakers' two pairs of binding posts, retaining the speakers' internal crossovers. Alternatively, for the intrepid amongst us, the internal (passive) crossovers can be removed and an active crossover added between the preamp and the power amps.

Horizontal bi-amping

A single-ended amplifier with 300B triodes has an output impedance of 2.1 ohms at 1,000 Hz. With an 8 ohm speaker, that amplifier would have a damping factor of DF = 8/2.1 = 3.8. Relatively speaking, that is an excellent result for a single-ended triode amp with no negative feedback. NFB would reduce the output impedance and therefore increase the damping factor but would make everything else worse.

The higher the damping factor, the better the amplifier will control the woofer's cone, the faster the woofer will respond to the changes in the signal. Solid-state amplifiers generally have a higher damping factor than tube amps, and among the tube amps, push-pull designs have higher damping factors than their single-ended cousins.

This intuitively leads to the idea of using a higher power, higher damping factor, solid-state stereo amp to drive both woofers (left and right), and then having a lower-powered stereo tube amp driving the tweeters.

This arrangement, horizontal bi-amping, has one stereo amp (the bass amp) driving the woofers, while the other amp drives the midrange drivers (if any) and tweeters.

Arguably, the tonal quality of the bass is not as critical as the smoothness, liquidity, and transparency of the midrange. Therefore, a lower cost, lower-quality solid-state amplifier would be fine for that purpose and would unburden the lower-powered tube amp from having to drive lots of current through the woofers' coils.

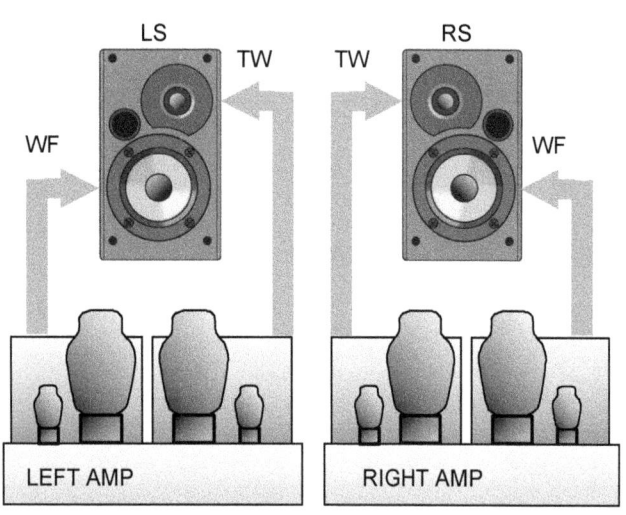

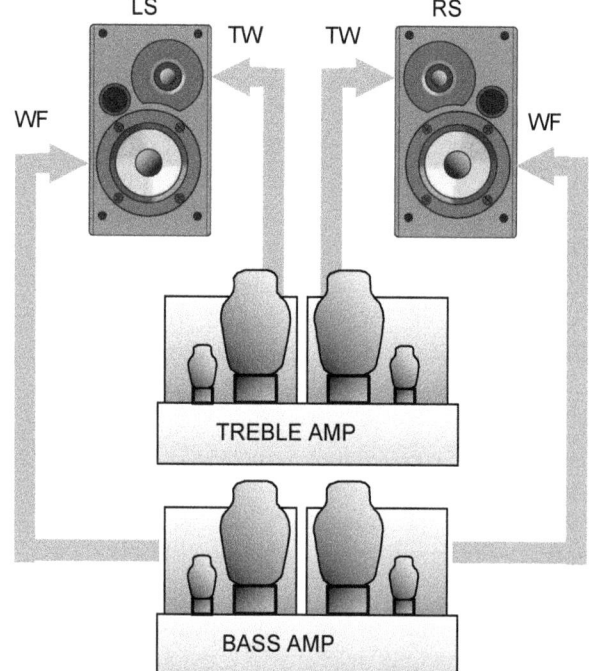

Fig. 10.14. (ABOVE) Vertical bi-amping
Fig. 10.15. (RIGHT) Horizontal bi-amping
NOTE: Although tube amps are shown, some or all of the amps could be solid state as well (see text). Both stereo amps are fed the same stereo signal from a preamplifier or an electronic crossover

While woofers are the power-hungry drivers, the midrange drivers and especially tweeters, need only a fraction of woofers' power. The tube amp will then operate at very low volume levels with very low distortion, almost effortlessly, which would also significantly prolong the life of output tubes.

The main drawback of this scheme is the tonal difference between the upper and the lower amp. This potential mismatch is accentuated by the fact that the common crossover frequencies usually lie within the frequency range to which our ears are most sensitive, that is, between 200 and 1,000 Hz. Of course, tonality is not the only aspect; the two amplifiers introduce different phase shifts and have different dynamics and distortion characters.

Vertical bi-amping

Since the amplifier channels for the tweeter and woofer are identical, there are no matching and tonal coherency issues that plague horizontal bi-amping. This arrangement ensures tonal cohesiveness, which may be lacking when two different amps are used in horizontal bi-amping. It is also easier to adjust the volume controls without disturbing the balance between the lower and upper frequencies as in horizontal bi-amping.

Also, because two separate amps are used for each channel, amplifier-generated crosstalk is eliminated. This will improve stereo imaging and result in a better soundstage.

Since most stereo amps have a common power supply (except those of dual mono construction), vertical bi-amping also works well in terms of power supply loading. Most of the power supply's capacity will automatically be assigned to powering of the woofer, with tweeters (high-frequency drivers) requiring much less power.

Vertical bi-amping also works better from the purchasing point of view. You could start by buying a stereo amp, and then, if you like it, when an opportunity presents itself (a clearance sale or a used amp bargain), you may buy a second, identical amp. Then, all combinations are possible, paralleling or bridging the channels (see pages 186-187) and vertical bi-amping.

However, the power margin and the driving capability are not maximized or optimized. The amplifier may be underpowered from the woofer's point-of-view and overpowered with respect to the tweeter. Using a 50-watt per channel push-pull tube amp to get a decent bass still isn't as good as the bass response of a 100 or 200-watt solid-state amp, and the midrange and treble will still not be as good as that of a SET amp.

The two bi-amping options discussed so far use the speakers' internal (passive) crossovers and are fed the same stereo signal from a source or preamp. Complexity-wise, the next option is to remove the passive crossovers from the speakers (or bypass them) and connect speaker drivers directly to amplifier outputs. However, in that case, active crossovers need to be added ahead (upstream) of the power amps. Let's look at that possibility.

ACTIVE CROSSOVERS

The main benefits of active crossovers

An active crossover is a frequency-selective preamplifier usually connected between the signal source (CD player, phono stage, or music streamer) or preamplifier and the power amplifier. Some provide voltage gain, so no additional preamp is needed, while others don't. Many even attenuate the input signal, thus requiring a line-level preamp. Most use integrated circuits (the cheapest and easiest solution, but the worst sounding), although vacuum tube-only designs (complex and expensive) are available for tube aficionados.

A two-way crossover has two filters, a low-pass filter, whose output feeds the low frequency or woofer amplifier, and a high-pass filter, whose output is connected to the input of an HF or tweeter amplifier.

These amplifiers can be but don't have to be identical, which is one of the benefits of this kind of bi-amping.

Since low powered single-ended amplifiers often struggle in this department, and since the spectral beauty and soundstage nuances of SET amplification are all but lost on a woofer, high powered but cheaper push-pull or solid-state amplifiers could be used for this purpose.

Tweeters require significantly less power so can be driven by (very) low power triode amps.

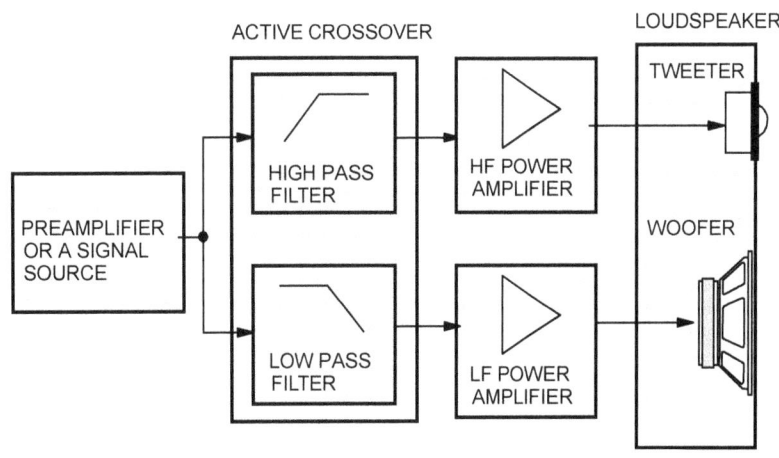

Fig. 10.16. Block diagram of a two-way active crossover setup (one channel)

This rationale is based on the most commonly used dynamic (moving coil) drivers. With full-range horn and electrostatic speakers (except hybrid designs such as Martin Logan), no crossovers are needed.

The LCR components (L-inductors, C-capacitors, and R-resistors) in passive crossovers must handle large signals and currents from the output of power amplifiers. As such, they introduce power losses which reduce speaker sensitivity and SPL levels of the whole system. Active crossovers operate at much lower voltage and power levels, so smaller (both physically and of lower voltage and power rating) and cheaper electronic components (resistors, inductors, and capacitors) can be used.

Active crossovers do not introduce power losses, so speaker efficiency is increased significantly. Amplifiers, especially low-powered SET ones, will work at reduced power levels and produce lower distortion, have more dynamic headroom, and sound better, with improved microdynamics and faster and cleaner transients.

Many passive crossovers use bipolar electrolytic capacitors and iron-cored inductors, which are nonlinear and cause significant distortion. This problem is eliminated in active crossovers, which don't use inductors or electrolytic capacitors, resulting in lower distortion.

Since the bandwidth of signals to each driver (woofer, midrange, if any, and tweeter) is strictly controlled and limited, the cross-modulation is, if not eliminated, then significantly reduced.

For instance, a distorting amplifier will pass a 2 kHz tone and a 2.5 kHz tone together but will also produce 500 Hz and 5.5 kHz intermodulation distortion tones. Only the two original tones will pass through the midrange crossover in a three-way active crossover, while the 500 Hz and 5.5 kHz distortion products will be significantly attenuated. The 500 Hz tone will not propagate through the LF fitter, and the 5.5 kHz tone will not even appear at the input of the HF filter since the filters are independent, resulting in significantly reduced IM (Intermodulation) distortion.

Active crossovers allow us to adjust the high- and low-frequency power levels separately and independently with volume controls for each frequency band, which are very easy to implement. With inbuilt, passive crossovers, that decision was made by the speaker manufacturer, and no user adjustment is usually possible. Active crossovers do not interact with speaker drivers' frequency-dependent (read "fluctuating") impedance as passive ones do, so a more precise and stable spectral division is achieved. The damping factor is also improved.

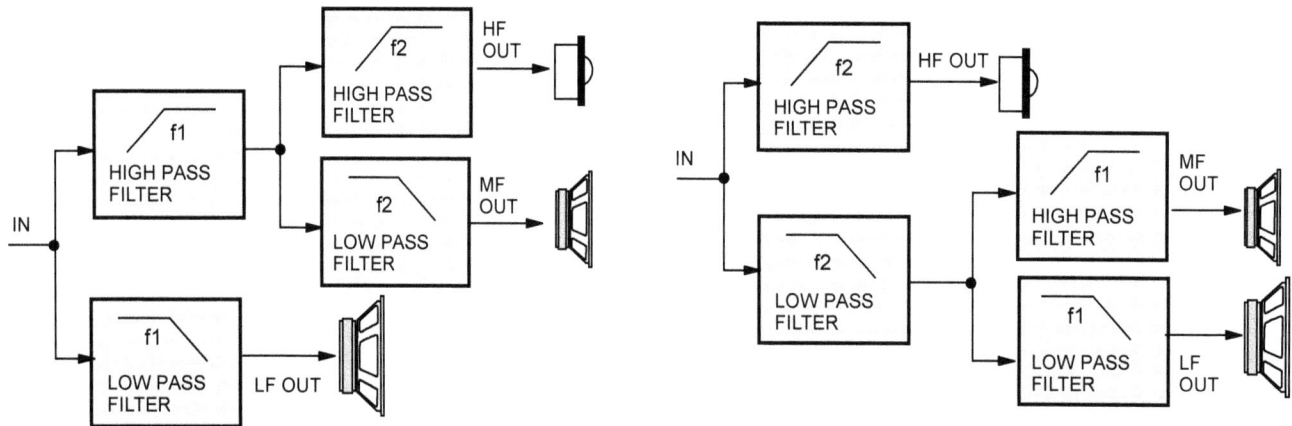

Fig. 10.17. Two ways of cascading the four filters required for a three-way active crossover

The main problems with and disadvantages of active crossovers

Since all speakers have a passive crossover built-in, building or buying an active crossover is a significant additional expense. The cost of an extra pair of interconnects (cables connecting the preamp to the crossover) also has to be factored in. If bi-wiring wasn't used before, this kind of bi-amping would also require an additional pair of speaker cables. The main cost, however, is in an additional power amplifier.

Apart from the significant cost, the main problem with active crossovers and bi-amping is system integration. Skillful speaker designers put lots of time and effort into voicing their speakers. They ensure the integration between the woofer, midrange driver, and the tweeter is as seamless as possible. That involves the judicious selection and fine-tuning of drivers, crossover frequencies and cabinet dimensions. With active crossovers, this demanding task has to be done by you or the person designing and building the active crossovers for you.

Just like bi-amping with passive crossovers, bi-amping with active crossovers violates that fundamental principle of all great audio systems - simplicity! The more complex the signal processing, the more things can go wrong. Unless you are time-rich and very patient, and unless you possess superior knowledge of electronics and acoustics, without the possibility of sitting on the fence and pontificating, my recommendation would be - proceed with caution.

BI-WIRING

The connection

Bi-wiring involves the use of two speaker cables, one for the high-frequency driver (tweeter) and the other for the low-frequency driver (woofer). Bi-wireable speakers have two separate sets of binding posts and two electrically separate crossovers.

Typically, when only one 2-core speaker cable is used per speaker, the two pairs of binding posts are bridged using external shorting links (the photo on page 161); these must be removed if bi-wiring is used.

Some 3-way speakers are designed so that the low-frequency crossover feeds both the woofer and the midrange driver (as in our Opera Terza speakers), while others have the midrange driver on the same crossover as the tweeter.

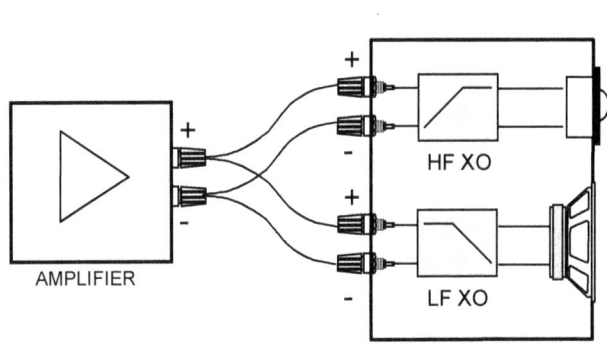

Fig. 10.18. A bi-wired 2-way loudspeaker. One channel only shown.

Why would bi-wiring make any sonic difference?

The rationale used by the proponents of bi-wiring is the separation of audio signals into two frequency bands, each propagating through its own speaker cable. Bi-wiring would thus separate high current, high-energy bass signals from the high-frequency, low power tweeter feed.

However, electrically, that should make no difference to the complex audio signal. Remember, we are not talking about two different signals, one HF the other LF, but one complex signal comprised of multiple harmonics that happily propagate together.

As with many other contentious topics in audio, most of which are discussed in this book, I believe the proponents of bi-wiring claim they've heard an improvement. As always, try it for yourself and make up your own mind.

Ignoring the super-expensive "all sizzle and no steak" abominations sold by rip-off merchants to audiophiles with more money than brains, decent speaker cables aren't that expensive, so getting an additional pair should not break the bank. Alternatively, borrow them from your audiophile mates for a day or two, or get together and organize a group listening evaluation of the bi-wiring concept (if you haven't already done so).

What factors could be in play?

First of all, many shorting links at bi-wireable speaker terminals are poor quality conductors and introduce two additional mechanical contacts into the signal path - each! Simply removing them from the signal path could result in noticeable sonic improvement. Soldering the two positive and two negative speaker terminals internally would achieve the same aim without bi-wiring.

To assess the quality of your shorting links, try this simple experiment using a single speaker cable (without bi-wiring). With shorting links in place, connect the speaker cable first to the woofer or LF binding posts and listen to a few test tracks (the ones you know intimately). Then connect the cable to the HF or tweeter binding posts and listen to the same tracks. Do you hear any difference? Many audiophiles do. In some cases, the difference is subtle, but often one connection sounds better than the other.

Just as some amplifiers excel in the midrange or treble but aren't that good in the bass department, some speaker cables sound better with higher frequencies. In contrast, others do a better job with the bass (low frequency) signals. If you wisely select the two cables for your bi-wired setup, you should get them to perform at their best.

Wire your speakers the other way around, and you'd get the worst of both worlds. That, of course, implies the use of different cables for HF and LF connections. Some audio experts warn that the two cables used for bi-wiring must be identical, but I am not so sure.

The amplifier's output impedance could also be a factor. The inductance, resistance, and capacitance of a speaker cable (its LCR parameters) could be of higher importance with solid-state amps due to their low output impedance. So a proposition that bi-wiring is more noticeable with SS amps isn't that far-fetched. I don't listen to solid-state amps, so I cannot comment any further here.

Inside tube amps, the audio signal flows through 20-40 meters of copper wire that make up the secondary windings of their output transformers. Another 2-3 meters of external speaker cable (or two in the case of bi-wiring) may not be such a significant factor, as they are in the case of solid-state amplifiers.

INTEGRATING SUBWOOFERS INTO AN EXISTING STEREO SYSTEM

Subwoofer controls

Although Polk Audio DSW PRO 440wi are considered budget subwoofers, they are by no means "cheap & nasty". We got a couple of them at a heavily reduced price. A 180 watt RMS Class-D amplifier drives an 8" down-firing woofer, but the four feet can be repositioned for front-firing operation.

The 13-3/4 "x 14-3/4 "x 16-1/4 "enclosure is of bass-reflex (ported) type and is magnetically shielded. The frequency response is specified as 30-125 Hz (-3dB), and the crossover frequency is continuously adjustable from 60 to 160 Hz.

With the "Power" switch in the "Standby" position (1), the subwoofer will only turn on when the user presses the power button on the remote control. In the "Auto" position, it will turn on automatically when it senses an audio signal at one of its inputs.

The sub can be fed a high-level signal from the main speakers (2) or a line-level signal (3) from a preamplifier (LINE IN). There are also LINE OUT (4) RCA outputs and "SPEAKER OUT" binding posts (5), which are then used to power the main speakers.

The unfiltered LFE (Low-Frequency Effects) input (6) is used only in home theater applications, with AV receivers with a dedicated "SUB OUT" output. In that case, the sub's internal filter is bypassed, and the AV receiver's filter is used. For hi-fi connections, the "Use as LFE" switch should be in the "NO" position.

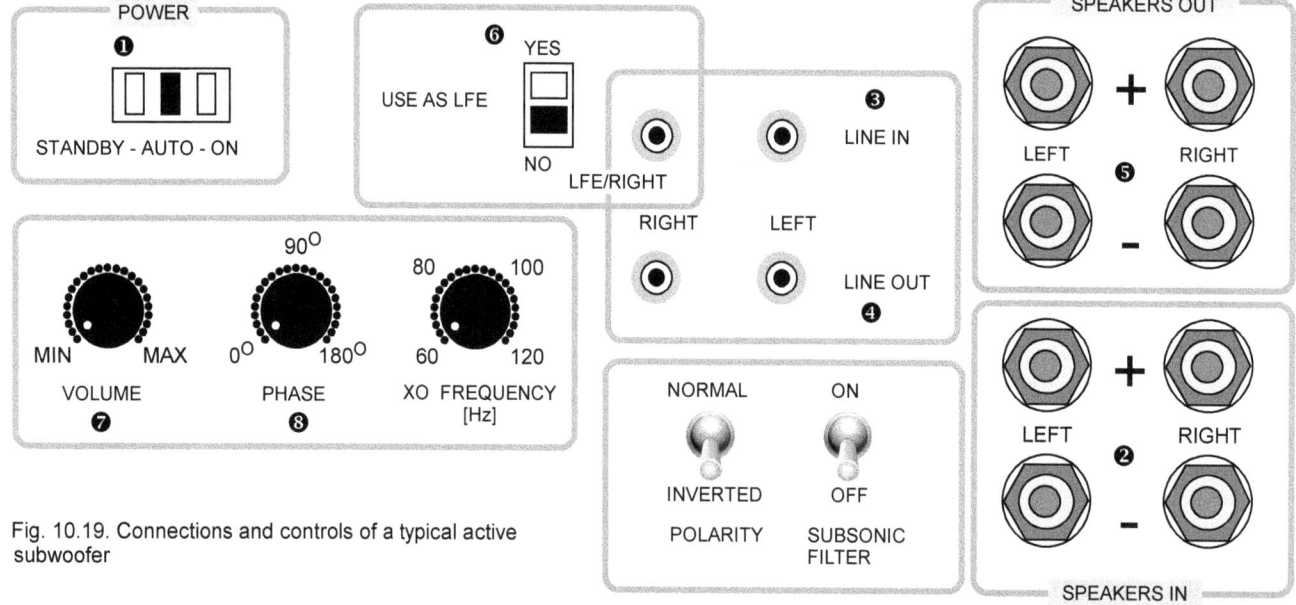

Fig. 10.19. Connections and controls of a typical active subwoofer

The Polk remote control has power on-off and mute buttons, volume up/down buttons, a "reset" button, and a button that engages the night mode dynamic range compression. A button that turns the power indicator LED on or off is always a good idea - some listeners may be annoyed by the bright light from the back of the woofer. The volume up and down buttons are also replicated at the back of the sub, although most others still have an analog volume control potentiometer (7).

Adjusting volume is easy, but the visual indication is awkward, done by a blue LED that blinks a 2-digit code, indicating volume setting from 0 to 40. The first digit is the number of long, and the second digit is the number of short blinks.

For instance, with volume set at "28", the LED will flash two long blinks followed by eight short ones. I found the need to count the long and short blinks annoying. Perhaps others felt the same, the reason for the subs' 50% discount fire sale?

Some active subs have a toggle switch at the back that selects 0 or 180 degrees phase, while others have a potentiometer that continuously varies the phase shift from 0 to 180 degrees (8). Polk DSW PRO 440w subs have neither. The only way to control the phase shift is by using the four buttons on the remote controller, 0-90-180-270 degrees (9), making it impossible to figure out the current setting by looking at the sub or the controller.

Fig. 10.20. Polk Audio remote controller

COMPONENT MATCHING AND AUDIO SYSTEM INTEGRATION ISSUES 243

The inbuilt Polk Room Optimizer buttons select one of four modes of operation, "cabinet," "corner," "mid-wall," and "mid-room," obviously relating to where the sub is positioned in relation to the cabinet, wall, and the room. Its aim, according to Polk, is to deliver "punchy bass no matter where you place the sub in your room."

Some subs have a wireless option included; others, such as Polk 440wi (and larger models 550wi and 660wi), are wireless-ready, meaning one only has to buy the wireless kit (Polk Audio PWSK1). The kit has only a pair of RCA "Line in" inputs.

Connecting active (powered) subs: "SPEAKER IN" and "SPEAKER IN - SPEAKER OUT" connections

The sub's high-level input (or "SPEAKER IN") is connected to the power amp's output or the main speaker binding posts. Both the sub and the main speaker's woofer are active, reproducing the same bass content. The main amp's connection to the preamp or the source is unchanged.

On the plus side, the subwoofer's input is the same as the main speaker's input, with no timing differences or phase shifts. However, the main speaker's woofer is also trying to reproduce the lowest frequencies that the sub is amplifying and producing. A careful setup is required, so the two acoustic outputs don't cancel each other or create a comb-filter effect.

Also, suppose the main amplifier's reproduction of the lowest frequencies is compromised - as it usually is, why would you need a subwoofer otherwise? In that case, the sub's reproduction of that frequency band will also be degraded! For instance, the low damping factor and the "muddy" bass of many lower-quality tube amps will be passed on to the sub, which cannot make a "tighter" or "faster" bass out of such a compromised signal.

Instead of doubling up the connections at one end, either at the power amp's or the main speakers' binding posts, most woofers offer "SPEAKER IN" and "SPEAKER OUT" connections, which are looped inside. With the sub positioned close to the amp or the main speakers, this usually results in shorter speaker cable runs.

Fig. 10.21.
ABOVE LEFT: The existing speaker cables between the amp and the main speakers are retained, with the sub's high level or "SPEAKER IN" input connected to the speakers' binding posts. The connections are thus "doubled up" at the speaker terminals.
ABOVE RIGHT: The existing speaker cables between the amp and the main speakers are retained, but both the sub's high level or "SPEAKER IN" input and the main speakers are connected to the power amp's output. The connections are "doubled up" at the amplifier terminals.

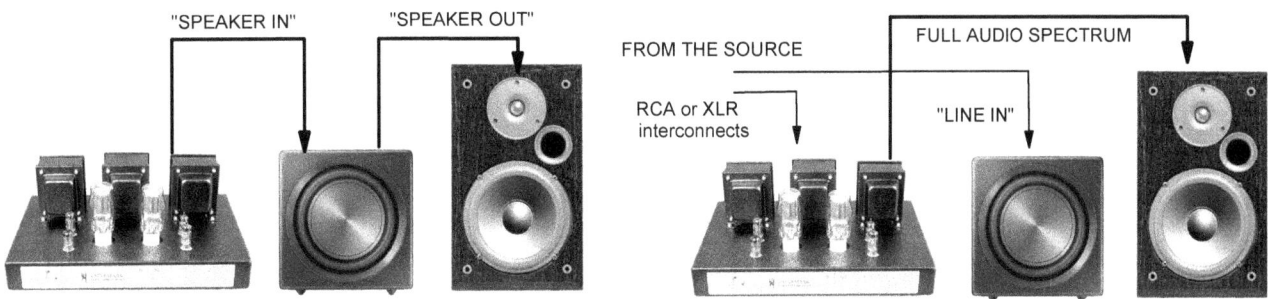

Fig. 10.22.
ABOVE LEFT: To minimize the length of speaker cables used, the amplifier's output is connected to the "SPEAKER IN" of the sub, and the main speakers are powered from the sub's "SPEAKER OUT" outputs.
ABOVE RIGHT: "LINE IN" connected to the output of the preamp, DAC, or music streamer ("source"). Some preamps have a second line-level output for such purposes. Both the sub and the main speaker's woofer are active.

Connecting active subs: "LINE IN" connection

The sub's "LINE IN" input is connected to the output of the preamp, DAC, or music streamer (line-level "source"). Some preamps have a second line-level output (wired in parallel to the main output) which can be used for this purpose. Alternatively, an RCA plug with two output RCA sockets (a "splitter") is needed, one for each channel.

If your preamp has only a single pair of outputs, another possibility is to use the "Tape out" or "Rec. out" feature on some line-level preamps to feed the signal to the active sub. Just as the previous setup ("SPEAKER IN" and "SPEAKER IN" - "SPEAKER OUT" connections), in this case, the main speaker's woofer is also reproducing the lowest frequencies that the sub is amplifying and reproducing. A careful setup is required, so the two acoustic outputs don't cancel each other or create a comb-filter effect.

Connecting powered subs: "LINE IN" and "LINE OUT" connection

The signal from the source enters the sub through the "LINE IN" connections and exits through the "LINE OUT" RCA terminals into the power amplifier. The main speakers are connected to the power amp in the standard way. On some subs, the "LINE OUT" is looped from the "LINE IN" input, passing the full frequency range onto the power amp. In such a case, both the sub and the main amp/speakers are reproducing the lowest frequency band. With others, the "LINE OUT" signal has the subwoofer frequencies filtered out (a high pass filter), so those lowest frequencies are not passed onto the power amp.

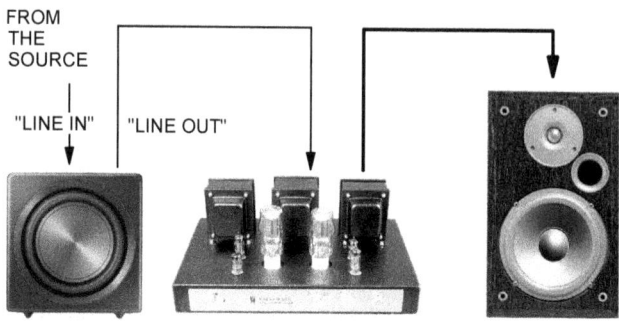

Fig. 10.23. The signal from the preamp, DAC, or music streamer ("source") enters the sub through the "LINE IN" and exits through then "LINE OUT" into the power amplifier. The main speakers are connected to the power amp in the standard way.

On the positive side, only the sub will amplify and produce those lowest bass frequencies, thus relieving the speakers' woofers of the most problematic and distortion-inducing frequency band.

However, on the debit side, in that case, the whole audio signal from the preamp passes through the high pass filter inside the subwoofer first, resulting in possible sonic degradation of midrange and treble frequencies.

With both of these setups using the "LINE IN" connection, the internal amplifier inside the powered sub receives the same signal as the main amplifier and amplifies it independently. This situation is equivalent to bi-amping using different amplifiers and suffers from the same potential problems. Different tonal and dynamic characteristics of the two amplifiers can result in a lack of coherence.

The myth of a single subwoofer

The myth that a single subwoofer is sufficient most likely originates from the fact that older AV processors (or "receivers") used in home theater applications had only a single subwoofer output. That was indicated by the 5.1 or 7.1 descriptors in their names, the "1" referring to the number of subwoofer outputs.

Although such a claim may have merits in home theater applications, better home theater setups use two or even four subwoofers to get a more evenly distributed bass response across the room. The differences in bass response and quality between different seats in the home theater are thus minimized. More modern and higher-specs AV receivers have two sub outputs, signified by 7.2, 9.2, and 11.2 descriptors.

A single sub is not sufficient in hi-fi setups. Intuitively, you may draw parallels with your main speakers, each of which has at least one woofer (bass driver), so each subwoofer (left and right) could be considered an extension of the main speakers. Subwoofers will affect your audio system's sonic and spectral balance, and with a single sub, such balance would be impossible to achieve.

Down- or front-firing orientation?

Most subs have a fixed, down-firing orientation, but some offer an additional front-firing option like Polk's DSWPRO series of subwoofers. Two sets of feet are provided. For forward-firing, the default feet are removed from the bottom of the speaker (which now becomes the front), and grilles can be attached. The feet are then attached to the sub's back (where the controls are), which becomes the bottom.

Should your subs have this option, it is worth experimenting; many audiophiles find the front-firing subwoofers easier to integrate with the main speakers and more conducive to a cohesive sound.

A sub will not sound equally good firing in any direction; there will be one or two directions where it will produce the best (fastest, tightest, best defined, and least boomy) bass. From my experience, those two directions are usually front-firing and, wait for it, side-firing.

Many subwoofers are of the down-firing type, down meaning into the floor, from the height of a few centimeters, as if their designers deliberately wanted to "excite" the floor and make it vibrate and resonate as much as possible.

Subwoofer placement

As if positioning of the main speakers, bi-amping and the use of active crossovers wasn't difficult enough, adding a subwoofer or two makes things infinitely more complex. The worst situation occurs when a single subwoofer is placed in one of the corners of the room.

This placement will excite (energize) all resonant modes across the width of the room. Note that only the first three modes are illustrated in the diagram (right). Room resonant modes are discussed in Chapter 13; you may wish to read that section first before continuing.

If two subs are placed in front corners, the odd-order resonances will be avoided, since such placement all but cancels the first, third, fifth and all other odd-numbered modes, leaving only the even-order modes across the width of the room (2). Note that while only the 2nd resonant mode is shown (for simplicity sake), the 4th, 6th and all higher order even modes will be present as well.

By moving the subs to the "nodes" or the locations of the 2nd mode (points where the sound pressure is zero or close to it), the subs will still cancel the odd-order modes and their ability to excite the even modes will now also be reduced!

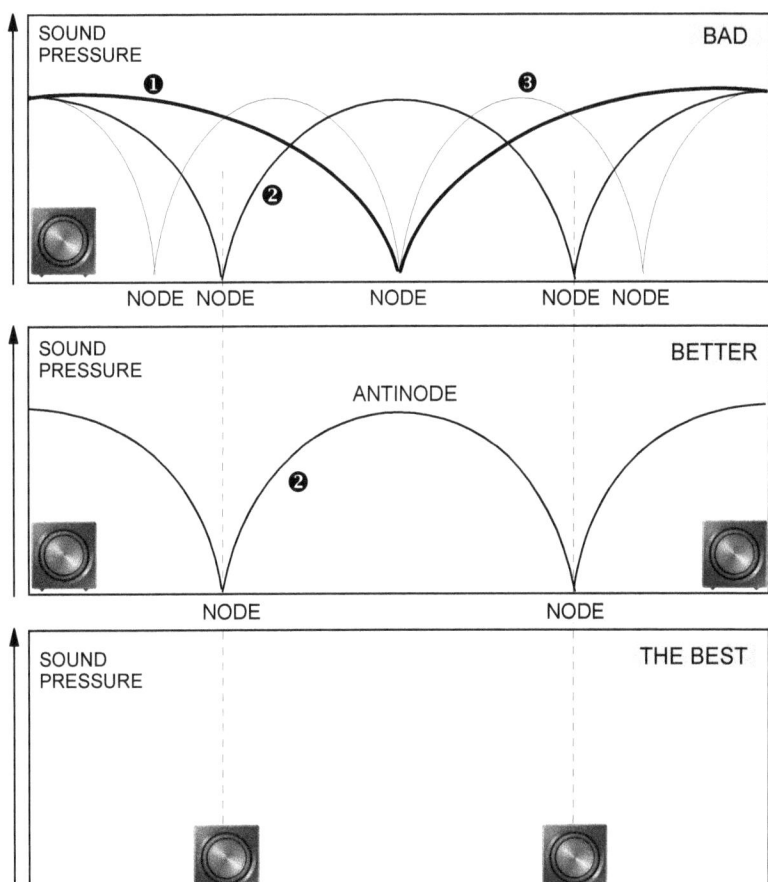

Fig. 10.24. Except for speakers designed for such a placement, Klipshorns, for instance, speaker placement in the corners is to be avoided. The same rule applies to subwoofers as well.

AUDIO SYNERGY

The frequency band balancing problem

"The frequency band balancing problem" is most noticeable in two instances. In amplifier design, amplifiers (tube amps in this instance) that have a firm and dominant (or rather "prominent") bass usually aren't very transparent or "airy" (there's no "air" around instruments), resulting in a muffled and "closed-in" treble and upper midrange.

On the other hand, those with "recessed" or unremarkable bass often have incredible transparency that not even the best solid-state amps can match (and that is supposed to be one of their strengths!), plus a refined and highly detailed top end. However, we thought that was "the nature of the beast," so to speak, the way tubes, especially power tubes, are. Some produce better, tighter bass, others are more detailed and transparent, but their bass reproduction is unremarkable.

We started connecting the dots when experimenting with bi-amping and tri-amping (not an exercise for the faint-hearted) and when trying to integrate active subwoofers into existing audio systems (could also be considered a case of bi-amping).

Increasing the volume of the treble amplifier impacts the bass, making it sound thin and weak. Likewise, adding a subwoofer does not just add to the bass but also changes the sonic character of the midrange and treble. Altering the overall tonal balance in such a way is usually unwanted unless it is an improvement, as it sometimes happens when it softens the harsh and sibilant treble and smoothes it out. This is most noticeable in the reproduction of female voices. I am not sure of the exact reason behind this phenomenon. It most likely lies in the human hearing mechanism, how the brain processes and integrates the sonic clues it receives from the hearing apparatus.

THE NEUTRALITY PARADOX a.k.a. THE INCREMENTAL REVELATIONS PRINCIPLE:

The more neutral or transparent each component of your audio system, the more the tonal character of other components becomes revealed.

The synergy principle

Synergy is defined as a phenomenon where the end result is greater or better than the sum of individual parts or components. It applies to both the macro level, the audio components that make up an audiophile music system, and to the micro-level of electronic components inside amplifiers, preamplifiers, and DACs.

You may remember "the worst sounding amplifier we've ever made" story, about one of our amps, bought by an audio dealer and speaker importer. He spent a considerable time experimenting, trying the amp with different speakers and sources until he found the right match.

The sonic problem wasn't in our tube amp; it was simply a poor match for the rest of the test setup we used at the time, our CD player and speakers.

Some aspects of synergy (or rather the lack of) are obvious, such as trying to drive inefficient, low impedance speakers by low-powered triode amps. Others are still a mystery. The absolutists claim that all of that can be explained by measurements and technical parameters; I don't think so. Damping factors, output power levels, frequency bandwidths, and distortion levels matter but are relatively poor predictors of how an amplifier will sound with particular speakers.

This makes component matching and the pursuit of audio synergy a dark art and not an exact, rational science. I wish I could give you here a checklist that would say, "follow these criteria-steps-principles, and you will achieve audio synergy"; I don't think such a list will ever be produced.

We have to "play it by ear" and learn through trial-and-error; there is simply no other way. To many of us, half of the fun in audiophilia is in such experimentation and auditioning.

Fig. 10.25. Simplicity and synergy are twin sisters. A common delusion amongst audiophiles is that the sound of their system would be improved by adding more (expensive) components. Usually, it is quite the opposite: simplify, simplify, simplify! The audio system above consists of only three components - a music streamer, tube amplifier (with volume control), and speakers.

> **THE SYNERGY RULE**
> Individual building blocks of an audio system need to be understood individually but matched and optimized together.

The ultimate test

I've noticed that women are often better judges of sound quality than men, who make up the majority of audiophiles. While male audiophiles pay attention to the bass, the midrange transparency, microdynamics, soundstage width & depth, and myriad other performance aspects of an audio system, women tend to listen holistically.

They notice first and foremost how the sound affects them emotionally. Perhaps we (male audiophiles) should, too, at least in the initial stages of evaluation. If a component or a system does not sound "involving" or "inviting," is there any point wasting your time on its more detailed aspects?

I tend to over-rationalize - I convince myself that a certain amplifier or audio component sounds good. Perhaps I subconsciously try to find something good in every design. After all, when you design and make something, you don't go trying to pooh-pooh it, do you? It's like deliberately finding faults in your children; normal parents don't do that. However, an impartial designer must sometimes be cruel and admit reality (but never defeat).

My music room (page 265) is right next to the kitchen/dining area, so my wife can hear how different amplifiers sound from ten meters away and through the closed French doors. On rare occasions, she'd comment how a particular amplifier sounded good (dynamic, tonally balanced, open or transparent, whichever way you want to define "good"). With others, she'd say nothing. To me, such comments have always been of immense value.

The ultimate test is how much enjoyment the audio brings you. I call those amplifiers, audio components, and systems "all-singing, all-dancing."

> **A GREAT MUSIC SYSTEM SUMMARIZED IN ONE SENTENCE**
> A great music system or component (for instance, an amplifier) evokes positive emotions in you, draws you in so you want to get up and dance, and makes you keen to listen more and more and never want to stop.

TUBE PREAMPS DRIVING SOLID-STATE POWER AMPS

The possible mismatch, how and why it happens

Some audiophiles consider pairing a tube preamplifier with a solid-state power amplifier to be the best of both worlds, a true sonic synergy. In other words, the sonic beauty of tubes is combined with the superior damping factor and high power of solid-state amplifiers. Even this idea is controversial.

While such pairing may improve things on the surface (better bass and dynamics), many audiophiles claim that solid-state kills the magic and introduces its fatigue-causing sonic artifacts.

Furthermore, tube preamps generally have a high output impedance, which may create problems with low input impedance solid-state amplifiers (in the order of 10-20 kΩ). Why is that so?

The output impedance of the Audio Research Reference 6SE line preamplifier is specified as 600 Ω (ohms) for the balanced and 300 Ω for the unbalanced (single-ended) output. 10 kΩ minimum load resistance and 2,000 pF maximum capacitance (of the following power amplifier) are also mentioned.

In its review in *Stereophile* magazine, the unbalanced output impedance was measured at 308 Ω at 1 kHz and 20 kHz, rising to 592 Ω at 20 Hz. This is not an anomaly or problem with AR preamps; it's normal in most tube preamps. However, no manufacturer that I know of mentions such doubling of the output impedance at high frequencies. Studying their output impedance-versus-frequency graphs in audio reviews shows how their output resistance R_{OUT} increases as the f_L (the lower -3dB or "cutoff" frequency) is approached.

Unless they use output transformers, all vacuum tube line stages have output coupling capacitors (C_{OUT} in the circuit diagram below). Such a capacitor forms a high pass filter with R_{IN}, the input resistance of a power amp. The lower the signal frequency, the more such a filter will attenuate that signal.

$\omega_L = 1/R_{IN}C_{OUT}$ and since $\omega_L = 2\pi f_L$, we have $f_L = 1/(2\pi R_{IN}C_{OUT})$ At the "cutoff" frequency (lower -3dB frequency) the signal is attenuated by 3 dB or approx. 30%. The larger R_{IN} and the larger C_{OUT}, the lower such frequency and the wider the bandwidth of the whole audio chain.

So why don't tube preamp designers and manufacturers simply increase the capacitance of the output capacitor C_{OUT} and avoid all possible problems with low input impedance power amps? The first couple of reasons are cost and size. High voltage film capacitors are very large (a larger chassis may be needed) and expensive, especially capacitors by "hi-end" brands (Psvane "Teflon" caps at $200 a pop? Anyone?)

Increasing the product cost by $100 at the production stage may result in a $500 increase of the retail price.

Sonically, the larger the capacitance, the slower the preamp's transient response (the charging time of a large capacitance) and the muddier the tonal presentation (the more they color the sound). Also, the larger (physically) its size, the better a capacitor works as an antenna in picking up unwanted hum and noise.

Many solid-state power amplifier designers and manufacturers have identified this trend of audiophiles driving their amps with tube preamps and have thus increased their amps' input resistance to 100 kΩ or even higher. This mismatch rarely happens with newer gear, but it does happen.

What about the 2,000 pF maximum capacitance specified by Audio Research? That requirement is in relation to another filter, this time a low pass filter formed by R_{OUT} of the preamp and the C_{IN} of the power amp, a RC circuit this time.

This filter determines f_U, the upper -3dB frequency of the combination. The formula is of the same format, namely $f_U = 1/(2\pi R_{OUT}C_{IN})$. This time the requirements are the opposite, the lower R_{IN} and the lower C_{IN}, the higher such frequency and the wider the bandwidth of the whole audio chain.

Strictly speaking, the capacitance of the interconnect cable C_C should be added in parallel to C_{IN}, but C_{IN} is usually much higher than C_C so that doesn't change the nature of the problem, just its magnitude.

The reduced frequency bandwidth results in a thinned-out bass and the loss of top end (treble). Also, the input resistance of the power amp "reflects" into the line preamp's tube circuit at frequencies where the impedance of C_{OUT} is very low and changes the operating point, which results in a reduced output voltage and increased distortion. Such "loading effect" is clearly undesirable.

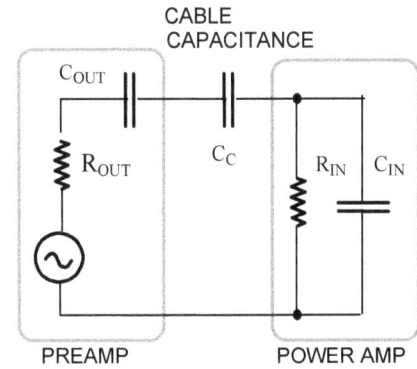

Fig. 10.26. A simplified circuit of a tube line stage (preamplifier) driving a power amplifier (tube or solid state)..

Three ways of achieving a low output impedance in tube preamplifiers

In the first version of Conrad Johnson's "Art" line-level preamplifier, each channel used ten ECC88 (6DJ8) triodes in parallel (five tubes, since each ECC88 tube is a duo-triode). According to a magazine review, the output impedance was measured as 475 Ω. Paralleling identical tubes can be considered a "brute force" approach.

The amplification factor of the paralleled tubes stays the same, but the overall transconductance increases N times, and the internal impedance reduces N times, N being the number of paralleled tubes.

Paralleled tubes must be matched to the highest degree possible. The standard practice of matching them only in one arbitrarily chosen quiescent point is necessary but insufficient. It doesn't ensure the tubes are also matched dynamically across their whole operating range. Their anode characteristics and transfer curves should be identical, which is seldom the case.

To make matters worse, even if they were perfectly matched when new, after 500 or 1,000 operating hours, they would not be matched any more. Tubes age differently and at different rates.

The next option is to use a cathode follower (CF) stage at the preamp's output. A cathode follower doesn't provide any voltage gain. Why would we use such a stage since it does not amplify the signal (it attenuates it 20-30%)?

The main reason is its very low output impedance Z_{OUT}. Depending on the type of the tube used, the output impedance of a common cathode stage with bypassed cathode resistor is usually in the 5kΩ - 10kΩ range. CF stages with the same tubes have Z_{OUT} in the 150 - 350 ohms range, lower than paralleling 8 or 10 triodes! Of course, triodes can be paralleled in the CF stage as well, for an even lower Z_{OUT}.

The second reason for cathode followers' superiority is their very high input impedance, meaning such stages do not load the previous amplification stage (usually a common cathode stage), which means very low distortion.

Some tube preamps use output transformers, just as most tube power amps do (except OTL designs). The line stage in the photo (right) has no capacitors in the signal path.

Each channel uses a single amplification stage with an E86C triode and a vintage USA-made STANCOR WF-35 output transformer. The step-down transformer's primary impedance is 15 kΩ while various output impedances are provided, from 600Ω down to 50 Ω, depending on which secondary terminals are used.

15 kΩ to 600 Ω means the impedance ratio of the output transformer is IR=25 and the voltage ratio is VR = √IR = 5, meaning the output transformer attenuates the primary signal voltage five times.

The triode circuit driving the output transformer has a voltage gain of around 50 times, so the overall voltage amplification factor is 50/5 = 10 times. With a 1 V input signal, up to 10 V are available at the output.

On the debit side, transformers increase the cost of line stages, introduce their sonic coloration (although it is much more sonically pleasing than the coloration introduced by coupling capacitors), and have a limited frequency bandwidth.

Fig. 10.27. A transformer-coupled triode line-level preamplifier using MSPL (Minimal Signal Path Length) approach, with RCA inputs (3) as close as possible to the input selector switch (5), which is only a few centimeters from the volume control knob (6) and the E86C tubes

Since the output transformers (2) are shielded, the proximity of the power transformer (1) makes no difference in terms of hum. The output RCA sockets are on the other side (4), with the on-off switch and mains IEC socket (for the power cable) at the back (7).

11 | LOUDSPEAKER POSITIONING

This chapter should be read in conjunction with the next one, "OPTIMIZING THE ACOUSTIC PERFORMANCE OF YOUR LISTENING ROOM," and the one after, "ACOUSTIC TREATMENTS," since they are all intrinsically linked.

I've pondered for quite some time which would be the best order to present the first two chapters and found arguments and counterarguments both ways. So, let's talk basic speaker positioning first, and then we'll go deeper into the issue of room acoustics, thus answering some questions you may have while reading this chapter.

Not all speakers are equally demanding in terms of their placement in the room. Some will sound good at various distances from the front and sidewalls. In contrast, the soundstage and tonal balance of others are so sensitive to their positioning that it will take you days of experimentation to find the optimal setup.

The sonic interaction between your speakers and your listening room is even more critical than the crucial amplifier-speaker match.

- EXPERIMENTAL (EMPIRICAL) SPEAKER PLACEMENT METHODS
- SPEAKER PLACEMENT CHALLENGES - ROOMS OF IRREGULAR SHAPES, FIREPLACES, SLOPING CEILINGS AND OTHER PITFALLS
- FORMULA-BASED SPEAKER PLACEMENT METHODS

> "If you want truly to understand something, try to change it."
> Kurt Lewin, applied psychologist, the founding father of change management

EXPERIMENTAL (EMPIRICAL) SPEAKER PLACEMENT METHODS

Where to place amplifiers, close to speakers or close to sources?

From the aesthetic point of view, monoblock amps look great when placed on the floor, each next (usually slightly towards the back) to the speaker they are driving. We often find such setups at prospective buyers' listening rooms but rarely at dealers' showrooms. And there are a few good reasons for that.

Loudspeakers vibrate, and these vibrations are transmitted to the floor and from there to anything on the floor, including equipment racks and amplifiers sitting on the floor. This is especially an issue with timber floors, which vibrate more than the concrete ones. Audio components should not vibrate (except electromechanical transducers such as loudspeakers and phono cartridges), so amplifiers should not be sitting directly on the floor.

There may be significant acoustic feedback from the speaker to closely placed tube amplifiers and preamplifiers; this is especially so with the rear-firing bass-reflex speakers. All tubes are microphonic to some extent, and if they are exposed to sound waves, an unwanted reproduction of the fed-back sound will result.

Without playing any music but with the system turned on, have somebody gently tap a tube in your amp or preamp with their fingernails while you listen with your ear on the speaker. You will usually hear a sound; sometimes, it will be faint, in other cases very loud.

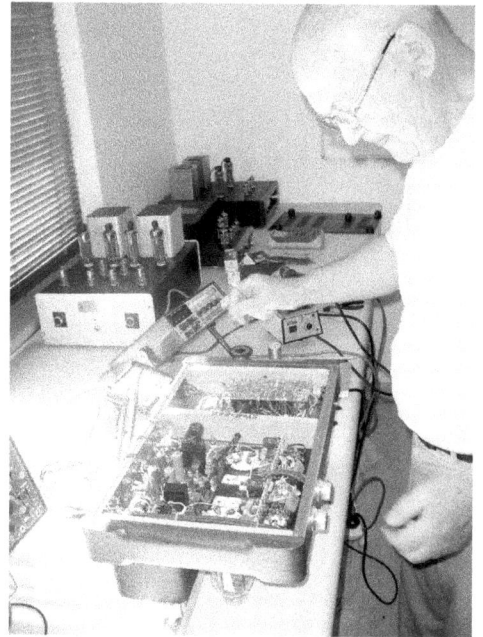

Before you even start listening to a tube component in the dealer's demo room or when buying privately, perform this simple microphony test. If the tube amp or preamp fails it, move on.

Fixing microphonic amplifiers is always tricky. Completely removing the microphony is usually impossible, especially if printed circuit boards are used. Messing around with such amps and preamps isn't worth your time and effort.

Some tube amplifiers we've analyzed over the years were incredibly microphonic. The worst was a DCC-90 preamp tube in a Music Angel XD-SE 300B amp. Its socket was mounted on a special rubber-suspended plate, meaning the manufacturer tried to mitigate the problem, obviously without success. The only solution was to rebuild the preamp stage completely.

Although you will not be able to isolate it acoustically or electrically to listen to it by itself, such acoustic feedback is always there to some extent. If not minimized, it will cause distortion, smearing, and loss of focus. It will impact the sound stage, accuracy, and transparency of the whole system.

The illustration shows a proper and improper way of equipment positioning. The source(s) and amplifiers (usually on an equipment rack) should be located as far back from the speakers as practical.

Fig. 11.1. Dr. Bob testing the extremely microphonic Music Angel 300B amplifier during the filming of our DVD workshop on upgrading and repairing audiophile tube amps.

The lower the signal amplitude, the shorter the signal-carrying cables should be. Thus, turntable interconnects should be as short as possible.

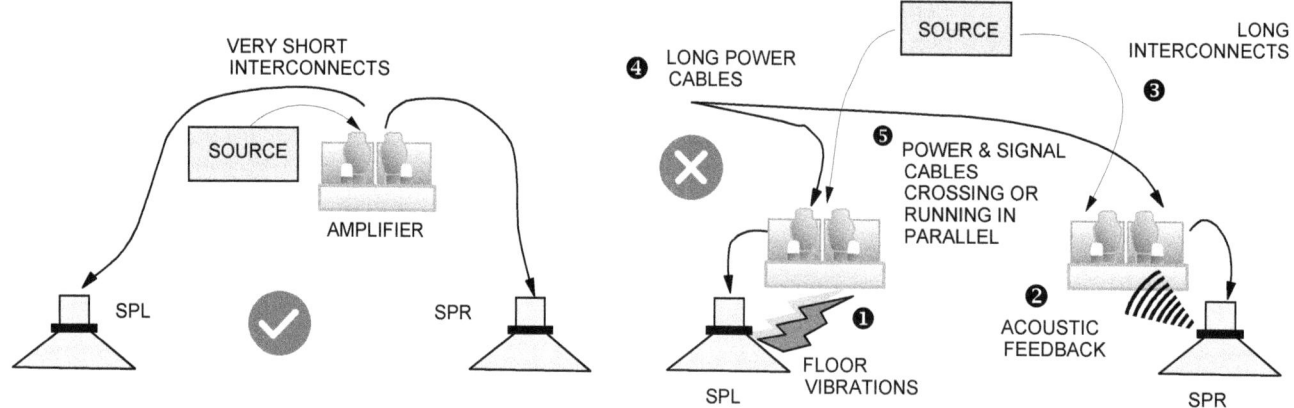

Fig. 11.2. Amplifiers sitting on a floor, close to speakers or subwoofers and far from the source introduce a few significant problems: (1) floor vibrations, (2) acoustic feedback from the speakers to the audio components (tube gear and turntables are especially susceptible), and (3) long interconnect cables between sources or preamps and power amps. In such cases, the long power cables (4) that feed the two monoblocks usually have to cross the long interconnects, increasing hum (5).

LOUDSPEAKER POSITIONING 251

Likewise, line-level signal interconnects should be shorter than speaker cables. Positioning monoblocks behind speakers shortens the speaker cables considerably, which is a nice benefit, but it lengthens the interconnect cables between the source and the amps significantly, typically from 0.5m to 2-3m, and that is bad, bad, bad!

Finally, there is also a problem of very long power cables (4) needed to feed the two monoblocks. Those usually must cross the long interconnects, increasing the prospect of hum development and injection.

> **DELICATE SIGNALS SHOULD NOT TRAVEL LONG DISTANCES**
> The lower the signal level, the shorter the cables between the components of an audio system should be.

How to find the optimal spacing between loudspeakers

The speaker positioning methods that we will analyze shortly are just suggested starting points for further experimentation and optimization. The final placement of your speakers and the listening position will depend primarily on your speakers and the size of your room.

Small or narrow speakers are easier to place than large or wide panel speakers, which could be overpowering in a smallish room.

With speakers too close together, the center of the stereo image is "crowded," the energy is concentrated in the middle, so there's no spatial information on the positioning of instruments and performers. It almost sounds like you are listening to a mono source or a single central speaker.

The overall dispersion pattern also depends on the design of each loudspeaker, so this is one aspect where superior speakers outshine the mediocre ones.

The smaller the difference between the spatial projections of various frequency bands, the more focused and cohesive the speakers will sound and the easier it will be to determine the optimal sweet spot for your listening position.

With speakers too far apart, the center of the stereo image is "sucked out," lacking energy and solidity. Instead of a realistically distributed sound stage, you are predominately hearing two speakers as easily identifiable sound sources, with no cohesion or integration between them and very little "center stage."

The proper spacing ensures that spectral energy is distributed optimally and uniformly, resulting in rich spatial information on the positioning of instruments and performers.

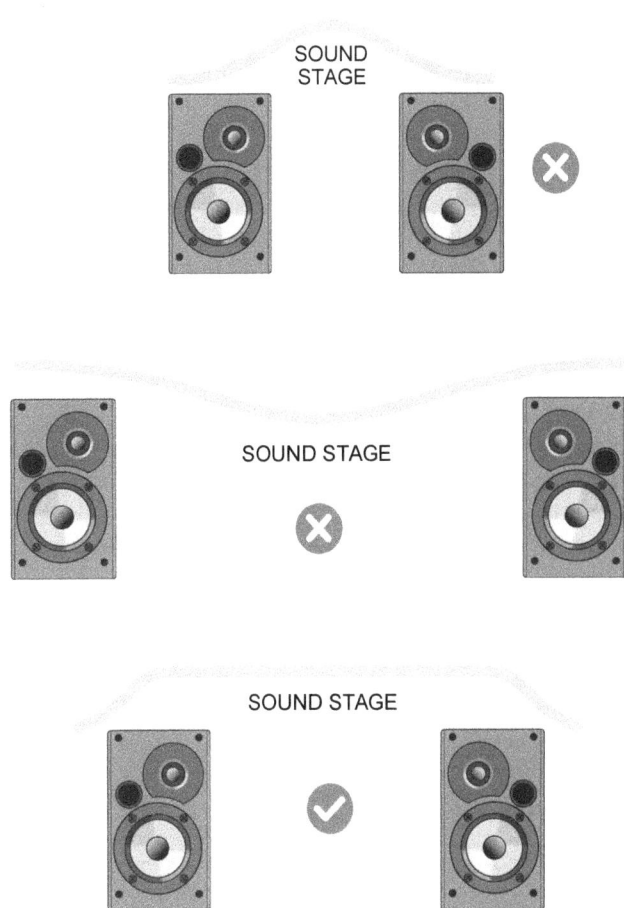

Fig. 11.3. Apart from the optimal longitudinal setup (distance between the speakers and the listening position), a transversal (between speakers) distance is crucial for a superior soundstage and imaging

Loudspeaker radiation patterns and their impact on the choice of the listening position

As you move your listening position forward and backward, away from the speakers, you may notice that the tonal balance of the sonic presentation changes. Loudspeakers project different frequencies differently. Woofers, midrange drivers, and tweeters become focused at varying distances from the speakers.

This is illustrated on the next page), with the whole audio spectrum is split into three main frequency bands, the highs (the treble), the mids (midrange), and the lows (the bass).

In accordance with their wavelengths, the lows (large wavelengths) are projected further than the mids (shorter wavelengths) and much further than the treble (the shortest wavelengths). Ideally, all frequencies would be "in focus" at the same distance from the speakers. Still, real-world loudspeakers are the least perfect of all audio components, except perhaps many listening rooms, which, acoustically, leave even more to be desired.

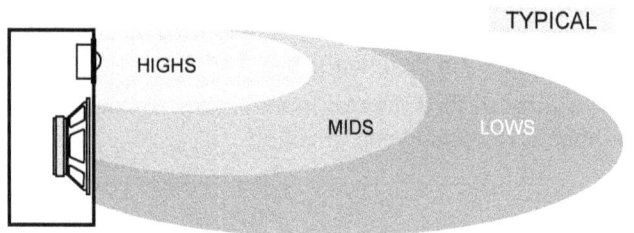
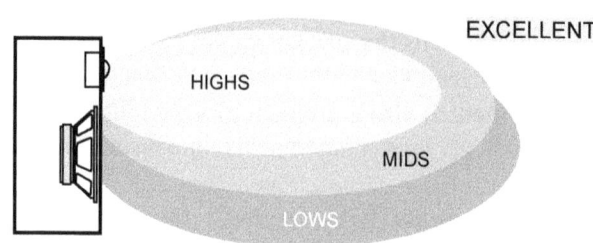

Fig. 11.4. A typical loudspeaker shows significant beaming and difference in projection patterns between various audio bands.

Fig. 11.5. In excellent loudspeakers, the differences between the projected distances of different frequency bands are minimized.

How to find the required height of speaker stands for bookshelf speakers

If you have just bought a pair of bookshelf speakers but have no stands for them, you shouldn't buy just any speaker stand. Apart from their looks, the material they are made of, and the price, the two most important factors are the stands' height and solidity.

"Solidity" means the stands should be stable, rigid, and as heavy as possible, so they don't easily transmit vibration from the speakers to the floor.

A proper height ensures that your ears are at the same level as the tweeters in your speakers when you are seated in the listening chair or sofa.

To determine the required stand height, you'll need an assistant and either a measuring tape or a laser distance gauge. You should be seated in your normal listening position so your assistant can measure the vertical distance from your ears to the floor (height H1 in the illustration).

Then, measure the distance H2 from the center of the tweeter to the bottom of the speaker box, including any feet or spikes that will be used. Subtract H2 from H1, and you have the required height of the speaker stands, HSS = H1 - H2.

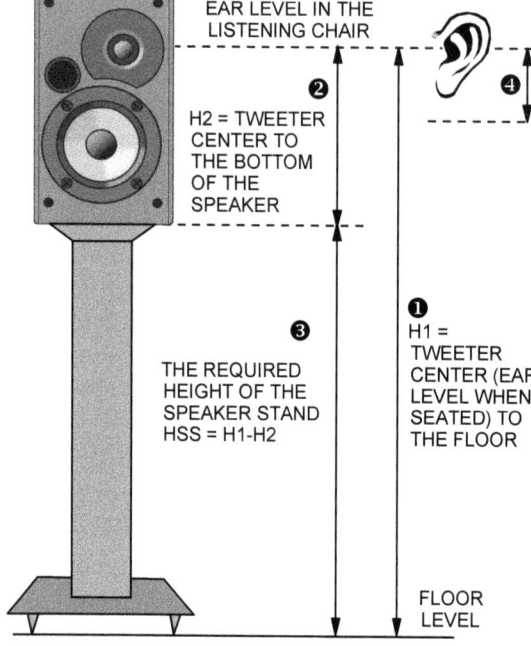

Fig. 11.6. How to determine the required height of speaker stands for a particular bookshelf speaker

Experiment with various speaker or listening chair heights

Just like most things in the audio field in general and acoustics in particular, the rule that the listener's ear should be at the tweeter level is not a "hard-and-fast" or strict requirement, rather a suggestion or a starting point for further experimentation. In quite a few cases, better results have experimentally been obtained by lowering the ear level below that of the tweeter's center, somewhere in the range marked with (4) in the diagram above, usually above the acoustic center of the woofer.

Such variation can be done by slightly changing the height of the speaker stands (some have adjustable feet or the height of the central column) or by trying out different listening chairs or chairs of varying heights.

Speaker time alignment experiments

Most multi-driver speaker boxes are of the straight front arrangement, as illustrated (next page). DT is the distance between the ear and the tweeter's center. We are assuming that the ear is at the same height as the tweeter, H. Because the center of the tweeter is some distance TW away from the center of the woofer, the distance between the ear and the woofer (DW) is not the same as DT! We have a right-angle triangle in play where $DT^2 + TW^2 = DW^2$ (Pythagora's Theorem).

The distance from the woofer should be measured not from the front plane of the speaker but the recessed point of the woofer's cone (DWR), its acoustic center. The triangle is not right-angled anymore, but for the time being, we will simplify our discussion and use DW, which is slightly shorter.

Since the tweeter is much smaller and thinner than a woofer or midrange driver, despite being mounted on the same vertical front surface of the speaker box, the tweeter's acoustic center is not in the same plane but forward of the woofer's acoustic center (distance δ).

LOUDSPEAKER POSITIONING

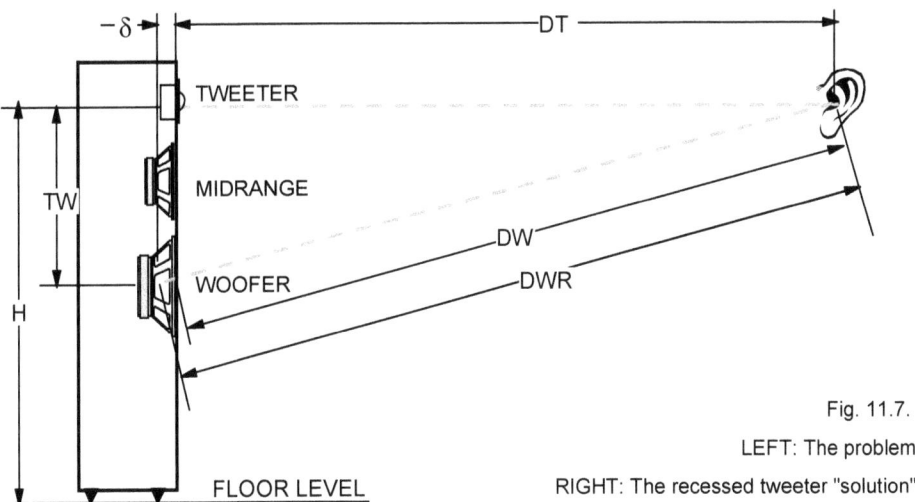

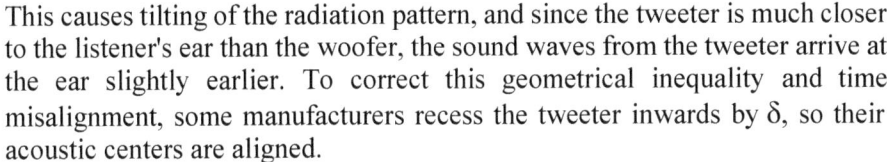

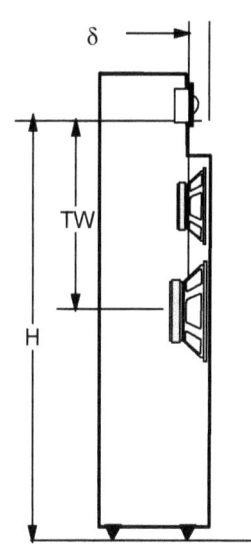

Fig. 11.7.
LEFT: The problem
RIGHT: The recessed tweeter "solution"

This causes tilting of the radiation pattern, and since the tweeter is much closer to the listener's ear than the woofer, the sound waves from the tweeter arrive at the ear slightly earlier. To correct this geometrical inequality and time misalignment, some manufacturers recess the tweeter inwards by δ, so their acoustic centers are aligned.

However, that creates new problems, such as possible diffractions around the sharp step. Others slant the whole front of the speaker box so that the tweeter is tilted backward and the woofer forward.

Tilting your speakers backward is relatively easy to do. Just insert a shim of some sort (a thin piece of plywood, cardboard, or plastic) under the front feet or spikes. Of course, you have to do it for both speakers - experiment with various thicknesses, usually from 5 to 20mm. If your spikes are adjustable, even easier, make the front ones longer than the rear pair and fine-tune them by ear.

However, now the acoustic axis is tilted upward, and the ear is off-axis relative to the acoustic centers of all drivers. You will notice a difference in the sound stage, in its height, and the overall "composition" of the sound, but the change may not be for the better. Sometimes you will like the change, other times you will not. Again, it all depends on other factors, on the design of your speakers and the possible electronic time alignment through the speakers' crossover, if and as implemented by their designer.

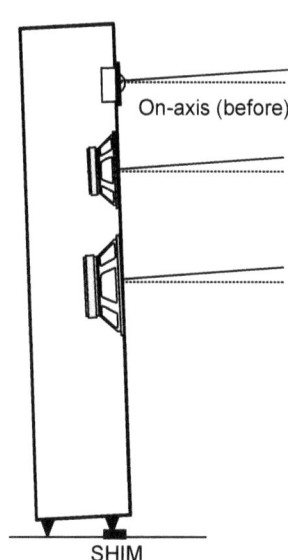

Fig. 11.8. Tiltback DIY solution. Experiment with shims of various thickness (5 mm - 20 mm).

Speaker toe-in experiments

The alignment of the speakers' center lines in the longitudinal direction is even more important than the vertical alignment just discussed. Once you position your speakers in the listening room and determine the best distance for the seating position (the sweet spot), experiment with the degree of speaker toe-in. Start with the no toe-in extreme and gradually keep adjusting it inwards. Eventually, the sound stage will enlarge, widen, and deepen at a certain setting, meaning you have found the optimal angle.

The first speaker arrangement (below) shows no toe-in, the speakers' center lines are parallel to each other and the side walls. Usually, this isn't the best arrangement from the soundstage point of view. The other extreme is to increase toe-in so much that speakers' center lines cross at the sweet spot. Usually, that is even worse. A healthy medium is generally optimal.

Obviously, you must toe in both speakers by the same angle. If you are fussy, buy a laser distance measuring device to position the speakers and the "sweet spot" precisely. Still, usually a ruler or a measuring tape is good enough.

Again, you will notice a change in the sound stage, especially its width, but its depth may also be affected.

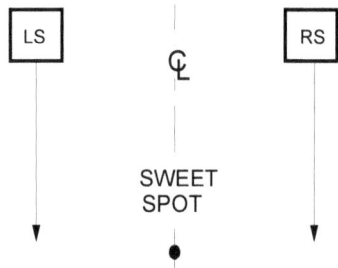

Fig. 11.9.
ABOVE: No toe-in
BELOW: Maximal toe-in

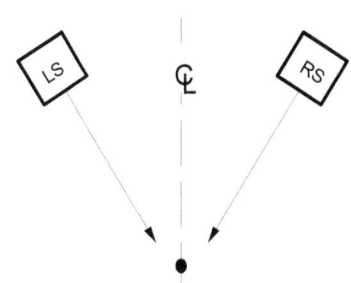

Once you find the toe-in that sounds the best to you, mark it somehow, either on a piece of paper ("the positioning plan") or on the floor. That way, if you have to move your speakers in the future for repair or just to clean the room or polish the floor, you don't have to go through the same process again. This experimentation takes time, so many audiophiles cannot be bothered, but the improvements can be significant.

SPEAKER PLACEMENT CHALLENGES - ROOMS OF IRREGULAR SHAPES, FIREPLACES, SLOPING CEILINGS AND OTHER PITFALLS

Fireplaces

A fireplace with a chimney is effectively a tuned resonant chamber with an open pipe on top of it. The issue isn't that serious with the fireplace centered in front, between the symmetrically positioned speakers, or at the back wall. A fireplace to one side of the room, however, will definitely unbalance the sonics in the room laterally.

The solution? Close the fireplace opening with a rigid panel (if you want to keep it functional) or brick it up entirely (and have it plastered over). That way, it will not add its resonant artifacts to the sound of your listening room.

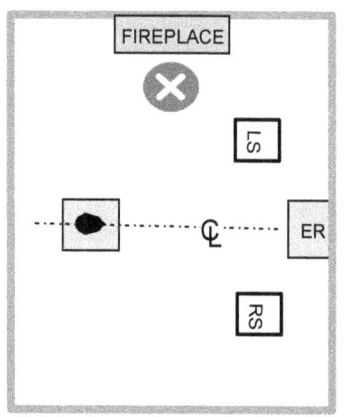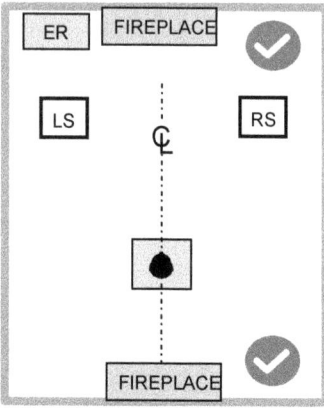

Fig. 11.12. To preserve the lateral (left-right) symmetry, try to have the fireplace either on the front or back wall.

Sloping and cathedral ceilings

Acoustically, flat ceilings are the safest and most predictable of all ceiling options. The other, more complex shapes are like boxes of assorted chocolates - you never know what you will get. In rooms with longitudinally sloping (slanted) ceilings, the loudspeakers should be placed at the end with lower ceiling height - the ceiling should be sloping down from you towards the speakers, not the other way around.

In rooms with symmetrical or "cathedral" ceilings, there are two possible orientations of the audio setup. In the longitudinal option, the ceiling is the highest at the center of the room, right above the listening position, and slopes down to the sides.

The transversal setup is similar to the sloping ceiling room previously mentioned, but in this case, after the initial sloping upwards from the speaker towards the sweet spot, the ceiling then starts sloping downwards. Each option has pros and cons. The ultimate choice may depend on other factors, such as the location of power points, windows, and doors. Again, experiment and make your own conclusions.

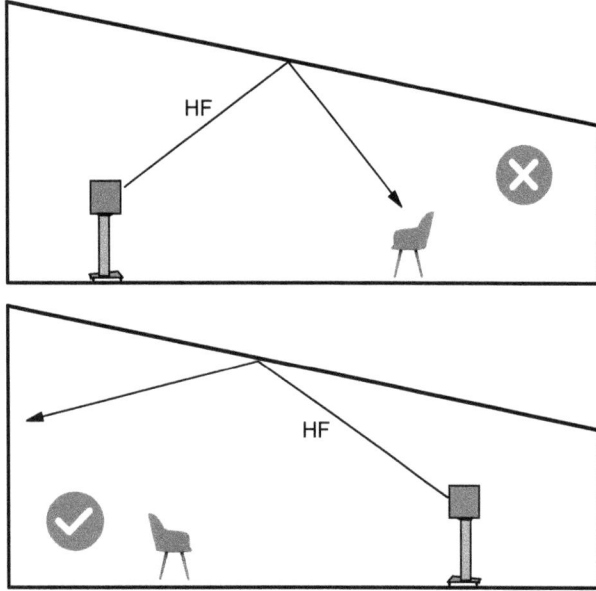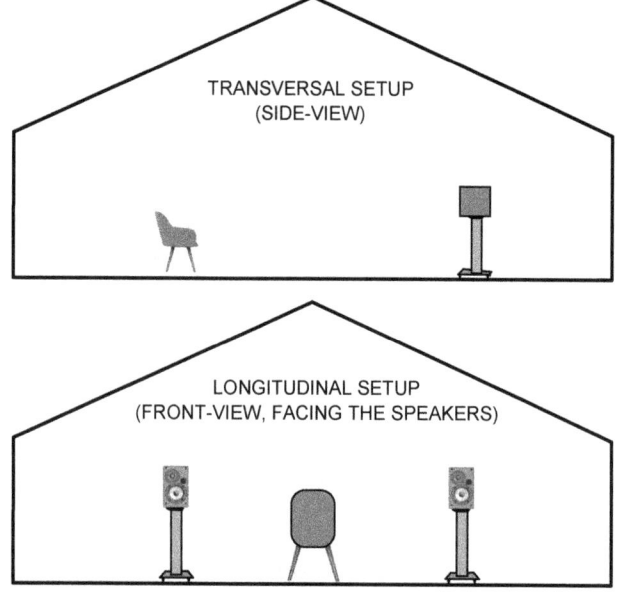

Fig. 11.10. In rooms with longitudinally sloping (slanted) ceilings, the loudspeakers should be placed at the end with lower ceiling height in order to manage early reflections off the ceiling towards the listening position.

Fig. 11.11. In rooms with symmetrical or "cathedral" ceilings there are two possible orientations of the audio setup. Each has pros and cons so the choice may depend on other factors, such as the location of power points, windows and doors.

LOUDSPEAKER POSITIONING

Very long rooms

Over the years, I've come across a few very long and narrow listening rooms. Despite being relatively narrow compared to the length, the width usually wasn't an issue; the room would usually be wide enough, at least four or more meters across. The problem was the length of eight or even ten or more meters in some cases.

Thus, their length to width ratio would fall way outside the acceptable proportions range. A solution is simple in principle but complicated and expensive in practice, a partition wall across the room's width.

Such a wall would bring the awkward proportions of the listening room back into the acceptable range, and thus improve its sonics, and bring a couple of additional benefits.

Shelving for the audio components could be inbuilt into the wall, thus obviating the need for an expensive equipment rack (ER). Such savings would go some way towards paying for the cost of the wall.

The partitioned space could be used as storage for audio equipment and could even incorporate shelves for CDs, LPs, and reel-to-reel tapes. Some audiophiles have even used the partition wall as a speaker baffle by installing their (almost always DIY-type) speakers onto it.

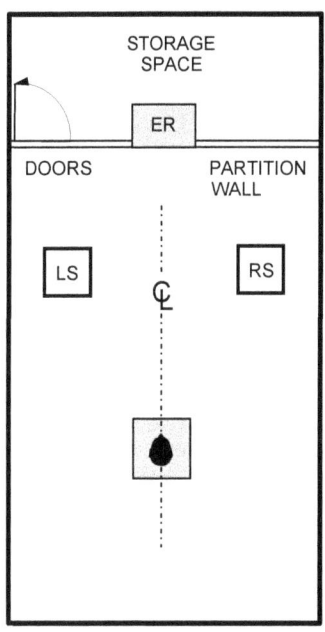

Fig. 11.13. (ABOVE) A partition wall can bring the awkward proportion of a listening room back into the acceptable range

Rooms of irregular shapes

It is possible to achieve a great sound even in highly irregular or asymmetrical rooms. Each case has to be analyzed and considered on its own merits, but certain common underlying principles hold in most cases.

If the irregularity (an alcove, a recess, or a bar, as illustrated) is at the back wall (behind the listener), the sonic impact will be minimal (just like the front-facing fireplace already mentioned).

Having irregularities on one or both sides is more serious and would be hard to compensate for by using acoustic treatments.

In irregularly-shaped rooms the speakers should be placed in the symmetrical part, with the listening position in the irregular end and not the other way around. No sharp corners or any other objects should be in front of either of the two speakers, as illustrated below right, to minimize the damaging early reflections.

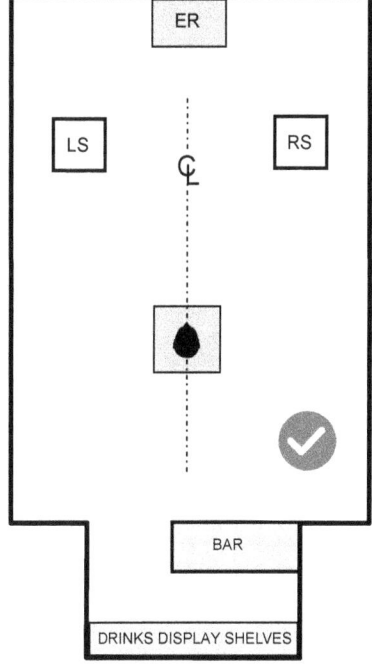

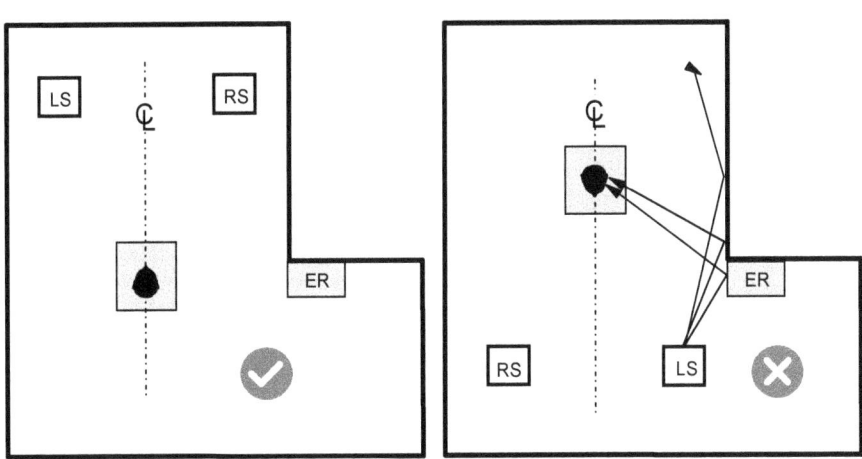

Fig. 11.14. (ABOVE) A typical case in older houses, "games" or "entertainment" room with an inbuilt recessed bar.

Fig. 11.15. LEFT) Irregularly- shaped listening room and two possible listening arrangements. The one with speakers placed in the "regular" part of the room usually sounds better.

A diagonal setup for square or close-to-square rooms

Rooms whose length and width are identical or less than half a meter either way are not irregularly shaped, but they create acoustical problems nonetheless. With its length-to-width ratio of 4.14/3.8 = 1.09 being so close to a square shape, our lounge room lends itself to another viable yet rarely used option, namely a diagonal setup.

One of the speakers (in this case, the left speaker LS) is closer to its adjacent wall (1), so the left-right symmetry is not achieved. However, the listening tests could not identify any impact on the sound stage, and the lateral (left-right) balance seemed fine.

Notice that the left speaker LS would probably sound slightly louder in a closed room due to its proximity to the adjacent wall, resulting in stronger side and rear reflections (rear-firing bass reflects ports on our speakers). However, while the RS is firing straight into the sidewall (2), the LS is aimed at the back of the room (3), which is open to the formal dining area, and as such, the effective area and the room volume it must cover is significantly larger, a compensating effect.

The equipment rack could be positioned alongside any of the three walls, with the illustrated position being the best. Our power points were on the front wall between the two windows, so the chosen rack location minimized the length of power cables, too.

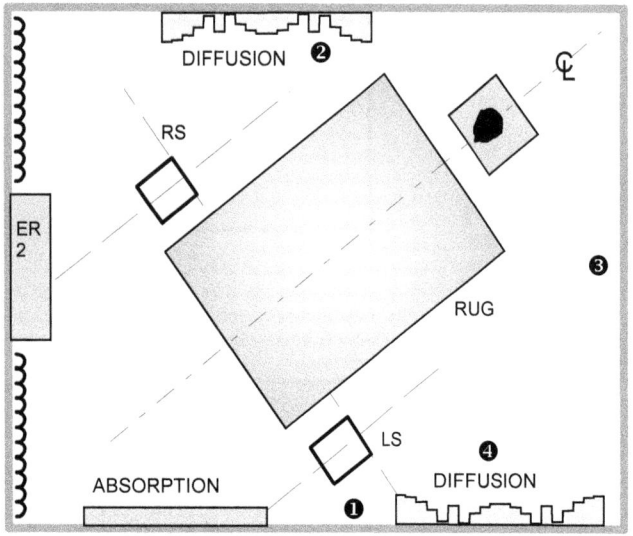

Locating the equipment rack against either side wall would bring it too close to the speakers and cause sonically damaging early reflections.

Also, one of the speaker cables would have to be much longer than the other, or, if made the same length, both cables would have to be unnecessarily long.

Diffusers on side walls (2) and (4) are usually necessary, as bass traps and absorption panels may be behind the speakers. We have been conditioned to symmetrical rooms and furniture placements within them.

So, while acoustically viable, the diagonal setup is disliked by many, most likely because it is also visually unbalanced; to me, it just does not feel right.

Fig. 11.16. A diagonal setup is a rarely used option. If not acoustically, it is undoubtedly visually unbalanced.

FORMULA-BASED SPEAKER PLACEMENT METHODS

How to avoid disastrous speaker placements

The worst position of a loudspeaker is an equidistant placement from the front wall (behind the speakers), the sidewall, and the floor. Since the most troublesome are the low-frequency artifacts such as resonances and standing waves. These distances are measured from the acoustic center of the woofer, not the tweeter or the whole speaker box.

Except for very large rooms (or those with low ceilings), the distance from the woofer to the ceiling is much larger than those three distances mentioned, so it is usually of no concern in this regard. The suggested speaker positioning ratios that follow comply with this rule of avoiding equidistant placement ($S = H = F$).

The equilateral ("Golden ratio") triangle speaker positioning method

Quite a few speaker positioning methods have been developed by speaker designers and manufacturers, audio experts, and audiophiles. We have found the two arrangements that follow to be close to optimal in most cases, especially when using relatively small speakers in small- to medium-sized listening rooms.

The two methods are very similar, in fact, identical in all aspects except the distance between the listening position ("sweet spot") and the front of the speakers (the "X-Y" distance).

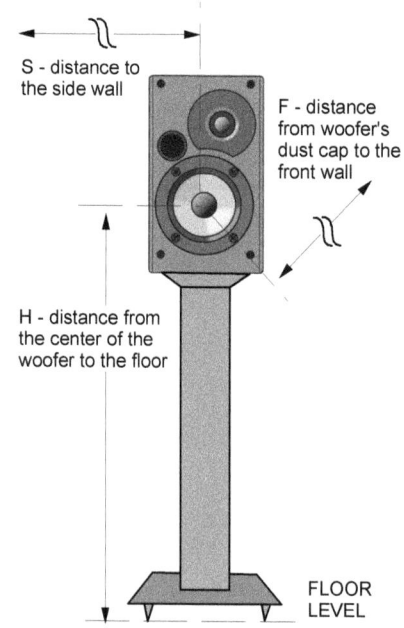

Fig. 11.17. Make sure the distances between the woofer's acoustic center and the three closest surfaces (the sidewall, the front wall, and the floor) are as different from each other as practically possible.

LOUDSPEAKER POSITIONING

We are assuming an equilateral triangle between the speakers and the "sweet spot" (all three sides A meters in length). This "A" corresponds to the "1" unit in the Golden Ratio. The distance between each speaker and the non-adjacent sidewall (the one further away) is then 1.618*A.

First, measure the width of your listening room (W). Since W=2.236*A, divide W with 2.236. That is your "A".

Our room is W = 3.93 m wide, so A = 3.93/2.236 = 1.758 m, meaning our speakers should be 1.76 m from the front wall. The sweet spot (the listening position) should be 0.866*A = 0.866*1.76 = 1.52 m from the front line of the speakers (the X-X line).

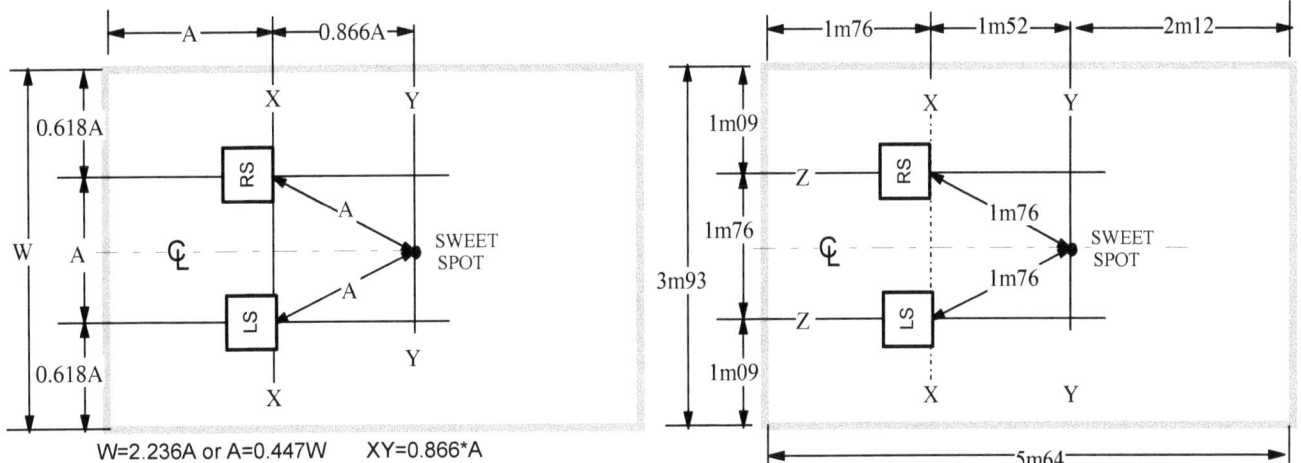

Fig. 11.18. The simplified shape of our listening room and the equilateral triangle speaker positioning distances

The isosceles triangle speaker positioning method (Cardas version)

Regardless of which method you use, the sweet spot is always on the centerline between the two speakers and the side walls (the setup must be symmetrical!). According to the method just outlined, its distance from the X-X line is 0.866A, so the three distances (between the speakers and between each speaker and the sweet spot) are all "A", thus forming an equilateral triangle.

Alternatively, as per the recommendation of Cardas Audio and many other audio experts, the distance from the X-X line should be 1.5A, so an isosceles triangle is formed.

The two X-Y distances, obtained by the equilateral and the isosceles methods, are extremes; the best placement is usually somewhere in between. Start with one or the other and then gradually move towards or away from the speakers (along the central line CL) until the best soundstage is found. Refer to "How to find the optimal spacing between loudspeakers" on page 255.

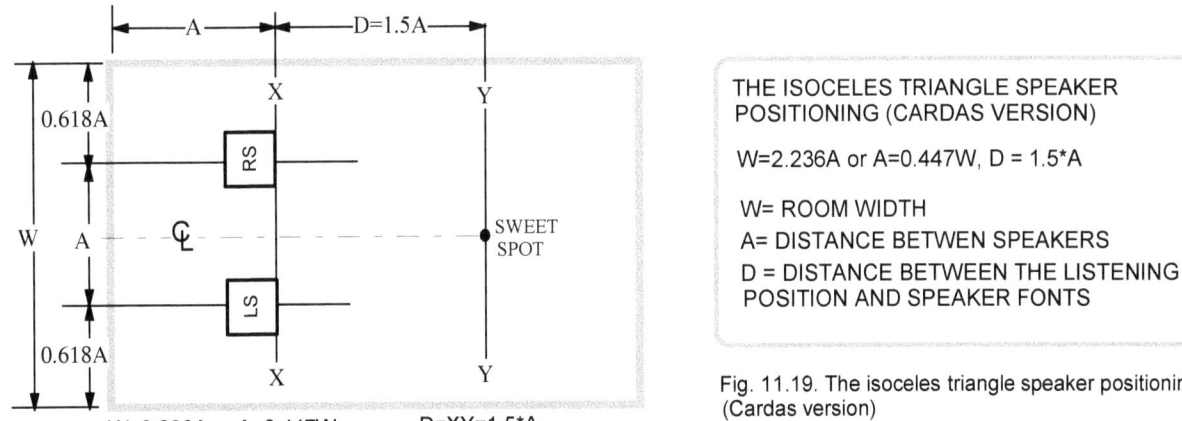

THE ISOCELES TRIANGLE SPEAKER POSITIONING (CARDAS VERSION)

W=2.236A or A=0.447W, D = 1.5*A

W= ROOM WIDTH
A= DISTANCE BETWEN SPEAKERS
D = DISTANCE BETWEEN THE LISTENING POSITION AND SPEAKER FONTS

Fig. 11.19. The isoceles triangle speaker positioning (Cardas version)

Bass-midrange-treble: what to listen for while positioning speakers

Now that we understand the basic tenets of speaker positioning, what should we listen for while fine-tuning their position? Different system integrators and audiophiles use different approaches. Most agree that the bass should be dealt with first. Once the speakers are positioned so the bass sounds clean and tight, we can experiment with fine-tuning to get the midrange to sound as good as possible.

Fig. 11.20. The room tunining sequence. Position the speakers to get the best bass response (1), then use the midrange to optimize the soundstage and imaging (2), finally turning your attention to treble (reflections and harsh sound)

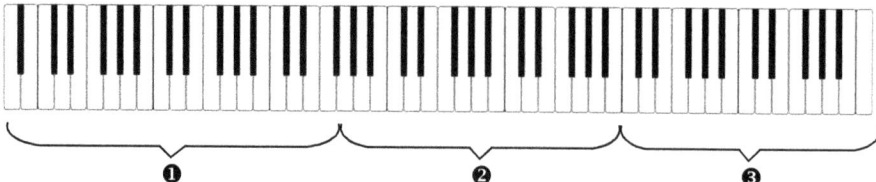

Here we pay attention to the soundstage dimensions, listener envelopment, instrument/performer localization, and tonal balance.

Generally, the treble is also fine if the bass and midrange are optimized, although there may be troublesome early reflections to deal with. The treble can also affect both the tonal balance and the soundstage (the sense of envelopment), so, unfortunately, the three steps are not independent but interactive. To the frustration of all involved, in some rooms and with some systems, the speaker positioning that yields the best bass does not make the midrange and treble sound good enough, so a compromise is required.

> THE SPEAKER PLACEMENT CURSE
> The optimal position for bass response is not necessarily ideal for the widest, most spacious soundstage and best imaging.

The elliptical speaker positioning method

Joachim Gerhard, the founder of the German loudspeaker company Audio Physic, advocates this relatively simple positioning method. The speakers are located on the longitudinal centerline CL1 in the said ellipse's two foci (focal points). The listener's head is on the vertical (transversal) centerline CL2, about 30 cm from the back wall, meaning you'd be sitting in a chair whose back is right against it!

All that's needed is to determine the location of the two loci of the ellipse, where the speakers will be placed, i.e., the distance "X" on the diagram. For an ellipse, $X = \sqrt{Z^2 - W^2}$ where Z is half of the room's length (the longer dimension) and W is half of the room's width (the shorter dimension).

For our listening room those figures were $Z = 5.64/2 = 2.82$ m and $W = 3.93/2 = 1.965$ m, yielding $X = \sqrt{(2.82^2 - 1.965^2)} \approx \sqrt{4} = 2.0$ m. This means each speaker is only $Z - X = 0.8$ m from the adjacent sidewall.

It is claimed this arrangement delays the first reflections (of the side walls), helping our auditory system to delineate between the direct sound and the early reflections, so "the best soundstage and tonal balance are achieved." Two reasons are given for placing the listening position near the back wall. All bass frequencies have an "anti-node" there (the maximum sound pressure), increasing the "feeling of deep bass."

Secondly, it is claimed, the reflections off the back wall, although still present, cannot be perceived by our brain since the short distance involved (between the rear wall and our ears) is smaller than the circumference of the human head. If the brain cannot measure such time delay, it can't locate the source of such reflections and just ignores them.

Our trials of this method and similar placements (we found that the "2X" spacing between speakers was way too wide) were not favorable, to say the least. Despite such a considerable depth "W" behind the speakers (almost 2 meters in our case), for some reason, the sound stage was flat, had no depth at all!

Also, sitting so close to the rear wall felt unnatural and made us somewhat uneasy and restless. We didn't notice any improvement in the bass levels, either.

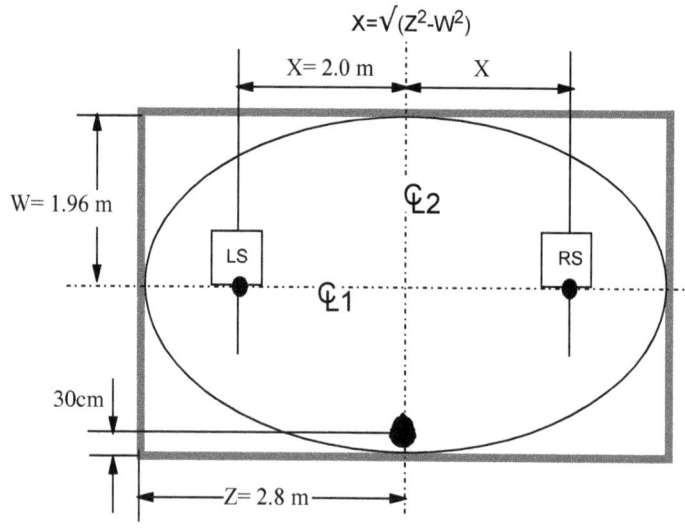

Fig. 11.21. The elliptical speaker positioning applied to the simplified shape of our listening room

12 OPTIMIZING THE ACOUSTIC PERFORMANCE OF YOUR LISTENING ROOM

The audiophile system is only as good as its weakest link, which is often the listening room. The acoustic behavior and thus the sonic signature of small listening rooms are very different from that of large concert halls.

This chapter will focus primarily on the room's bass (low frequency) response and the associated problems such as room modes (resonances) and standing waves. The next chapter, "Acoustic treatments," will deal with bass traps used to tame such unwanted bass artifacts and with other types of acoustic panels.

Two practical examples of listening rooms are analyzed together with two different setups, the longitudinal and transversal arrangement.

- THE FINAL TRANSDUCER OR THE LAST LINK - THE LISTENING ROOM
- ROOM DIMENSIONS, PROPORTIONS AND RESONANT MODES
- CASE STUDY: OUR LISTENING ROOM
- CASE STUDY: FORMAL LOUNGE AS A LISTENING ROOM

> "Stradivarius, in particular, was the most amazing craftsman and one of the great artists and scientists that ever lived because he figured out something with the sound and the science of acoustics that we still don't understand completely."
> Joshua Bell, American violinist

THE FINAL TRANSDUCER OR THE LAST LINK - THE LISTENING ROOM

Hard and soft rooms - the two sonic extremes

From experience, most listening rooms either sound "too harsh, "cold," and "clinical," or they sound dull and "dead." The first symptom indicates too many solid surfaces in the room, for instance, a tiled floor, bare walls or exposed windows (without curtains or blinds), or any solid objects or surfaces that reflect sound waves. This is often the case in warmer parts of the world, such as in the Mediterranean and in the tropics.

The second extreme indicates the opposite situation; too many soft surfaces, furnishings, carpets, curtains, and other plush objects absorb sound waves, especially in the mid-and high-frequency bands. This usually happens in colder climates, where, for instance, it's less likely for "cold" floor surfaces (tiles) to be used, or, if they are, they are almost completely covered with carpets and rugs.

Generalizing is always a difficult and often a dangerous practice. If I had to assign numbers, around 80 percent of the problematic rooms I've seen and listened to in Perth, Australia, where I live, are bright and boomy. Noticeable bass resonances resulted in muddy, undefined, "one-note" bass, and uncontrolled higher frequency early reflections made the treble harsh and unpleasant.

These early reflections also impact the timing and spatial clues, resulting in an unstable and poorly defined soundstage. Obviously, there were too many hard, reflective surfaces in such rooms and not enough absorption on the floor and the walls. Some larger rooms even had a prominent echoing.

About 20% of the listening rooms had the opposite problem: too many soft furnishings absorbing mid and high frequencies, making those rooms sound dull and lifeless. In all such cases, the whole floor was carpeted, and thick curtains covered many wall surfaces, often an entire wall or two. Throw in a couple of large plush 3-seater sofas with high backs, and you have an acoustically "dead" listening room.

The weakest link syndrome and finding a balance between reflective and absorptive surfaces

Therefore, from the dead room - live room perspective, your first task in any room is to strike an optimal balance between the reflective and absorptive surfaces. That does not have to involve installing expensive professional acoustic panels and bass traps, although it may be required in some problematic cases.

We have a floating timber floor on top of a soft underlay and concrete floor in our music room. Initially, the sound was too harsh for my liking. Once we added a $300 (3 x 2 meters) thick wool rug in front of the speakers, all was well again.

The sound became soft and warm; the rug completely tamed down the offending reflections of the floor. The DIY absorptive panels to the sides of the speakers (to tame the first reflections), as featured in a few photos in this book, were not needed. They are in those photos simply for illustrative reasons.

Another automotive analogy is in order. An amplifier is akin to the engine of a car. The speakers can be likened to the transmission, wheels, and tires. You may have the best engine, but the car will not drive well if its transmission isn't optimized. As the ultimate acoustic transducer, the room is the equivalent of the road on which the car travels. Even the best car and the best tires will lose their grip on a lousy road. I call it "The Weakest Link" syndrome:

> THE WEAKEST LINK SYNDROME:
> The audiophile system (and that includes the listening room) is only as good as its weakest link, which is very often the said listening room.

The difference between small listening rooms and large rooms (concert halls)

The diagram (next page) illustrates a typical echogram of a large room or a concert hall. Direct sound and its amplitude D undergoes no reflections. Still, in comparison to the SPL at the sound source (amplitude S), it is attenuated, so the sound at the listening position has a reduced amplitude (lower SPL). This attenuation follows the so-called "inverse square law." Doubling the distance between the source and listener results in a square drop in amplitude - twice the distance, four times lower the SPL.

Sometime after time t_T (transit or travel time needed for the sound to travel from its source to the listener), the first reflection R1 arrives. The time delay between the direct sound and R1 is a critical parameter of a room, called the Initial Time Delay Gap or ITDG.

In great-sounding concert halls, ITDG is in the order of 20 or so milliseconds (0.02 seconds), but in small listening rooms that we audiophiles use, it is pretty short, typically only 1-3 ms.

The exception is the LEDE approach (Live End - Dead End), used in recording studios and some listening rooms, where higher ITDG figures can be achieved, approaching those of live concert halls. The front part of the room (where the speakers are positioned) is made as "dead" or absorbent as possible to minimize reflections. The back of the room, behind the listener, is made as reflective and diffusive as possible ("live").

The first reflection is followed by a series of discrete reflections of various surfaces, the walls, floor, and ceiling. This is followed by a third component, the late reflections that compose the so-called reverberant field. The amplitudes of such echoes reduce or decay exponentially.

In the diagram, the reduction seems linear, and it is because the vertical axis is amplitude in dB (decibels) which is a logarithmic unit.

The direct to reverberant sound ratio will depend on the distance between the listener and the sound source. The distance at which the direct sound and the reverberant sound are of equal level (loudness) is called the critical distance d_C.

This parameter is of interest in large rooms only, such as auditoria and concert halls; it is usually larger than the longest dimension of a typical home listening room.

In the rooms of interest to us in this book, the reverberant field (echoes) doesn't develop; the rooms are too small for such a relatively uniform field to exist. Thus, we are left with the discrete early reflections, which are behind many acoustic problems of listening rooms.

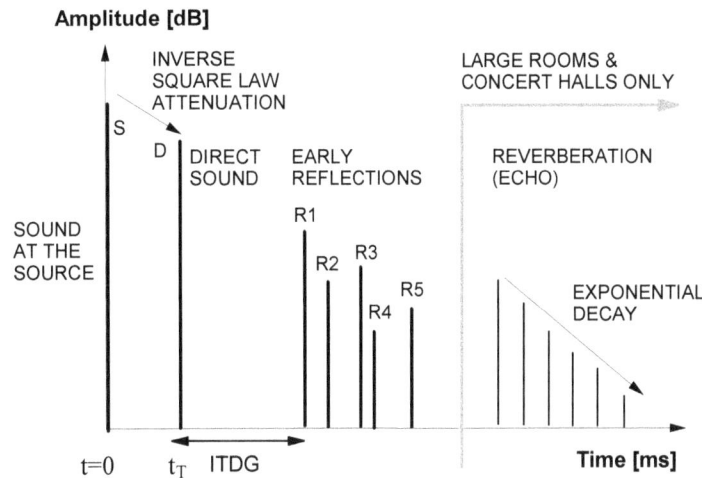

Fig. 12.1. The idealized (in principle) acoustic response (echogram) of a large room

ROOM DIMENSIONS, PROPORTIONS AND RESONANT MODES

What makes a well-proportioned listening room?

Humanity has been grappling with the question of the optimal room dimensions for millennia, not just in terms of their acoustics but also regarding the health benefits they bring their occupants and other unrelated criteria. Various "optimal" designs have been suggested. Let's have a brief look at three such examples.

The simplest has proportions 2 : 1 : 1 (length : width : height). These were often used in Egyptian and ancient Greek temples, also later on in Romanesque and Gothic churches.

For a 6 m long listening room, the room width and ceiling height would be 3 m. However, due to standing waves and room modes (more on which soon), it is best to avoid having two identical room dimensions, for instance, the width equaling the height or the length the same as the room width!

The second set of numbers is 2:1:1.118, used for the shape of "King's Chamber" in the Great Pyramid of Giza. For a 6 m long listening room, the width would be 3 m, and the ceiling height would need to be 3.354 m.

"Golden Solid" or "Golden Cuboid" proportions were used for Egyptian tombs and medieval churches. Φ (Greek letter "phi") is the "golden ratio." The dimensions progress in a 1 - Φ - Φ² order, also called the Fibonacci sequence. For a 6 m long listening room, the width would be 3.71 m, while the ceiling height needs to be 2.29 m. This is close to the often-used games room size in Australia of 6 x 4 x 2.2 m.

These exact ratios have been successfully applied to the dimensioning of musical instruments, loudspeaker boxes, and even the placement of loudspeakers in a room.

Apart from the very affluent and those of you planning to build your new home, most audiophiles have little or no control over the shape and the size of their existing listening room. However, that does not mean you cannot improve its sonics. This chapter and the next will show you how.

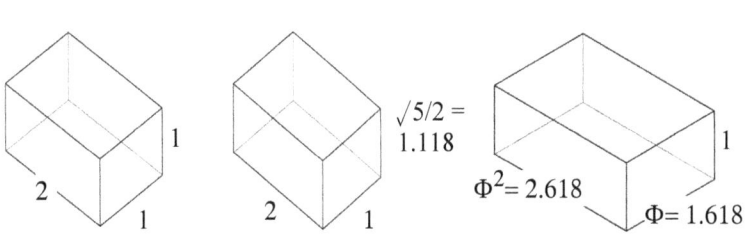

Fig. 12.2. Three chamber proportions commonly used throughout history

The importance of room height

The most unfortunate building trend has been the rapid decrease in room height. In Australia, the most common room height for houses built in the 1930s was around 3 meters. In the 60s, that decreased to 2.6 m, and currently, new homes have a ceiling height of only 2.4 m. This seems too low and depressing for living areas, but is it OK for a listening room? After all, you don't listen standing up, but in a sitting position? Let's consult the Golden Rule!

For a room with golden cuboid proportions, a width of 4 m and a length of 6.5 m, the ceiling height would need to be 4/1.618 = 2.47 m. Perhaps, when it comes to acoustics, modern builders do know a thing or two.

For our listening room (case study on pages 264-268), the width to height ratio is 3.93/2.914 = 1.35, while the length to height ratio is 5.64/2.914 = 1.94.

For our lounge room (case study on pages 268-270), the width to height ratio is 3.8/2.66 = 1.43, while the length to height ratio is 4.14/2.66 = 1.56.

Both fall just outside of the Bolt's "blob." The music room's length to height ratio is acceptable, but the width is too narrow compared to the height. Another meter or so (widening it from 4 meters to 5 meters) would bring its proportions into the recommended area.

The lounge room's case is the opposite, the width to height ratio is perfect, but the room is way too short; an additional meter or two in length would improve things.

Notice that even a "Golden Cuboid" proportioned room would fall outside Bolt's recommended ratios, a reminder not to lose too much sleep over these "recommendations."

Room's modal resonances and standing waves

Those with quality education and a good memory will remember high school physics lectures on standing waves. These mechanical or sound waves result from reflections from the walls, ceilings, and floors and are particularly pronounced or noticeable at low frequencies, below 300 Hz or so.

"Standing" means that the peaks and zeros (or "nulls") of a wave are not moving around the room but are always "standing" or appearing in the same spots. You may have noticed that if you move your listening chair one foot or even less forward or back, the tonal balance of the sound, especially the bass, will change.

If you were sitting at the spot where the incoming and reflected waves cancel one another, you experienced a weak bass. By moving slightly forward or back, you moved away from the "zero" spot, and the volume of the certain frequencies increased; the bass was not weak anymore.

If you were sitting in the spot where the incoming and reflected waves aided one another, you experienced a high volume of specific bass frequencies; you were sitting at the "peak" spot resulting in boomy, "one-note" bass. Moving slightly forward or backward "tamed" the bass.

However, you need to understand that no matter where you are sitting, there will always be one or more frequencies with a "zero" in that spot, and there will always be other frequencies that will have the peak at that exact spot.

Different frequencies have different wavelengths and, therefore, different distributions of their amplitudes in a room.

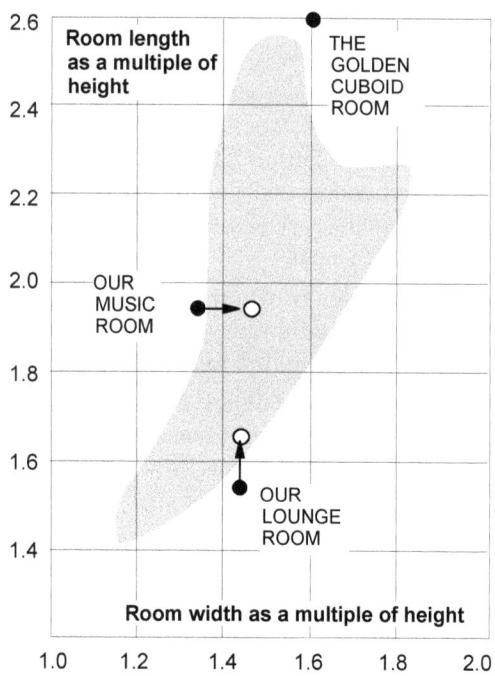

Fig. 12.3. The Bolt "blob" is a recommended specification of small room ratios, expressed here as length and width compared (normalized) to the room's height. The dimensions that fall within the shaded area are claimed to yield the smoothest frequency responses at low frequencies.

Fig. 12.4. The validity of the recommended ratios depends on the frequencies of interest and the room volume. For our listening room (65 m^3 in volume), the frequency range the recommended ratios relate to is 52 - 132 Hz. Adapted from Bolt, 1946.

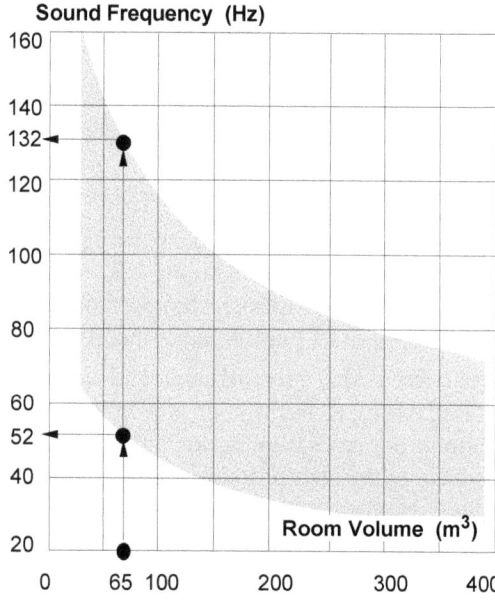

Axial, tangential and oblique reflections

The axial modes are reflections and standing waves between opposite walls and between the floor and ceiling. They are dominant (of highest magnitudes) and, as such, are of primary concern.

Tangential modes involve four surfaces in one plane (two pairs of opposite surfaces), such as all four walls (1-3-2-4), two walls (front & back), ceiling and floor, or two side walls, ceiling and floor. They are less significant but may become an issue in rooms with rigid walls, such as those made of brick or concrete, as is the norm in Western Australia, where houses are constructed using the double brick outside and single brick inside walls. The "brick & veneer" homes, as used in the rest of Australia and many parts of the USA, have only outside walls made of brick, while the internal walls use timber frame and "veneer" or plasterboard cladding.

The oblique modes are seldom an issue and can be ignored in almost all domestic listening rooms.

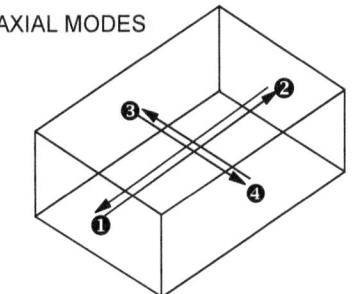
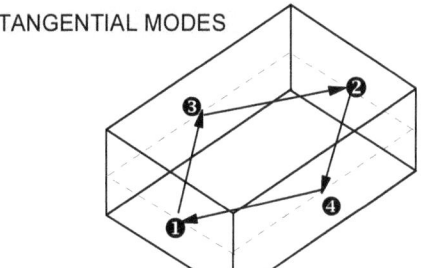
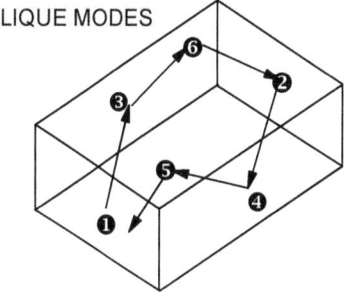

Fig. 12.5.1. Three axial modes (standing waves between two opposite walls): Between the front wall (1) and the back wall (2), sidewalls (3) and (4), and the ceiling and floor (not drawn to preserve clarity)

Fig. 12.5.2. Tangential modes involve two pairs of opposite surfaces: all four walls (1-3-2-4), two walls (front & back) + ceiling + floor, and two side walls + ceiling + floor. All reflections are in one plane (dotted line).

Fig. 12.5.3. The oblique modes involve all six surfaces, the four walls, floor (5), and ceiling (6). Only one of the many possible paths is illustrated.

Calculate and analyze your room's axial modes

Providing the room dimensions are known, calculating axial modes and resonances in a room isn't difficult. It can be done manually or in a spreadsheet. There are even online calculators, but only for regularly shaped rooms.

The fundamental formula is $\lambda = v/f$. λ (Greek letter lambda) is the wavelength (the distance between two identical peaks or zeroes of a periodic wave), v is the speed of sound in free air, and f is the sound frequency in Hz.

Since every room has three orthogonal dimensions (length, height, and width), standing waves appear in all of those dimensions. So, there are vertical standing waves, horizontal standing waves, and those propagating sideways (lateral or transversal standing waves).

Let's study an example of a 6.0 m by 4.0 m listening room which is 2.2 meters tall. The first three "modes," or the standing waves of the three lowest frequencies, are illustrated (next page), but there are higher-order modes, of course, the 4th, the 5th, and so on. We'll talk about longitudinal waves; the situation with the other two axes is analogous.

The thick line (marked "f_L") shows a sine wave that starts with an air pressure maximum (or peak or "anti-node") at the left wall, then has a null (or "node") in the middle of the room, and again raises to the maximum at the left wall (back wall from the listening position).

The wavelength of that wave L1 equals twice the length of the room or 12 meters. What is the frequency of that wave? Well, since $f = v/\lambda$ and the speed of sound at sea level in free air is around 340.3 m/s (meters per second), we get f = 28.4 Hz (Hertz is also known as cps - cycles per second or 1/s).

If you were sitting in the middle of that room, the frequency of around 28 Hertz would be seriously attenuated. Theoretically, in that spot, you would not hear that frequency at all. Such points are called "nodes."

The second mode in the longitudinal axis is depicted using a medium line and is marked "$2f_L$"). Two quarters (a quarter from each wall) and one-half of its wavelength fit the room's length L, so one whole wavelength equals 6 meters. Using the same formula, we calculate its frequency of 56.7 Hz or double the previous one, hence "$2f_L$". Instead of two peaks and one null, this standing wave has three peaks and two nulls.

> **THE FUNDAMENTAL ACOUSTIC FORMULA**
>
> $\lambda = v/f$
>
> λ = the wavelength (the distance between two identical peaks or zeroes of a periodic wave)
> v = the speed of sound in free air (meters per second)
> f = sound frequency in Hz.

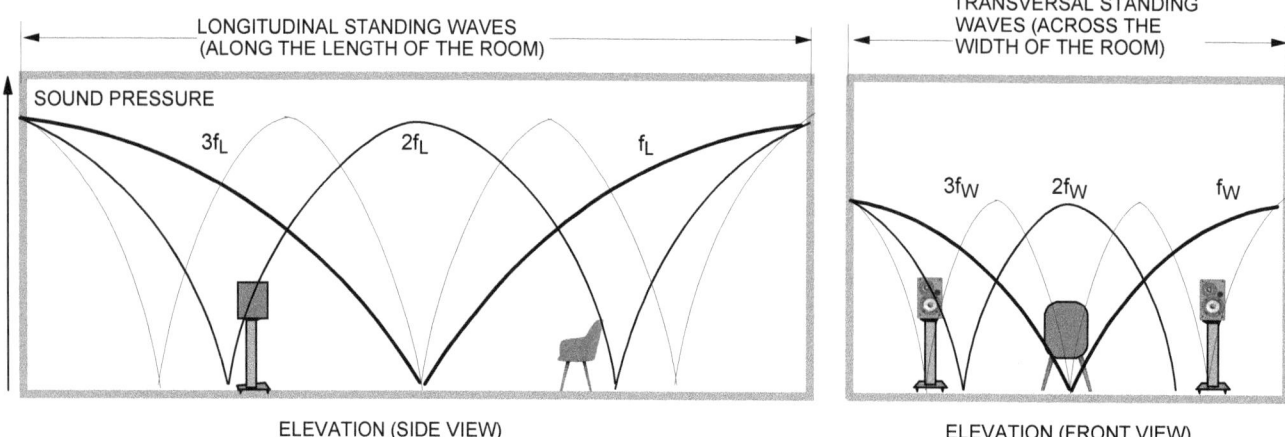

Fig. 12.6. The first three axial modes in the two of the three dimensions of a typical listening room (the length "L" and the widtgh "W". A similar situation happens in the 3rd dimension, the heaight ("H"). The speaker placement distances and the listener's position are for illustrative purposes only and have not been optimized.

The third wave (marked "$3f_L$") has four peaks and three nulls; 1.5 wavelengths fit into the room longitudinally, so the frequency of this standing wave is 85.1 Hz or three times the frequency of the first node.

So, you can now quickly and easily calculate any resonant frequency, determine how many peaks (anti-nodes) and nulls (nodes) it will have ("n+1" peaks and "n" nulls, where "n" is the node number), and at which points in the room those peaks and nulls will appear. So, the 9th node will have ten peaks and nine nulls, and its frequency will be nine times higher than the frequency of node #1, or $9*28.4 = 255.6$ Hz.

Let's apply this knowledge to a couple of real-life cases. Both examples are of listening rooms of small to medium size, but the issues discussed and the physical principles behind them apply to any size room.

CASE STUDY: OUR LISTENING ROOM

The shape and size

Originally part of one sizeable irregularly-shaped kitchen/dining/family room, this "entertaining" room was closed off by adding a brick wall (2) with a diagonal section featuring double French doors (1). Since the main storage area is in the corridor, on the other side of the sidewall (3), there was nothing we could do about the fact that two large recesses were created on each side, (4) and (5).

Together with the diagonal wall (1), they made the room asymmetrical, both in terms of the material used (glassed areas on the left and brick walls on the right side) and the shape.

This is a far from arrangement for a listening room since it will always remain sonically unbalanced to some degree, but is indeed a very common situation.

Most audiophiles don't have the luxury of designing an ideal listening room and have to live with whatever is available and make the best out of it. Many don't even have the luxury of a dedicated listening room and have to accept even more restrictions - the inability to add bulky acoustic treatments such as bass traps and less-than-ideal arrangements (bulky furniture, suboptimal and even awkward furniture layout, etc.)

While writing this book, I went through all past issues of the most prominent USA audio magazines, *Audio*, *Stereophile*, and *The Audio Critic*.

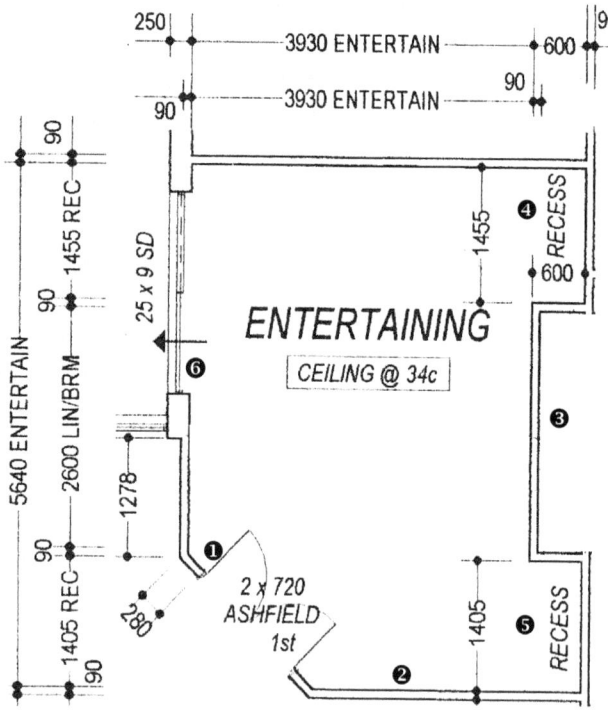

Fig. 12.7. The architectural plan of the "entertaining" room. "Ashfield" is a trade name for French doors with ten glass panels. "25x9" are sliding glass doors leading to the back yard (6). The internal walls are single brick with plaster (90 mm thick), the outside walls (only two short sections on either side of the sliding doors) are 250mm wide, double-brick construction with an air cavity.

OPTIMIZING THE ACOUSTIC PERFORMANCE OF YOUR LISTENING ROOM

In the early 1990s, *Stereophile* published a series of articles about their editors, equipment reviewers, and contributors. Each article included the sketch of the floor plan of the person's listening room, which I found of most interest. I anticipated a certain degree of criticism from audiophiles reading this book as in "You were supposed to be an audio expert, yet your listening room is far from ideal!"

To my relief, among a dozen or so setups, all the rooms were of irregular shape. Not a single room was completely symmetrical! Many featured a fireplace, awkwardly positioned doors and windows, and other features usually considered acoustically undesirable. A few were even of the dreaded L-shape, the bad boy of listening spaces, due to their asymmetrical sound field and propensity to add further standing waves to the equation.

Room resonance modes

To estimate the resonant frequencies of this room and the wavelengths of the standing waves across its three dimensions, we simplified things and assumed a rectangular shape with the length L=5.64 m, width W=3.93 m and the ceiling height of H=2.91 meters.

Subjectively, the short diagonal wall in one rear corner of the room does not make much of an acoustic difference and is not troublesome in any way; since the two recesses are almost entirely filled with inbuilt bookcases, the recesses seem to be innocuous, too.

Again, that fundamental wave L1 wavelength equals twice the room's length (2*5.64 m) or 11.28 meters, so its frequency is $f_{L1}= v/\lambda_{L1} = 344/11.28 = 30.5$ Hz. Now we can calculate the second mode $f_{L2} = 2f_{L1} = 61$ Hz, and all the higher ones.

Due to the room's relatively short length of 5.64 meters, there are no resonant frequencies below 30 Hz. There are no coincident frequencies either, which is good news. However, there are three cases of potentially problematic close pairings of modes, marked with an asterisk (*), 59.1 & 61 Hz, 175.3 & 177.3 Hz, and 305.3 & 306.7 Hz.

The spacing of the combined modes is relatively uniform, which is another piece of good news, but, as with all rooms, there are some spacings between the modes that are larger than 20 Hz (marked in bold), in total 7 out of 18 modes below 307 Hz, or 39%.

Room parameters at a glance

With height H = 2.91 m, width W = 3.93 m and length L = 5.64 m, our room ratios are 1: 1.34 : 1.93.

In the 1993 BBC Research and Development report titled "Optimum dimension ratios for studios, control rooms and listening rooms," R. Walker proposes an evaluation criterion of acceptable room proportions.

Rooms that satisfy such a rule-of-thumb would achieve reasonably even distributions of low-frequency modes.

Let's check ours: 1.1W /H < L/H < (4.5W/H - 4) or 1.1*3.93/2.91 < 5.64/2.91 < (4.5*3.93/2.91 - 4), and finally 1.49 < 1.94 < 2.08, so all seems well.

The (simplified) room's volume is 64 m³, and its surface area is 98 m² (neglecting the two bookshelf-filled recesses), out of which the floor and the ceiling are 22 m² each, and the four walls are 54 m² in total.

Axial modes [Hz] Longitudinal L □	Axial modes [Hz] Transversal W ○	Axial modes [Hz] Vertical H △	All axial modes	Difference	
			30.5		
30.5	43.8	59.1	43.8		13.3
61	87.6	118.2	59.1		15.3
91.6	131.5	177.3	61	*	1.9
122	175.3	236.4	87.6		**26.6**
152.7	219.1	295.5	91.6		4
183.2	262.9		118.2		**26.6**
213.7	306.7		122		3.8
244.3			152.7		**30.7**
274.8			175.3		**22.6**
305.3			177.3	*	2
			183.2		5.9
			213.7		**30.5**
			236.4		**22.7**
			244.3		7.9
			262.9		18.6
			274.8		11.9
			305.3		**30.5**
			306.7	*	1.4

Fig. 12.8.1. The room resonant frequencies (axial modes) for the simplified rectangular room

Fig. 12.8.2. All three axial modes below 300 Hz combined, with the difference between adjacent modes calculated in the right column.

Fig. 12.8.3. All three modes are represented graphically on the continuum of frequencies (linear scale). The symbols (shapes) of the very closely spaced modes are filled (in black).

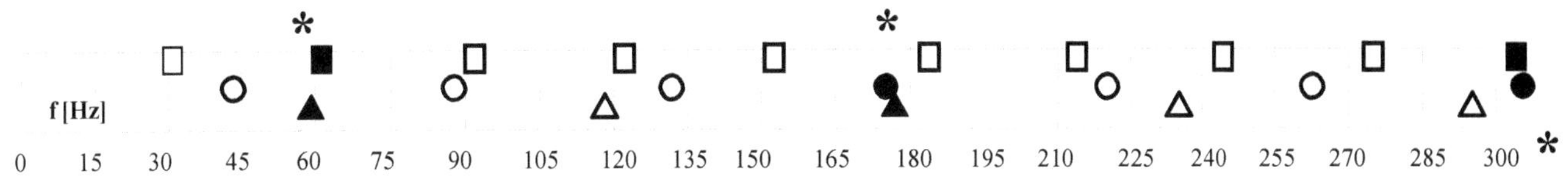

Some online calculators defined Schroeder's critical frequency of this room as $f_C = 109$ Hz and calculated the critical distance $d_C = 4.40$ m.

However, these results must be taken with a great deal of reservation, even skepticism. The use of RT60 (the time taken for the reverberant sound to drop 60 dB) required to calculate Schroeder's f_C, MFP (Mean Free Path), and other such statistical equations is invalid in such a small room. All such parameters are based on the premise of validity of the Sabine equation, which only applies to large halls.

If you are not familiar with MFP, the Sabine equation, Schroeder's critical frequency, the critical distance, and RT60 reverberation time, don't despair; you are not missing out on anything. You don't need to learn acoustics to such depth to optimize the acoustics of your room.

The longitudinal arrangement

This arrangement has the speakers in front of a solid brick wall (with two double power points PP), away from the entrance door to the room (bottom left corner), and as such, is the most intuitive option. There's plenty of depth behind the speakers (1.76 m in this setup), resulting in a deep and spacious sound stage. The listening position is relatively far from the back wall (1.3 m), so there are no problems with rear reflections.

A large percentage of the back wall's area is covered by a large bookshelf, which acts as an absorber and a midrange and high frequencies diffuser.

On the debit side, due to the relatively narrow width of the room (just under four meters), the speakers are fairly close to the side walls (about a meter), so early lateral reflections arrive only slightly delayed from the direct sound, "bundled" with it.

This is more of an issue with the right speaker (RS), which is close to the hardwood-clad sidewall, and less so for the left speaker. Its mid- and high-frequency reflections are significantly absorbed by the thick curtains adjacent to it.

The equipment rack (ER) is behind the speakers, possibly interfering with the sound stage to some degree. Ideally, there should be nothing behind the speakers.

However, the sound stage was much more spacious in this case compared to the transverse arrangement (next page), where the equipment rack was at a far side, away from the central area between the speakers.

So, it seems that the myth of the equipment rack behind the speakers being detrimental to the soundstage size is just that - a myth.

Concerning sidewalls, both the shape of the room and the furnishings are asymmetrical. On the left (sitting in the listening position), there are curtains with a glass wall and sliding door behind them, a brick wall (with partial wood paneling), and French doors.

On the right are two recesses with inbuilt bookshelves and a brick wall with wood paneling.

However, no detrimental sonic effects were ever detected as a result of such asymmetry. It seems that other factors are far more critical, namely the considerable distance between the front wall and the speakers (1.76 m in this case) and the significant space behind the listener (between the rear wall and the listening position), about 1.3 m.

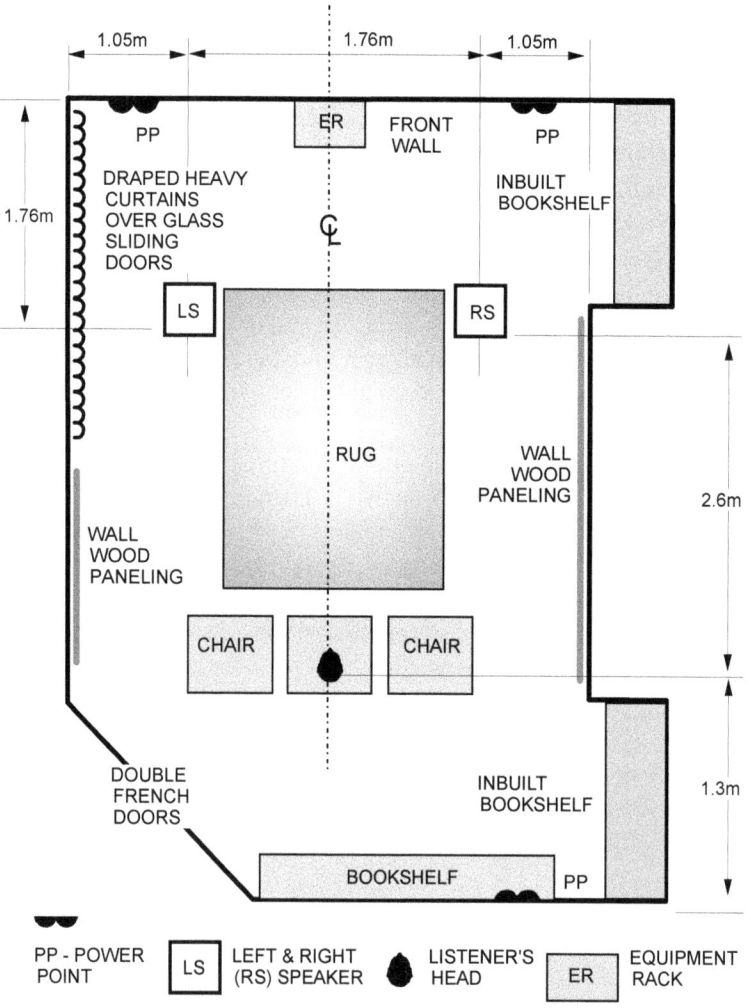

Fig. 12.9. The longitudinal arrangement of our listening room, indicating speaker and "sweet spot" positions and distances

Fig. 12.10. The longitudinal arrangement of our listening room, with speakers firing across its length

I had a feeling the four glass-covered pictures that used to be on the front wall (1) were too reflective. I found these "horse" prints (solid timber frame and stretched canvas) on sale at Kmart for AU$10 each, filled their backs with glass fiber and voila - four homemade absorption panels, resulting in a sonic improvement. Commercial acoustic panels of a similar size are AU$500+ each.

The four glass-covered pictures were moved to the right wall (2); you can see them in the photo of the transversal arrangement on the next page.

Speaker positioning using the isosceles triangle method (Cardas version)

Over the years, we have tried various speaker placing formulas with different model speakers, but the one that generally worked the best in our room was the Cardas formula (see page 257). In some cases, that exact setup would sound the best; with other speakers, it would serve as a reasonable starting point.

With the room W = 3.93 meters wide, the speakers are A = 0.447W = 1.76 m apart and the same distance (1.76 m) from the front wall. They should be X = 0.618A = 1.05 m from the side walls. The distance from the frontal speaker plane X-X to the listening position is 1.5A = 1.5*1.76 = 2.6 m.

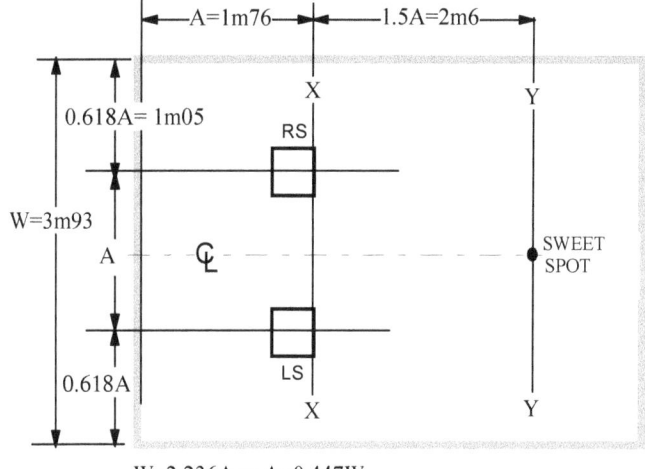

Fig. 12.11. The isosceles triangle method (Cardas version) applied to our listening room

The transversal arrangement

Transversal in this context means that the centerline between the speakers through the listening position is transversing or perpendicular to the room's length. Thus, the listening arrangement is "across" the shorter dimension of the room, its width.

The advantages:

- Wall recesses with inbuilt bookshelves are now symmetrical with respect to the listener.
- Speakers are far from sidewalls, so early lateral reflections aren't significant and take a long time to reach the listener, so they aren't "bundled" with the direct sound.

- The equipment rack is away from the speakers on one side, so there's nothing behind the speakers to interfere with or skew the sound stage.

The disadvantages:

- The listening position is close to the rear wall and the sliding door behind it.
- The asymmetrical rear wall surface, with lots of reflection (wood paneling and French doors) and some absorption (curtains), but without any diffusion.
- Speakers are relatively close to the front wall, resulting in a shallow sound stage.

By now, you are probably wondering which of the two arrangements sounded better. Well, without the slightest doubt, it was the longitudinal one. Despite the equipment rack being behind the speakers, it had a deeper and more expansive sound stage. The bass was also better compared to the transversal setup, so right after the photo below was taken, we reinstated things as they were.

The two subwoofers were not used in the transversal arrangement; that's why we left them in their original positions (as in the longitudinal setup).

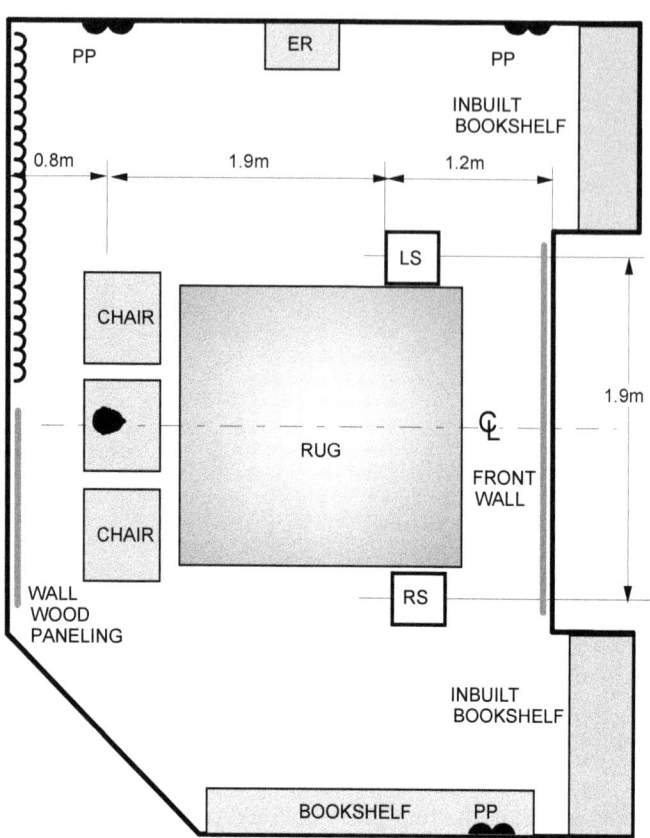

Fig. 12.12. The plan of the listening room with the transversal arrangement, including the significant distances

Fig. 12.13. The transversal arrangement of our listening room, with speakers firing across its width

CASE STUDY: FORMAL LOUNGE AS A LISTENING ROOM

Formal lounges and dining rooms were a standard feature in Australian houses built in the last century. Since they hardly get used in most families, the early 21st century has seen designers abandoning the idea of having these two rooms. Instead, they use that floor space to enlarge the family and informal dining rooms, usually connected with an open-plan kitchen.

Our formal lounge is a rectangular room 4.14 m by 3.8 m, with a height of 2.914m (34 standard brick courses). We will overlook the coffered ceiling, whose perimeter (0.48m wide) is lowered down to 31 courses or 2.66 m.

However, the issue of the back wall remains. All that is left is a 210 mm wide column on one and a 710 mm wide column on the other side. The rest is open to the formal dining room. We will ignore the two 0.5 m long low walls (six brick courses or 514 mm).

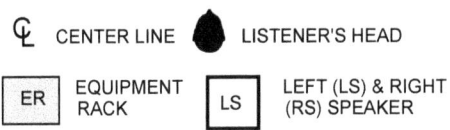

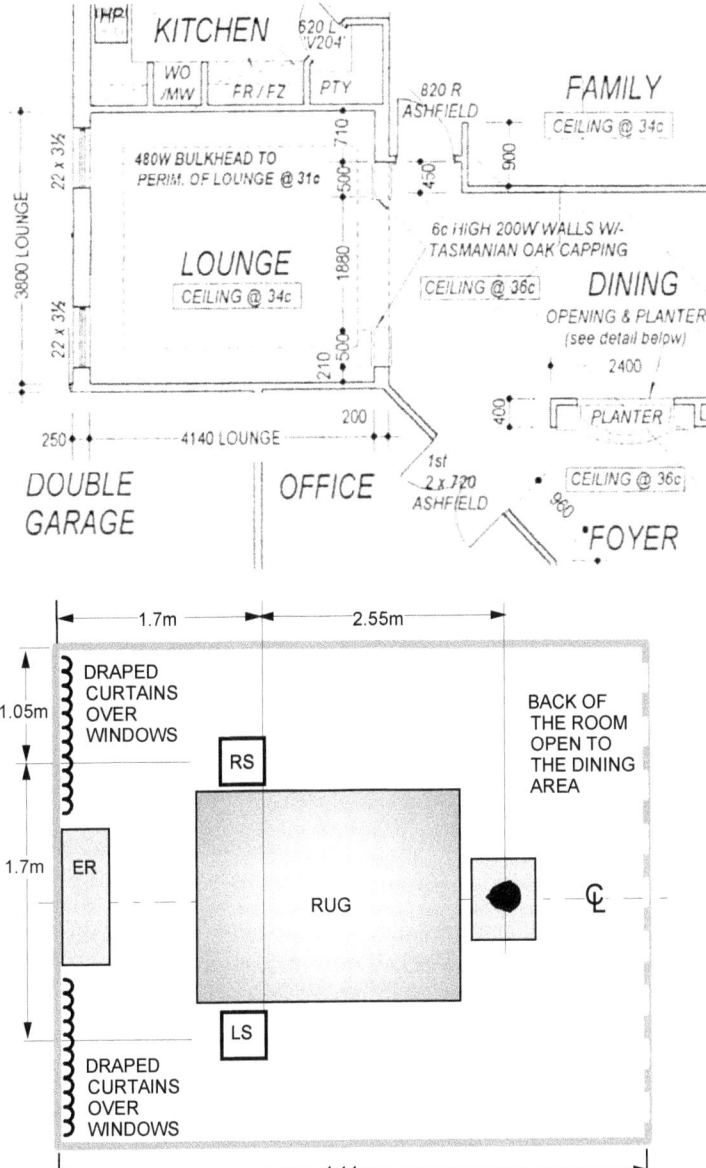

The dining room is 4.75 m long, so for the estimation of longitudinal modes, the two rooms are acoustically considered to be one. We will use the 4.14 m lounge length plus 4.75 m dining room length plus 0.2 m nib wall thickness, or 9.1 meters in total.

Fig. 12.14. The architectural plan detail showing the relation of the formal lounge room to the interconnected dining area and the foyer (entrance hall) to the house.

Fig. 12.15. The audio setup found to be optimal. Not to scale.

℄ CENTER LINE   ● LISTENER'S HEAD
ER EQUIPMENT RACK   LS LEFT (LS) & RIGHT (RS) SPEAKER

Speaker positioning using the isosceles triangle method

Since the width of the lounge is W=3.8 m, the speakers will be A = 0.447W = 1.7 m apart and the same distance (1.7 m) from the front wall. They should be X = 0.618A = 1.05 m from the side walls.

The distance from the frontal speaker plane X-X to the listening position is 1.5A = 1.5*1.7 = 2.55 m.

Room resonance modes

Due to the much higher combined length of the lounge and dining areas (around 9.1 meters), there are no resonant modes below the relatively high frequency of 41.6 Hz, which is excellent news. Again, there are no coincident frequencies at all.

The spacing of the combined modes is quite uniform, which is another piece of good news, but, as with all rooms, there are five spacings between the modes that are larger than 20 Hz (marked in bold) and two pairs of modes that are just under the 20 Hz limit (2), at 208 & 226.7 Hz and 291.3 & 296 Hz.

A very good result, confirmed by favorable listening tests.

Axial modes [Hz] Longitudinal L ■	Axial modes [Hz] Transversal W ●	Axial modes [Hz] Vertical H ▲
41.6	45.3	59.2
83.2	90.67	118.4
124.8	136	177.6
166.4	181.3	236.8
208	226.7	296
249.7	272	
291.3	317	

L+W+H		Difference
41.6		
45.3		3.7
59.2		13.9
83.2	❶	24
90.67		7.47
118.4	❶	27.73
124.8		6.4
136		11.2
166.4	❶	30.4
177.6		11.2
181.3		3.7
208	❶	26.7
226.7	❷	18.7
236.8		10.1
249.7		12.9
272	❶	22.3
291.3	❷	19.3
296		4.7

Fig. 12.16.1. The room resonant frequencies (axial modes) for the rectangular lounge room

Fig. 12.16.2. All three axial modes below 300Hz combined, with the difference between adjacent modes calculated in the right column.

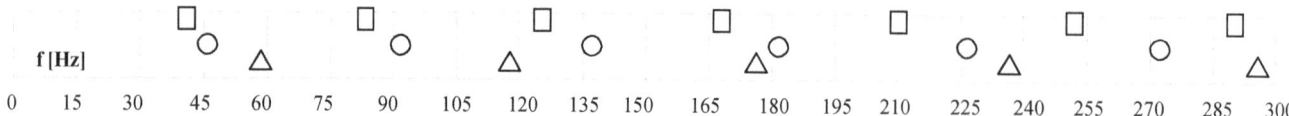

Fig. 12.17. All three modes represented graphically on the continuum of frequencies (linear scale)

Fig. 12.18. Setup #1, with the listener's back to the large dining room and entrance hall.

Two alternative setups were tried. There was a power point centrally positioned on the lounge room's external wall, between two windows, so we placed the equipment rack there, with the listener's back to the large dining room and entrance hall.

Initially, the lounge was bright sounding with somewhat boomy bass and a long echo, probably due to the passageway and the large dining room behind the listening position. Adding broadband absorbers behind the speakers (1) improved the bass, while a large and thick rug (2) and movable side absorption panels (3) tamed the treble and shortened the reverb time.

Drawing down curtains on two windows on the front wall (4) improved the sonics slightly. The setup was neat and practical, but I could never relax completely with my back towards open areas and the passageway.

Reversing the setup resulted in a more relaxed seating position, with the sweet spot about a meter in front of the solid wall (5). However, having monoblocks on the floor between the speakers blocked the entrance to the lounge room, so one had to navigate carefully through the narrow space on the left (6).

None of our interconnects were long enough to reach the monoblock further away from the equipment rack (7), so the rack had to be moved right behind the left speaker instead of being placed against the sidewall, where the power point was (8). Using a rack-mounted amplifier would solve all those issues, except the inevitability of one speaker cable transversing the entrance to the room (9).

This setup resulted in much better sound, the echo and boominess were completely gone, and the treble was more precise and refined. The incredible depth of the soundstage extended all the way down into the dining room.

Fig. 12.19. Setup #2, with the speakers' back facing the large dining room and entrance hall.

13 | ACOUSTIC TREATMENTS

Acoustics is such an involved, maths-laden technical and architectural discipline that most audiophiles would rather subject themselves to a detailed tax audit for a whole day or colonoscopy (without anesthesia) than study such a demanding subject.

Luckily, once you understand the fundamental formula that links the frequency, speed, and wavelength of sound waves and the fundamental behavior of sound waves (diffraction, reflection, and absorption), you will be able to figure out how various types of acoustic treatments work.

We look at bass traps (bass absorbers), midrange and broadband absorbers, of both panel and cavity resonator type, and finally, another product category, diffusers, primarily of the QRD type (quadratic residue diffusers).

The rapid advance of digital technology has made digital room correction (equalizing) a reality. Fundamentally sound (no pun intended) in principle, just as with all digital systems and manipulation techniques, the jury is still out as to its sonic merits and shortcomings.

- THE PROPERTIES AND BEHAVIOR OF SOUND IN CLOSED SPACES
- SIMPLE ACOUSTIC TREATMENTS FOR LISTENING ROOMS
- POROUS MATERIALS AS SOUND ABSORBERS
- PANEL ABSORBERS
- CAVITY RESONATORS
- BASS TRAPS
- DIFFUSERS
- DIGITAL ROOM EQUALIZATION SYSTEMS

> "If the midrange isn't right, nothing else matters."
> J. Gordon Holt, founder of *Stereophile* magazine

THE PROPERTIES AND BEHAVIOR OF SOUND IN CLOSED SPACES

How sound travels through air and other materials

To understand both the acoustic behavior of listening rooms (in contrast to large concert halls) and various sound treatments and their effectiveness, we need a speed course on the fundamentals of acoustics. While this is by no means an extensive coverage, it should serve as a convenient starting point for your further study, should you wish to deepen your understanding of why things are the way they are and why they sound the way they do.

By now you shoud already be familiar with $f=v/\lambda$, the fundamental formula that governs the propagation of sound waves through any medium is. We can calculate the speed of the sound in a solid material from its density ρ (Greeek letter "rho") and its Young's modulus (elasticity) E: $v=E/\rho$. So, the denser the material, the slower the speed of sound. The higher the material's modulus of elasticity, the faster the sound travels through it.

The formula is not valid for gases, including air. For instance, the speed of sound varies from around 5,189 m/s in steel and 3,536 m/s in concrete to 1,517 m/s in water and 344 m/s in air. In gases (such as air), the speed of sound depends also on the temperature and the molecular weight of the gas.

An interesting case is a solid timber, whose elasticity module is much higher along the grain than across it. For instance, Young's modulus of beech wood is approx. 14×10^9 N/m^2 along and only 0.9×10^9 N/m^2 across the grain. With a density of 680 kg/m^3, the speed of sound along the grain will be about four times faster than across it!

This fact is important in the construction of musical instruments and the design and construction of loudspeakers made of solid wood, such as our Opera Terza, made of solid mahogany (elasticity modulus of 8.7×10^9 N/m^2 along the grain). This difference in speed propagation will also impact speaker cabinet vibrations in various directions and the damping of its resonances. In contrast to solid timber, the elasticity modulus of man-made timber products such as medium density fiberboard (MDF) and plywood does not vary with direction.

Sound diffraction, reflection and absorption

When a sound of specific frequency f hits a solid object such as a wall or acoustic panel, depending on the density and other properties of that material, part of it reflects back, and part of it is absorbed by the material (3). That portion that enters the material also reflects back at the other (rear) boundary, with only part of it exiting back into the air on the other side (4).

Every audiophile should understand a few fundamental aspects of this behavior. First of all, the incoming or incident angle with reference to the vertical (dotted line in the illustration below) is always equal to the reflected angle. Both are thus marked α (Greek letter alpha). Secondly, each material has its sound absorption coefficient (SAC). SAC is not a single figure but is frequency-dependent; there are SAC-versus-frequency curves for various materials. You can find them online for building materials (concrete, timber flooring, etc.), home furnishing materials (carpets, drapes, sofas, etc.), and various acoustic treatments such as panel absorbers and bass traps.

Notice that the frequency of the incoming, refracted, reflected, and the transmitted sound wave is the same. The frequency does not change, but the wave's wavelength and speed do change in inverse proportions. Sonically, the most damaging is the re-radiated 'sound" or vibrations of unrelated frequency (5). Each object or structure (a wall, a panel, a piece of furniture, a door or a window, etc.) absorbs the sonic energy from the speakers. The sonic waves "excite" the structure and cause it to vibrate at its own natural frequency. Such vibrations are then re-radiated back into the room as a discordant, unpleasant sound since its frequency is unrelated to the frequency of the original sound.

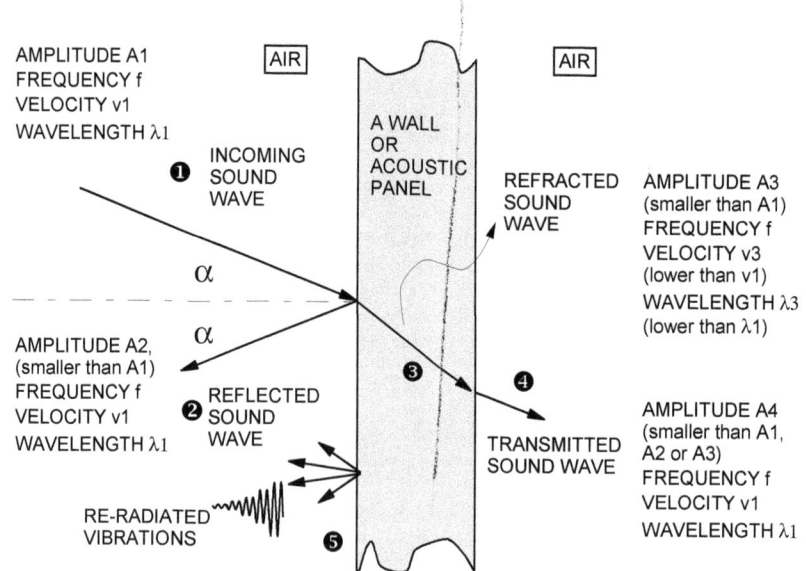

Fig. 13.1. Reflected, refracted, absorbed, and transmitted sound, plus the most sonically damaging component, the re-radiated "sound" (vibrations of unrelated frequency)

Why the size of objects matters and how it affects the sonics

Using the fundamental speed formula $f=v/\lambda$, we can calculate the wavelength corresponding to a few selected frequencies across the audio range of human hearing. At the upper end of 20 kHz, the wavelength is 1.72 cm. Thus, draped curtains and thick carpets are of relatively large dimensions compared to such small wavelengths (high frequencies), and such frequencies get absorbed well.

Moving down to the upper bass-lower midrange (200 Hz), the wavelengths are in the order of a couple of meters, and this is where the trouble starts. To be effective, room treatments in that frequency range need to be very large (2 meters at 200 Hz and more than 10 meters at 20 Hz), and very few listening rooms can accommodate acoustic panels and treatments of that size or even 10x smaller (1.7 m).

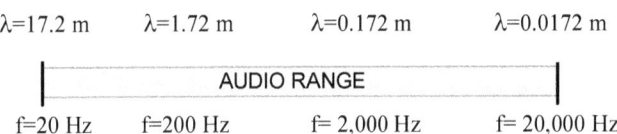

Fig. 13.2. With a speed of sound of v=344 m/s (at 20 deg C), its wavelength varies from 17.2m at 20Hz to 1.72 cm at 20 kHz!

> **THE "FREQUENCY DETERMINES THE FIX" PRINCIPLE**
> There's no universal acoustic treatment. What works and what doesn't depends on the frequency in question. Generally, the higher the offending frequency, the shorter its wavelength and the easier it is to deal with acoustically. Bass problems are the most difficult to treat!

What is the problem with room's skewed spectral response

Amplifiers, preamps, and most other audio components have relatively flat frequency characteristics. The first significant change to the recorded and reproduced signal spectrum happens due to loudspeaker action, where their frequency-dependent and often highly variable impedance causes some frequencies to be reproduced at higher and others at much lower loudness (SPL) levels.

The final transducer, the listening room, changes the spectral distribution even further. Again, depending on the room's size, the speakers' placement in relation to the listening position, and on the reflective/absorptive properties of the walls, floor, and ceiling, some audio frequencies will be amplified (boosted) while others will be attenuated. The listening room acoustic properties will change reproduced spectra of different musical instruments in different ways. Some will be reproduced more faithfully than others.

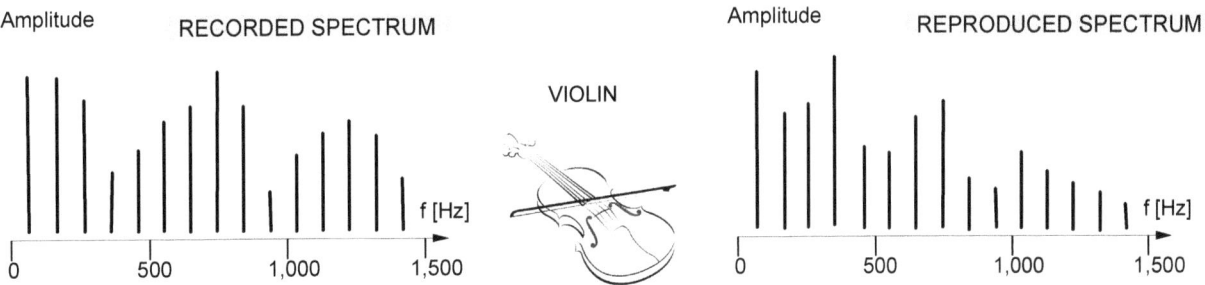

Fig. 13.3. Listening rooms change the spectral distribution of the reproduced sound even more than loudspeakers.

The ideal of the uniform and continuous sound field

Ideally, all audio frequencies should be distributed uniformly throughout the room. However, since sound is a longitudinal mechanical wave motion (pulsating changes in air pressure), its distribution in a room is affected by the room's dimensions, shape, and curvature of its boundaries.

The materials the walls are made of (brick, concrete, plasterboard) and covered with (wallpaper, timber cladding, curtains, tapestries, mirrors, large pictures & paintings, etc.) also have a marked impact on the room's sonic "signature."

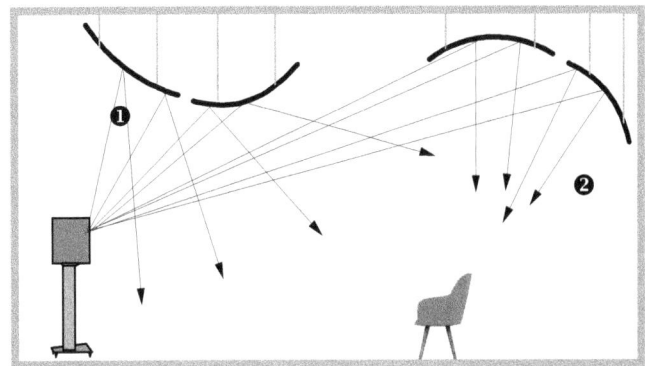

Fig. 13.4. Convex surfaces (1) diffuse or disperse sound waves while concave surfaces (2) focus or concentrate them.

Convex boundaries and objects (large round pillars, for instance) disperse sound, redirecting and scattering it over a wider area. Others, of convex shape, behave akin to a parabolic lens (like the ones often seen on satellite dishes), focussing the sound waves into narrow beams. In acoustic terms, convex is generally good (beneficial), while concave is detrimental to sound quality.

The listening room construction methods and their impact on its sonics

Single-story (one-level) houses have light ceilings made of gypsum board, drywall, or plasterboard sheets attached (screwed & glued) to ceiling battens. As such, the whole ceiling acts as a sizable resonating membrane, a passive radiator above your head. Older houses use timber battens and ceiling joists, while some newer ones have the entire roof construction made of steel profiles. This could change such ceilings' resonating frequency and other vibrational characteristics but fundamentally make no difference to their undesirable sonic effect.

Some double-story houses have a suspended concrete slab between the two stories, so if your listening room is on the ground floor, it will have a solid concrete ceiling, an acoustically preferable option.

Most homes in Western Australia are of double brick construction (outside walls), with single brick plastered internal walls. Other Australian states use "brick veneer" construction. The exterior walls are single brick with timber-framed and plasterboard-lined inner shell, while the interior walls are also timber-framed and plasterboard-lined. In the last few decades, steel-framing has been slowly replacing the use of timber. Timber and steel framed homes are prevalent in the USA and many other parts of the world.

Due to the significant differences in their resonant nature, rigidity, and sound absorption/reflection coefficients, timber and steel framed rooms have different acoustic properties from the ones built using brick or concrete blocks.

Reducing background noise levels

We've already mentioned the nighttime listening effect when the mains power supply is usually cleaner and the background noise level noticeably lower. Reducing the background noise level isn't of paramount importance just to recording studio owners and designers; it will significantly improve both the dynamic range and the microdynamics and low-level resolution of your audio system.

While 18 dB is achievable in an average recording studio, a typical listening room's background noise level may be as high as 40 dB. A 22 dB reduction may not be financially feasible or even possible from the construction point of view (short of rebuilding the whole house or listening room), but 10-20 dB is an achievable goal.

In terms of sound pollution, the main offenders are usually the air-conditioning system, the ceiling (most aren't insulated at all or only have thermal insulation bats, not sound insulation treatments) and the ingress of outside noise through large doors and windows, which are often only single-glazed.

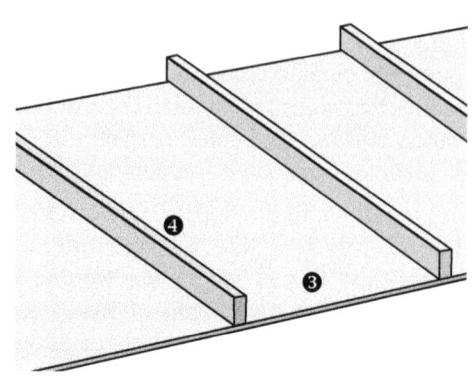

Fig. 13.5. Gypsum board, drywall, or plasterboard sheets (3) attached (screwed & glued) to wall framing and/or ceiling battens (4) form a large resonating panel.

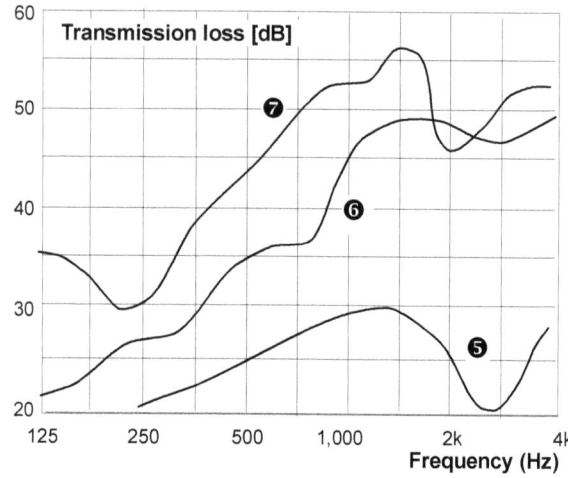

Fig. 13.6. Sound transmission loss for 3 mm single pane window (5), double glazed 3 mm window with 10 mm internal air space (6), and double glazed 3 mm glass window with doubled (20 mm) internal air space (7).

Fig. 13.7.: By lowering the background noise level in a music room from a typical 40 dB to 20 dB, the achievable dynamic range increases from about 70 to 90 dB!

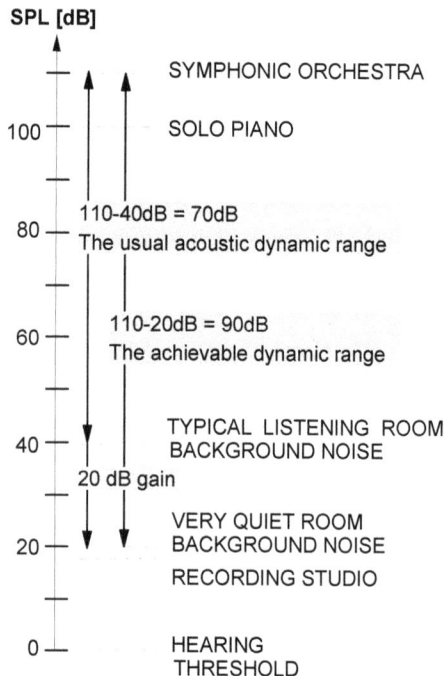

6.38 mm thick laminated glass gives you only a 2 dB advantage over the standard 4 mm float glass (STC29 versus STC27), so such an upgrade to the existing window(s) on your listening room would hardly be worth the expense.

Even investing in double-glazed windows may not bring drastic improvements. Not all double-glazed windows are equally good sound attenuators.

For instance, two standard 4 mm glass sheets with 14 mm of air space between them may have an STC31 rating (4 dB better than typical 4 mm float glass). A combination of 4 mm float glass and 4 mm E-glass with a 16 mm Argon-filled cavity gains you only another 3 dB (STC34), or 5 dB over what you most likely have now.

STC (Sound Transmission Class) ratings are a generic measure of the sound attenuating properties of a particular material or product. It indicates a transmission loss in decibels (dB) at 16 different frequencies across the audio range, in accordance with ASTM E413 standard, and then boiled down to a single figure.

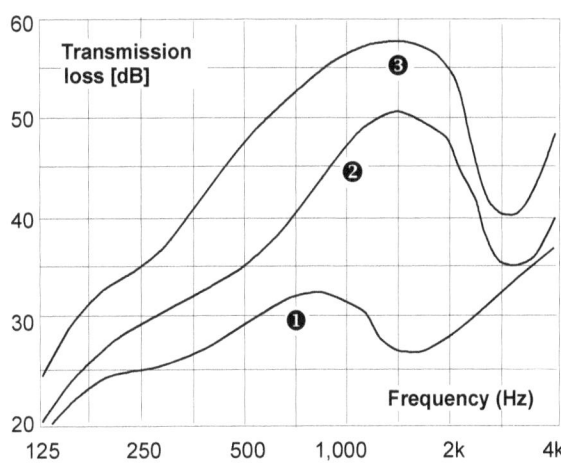

Fig. 13.8. Sound transmission loss for 1) two ½" gypsum board (Gyprock) sheets joined together (STC = 31), 2) wall constructed using two ½" Gyprock sheets on 4" steel channels (STC = 37), and 3) the same wall construction as in 2), but with an added 2" sound insulation between the two sheets (STC = 45)

It is essential to understand a few simple sound transmission principles. Thicker and higher mass partitions attenuate sound more. For instance, a 4" cavity concrete block wall would have an STC of around 40. Plastering both sides increases its STC to 48. Using blocks of a double thickness (8") would raise its STC from 40 to 45 or from 48 to 56 if both sides were plastered.

Joining two identical partitions increases STC only marginally; introducing a cavity is much more effective. For instance, as in (1) above, bonding two 2" gypsum boards (STC=28) together only gains 3 dB (STC = 31). However, using the same two sheets to construct a cavity wall, as in (2) above, raises the STC to 37. Introducing sound-absorbing insulation into the cavity (3) would increase the STC rating even further, in this case to 45!

Soundproof your doors

Most internal doors in the USA and Australia are hollow-core doors, made as cheap as possible. As such, they are the weakest link in your fight against outside noise. Molded MDF skins cover the front and back of a glued-together wood or MDF frame. The void is filled with a cardboard honeycomb structure to prevent flexing, or there's no fill in the void at all.

Better doors have an OSB (Oriented Strand Board) core, a material manufactured from wood shavings and chips, similar to MDF board, or their void is filled with a polyurethane foam core to reduce the weight.

If you have plenty of time on your hands and don't mind messing around, you can try filling your listening room's hollow doors with expanding foam or sand. Otherwise, contact door specialists and ask for information on their acoustic door range and get them to install a "soundproof" door.

Alternatively, buy an external solid-core door and install it as an internal door. If you are really keen, cover the internal surface (facing your music room) or both door sides with an acoustic (absorption) panel.

To give you a feel for the numbers, a hollow door without a seal achieves an STC rating of around 17, which is pretty dismal. Adding a gasket increases the STC rating to about 24 (a gain of 7 dB), which is significant. A solid-core door (without a seal or gasket) is typically rated around STC = 24; adding a gasket improves it by 2 dB to STC = 26. External aluminium sliding doors, such as those in our listening room, are usually soundproofed by installing a second internal sliding door, thus creating an insulating space between the two.

When I said the doors are the weakest link, I wasn't entirely truthful; the most problematic are the gaps around the doors, especially the space at the bottom of the door, which is usually a much bigger gap than the space around the sides.

Applying adhesive weather-stripping around the door is the minimum you should do, but that yields only a marginal sound insulation improvement. Proper rubber seals and gaskets are a much better option. You could even install a magnetic seal or gasket, such as those used on refrigerator doors. A magnetic material fills the PVC rod inside a rubber seal.

Whichever type of seal you opt for, a soundproof door sweep is a must. Sold in hardware shops, they take some time and effort to install. You have to precisely measure the required length and cut the strip down to fit, and then you need to fix it to the bottom of the door using self-tapered screws.

SIMPLE ACOUSTIC TREATMENTS FOR LISTENING ROOMS

Remove all objects that can vibrate or rattle

Every object in the listening room contributes to its acoustics, so before you do the final evaluation and consider the additional acoustic treatments, make sure all the equipment and furnishings are in place.

Keep in mind that each additional chair, sofa, rug, screen or mat constitutes primarily an absorber and possibly a diffuser of mid and high frequencies. In many cases, where a few such large furnishings and furniture pieces feature in a listening room, no further acoustic treatments are needed.

Since they are usually made of porcelain, glass, and similar sonically unfriendly materials, figurines, small statues, vases, bric-à-brac, and other small "hard" objects that can vibrate and rattle have no place in a listening room and should be removed.

Remove all objects between your speakers and the listening position

Early reflections don't just bounce off the sidewalls, the floor, and the ceiling; any object placed between the speakers and the listing position will reflect the mid and high frequencies.

In the case illustrated in the photo, the glass-top coffee table, while useful for keeping the remote controllers, CDs, and drinks in front of you, will usually have a noticeable negative sonic impact.

While the wool carpet toned down the high-frequency glare and harshness of our room by reducing the early reflections of the floating timber floor, the glass top brought some of those unpleasant treble reflections back.

Glass is not just a highly reflective but also a resonance-prone material. It has no place in your listening room. Remove glass shelves from the walls, your hi-fi rack, and even glass turntable platters, and replace them with other, sonically-friendlier materials, such as timber, marble, bamboo, or carbon fiber.

Common combinations of acoustic treatments

Bass reflex speakers with rear-firing ports and audio systems with subwoofers behind the main speakers result in a bass-heavy front of the listening room. Bass traps can be placed between the two vertical walls, in corners, as shown (1), but also in the top (between the front wall and the ceiling) and bottom corners (the front wall and the floor).

In other rooms, bass traps may be more effective in corners at the rear, behind the listening position (2). In some cases (troublesome bass resonances), bass traps in all four corners and wall/floor and wall/ceiling interfaces may be needed.

With highly reflective sidewalls, absorption panels will tame down the early reflections, while diffuser panels are best used on rear walls.

Some setups and rooms benefit from ceiling-mounted diffuser panel(s). Depending on the surface area of other absorption panels in the room, most rooms with hard floors (concrete, tiles, or parquetry) will definitely benefit from a large, thick rug (3) to provide floor absorption.

Fig. 13.9. Remove any objects between loudspeakers and the listening position, especially those using hard reflective tops and surfaces, such as this glass and steel coffee table!

Fig. 13.10. The plan (top view) of two typical listening room acoustic treatment schemes (arrangements).

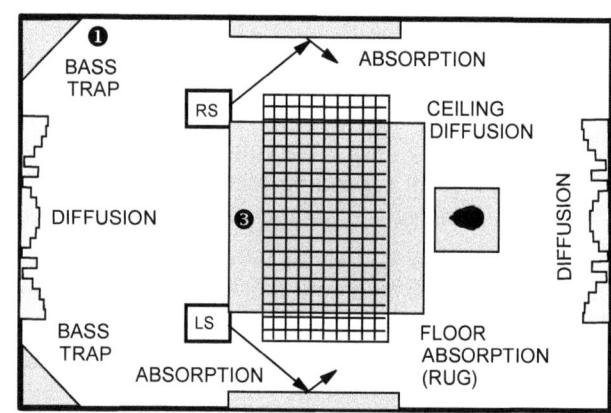

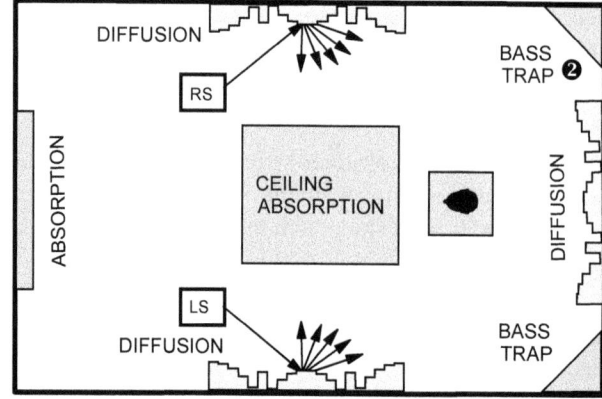

ACOUSTIC TREATMENTS

If a rug isn't viable or desirable for some reason, ceiling absorption can be used (4) instead, although (since it's closer to the speaker drivers), floor treatment is more effective. Likewise, if sidewall treatments are providing diffusion, usually some ceiling absorption is needed for balance. The most effective treatment for the rear wall is usually a diffuser of some sort, just as in the previous case.

Dealing with early reflections

We have defined axial, tangential, and oblique reflections in the previous chapter. The axial and tangential reflections are most prominent, so most room treatment strategies focus on taming them down using various approaches, the most common of which is absorption, followed by diffusion.

Diffusion treatments (acoustic panels) are generally the largest and costliest and not always beneficial, so they should be approached with a degree of caution. Even too much absorption can be counterproductive, resulting in an acoustically "dead" room.

The first reflections off the side walls are sonically the most damaging since they arrive very soon after the direct sound from the tweeter, this affecting the imaging and soundstaging.

The same applies to highly reflective or low ceilings. If the ceiling or side walls of your listening room feature a hard (and that means "sound reflecting") surface, as ours does (bare walls, wood paneling, glass doors, etc.), read on.

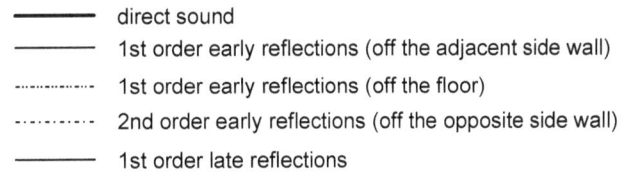

Depending on your loudspeakers' directionality and off-axis frequency response, these first reflections of the treble frequency band may or may not be a problem.

There are two methods of finding the point of the first sidewall reflections. The experimental method requires you to sit in the sweet spot and hold a laser pointer or laser distance measuring device right next to your right ear.

Have your helper place the mirror at the tweeter height, flush with the sidewall. Place the light source on your right ear and aim it at the mirror, which should be moved forward and back along the wall until you see a spot of light on the tweeter's dome (1).

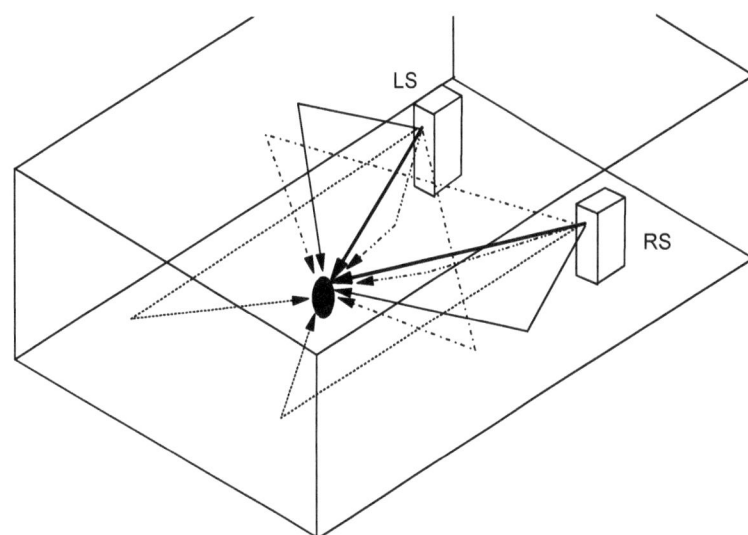

Fig. 13.11. Some typical lateral reflection paths in a listening room. To preserve clarity, reflections off the ceiling and front wall are not shown but can be significant and troublesome. The same applies to 2nd order late reflections.

Fig. 13.12.
BELOW LEFT: The analytical method ("reflected image") uses geometric proportions and calculations after a few distance measurements.
BELOW RIGHT: The experimental, or "Mirror method" requires no measurements or calculations.

Mark that position on the wall (2) - that is the "R" spot, at which you should place an acoustic panel, a wideband absorber.

Alternatively, move the mirror (without using any light source) until you see the tweeter's dome in the mirror.

The analytical method uses a ratio of two right-angle triangles to arrive at a simple formula that can be used instead of the experimental mirror method just described.

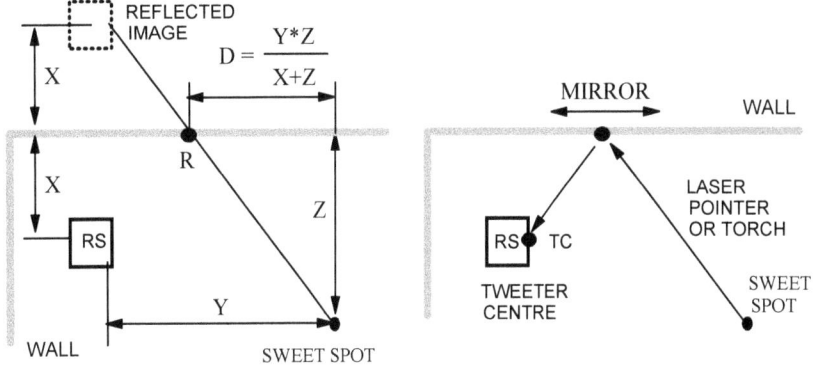

 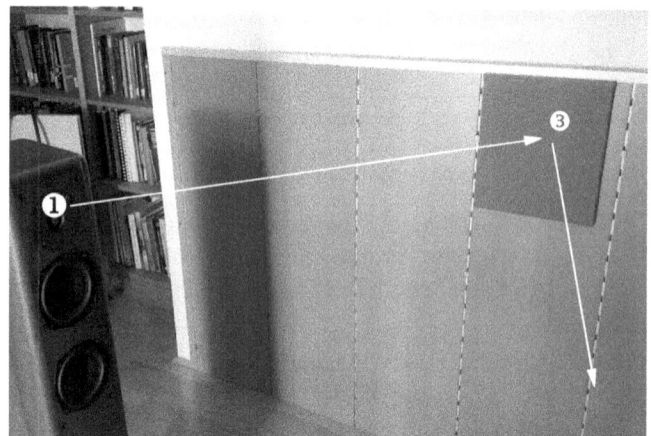

Fig. 13.13. (LEFT) The mirror (2) method to determine the point of tweeter's (1) first reflections off the side walls
Fig. 13.14. (RIGHT) The wideband absorber panel (3) fixed into that position

It can be shown that the distance between the sweet spot's projection to the sidewall and the "R" spot is D = (Y*Z)/(X+Z)!

The absorber (3) doesn't have to be large; we used 12" x 12"canvas frames, sold in pairs in Kmart, costing AU$7. On top of them, we stretched the same upholstery fabric used to cover the larger panel absorbers (page 282) for a uniform look. The frames were shallow, about 2 cm, but we still managed to stuffed them with glass fiber filling.

Repeat the same procedure and treatment with the ceiling reflections.

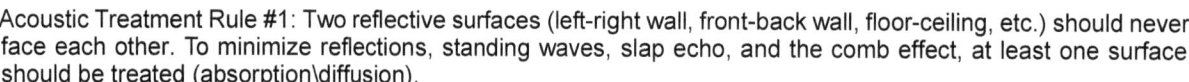

THE ACOUSTIC TREATMENT RULES

Acoustic Treatment Rule #1: Two reflective surfaces (left-right wall, front-back wall, floor-ceiling, etc.) should never face each other. To minimize reflections, standing waves, slap echo, and the comb effect, at least one surface should be treated (absorption\diffusion).

Acoustic Treatment Rule #2: Treat the bass frequencies first. Due to their large size, bass traps will also significantly attenuate higher frequencies, so you may find that you need very little additional mid- and high-frequency absorption, if at all.

Acoustic Treatment Rule #3: Early reflections from the sides of your speakers are the most sonically damaging and should be treated once the bass boom is sorted out.

POROUS MATERIALS AS SOUND ABSORBERS

Four types of sound absorbers are of interest to audiophiles: porous materials, panel absorbers, cavity resonators, the often forgotten "everything else" category, or, in other words, people and furniture.

Carpets and other commonly used materials that provide sound absorption

Generally, the more porous the material (the more openings, cavities, nooks, crevices, or fibers it has), the higher its sound absorption coefficient (SAC). On the other hand, the harder and smoother (more polished) the surface, the lower the material's absorption coefficient.

Another phenomenon that accounts for sound absorption is the vibration and eventual resonance of panels and similar items. More on that important issue soon. In all cases, the absorbed sound energy is converted into heat.

The SAC curves on the next page show how the absorption coefficient of all soft materials (drapes, carpets, and even fiberglass and acoustic panels) drops sharply at lower frequencies. This means that such materials are only effective in absorbing midrange and high frequencies in a listening room and do very little or nothing at all if the bass is boomy and needs to be tamed down.

A wool or synthetic carpet is quite an efficient high-frequency absorber. Its sound absorption coefficient is around 0.5-0.7 at higher frequencies (above 2 kHz) but drops off rapidly in the 500 Hz - 2,000 Hz region. It is practically negligible for bass frequencies (below 250 Hz). The curve resembles that of a low pass filter in electronics, and that's exactly how it behaves acoustically; the low frequencies are not affected much, if at all.

For that reason, fully carpeted floors (wall-to-wall) usually aren't desirable; they introduce an imbalance in the vertical plane, with too much high-frequency absorption off the floor and very little off the ceiling.

Likewise, hanging large rugs and tapestries on walls tilt the balance too far and make the room sound dead or dull. The only application that proves beneficial most of the time is placing a thick rug in front of the speakers, between them and the listening position. On hard floors such as bare concrete, ceramic tiles, parquetry, or floating timber floors, a thick rug tames the treble brightness and improves the tonal balance of the room.

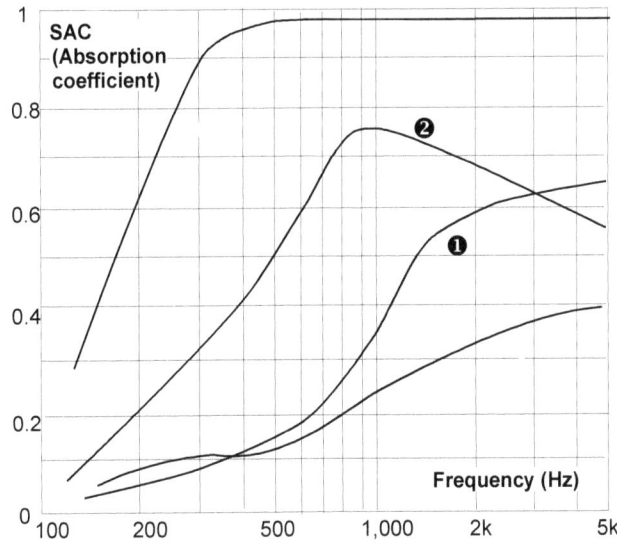
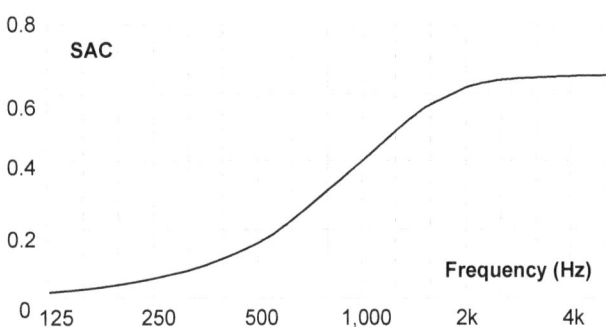

Fig. 13.15. The absorption coefficient versus frequency curves of various high-frequency porous absorbers: 1) heavy wool carpet laid on a concrete floor 2) Cotton curtain, medium weight, draped to half of its coverage area 3) Cotton curtain, medium weight, draped to the whole of its coverage area (almost flat, no draping) and 4) 2" thick fiberglass panels.

Fig. 13.16. The sound absorption coefficient of a typical domestic carpet with underlay.

What should I use for my DIY panels, acoustic foam or fiberglass?

In terms of acoustic properties and, ultimately, their usefulness as acoustic absorbers, not all types of foam were created equal. Acoustic foams have open cell structures, where the pores are interconnected, resulting in significant sound absorption properties and high SAC figures. On the other hand, closed-cell foam materials don't provide much absorption.

The acoustic properties of foam and fiberglass are fairly similar, although, in some tests, the bass traps made of acoustic foam slightly outperformed those using fiberglass. When disturbed (drilled, cut, or handled), fiberglass releases fine dust into the air, which is a known irritant.

For that reason, some manufacturers use polyester or polypropylene fibers instead in their commercial panels and traps since those materials don't require protective equipment while handled, as fiberglass does.

Acoustic foam can easily be manufactured into different profiles and shapes or even cut and shaped by the DIY user, giving it a distinct advantage over the relatively inflexible fiberglass batts.

Acoustic properties of fiberglass panels

At frequencies above 1,000 Hertz or so (3), there's no practical difference in the absorption coefficient between the 25 and 50 mm thick panels. Thus, for midrange and high-frequency absorption, thinner (1") panels are sufficient.

Between 400 and 800 Hz (the lower midrange), the 2" panel has a very high SAC (4), almost 1.0, meaning total absorption. There is a noticeable increase in SAC of 50 mm panels at bass frequencies compared to the thinner, 25 mm ones. For instance, at 125 Hz (5), the doubling of thickness results in doubling the absorption coefficient, from approx. 0.15 to 0.3!

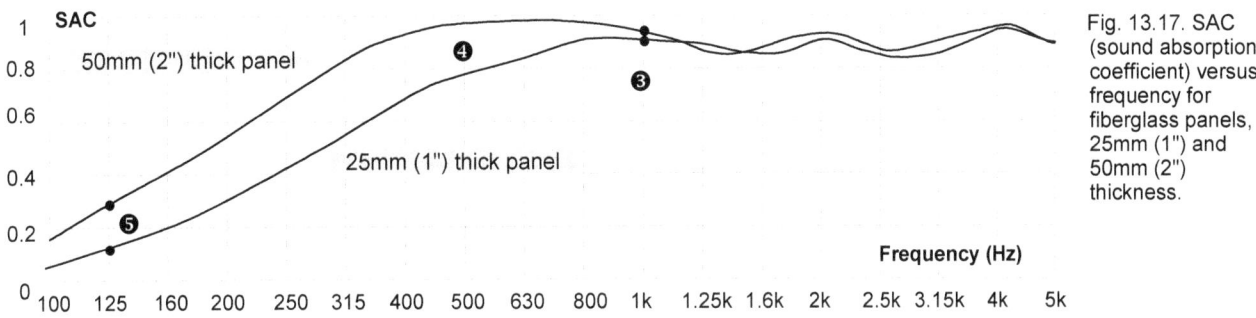

Fig. 13.17. SAC (sound absorption coefficient) versus frequency for fiberglass panels, 25mm (1") and 50mm (2") thickness.

PANEL ABSORBERS

How panel absorbers work

Panel absorbers (also called membrane or diaphragm absorbers) are low- to mid-frequency sound treatments that work as mechanical resonators. The air space behind the panel acts as a spring, with the board or panel being the mass, thus forming the spring-mass resonator. Sound waves cause the panel to vibrate, with the maximum vibrational amplitude occurring at the panels own resonant frequency f_R, which is inversely proportional to both the mass of the panel and the distance behind the panel:

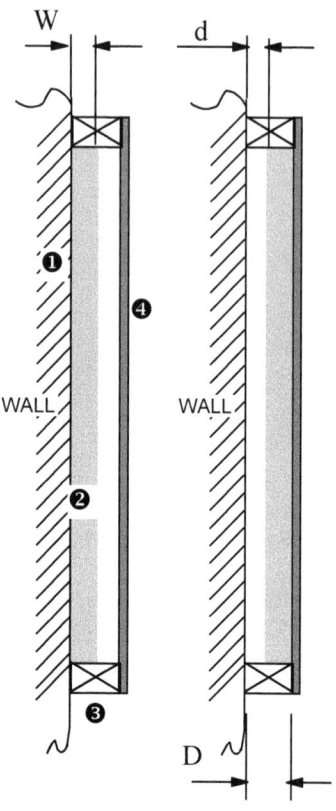

$f_R = \dfrac{600}{\sqrt{m_S * D}}$ [Hz], where m_S is the specific mass of the panel in kg/m², and D is the internal distance to the wall or ceiling in cm (centimeters)

The panel must be mounted on a timber frame a specific distance D away from the wall or ceiling it is fixed to. The structure must be airtight, so it forms a sealed chamber between the wall and the panel.

Notice that the resonant frequency does not depend on the actual area of the absorbing panel. However, it depends on its thickness, which is implied in the specific mass m_S - a thicker panel will have a higher specific mass.

As with any oscillatory system, mechanical or electrical, the panel's inertia and damping cause its vibrations to convert some of the sound energy into heat, thus providing attenuation in the region centered around the resonant frequency, meaning the reflected sound is of diminished energy compared to the incident sound. By carefully choosing the two variables (m_S and D), we can tune the panel to the desired frequency, usually one of the modal frequencies of the room. By reducing the amplitude of the standing waves, such tuning makes that particular room resonant mode less troublesome (less audible).

Fig. 13.18. DIY panel absorbers are relatively cheap and easy to make. 1) the wall 2) fiberglass or similar damping material 3) Mounting timber frame 4) Front panel

Fiberglass bats can be mounted against the wall or the front panel. "d" is the depth of the air cavity, the difference between "D" and the thickness of the fiberglass bat "W".

How panel absorbers' SAC varies with the back spacing and the thickness of the fiberglass batts

The two graphs below, redrawn from those published by Owens-Corning, will give us a better feel for the impact of "d" and "W" on panel absorbers SAC figures.

With no free space behind the fiberglass batt (d=0), the reasonable level of attenuation, say, SAC=0.3, is around 335 Hz (3). With a 3" deep backspace, the SAC=0.3 point is lowered to about 185 Hz. Likewise, increasing the fiberglass batt thickness from 1" to 3" lowers the SAC=0.7 frequency from about 510 Hz (5) down to 155 Hz (6). Both will result in significant improvement in the upper bass/lower-midrange attenuation.

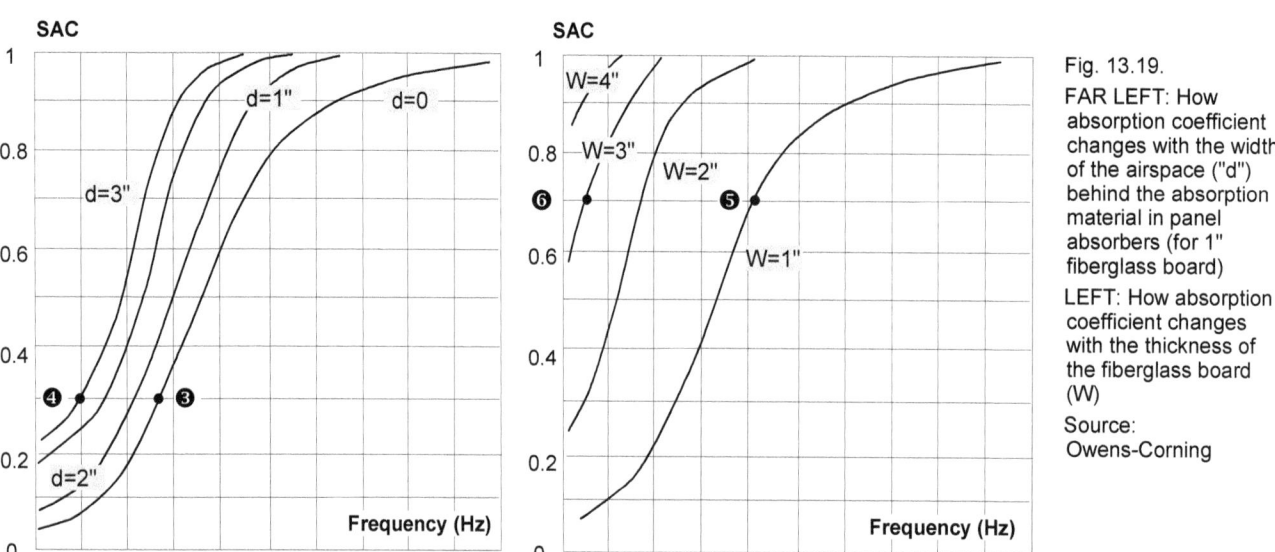

Fig. 13.19.

FAR LEFT: How absorption coefficient changes with the width of the airspace ("d") behind the absorption material in panel absorbers (for 1" fiberglass board)

LEFT: How absorption coefficient changes with the thickness of the fiberglass board (W)

Source: Owens-Corning

CASE STUDY: "Wave" absorption-diffusion acoustic panels

Sold online, factory-direct from UA Acoustics, these 500 x 500 x 53 mm "Wave" absorption-diffusion panels will cost you US$16 each plus shipping. Each panel weighs around 1.5 kg and covers 0.25 m².

The panels are made of 50 mm thick M1 acoustic polyurethane foam. Cheaper, 30 mm thick panels are also available. M1 refers to their fire rating, meaning that they are nonflammable, the only kind you should ever use inside your house. The foam is then covered with a decorative top, made of laminated MDF, and available in seven finishes, including white, black, cherry, and oak.

The hardtop surface includes grooves of various lengths, which allow some sound to pass through while also providing a degree of diffusion.

The sound absorption coefficient versus frequency graph, tested according to EN ISO 354:2003 standard, shows that between 300 Hz (1) and 1 kHz (2), the SAC is above 0.8, a very high absorption rate. It then drops in an almost linear fashion towards 0.25 at 100 Hz (still a reasonable attenuation of the bass frequencies) and, at the high-frequency end, towards 0.4 at 5 kHz (4).

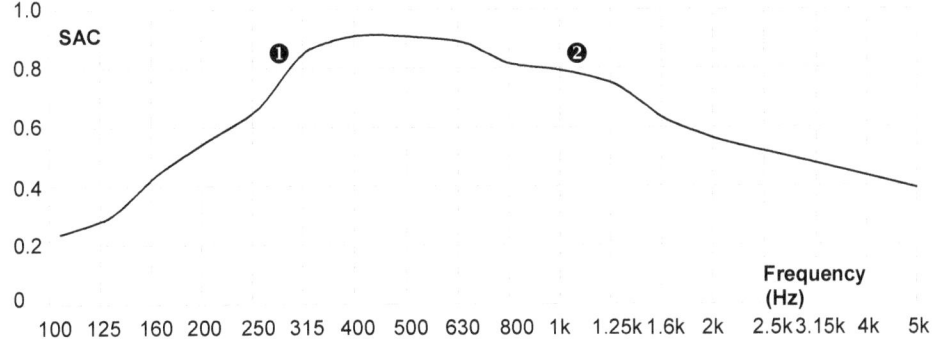

Fig. 13.20. (ABOVE) Eight "Wave" panels (covering two square meters) on the front wall of our listening room

Fig. 13.21. (LEFT) The SAC (sound absorption coefficient) versus frequency curve for 50mm thick panels sold by UA Acoustics, tested according to EN ISO 354: 2003 standard

Fig. 13.22. (BELOW) "Foamfix" mounts for the installation of acoustic foam panels on walls.

We tried using a strong double-sided self-adhesive tape, but without success. The tape would stick to the walls but not to the foam at the back of the panels.

For US$18.- (free worldwide shipping), the same eBay seller offers 10-packs of "Foamfix" mounts (right) for the installation of acoustic foam panels on vertical surfaces such as walls.

Commercial example: Vicoustic Wavewood panels

Vicoustic is a Portugal-based acoustic solutions company. Its Wavewood panels are made of acoustic foam (M1) and wood, based on the "nonlinear sequential cavities" principle, making this product act as both an absorber and diffuser for medium and high frequencies.

Allegedly, it could also be used as a "very efficient" bass trap in the room's corners. However, the panels' depth is only 16 cm, so it's hard to see how it could be effective below a hundred or so Hz. The published absorption coefficient curve shows a SAC of 0.4 at 200 Hz, which is significant, but drops to only 0.1 at 100 Hz.

Weighing about 1.5 kg per panel, Wavewood can be attached (glued) to both walls and ceilings.

In 2021 a pack of ten panels sold for AU$1,250, and since each panel is 595 x 595 mm, covering 0.354m² (3.54 m² per pack of 10), the cost is AU$1,250/3.54 = AU$353 per m²!

Larger but thinner 120 cm x 60 cm x 6.4 cm panels are also available, sold as "Wave Wood 1200".

DIY PROJECT: *Absorption panels*

Just as we were planning to make a few panels from scratch, a local seller advertised these panels at a meager price, so we bought them. The timber used was treated radiata pine, 19 mm by 70 mm profiles. Meant for outdoor use, they are much cheaper than "dressed" pine profiles for indoor use (furniture grade). A 2.7 m length costs only AU$6.75 in hardware shops. The panels were stuffed with fiberglass insulation batts, about 90 cm long and 45 cm wide.

Neither the timber frame finish nor the fabric covering was of an acceptable standard - they used it in a makeshift recording room, so the wife-approval factor was not an issue. A daggy-patterned fabric was tacked to the black-painted frames in a haphazard manner (4), so we removed it and completely upholstered the panels. The expensive upholstery-grade material in golden hues looked very classy and matched the decor of the listening room (pastel colors).

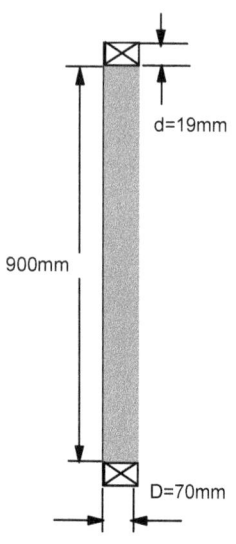

Fig. 13.23. The cross-section of the panels. All were of the same height to accommodate 90 cm long fiberglass insulation batts. Two were 450mm wide (the width of one batt), and two were double-width, 90 cm internally.

Fig. 13.24. ABOVE LEFT: The DIY panels we bought had a poor finish but were solidly made and cheap, below the cost of the parts, let alone the time involved in making them. The self-tapering screws were protruding, so we had to take them out, enlarge the holes so their heads can fit, and screw them back in (1). ABOVE RIGHT: The fabric is trimmed to the required size, ready for gluing and stapling. The spray adhesive (2) isn't essential; the material can just be fastened to the frame using a stapler (3).

Fig. 13.25. One of the two larger panels in position as a treatment for lateral reflections of the sidewall. It is a bit lower than the tweeter level, but it still resulted in a noticeable improvement - softer and more refined treble, less top-end glare, and better resolution.

Fig. 13.26. Two smaller absorption panels hung on the front wall instead of the two middle "horse" frames.

CAVITY RESONATORS

The Helmholtz resonator

Named after German physicist Hermann von Helmholtz, this resonator consists of a cavity with rigid walls and volume "V," with a neck area "S" and neck length "L." The acoustic pressure in the cavity provides the stiffness element (acting as a spring), and the air in the tube (neck) provides the resistance, the mass element. The mechanical equivalent of such a resonant system would thus be a mass suspended on a spring.

The fundamental resonance occurs at the angular frequency ω_R (formula on the right). Cavity resonators with perforated panels, such as our DIY absorber below, are arrays of Helmholtz resonators with individual necks and a joint cavity. The perforations act as necks, while the backing volume works as the cavity.

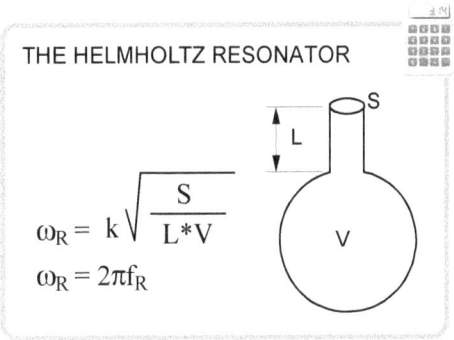

How SAC changes with the perforation ratio

The SAC graphs (left) illustrate an important principle. High perforation ratios (bigger holes or more of them) result in the highest SAC values, occurring at higher frequencies.

As the perforation ratio (PR) gets smaller, the peak SAC is reduced; the frequency of the peak is also lowered.

In this typical example, the peak for 25% PR happens at around 1 kHz and is just over 1.0. In comparison, the peak for 5% PR is at 500 Hz and is lower, around 0.85. Perforation ratio of 0.5 % results in a panel whose peak SAC is "only" 0.7, at an even lower frequency of 250 Hz!

Also, as the perforation ratio is reduced, the curve becomes narrower and "sharper," meaning the panel's Q-factor (quality factor) is higher. High perforation ratios result in the broadest curves and lowest acoustic Q-factor.

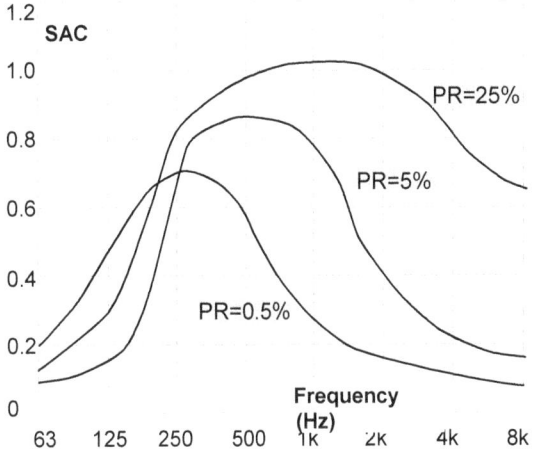

Fig. 13.27. SAC curves of wideband absorbers with hardboard perforated panels, 25%, 5% and 0.5% perforation

DIY project: IKEA panel resonator

Stuva is IKEA's modular furniture system. This frame sells for AU$40.- and comes only in a white plastic foil finish. It's 60 cm wide, 64 cm tall, and 50 cm deep. Made of particleboard, its honeycomb structure filled with recycled paper is used to make it lighter, resulting in a weight of only 8 kg.

We've used the Skadis pegboard for the front panel, also from IKEA, made of fiberboard; the largest size is 76x56mm.

The cabinet should be filled to at least half (250mm) by fiberglass bats, nonflammable acoustic foam, or similar absorption material. For instance, dense cotton and feather pillows work well too. The front edges should be sealed; the air should be allowed to enter the structure only through the perforations.

Fig. 13.28. DIY cavity resonator panel positioned in the front corner of the listening room

Commercial example: Vicoustic Visquare 60.4 V2 Premium

Vicoustic Visquare 60.4 V2 Premium comes in packs of eight panels and sells for AU$720.- Each panel is 595 x 595 x 40 mm (covers 0.354 m²) and weighs 650 grams, so the cost per square meter is around AU$255.-

Constructed of open-cell polyurethane foam with fabric covering, these mid- and high-frequency absorber panels are only 4 cm thick but of equivalent performance to 70 mm ordinary foam panels. The "secret" is their perforated foam construction. The perforations increase surface absorption and also act as a sound trap.

The sound absorption coefficient is more or less constant (1) between 630 and 5,000 Hz, with a slight dip around 2,500 Hz, then reduces linearly (2) as the frequency falls into the bass region. There is some absorption (SAC=0.3) in the 200-250 Hz upper bass region (3), which could be helpful in some rooms.

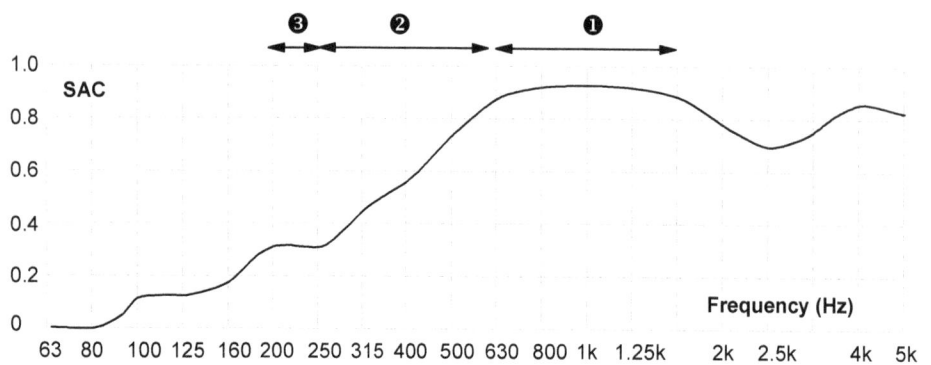

Fig. 13.29. Sound absorption coefficient (SAC) versus frequency for Vicoustic Visquare 60.4 V2 Premium panels

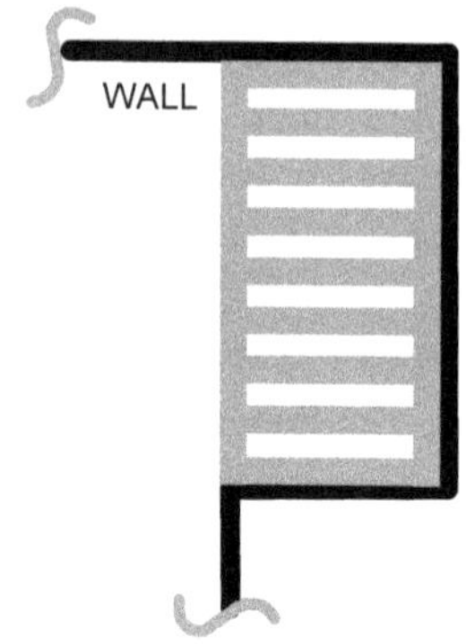

Fig. 13.30. Turning an unused corner recess into a bass trap

BASS TRAPS

Low-frequency sound waves (or "bass") builds up in the corners of a room, and especially in "trihedral" corners where two walls meet the ceiling and where two walls intersect the floor. Thus, these are the primary placement points for "bass traps," acoustic panels designed to provide attenuation at such low frequencies. The name is unfortunate since bass waves are not trapped; they are simply attenuated. Unfortunately, due to the long wavelengths of bass waves, most bass traps provide a very modest degree of attenuation.

Remember, a 50 Hz sound wave has a wavelength of $\lambda = v/f = 340/50 = 6.8$ m, so using the $\lambda/2$ principle, a very effective bass trap at such low bass frequency would have to be almost 7 meters deep, which is, of course, absurd.

Wall recesses as bass traps

Our listening room features two 600 mm deep recesses in the corners of the same wall (page 265), the left wall in the longitudinal listening arrangement, or the front wall in the transversal setup (page 268). They aren't of the same width (1.405 m, the other 1.455 m), but that is of no importance. What matters is that both are 600 mm deep.

In problematic rooms with boomy bass, such alcoves can be made into very efficient bass traps. The peak bass absorption is at the frequency for which the depth is a quarter-wavelength. So, in our case, $\lambda/4 = 0.6$m and $\lambda = 2.4$ m. The peak absorption frequency would thus be $f = v/\lambda = 343$ [m/s]$/2.4$ [m] $= 143$ Hz.

You would need to cover the front of the recess with fiberglass acoustic insulation boards (such as those featured on page 287) and at least partially fill the cavity with similar boards or absorption materials (see the illustration). To maximize the absorption, line the recess walls with the insulation boards and fill it with an array of equally-spaced parallel panels, side facing the front of the bass trap (as illustrated).

Wall cladding or paneling as upper bass traps

Our timber wall cladding partially covers two side walls of the listening room. 11 mm thick, the panels (6) are held in place by an array of vertical aluminium profiles (5) bolted to the brick wall (4) and locked in by round metal pegs (7) that slide into the slots in the distancing channels. The panels are 1.09 m tall and 0.356 m wide and weigh 2.7 kg each, meaning their mass per square meter is $m_S = 6.96$ kg/m^2.

Since $D = 1.9$ cm, they are tuned to the frequency of 165 Hz. However, the formula used assumes flexible vibrating membranes, and our panels are pretty stiff, resulting in a higher resonant frequency, 200 - 250 Hz.

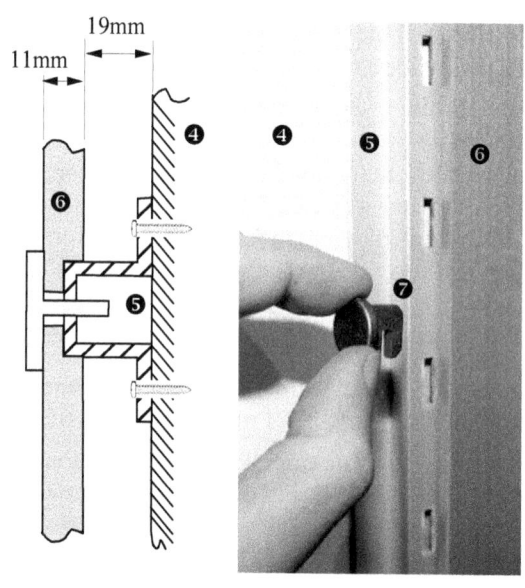

Fig. 13.31. The cross-sectional drawing showing the mounting details of our IKEA wall panels

Electronic or active bass traps (ABTs)

Passive LF absorbers (bass traps) are physically large and tuned to a single frequency only. Furthermore, with insufficient damping in the resonant cavity, the so-called "mode splitting" occurs, the pressure amplitude of the target mode is slightly reduced, and two lower pressure peaks appear. Most of the troublesome energy isn't dissipated, just redistributed. Active or electronic bass traps solve all those problems. They are small, portable (can be taken to another room and re-tuned), and can act on more than one resonant mode.

The Bag End's active bass trap is called E-Trap™, uses a 10" woofer, and sells for U$1,500.- It is not an automatic system. Instead, PC measurement software is included, which means the user first needs to perform the measurements to determine the frequencies that require damping and then set up the controls manually. Since two modes can be targeted simultaneously, the unit has two identical rows of controls to adjust the frequency and the amount of damping by changing the amount of feedback and the Q-factor of the filter.

The AVAA C20 (Active Velocity Acoustic Absorber) by PSI Audio also looks like a subwoofer but emits no sound. Its transducer (woofer) membrane is an active absorber with an efficiency equivalent to that of passive absorber up to 25 times its size. It also features a microphone, completely analog signal processing, and a driven woofer. The latency (delay) of DSP-based systems is thus eliminated for a much faster and tighter response.

Using two or three AVAA C20 bass traps, the before and after room response measurements in the frequency domain indicate a minor reduction in peaks (1-2 dB) with even lower "lifts" in the dips. At some frequencies, the peaks and dips got worse! Spending US$6,000.- for such a minor improvement would seem a total waste of money, but listening evaluations are unequivocally positive. It appears that the corrected time-domain response is a much better predictor of active bass traps' success, another fascinating example of the fact that what we measure and what we hear are two very different things.

For more information on the operational principle behind AVAA C20 see patent WO 2016/083971 Al, titled "Low frequency active acoustic absorber by acoustic velocity control through porous resistive layers." Search online and download US patent US7190796 B1 for the Bag End E-Trap background details, titled "Active feedback-controlled bass coloration abatement."

COMMERCIAL BENCHMARK: "Super Bass Extreme" bass trap by Vicoustic

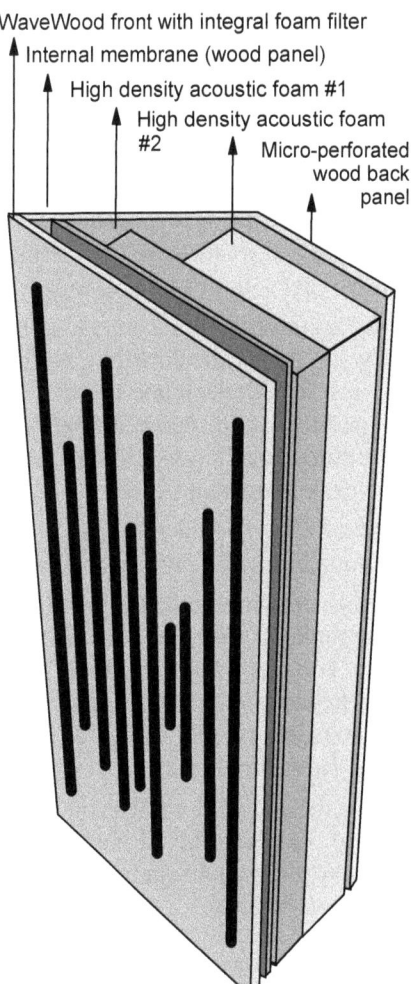

This smallish (595 x 595 x 155 mm) corner-mounted bass trap was designed to provide absorption between 60 and 125 Hz, with maximum absorption in the 75 -100 Hz band (upper bass). It consists of half a dozen or so layers of various materials.

The wooden front is based on Vicoustic's WaveWood panel and acts as a diffuser and a mid-high frequency absorber. The rest of the layers (a wooden membrane or resonator panel, two high-density foam layers, and a micro-perforated back panel) provide bass attenuation. At bass frequencies, porous absorbers such as foam or fiberglass suffer from reduced effectiveness (a rapid drop in SAC).

The manufacturer summarizes the operation of the bass trap as a resonant absorber where "... internal membrane transforms this high pressure by converting the pressure fluctuations into air motion. The membrane sympathetically vibrates over a frequency range of 75-85 Hz, causing the air to pass through a layer of high-density foam that absorbs the low frequencies."

Seen from its back, the micro-perforated rear wood panel (1 mm diameter holes) acts as a tuned cavity or Helmholtz resonator, further adding to the bass absorption.

Vicoustic claims that its tests show a 20% increase in efficiency when a perforated panel is used instead of a simple non-perforated panel.

In 2021, a pack of two bass traps sold in Australia for AU$895.-

Fig. 13.32. The cross section showing the main construction details of "Super Bass Extreme" bass trap by Vicoustic

COMMERCIAL BENCHMARK: *LF70 bass trap by SoundAcoustics*

Australian designed and made LF70 bass trap can be ordered with or without the front-facing reflector inserts, which come in two varieties, a clear acrylic glass (Plexiglas) sheet or an unfinished timber panel. SoundAcoustics developed the BEAR or "Bass Enhanced Absorption Reflector" concept at the RMIT University (formerly known as Royal Melbourne Institute of Technology) in 2014.

The 1m high and 52 cm wide trap is made of high-density acoustic foam. In 2021 they sold for AU$119.- each. Looking at the published SAC versus frequency graphs, it is evident that this "trap" is actually a broadband absorber. Without an insert, its SAC is very high (above 0.8) between 160 and 630 Hz, and then it drops slowly in an almost linear fashion to still relatively high SAC of 0.67 at 5kHz.

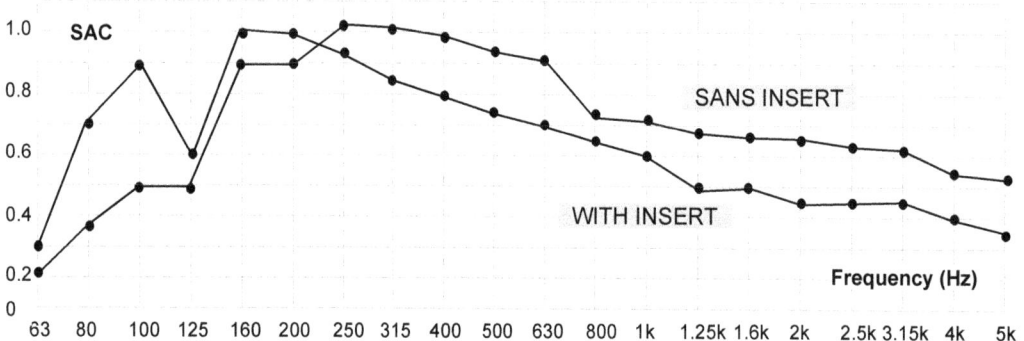

Fig. 13.33. Sound absorption coefficient versus frequency for LF70 bass trap (Source: SoundAcoustics website)

Fig. 13.34. The cross-sectional profile of LF70 bass trap by SoundAcoustics

0.21 at 63 Hz and 0.38 at 80 Hz are quite useful absorption coefficients, but with an insert, they jump significantly to 0.3 and 0.7, respectively. Thus, for these panels to perform as a bass trap, an insert is a must.

Above 200-250 Hz, the insert partially reflects higher frequency waves, thus reducing the overall SAC by about 0.2 across the 315 Hz-5 kHz band.

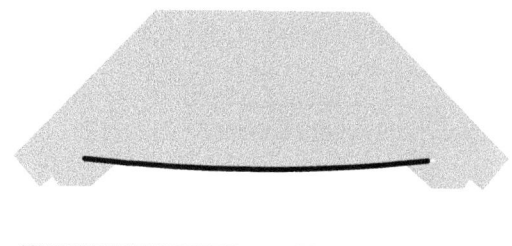

DIY PROJECT: *Turn bookshelves into bass traps & diffusers*

If you have a storage issue in your listening room and need a few deep & wide shelves for audio components you aren't using, this DIY bass trap kills two birds with one stone. Buy the tallest shelf unit (not necessarily a bookshelf, it doesn't matter what the shelf is for, a kitchen, bathroom, etc.) with at least three shelves (1), so you will get four or more "compartments."

Bass "accumulates" in corners, so bolt the shelves to the front or sidewall against the corners (2). There should be no space (gap) between its sides/top and the two walls. Fill the top and bottom compartment with acoustic material, fiberglass, acoustic foam, even pillow stuffing would do. You can then use the middle two shelves to store audio gear.

Alternatively, should you have a severe bass problem, fill all of the 'compartments" and cover them with decorative fabric.

Another option is to add some diffusion or reflection to the mix. Some modular shelving units come with optional glass or timber doors. You could add such doors to cover the two middle sections (4). Then cut-to-size and attach two-dimensional EPS diffuser panels (such as Multifuser DC2 featured on page 289) to the doors. How cool is that?

One of the standard widths is 60 cm, but narrower shelving units are also fine; even the 40 cm wide units will work well. As for the depth, again, with 40 cm being standard, anything between 30 and 60 cm can be used.

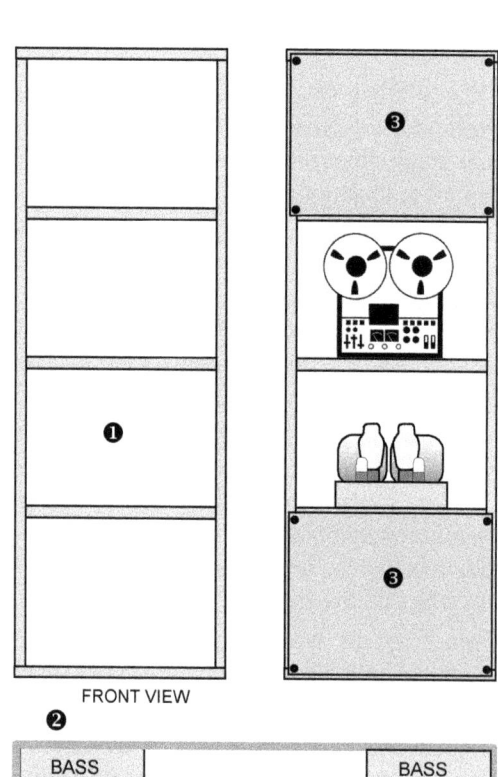

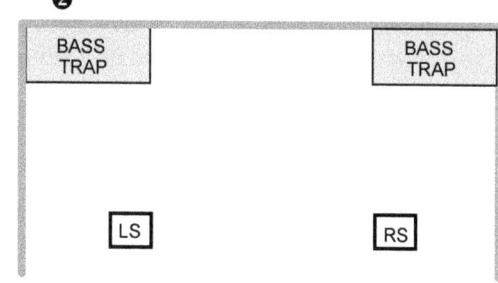

Fig. 13.35. A 3-shelf 4-compartment bookshelf and its conversion to a bass trap. Two such bookshelves should be placed in front corners.

ACOUSTIC TREATMENTS

DIY PROJECT: *Triangular bass trap*

Owens Corning type 703 and 705 acoustic insulation boards are made of inorganic glass fibers (Fiberglass™) bound with a thermosetting resin binder and formed into semi-rigid (703) or rigid (705) boards of various sizes.

The best to use for this purpose are 4' x 2' x 2" boards, which come in a pack of 5 and sell for US$95-120. The density of type 703 is 48 kg/m^3, with type 705 being twice as dense (96 kg/m^3), which means it will perform better in terms of sound absorption. Each type 705 2" board weighs around 8 lb. (3.6 kg).

The beauty of type 705 is that it is rigid enough to retain its shape, even when stacked relatively high, while still easy enough to cut.

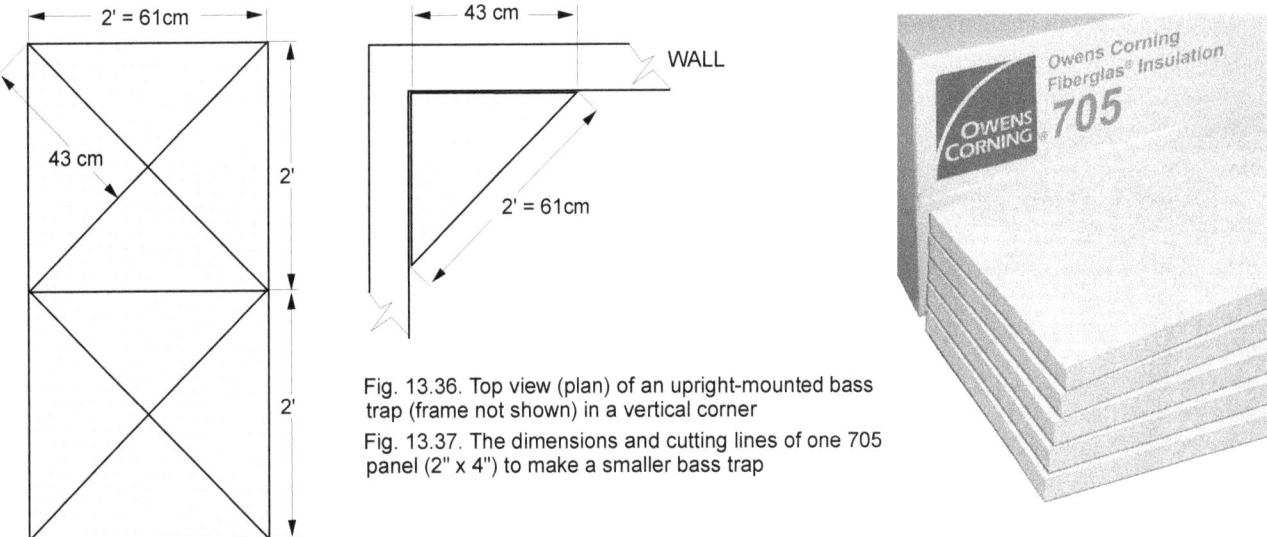

Fig. 13.36. Top view (plan) of an upright-mounted bass trap (frame not shown) in a vertical corner

Fig. 13.37. The dimensions and cutting lines of one 705 panel (2" x 4") to make a smaller bass trap

Depending on your budget, you can opt for a smaller size bass trap, where you cut each sheet in half and then each half into four smaller triangles, or a larger variant, where you cut each sheet in half and then diagonally into two larger triangles. The smaller variant will measure 43 x 43 cm, or 61 cm diagonally; the larger version will be 61 x 61 cm or 86.3 cm diagonally.

It's up to you how high you want your stack to be; each triangle is 2" thick (tall), so to cover 122 cm (48"), you'll need either three or six sheets. The effectiveness of the larger bass trap will extend into lower frequencies, but since its cost is double that of the smaller version, keep in mind that it won't be twice as effective.

With 24 triangles stacked on top of each other, there will be some compression of the lower ones. To avoid that, you could build a triangular frame out of thin plywood, with horizontal dividers after every 6 or 8 triangles (segments).

A rectangular frame can be built around the stack to hold the segments in place, and a decorative cloth may be used to cover the front and top surfaces.

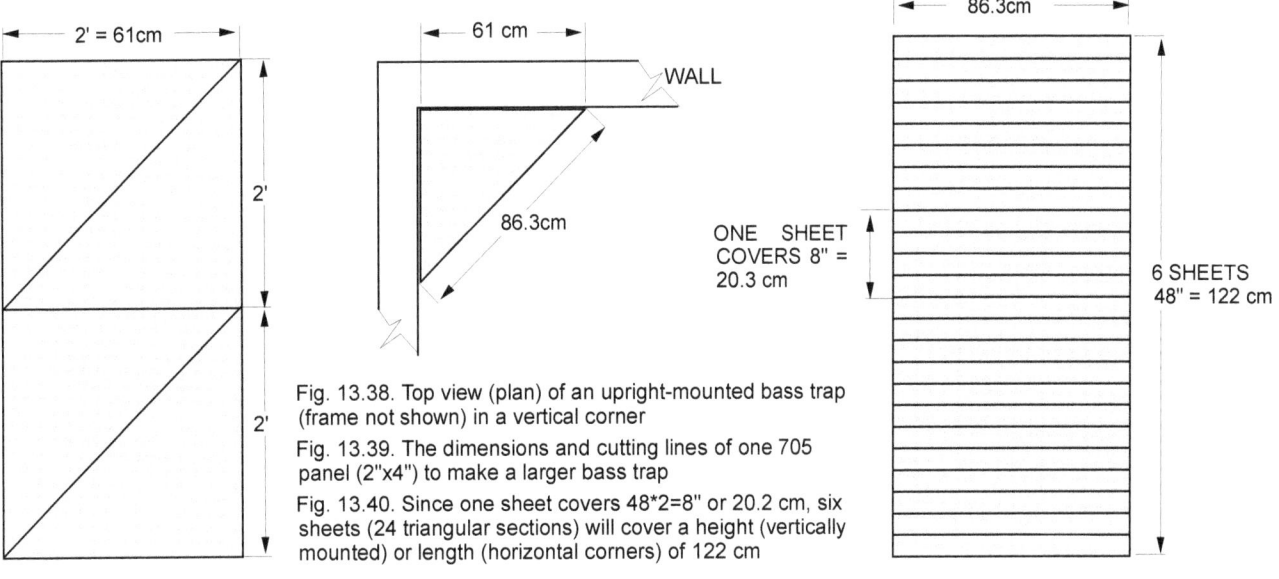

Fig. 13.38. Top view (plan) of an upright-mounted bass trap (frame not shown) in a vertical corner

Fig. 13.39. The dimensions and cutting lines of one 705 panel (2"x4") to make a larger bass trap

Fig. 13.40. Since one sheet covers 48*2=8" or 20.2 cm, six sheets (24 triangular sections) will cover a height (vertically mounted) or length (horizontal corners) of 122 cm

DIFFUSERS

Diffusers disperse acoustical energy throughout the listening room without removing acoustic energy (as absorbers do) while enhancing the room's sonic ambiance and the listener's sense of envelopment. In hi-fi listening rooms, diffusers are typically used on both the rear and front walls. This gives an aural illusion of pushing the walls out, making the room feel acoustically bigger.

How Quadratic Residue Diffusers (QRDs) work

A quadratic diffuser is an acoustic panel made up of a series of wells or troughs of different depths. The largest (deepest) well depth determines the highest wavelength (lowest frequency f_L) that will be diffused, while the width of the wells is around half of the shortest wavelength (upper-frequency f_U) to be diffused.

This applies only to incoming sound waves striking the diffuser straight-on. This high-frequency cutoff reduces for the waves with smaller incidence angles and is zero for waves hitting the QRD panel sideways (at 90°). Different well depths stagger the reflections and make them less coherent.

The wells run in parallel to one another, with thin dividers between them. A vertically oriented QRD will spread the sound waves in a horizontal plane, and a QRD placed, so the wells run horizontally will diffuse the sound energy in a vertical, fan-like array. Thus, a two-dimensional diffused sound field can be achieved by combining vertically and horizontally placed diffusers.

How to calculate well depths

The well depths (WDs) are calculated using the formula WD $= k*n^2$ modulo p, where p is a selected prime number (numbers not divisible by any other integer), such 5,7,11,13,19,23, and so on, and where n are integer numbers. Factor "k" is the proportionality factor.

The "modulo" function means residue. Although it all seems complicated, an example should clarify the calculation procedure, which is quite simple. For instance, let's choose p=19 and n=10, so the well depth will be WD = k*100modulo19, meaning we keep subtracting 19 from 100, which we can do five times (5*19=95) and the residue to 100 is 100-95=5.

Calculating all residues for p=19 and n ranging from 0 to 19 gives us the quadratic residue WD sequence 0, 1, 4, 9, 16, 6, 17, 11, 7, 5, 5, 7, 11, 17, 6, 16, 9, 4, 1, 0. Notice that after ten wells (in this case, p=19), the same sequence is repeated backward. In other words, the first and the second half of the entire sequence are mirror images of each other with respect to the centerline CL.

Multiplying well depths by the proportionality factor "k" results in the actual physical depth "d," so your well depths will be 0, k, 4k, 9k, etc. In this case, the deepest well is 17k deep, plus you may need one depth "d" at the bottom, resulting in an overall panel depth D=18d.

Remember, the larger "d" and thus also "D," the lower the f_L cutoff frequency, the diffuser will scatter. As always we see that lower frequencies (longer wavelengths) require physically larger acoustic treatments! If you only want to scatter upper midrange and treble frequencies, you can make your QRD panels very shallow (small "d" and "D").

If you choose your well width W to be say 5cm (around 2"), keeping in mind that $W=\lambda_U/2$, we get $\lambda_U=2W=10cm = 0.1m$, so $f_U = v/\lambda_U = 343/0.1 = 3,430$ Hz. Halving the well widths would double the upper scatter frequency to around 6,860 Hz!

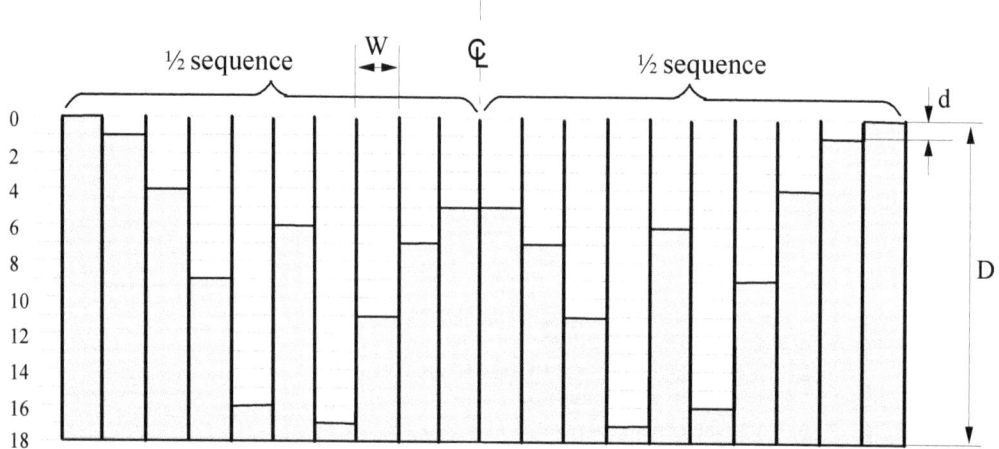

Fig. 13.41. The QRD profile based on quadratic residue sequence for *p=19*. The vertical lines are thin dividers between wells.

ACOUSTIC TREATMENTS

COMMERCIAL BENCHMARK: *Acoustic Diffuser QD-13 by Acoustic Fields*

Priced between US$1,500.- and US$2,000.- and weighing a hefty 105 lb., USA-made QD-13 is a wooden quadratic diffuser, 27¾" wide, 48" high and 12½" deep. Since it is based on prime number 13, it has 12 wells and a diffusion frequency range between 280 Hz and 3,400 Hz.

Let's check that lowest 280 Hz frequency claim. The overall depth of 12½" is approx. 0.3175 meters, so assuming that is the depth D of the deepest well, since $D=\lambda_L/4$, we get $\lambda_L = 4D = 1.27$ m. That corresponds to the lowest diffused frequency $f_L = v/\lambda_L = 343/1.27 = 270$ Hz. Close enough.

DIY PROJECT: *How to make your own QRD*

A 105 lb. (almost 50 kg) weight for a smallish commercial modulo-13 diffuser means almost prohibitive transport costs to overseas buyers and makes a DIY option even more attractive.

The space behind the wells' bottoms does not need to be solid or even filled; simply use cross-sectional dividers, which will all be of the same width and length.

The vertical well dividers will also all be of the same size. If you are constructing your QRD out of timber, use 1/8'-1/4" inch thick plywood panels.

Commercial QRDs omit the end two wells (1) and (2), so, in that case, you'll need 18 cross-sectional dividers (for the bottom of the wells) and 19 vertical dividers between wells.

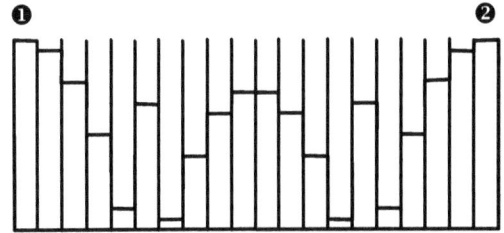

Fig. 13.42. Apart from the rectangular frame and the backing panel, for our p=19 QRD you will need 20 cross-sectional dividers (for the bottom of the wells) and 21 vertical dividers.

Fig. 13.43. QRD panels are large, heavy, expensive, and, most importantly, scatter in only one plane. Combining them in an alternate fashion (upright - sideways) is necessary.

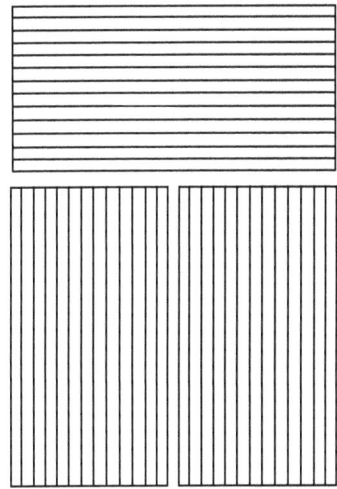

Two-dimensional "primitive root" diffusers

QRD panels are large, heavy, expensive, and, most importantly, scatter in only one plane, so they must be combined alternately (upright + sideways).

2D "primitive root" diffusers scatter in two dimensions and are generally lighter and cheaper (per area covered). Some are made of plastic, others of wooden or bamboo blocks, but generally, the cheapest are those carved out of polystyrene (EPS) foam.

Just as with their QRD brethren, there is lots of information on 2D diffusers online. For their operation, see US patent US5401921A, now expired, so make your own without fear or guilt. You'll have lots of fun, I'm sure.

COMMERCIAL BENCHMARK: *Vicoustic Multifuser DC2*

Multifuser DC2 is a two-dimensional "primitive root" diffuser panel made of high-density expanded polystyrene (EPS). It is glued to walls and/or ceilings, providing diffusion and scattering in vertical and horizontal planes.

Each panel measures 590 x 590 x 147 mm, six panels come in a pack that in 2021 sold in Australia for AU$895.- Since a 6-pack covers just over 2m², the cost is very high, around AU$447/m²!

While each panel's back and sides are flat, the front is an array of posts (196 of them in total) of the same footprint but varying heights. The heights are determined by the so-called "primitive root" numeric sequence. Some posts have flat front surfaces, while others are angled, further improving the panel's diffusion performance.

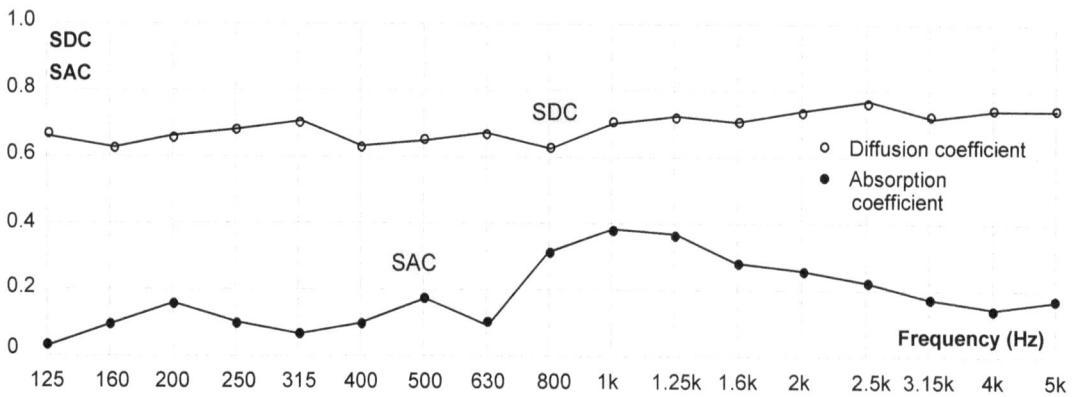

Fig. 13.44. Multifuser DC-2 diffusion and absorption coefficients versus frequency

Large bookshelves as broadband diffusers

If you have an extensive book collection and wonder where would be the best place for a large bookshelf or two (or three, or four, as in our case), the answer is simple: against the back wall of your listening room.

Books come in different sizes and thicknesses, so when stacked vertically (spine-out), they provide an almost random surface profile, which is a surprisingly effective broadband absorber and diffuser.

If your bookshelves came with a thin craft board or plywood backing, make sure you remove such panels before stacking the shelves with books. Such thin and usually flimsy backboards resonate and act as panel or diaphragm absorbers, which in this application is undesirable. They will also release such absorbed broadband energy with a delay, negatively affecting sound coherence and imaging.

COMMERCIAL BENCHMARK: SpaceCoupler diffusion panels by Auralex Acoustics

The SpaceCoupler looks like a wall shelf for figurines and trinkets. Three inches deep (7.6 cm) and measuring 2'x2' (61x61cm), each panel is an array of 64 square openings (8" x 8" each) and covers 0.37 square meters. Released in 2005 with MSRP of US$598.-, it mostly sold online for US$398.- for a pack of two panels. The cost is thus US$199.-/0.37 = US$538.- per square meter, a significant expense.

Fig. 13.45. Large bookshelves filled around 80% with books of various sizes work well as broadband diffusers and absorbers.

The effect of this type of acoustic treatment depends on the incidence (incoming) angle of the sound. At 90 degrees, the sound waves will pass straight through the openings.

As the incident angle reduces, more and more of the sound hits the inner surfaces and reflects at various angles (diffuses). Thus, by changing the location (wall-mounted, ceiling-mounted, suspended from the ceiling) and the positioning of the panels, the listening room's acoustic response can be altered.

DIY PROJECT: Wall- or ceiling-mounted diffuser/acoustic lens

These sets of three solid timber (pine) wall shelves (next page) gave us an idea to combine a dozen of them into a simple yet effective diffuser and acoustic lens. All three frames in a set are 10 cm deep and measure 30 x 30, 25 x 25, and 20 x 20 cm.

The finished assembly is relatively light, so it can easily be hung on a wall, as pictured, or suspended from the ceiling between the speakers and the listening position. This would provide a welcome diffusion of the early ceiling reflections and act as an acoustic lens, along with the principles used in Auralex Acoustics SpaceCoupler panels.

The only time-consuming task is drilling the frames and joining them, which can be accomplished using different methods. Hanging the finished diffuser on the wall only took ten or so minutes, mostly spent drilling three holes in the brick wall and inserting wall plugs. Suspending it from a ceiling was a longer and more frustrating task since ceiling joists had to be located and drilled into.

Fig. 13.46. A set of solid pine shelves sold in Kmart. Light and cheap, they can be joined in various ways to create acoustic lens panels.

Fig. 13.47. Four sets (12 shelves) were needed for this simple wall-mounted diffuser. An identical one was hung from the ceiling between the speakers and the listening position. Total cost for two? AU$100.-

DIGITAL ROOM EQUALIZATION SYSTEMS

The ultimate aim of acoustic treatments is to flatten (or "smooth out") the room's response curve by lowering the peaks and raising the dips without affecting any other aspect of its sonics. With acoustic panels and bass traps that is not generally possible, most have undesirable effects on other frequency bands.

Electronic or active bass traps and subwoofers with inbuilt room correction or equalization generally succeed at that task since their operating range is comparatively narrow, generally below 200 Hz. Digital room equalization systems operate on the same principle but over a much wider frequency band, although their effect is also most noticeable in the bass region.

Digital room equalization systems come either as an add-on feature of AV receivers or preamplifiers or as stand-alone units. These self-contained ones range from the very expensive (Accuphase DG-68 for instance, selling for US$24,000!) to small, cute, and affordable, such as DSpeaker Anti-Mode 2.0 digital room equalizer (US$1,099.-), and many other similar models in their lineup.

How they work

At the heart of any DRE system is a DSP, a digital signal processor, in which various parametric filters (equalizers) are digitally programmed. The parameters of such filters and other aspects of its operation are accessible and programmable through the software installed either internally (for stand-alone DRE systems) or on an external device such as a PC, smartphone, or tablet.

The system uses a calibrated microphone in various positions (the user is guided step-by-step in its positioning) to measure the room's response to multiple frequency sweeps and impulses generated by the DRE, amplified by a power amplifier, and reproduced by the speakers.

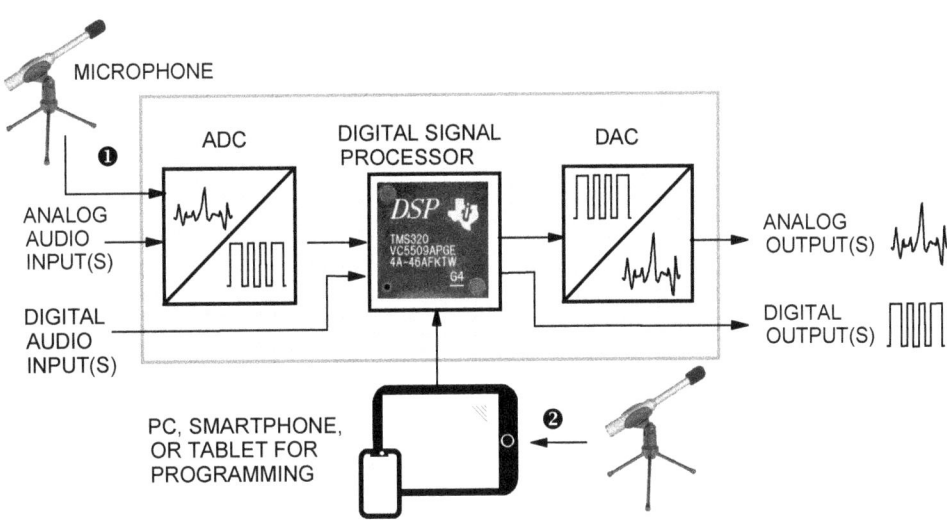

Fig. 13.48. Block diagram of a digital room correction system (a stand-alone unit, externally programmable).

Some feature a direct microphone connection (1); others connect the measuring microphone to the external programming unit such as a PC, tablet, or smartphone (2).

Once the irregularities are identified, the DSP calculates the optimal configuration of its software filters to minimize them. Some DRC systems are fully automatic; with others, the parametric equalizers must be manually set. The more flexible automatic ones still allow users to manually fine-tune the response by editing the target EQ curves to suit their tastes.

Digital audio sources can be directly connected to the DRE's digital audio inputs, but analog sources first must undergo the analog-to-digital conversion in the ADC converter. Likewise, after DSP processing, the DRE system must perform a digital-to-analog conversion, and such modified analog signal is then fed to a power amplifier from its analog outputs.

And this brings us to the fundamental objection of the puritan brigades, and that is the claim that such dual-conversion from the analog to a digital domain and back cannot be but detrimental to the sound quality. Again, we see the "There's no free lunch" law in action: we fix one problem but create two or more new ones!

Passive acoustic treatments and subwoofers with their own active equalization - to keep or not?

Generally, if sound pressure dips and peaks resulting from standing waves and other anomalies in the room's response (such as those illustrated on the right) can be reduced by *passive* acoustic treatments, that should be attempted first. Such treatments should then be kept in place before the application of DRE systems.

As for subwoofers with inbuilt active room correction, in some cases, their own equalization should be kept operating; in others, it should be disabled, so some experimentation is needed.

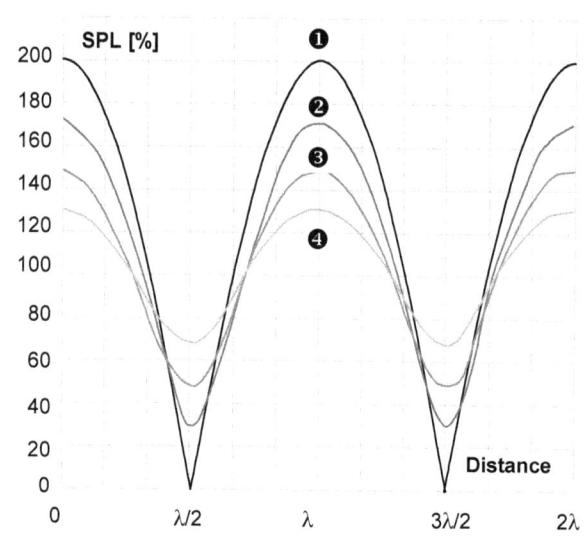

Fig. 13.49. Single sinusoidal frequency standing waves (sound pressure levels) as a percentage of the original sound amplitude (100%) for 1) 100% reflective surfaces 2) 70% reflective ends 3) 50% reflective surfaces, and 4) 30% reflection

---

THE "PEAKS & DIPS" RULE
Digital room correction can reduce or even completely flatten the peaks of standing waves, but it cannot eliminate or even substantially reduce the dips.

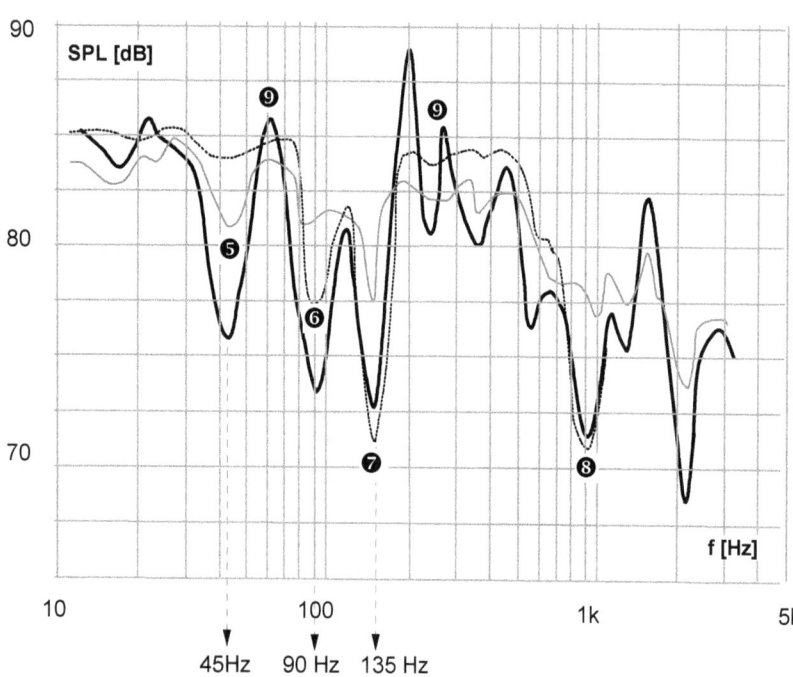

There are dozens of DRE systems in the marketplace, ranging from next-to-useless to very good. None will correct *all* the peaks & dips, and none will flatten them *completely*.

In the example (left), system A (dotted line) corrected dips (5) and (6) better than system B (solid thin line) but made dips (7) and (8) slightly worse, a pervasive issue. As for the peaks, for instance (9), system A flattened them better than system B.

Fig. 13.50. The uncorrected room's frequency response (thick line) and the frequency response after equalization by two different digital room correction systems (thin line and dotted line).

14 | MINIMIZING UNWANTED VIBRATIONS & OSCILLATIONS

Electromechanical transducers such as microphones (used in the sound recording) and phono cartridges and loudspeakers (on the sound reproduction end) are the only parts of the audio chain that should vibrate freely. Others (amps, preamps, cables, audio racks) should not vibrate at all.

Unwanted vibrations of larger objects in a listening room (glass doors and windows, pictures or mirrors on a wall, even sofas) produce sound waves that bear no harmonic or temporal relationship to the original sound, negatively impacting the sonic presentation in various ways.

The vibrations of smaller objects, such as a turntable tonearm or a power transformer in an amplifier, may not be directly sonically detectable but interfere with and degrade their performance, increasing distortion and changing the frequency response and tonal balance of an audio system.

Vibration reduction and minimization measures usually don't involve significant expenses. The sonic improvements are similar to those achieved due to cleaning up the power supply: cleaner, tighter bass, more precise and stable soundstage, the removal or fuzz, grit and haze around notes, and, ultimately, a more relaxed and "natural" sound.

- UNWANTED VIBRATIONS: CAUSES AND REMEDIES
- HI-FI RACKS AND EQUIPMENT PLATFORMS
- CABLE LIFTERS (RAISERS)

> "If you want to find the secrets of the universe, think in terms of energy, frequency and vibration."
> Nikola Tesla, Serbian inventor

UNWANTED VIBRATIONS: CAUSES AND REMEDIES

One of the most fundamental rules in electronic design is that specific points in a circuit (such as an amplifier) should be at a rock-steady voltage level, while others should be free to move up and down with the signal. All electrical rules have their mechanical and acoustical equivalents, and this rule applied to acoustic is that some vibrations are good and desirable while others are less so. Only two things in your audio system should vibrate freely, namely the cones of your speaker drivers (or the membranes if you are using electrostatic panels) and the stylus/cantilever in your phono cartridge. Anything else, ideally, should not vibrate at all.

> **THE VIBRATION RULE**
> Only a couple of things in your audio system should vibrate freely, namely phono styli and loudspeaker cones. Nothing else should vibrate at all.

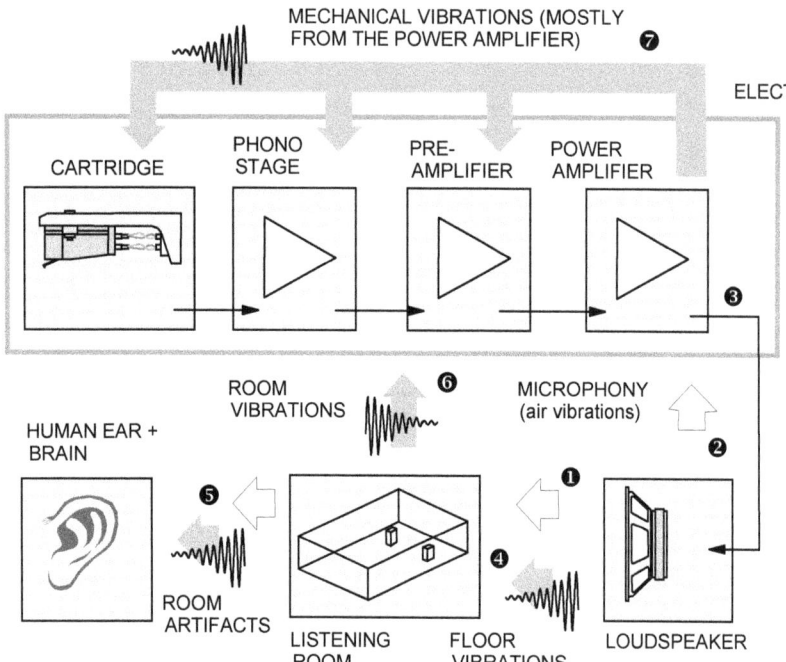

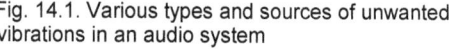

Fig. 14.1. Various types and sources of unwanted vibrations in an audio system

Fig. 14.2. Floor vibrations combine with vibrations from audio components (mainly power amplifiers) and propagate through the hi-fi rack to vibration-sensitive sources such as CD players, audio streamers, and turntables.

The block diagram above summarizes the various types and sources of unwanted vibrations in an audio system. While speaker drivers' cones vibrate due to electrical signals from the power amplifier (3) and produce sound (1), the speaker's enclosure also vibrates, and such vibrations (4) travel through its coupling with the floor and propagate throughout the listening room.

They get mixed with various room artifacts (standing waves, room modes, comb filtering, slap echo, various types of reflections, etc.) and affect the listeners' perception of the sonic presentation (5).

The speakers' direct sound and the distorted sonic artifacts added by the room combine; such vibrations hit the electronic part of the audio system (turntable, CD player, DAC, streamer, preamplifiers, amplifiers) and cause their chassis and internal components to vibrate (6).

These external vibrations combine with mechanical vibrations (7) from audio components (primarily power amplifiers) and propagate through the hi-fi rack to sensitive sources such as CD players, audio streamers, and turntables.

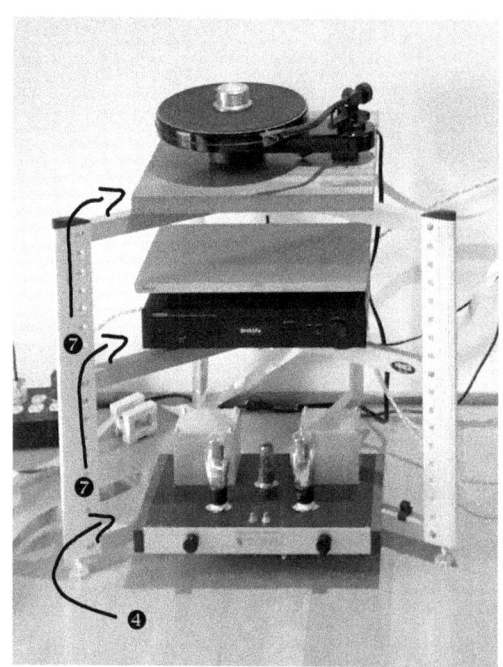

Transformer vibration checks

Due to the magnetostrictive effect, the laminations of power and output transformers always vibrate; it is only a matter of degree. The vibrations of better quality and potted (encapsulated) transformers are almost entirely dampened.

On amplifiers whose transformers are clearly identifiable (either of the "open" type with end-bells or enclosed (under metal covers), with speakers connected but no music playing, listen for any buzzing from the power transformer (put your ear against it). This test can be done on both tube and solid-state amps.

The sound that output transformers produce is generally masked by the music playing through the speakers. This is undesirable - you don't want them to act as miniature parasitic speakers. To check your tube amp, with music playing, briefly disconnect both speakers and listen with your ear close to the transformers.

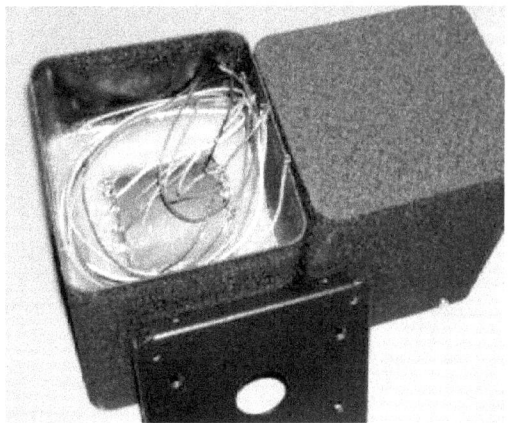

Fig. 14.3. Some transformers are potted (encapsulated in a hardened resin) inside their metal covers.

They are usually much smaller than their "cans" or enclosures, so don't judge tube amps by the size of their transformer enclosures.

Do this test very quickly since operating tube amps without connected speakers can damage output transformers. Do not disconnect the speakers from solid-state (transistor) amps. They don't have output transformers anyway, so such a test would be meaningless.

Why are low level parasitic vibrations so sonically damaging?

Although most parasitic vibrations so far mentioned are of relatively low level or magnitude, they still affect the signal produced or amplified by audio components. They add electrical distortion to the speakers' inputs and produce unwanted acoustic output (sound).

The main problem is that the added electromechanical distortion is not just the (relatively innocuous) harmonic distortion. Most such parasitic vibrations happen at frequencies unrelated to the original audio signals and are also delayed in terms of time and phase shift.

Various materials and components absorb some of the incoming sonic energy and then release a portion of it sometime later. Such phase distortion and time delay artifacts are the most sonically damaging since they are harmonically unrelated to the original sound. Critical aspects of sonic fidelity such as coherence, sound stage, transparency, and transient response are impacted.

Hunting down and removing various parasitic resonators from your listening room

As a rule in life in general, and in electronics and acoustics in particular, preventing problems from happening is always better than trying to deal with them later on. Hunt down the sources of parasitic mechanical vibrations and eliminate them mercilessly. If they cannot be eliminated, vibrations should be either absorbed (reduced) or isolated, so they cannot propagate and reach sensitive components where they can do sonic damage.

Again, the higher up the sound chain a component is, the more it's susceptible to vibration. That means that turntables, phono stages, and CD players are your prime candidates for anti-vibrational treatments. Preamps and amps are the lower priority, although damping their enclosures and thus minimizing their mechanical vibrations does often result in significant sonic improvements.

Turntable lids, floating timber floors, bookcase backing panels, sofas and other parasitic resonators

Due to its considerable size, a turntable lid, usually made of acrylic material, is a large resonator panel. Due to the microphonic effect, sound waves emitted by loudspeakers will cause it to vibrate. The lid will transmit such vibrations to the turntable plinth it's attached to and thus affect its sonics.

One solution is to remove (detach) the lid while listening. However, with many turntables, that is not possible. If the lid of your turntable is bolted onto the plinth, remove it permanently and use a soft dust cover, such as a piece of thick cloth.

Other household items that will change the sonics of a listening room are sofas and oversized upholstered chairs with tightly stretched backs. Large oil and acrylic paintings (without glass front covers), whose canvasses are also tightly stretched, are also problematic. These vibrate and resonate but also act as unwanted midrange and high-frequency absorbers.

Even surfaces which we normally consider "hard" or "solid" vibrate. They store and release acoustic energy in unpredictable ways, smearing transients and coloring the tonal balance.

Floating timber floors, non-solid external and internal walls, and ceilings (gypsum boards on studs or any cladding used in timber-framed houses) are common in many parts of the world.

Large bookshelves often have closed backs, usually a thin panel made of craft board, fiberboard, or similar material. Even if hard-pressed against the wall, there are still a few millimeters left for those back covers to vibrate, and vibrate they do. They could be stiffened by additional bracing, but the best and easiest solution is to remove such backing panels altogether.

While multiple pairs of speakers are almost always present in audio dealers' showrooms, they are less likely to be found in domestic listening rooms. Nevertheless, since they are free to move and vibrate due to the varying air pressure from the active speakers, unused speakers' cones act as microphones and passive resonators. This will affect the sonics in subtle and, depending on the size of the bass and midrange drivers, not so subtle ways, so remove all unused loudspeakers from the listening room and store them safely somewhere else

Power transformer mounting

In this vintage Japanese hi-fi tube amplifier, the whole power supply section is on a separate steel base, "floating" on rubber mounts (1). This minimizes the transmission of vibrations from the mains transformer (2) onto the chassis, onto which the audio tubes' sockets (3) are directly bolted.

None of the modern audiophile amplifiers that I've seen take such precautions, presumably to minimize costs, or their designers don't care about this issue.

Ventilation holes around the rectifier tube (4) improve airflow to help with cooling.

Fig. 14.4. Suspension-mounted power supply of a vintage Japanese tube amplifier.

How to check tube amps and preamps for microphony

Many tubes and tube amplifiers are incredibly microphonic. Stay away from those. Any tube can be or become microphonic as it ages, but some tube types are known for their microphonic propensity. Generally, high gain pentodes are the worst offenders, although some triodes can also be troublesome.

Internal parts of a microphonic tube vibrate due to the sound waves hitting the tube or vibrations that reach the tube through its socket. Before you even start listening to a tube component (phono stage, line preamp, power amp) in a dealer's demo room or when buying privately, perform this simple microphony test.

If the tube amp or preamp fails it, move on; don't even listen to it. Sure, the particular tube the seller has installed in the amp may be microphonic. By replacing that tube, you may reduce or even eliminate the microphony, but would you take the gamble?

If it's the printed circuit board or tube sockets (ceramic tube sockets are the worst offenders!) that are microphonic, there's nothing much you can do except avoiding such amps and preamps. We've had quite a few cases of Chinese-made preamplifiers and amplifiers that were extremely microphonic and, as a result, sounded terrible.

Assuming the amp or preamp is powered up and connected to the rest of the audio system (but without any music playing), to test tubes for microphony, tap the tube's glass bulb and listen for any sound from the speakers. A microphonic tube will produce a distinctive sound in one or both speakers (depending on where in the circuit it is). You can use a pen, a pencil or a small stick (chopsticks, for example). You could even flick tubes with your fingernails. Since the contact is of such a short duration, you will not get burned even with the hottest of tubes.

The same acoustic phenomenon will happen with acoustic feedback from the speakers to the amp or preamp; it's just that the primary signal of the amplifier will mask it. Although you will not be able to isolate it acoustically or electrically to listen to it by itself, it is there and will cause distortion, smearing, and loss of focus. It will impact the sound stage and accuracy and transparency of the whole system!

Tube dampers

Although they work as electronic amplifiers, just like solid state transistors, vacuum tubes are mechanical structures built by hand and are of the considerable physical size. Thus, they are much more susceptible to vibrations, which can be passed from transformers through the chassis-mounted sockets or from loudspeakers through the air directly onto the glass bulb and the electrodes inside it (microphony).

In demanding audio- and radio-frequency applications, tubes are mounted on suspended anti-vibration mounts. You can see them in vintage tube tape recorders and other audio equipment, but rarely in contemporary audio gear. Today's tube equipment designers are either ignorant (don't understand how tube vibrations can significantly degrade the sound), or they are overruled by the accountants and managers in their relentless race for profits.

While tube dampers aren't as effective as anti-vibration mounts, they improve clarity, microdynamics, and focus, and reduce the noise floor. A very simple accessory that is worth investigating and trying out in your tube-based audio components, tube dampers use various materials and come in different shapes and sizes.

Silicone rings, resistant up to 240°C, are generally the cheapest yet quite effective in minimizing microphony and vibration. Illustrated on the right are tube dampers of various sizes.

The small ring on the left is for Noval (9-pin miniature) tubes such as 12AX7, 12AU7, 12AT7, and EL84, which require 20 mm diameter dampers. In the middle is a 28 mm damper for small octal tubes (preamp tubes) such as 6SN7 and 6SL7, power tubes such as 6V6, and small rectifiers such as 5Y3.

Medium-sized power tubes such as F2A (pictured) or EL34 need 32 mm rings. Use 39 mm rings for power tubes such as 6L6 and 44 mm dampers for KT88 and KT120.

Two rings of different diameters will be needed to be used with "coke bottle" tubes such as 300B (tubes whose glass bulb is not a straight tube and changes diameter).

Fig. 14.5. (L-R): various size tube dampers: for Noval (9-pin miniature) tubes, small octal (preamp) tubes and for medium sized power tubes (F2a pictured).

Prevention is always better than cure

Vibration treatments or "anti-vibrational" devices work in two ways, they either isolate the vibrations, preventing their transmission or dampen them by reducing the amplitude and duration of the vibrational wave. Some of the vibrational energy is absorbed and converted ("dissipated") into heat. Preventing vibrations from happening is, obviously, preferable to dealing with them once they've already developed.

The speaker-floor interface

When it comes to speaker mounting methods, there are two schools of thought. One is to mechanically couple the speakers to the floor via spikes, for instance. Such rigid "grounding" increases the effective mass and reduces speaker cabinet movement and vibrations.

However, now such vibrations are transferred into the floor, and if the floor has even one degree of freedom, it will become a source of sound itself. Such absorption and release of sonic (vibrational) energy take time, so the sound retransmitted by the floor is delayed, causing time smear and loss of transient detail.

This is more of an issue with floating timber floors, for instance, which can move and flex, less so with solid parquetry floors that are glued to the concrete slab and much more rigid. A suspended timber floor in many American houses can also be problematic, for it constitutes an enormous resonating panel.

The type of flooring in your listening room will further complicate this issue since spikes are not equally effective on all surfaces (carpet, timber floor, tiles, etc.) Using speaker isolation feet, such as those produced by Isoacoustics and others, reduces the speakers' mechanical coupling with the floor, thus minimizing the retransmission effect.

However, Newton's third law of motion, the law of action and reaction, states that when two objects interact, they apply equal forces to one another but in opposite directions. Thus, when large speaker cones move to produce sound, the reactionary force is exerted onto the speaker cabinet, which must move in the opposite direction.

With a rigid floor coupling (spikes), most such energy is transferred to the floor. Now it will remain trapped inside the speaker enclosure, which will absorb some of that energy and release it with a delay, a recipe for sonic smudge.

Even loudspeaker designers and acoustic experts disagree on which mounting method is better, so experimentation is in order. Try both and make up your own mind. The two setups will sound different, but the question of which sounds better remains.

However, every time you install or remove the spikes and speaker supports in the course of listening evaluations, you have to be very careful to place the speakers back in the exact original position.

In some audio setups, changing speakers' position or direction (toe-in) just 10 or 20 mm will result in noticeable sonic changes, which could erroneously be ascribed to the change of speakers' mounting method. Just as with any scientific experiment, audio comparisons have to be done by only changing one parameter or aspect at a time.

The Rolls Royce engine coin test on loudspeakers

The famous Rolls Royce engine coin test has been used for decades to demonstrate how vibration-free a well-designed and balanced car engine is. A coin placed on its edge on top of the engine block should not fall even during engine start and revving.

The test is also helpful in evaluating loudspeakers. Speaker cabinets should not resonate (vibrate) when the music is played, but no speaker cabinet is free of resonances; it is just a matter of finding the right frequency of high enough amplitude to excite it. Most cars and many loudspeakers will not pass this test.

Fig. 14.6. The Rolls Royce coin test on Opera Terza loudspeakers using an Australian 50 cents coin (dodecagon or 12-sided polygon) They passed, the coin did not fall, most likely due to the rigid, solid mahogany cabinet and the spikes that couple the speakers rigidly to the timber floor.

The knock test

You may have been to a doctor who tapped your back or abdomen with their fingers and wondered what possible use such a seemingly primitive check could be. Known as percussion tapping, it's one of the "hands-on" methods used to discern the condition of internal organs such as the liver, lungs, and even muscles.

Each part of the body should "sound" a certain way, and if it doesn't, it indicates an abnormal condition that needs further investigation. For instance, a resonant sound should be produced by a hollow, air-filled cavity or structure such as lungs; a dull ("flat") sounding lung may indicate a presence of water or a solid mass (tumor).

However, a liver or muscles should sound dull and flat. Likewise, abdomen tapping should produce a "tympanic" type of sound, relatively loud and higher-pitched. The skin over the stomach area, the muscle layers under it, and the abdomen cavity itself form a resonating structure, just like timpani or kettledrums.

An analogous test can be done on loudspeakers, turntable plinths, and acoustic panels. The test itself is super easy, but interpreting its results is another matter. Like many issues discussed in this book, it requires intuition, experience, and knowledge of physics and engineering principles.

Speaker builders know that as various material choices (solid timber, MDF) sound different when tapped while still in the form of boards, planks, and strips, they will sound different once made into speaker enclosures. The knock test is most helpful in evaluating the resonant character of electromechanical transducers such as turntables and loudspeakers.

A turntable plinth or base should not sound resonant or "hollow"; the vibrations caused by your tapping should die down as quickly as possible (a high degree of damping).

When tapped, the sound produced by a loudspeaker will depend not just on its cabinet construction and damping but also on few other factors. The interface with the floor or speaker stands (spikes, cones, or different types of supports), the type and resonant properties of the floor, and the design and acoustic properties of the speaker stands are all part of the sonic picture.

That being the case, it can be concluded that speakers' sound quality also depends on all these factors.

Just below the two rear-firing bass-reflex ports (3), our speakers have a horizontal divider that enables their lower section to be filled with sand or any other type of granular material. The denser (heavier) the filler, the better, since it gives the speakers stability and acts as a vibration damper. Remember, the speaker cabinet itself should not vibrate or move; only the drivers' cones and the air inside should do so.

Tapping the speakers' top, back, and upper sides (1) produces a different sound, the longer and larger sides usually being much more resonant than the smaller and thicker top.

That is all normal and to be expected. Tapping the speaker's side towards the bottom, where the filled compartment is (2), produces a deeper, duller, less resonant sound than the upper part of the same side panel (1) behind the woofer.

In short, look for speakers whose cabinets (when knocked on) resonate as little as possible; the flatter and duller the sound, the better.

Fig. 14.7. The knock test on a loudspeaker

HI-FI RACKS AND EQUIPMENT PLATFORMS

Apart from the impact of the room acoustics on the overall audio system's performance, another critical part of an audio system that is often ignored or paid very little attention by audiophiles is the equipment rack. The rack is usually bought *last* as an afterthought, but it is truly a fundamental cornerstone of every superior system. Buy your audio rack *first*, so you have a stable, vibration-free platform for evaluating all subsequent audio components, such as turntables, CD players, phono stages, line preamplifiers, and power amplifiers. And buy the best, most expensive rack your budget will allow, so you don't have to upgrade it as your system improves.

As always, there's also the DIY option, or a combination of the two, as in buying a quality bargain-priced audio rack (used or new at a clearance sale) and upgrading its shelves or anti-vibration supports later on.

What to look for in an audio rack

There's a mind-boggling variety of hi-fi racks on the market, covering a vast range in both price and features: timber, steel, aluminium supports ("legs"), glass, wood, solid stone, or carbon fiber shelves, plus various "proprietary" features, such as grooves, spikes, dampening gaskets and other vibration-reducing treatments.

As for configuration options, there are only two types - fixed and modular racks. With the modular ones, you can choose the number of shelves and the distance between them (the height of the uprights), which adds versatility. Some racks even offer you a choice of bolted or spiked uprights. Due to transport costs and to minimize the chance of transport damage, even fixed ones usually come flat-packed, so that most racks will require some assembly.

Cheaper racks usually have tubular steel uprights ("legs") and glass or MDF shelves. Some legs can be filled for added mass, reducing resonance and vibration. More expensive varieties have solid timber, bamboo, marble, or reconstituted stone shelves (Corian® and similar materials).

Our equipment racks don't use any shelves; the audio components were supposed to sit on rubber pads that slide across the top of the four solid timber crossbars (1). The crossbars hold the four machined aluminium legs or "uprights" (2) together and are joined in the middle by a star-shaped solid machined aluminium ring (3). Each leg has a screw-in spike at its bottom (4) which can sit directly on the floor (but will, just like speaker spikes, damage a timber floor) or in a supplied dish, also machined out of aluminium.

The 18 holes, pre-drilled into the uprights, make it possible to adjust the height between the crossbars by moving them up and down.

Fig. 14.8. Our two audio racks. The width between front legs ("uprights") is much larger than between the two rear ones.

Upgrading the shelves of your hi-fi rack

Shelves made of different materials will imprint their sonic properties onto the sound of the audio components sitting on them and thus onto the whole audio system. Some stylish hi-fi racks come with glass shelves. I found them inferior to pretty much all other materials. They add a glassy and brittle edge to the sound, making the system sound cold and fatiguing.

Shelves or slabs cut from natural or engineered stone are a better option. The most common choices are marble, a natural stone, and various engineered materials for kitchen benchtops and bathroom vanities, such as quartz stone and Corian®. Corian® is DuPont's brand name but is often used for the whole class of similar materials from other manufacturers. Its composition is roughly 2/3 minerals and 1/3 synthetic polymer (filler).

Not all slabs made of natural stone are equally good sonically. Many audiophiles have found granite sonically unsuitable, imparting a harder edge and grainy character on the sound.

Shelves made of natural or manufactured stone are stable and add a welcome mass and mechanical inertness to audio racks. However, they are hard surfaces, and to some audiophiles, that's how they sound - hard! So, do your research by auditioning such shelves (if you can find a dealer that uses them) before you fork out the money.

Another option is to approach a few of your local kitchen and bathroom cabinet-making businesses that use marble vanity and kitchen tops and ask them for any offcuts or odd sizes they cannot sell or use. You may get a bargain. However, in my case, the quotes I got for natural or artificial stone shelves were all between $300 and $600 per shelf. Since I needed six of them, the total cost would be $1,800-$3,600! Perth, Australia, must be one of the most expensive places globally; local tradespeople charge exorbitant hourly rates, and materials of any kind are costly.

Dealing with such an extortionist attitude certainly makes you more motivated to develop and improve your DIY skills and turns your attention to finding an imported product that could fit the bill. These are always much cheaper due to their countries of origin significantly lower labor costs (China, Vietnam, Indonesia, and many other Asian and even some European countries are in this category). And that's exactly what I did. I forgot about the hard stones and turned my attention to prefabricated timber products.

DIY PROJECT: *Make your own timber shelves and equipment platforms*

Many musical instruments are made of timber (I cannot think of one made of hard rock or glass), meaning that timber shelves should impart a softer, more natural, and more euphonic sonic imprint onto the music.

The Pro-Ject turntable, due to its construction (not of a standard, rectangular shape with a box-type plinth), could not sit on the crossbars of our shelf-less audio racks and required a solid shelf.

"Mondella Rococo" range of bathroom vanity tops in a hardware store was just what the doctor ordered. Made in Indonesia from laminated pieces of meranti wood, with 30 mm thickness and 465 mm width, it comes in various lengths, 60, 75, 90, 120, 150, and 180 cm.

The 90 cm top (AU$67 or US$48) was chosen and sawn in half, making two shelves 465 x 450 mm each. We lost 10 mm due to uneven cutting (using a wonky circular saw) and filing. After sanding and a few coats of satin polyurethane finish (lacquer), the shelves looked tremendous and undoubtedly improved the sound of audio components placed on top of them.

Timber chopping boards - cheap equipment platforms and rack shelves

Solid stone or timber shelves are not the only options. Strictly speaking, bamboo is not wood but fast-growing grass. Due to its thin and tall (elongated) shape, laminated bamboo boards are made from flat rectangular bamboo strips cut from their stems. These are glued together under pressure either horizontally or vertically (side pressed).

Bamboo is used to make floorboards, chopping boards, shelves, and furniture. Lightweight material with a high strength-to-weight ratio, bamboo is ideal for equipment platforms and hi-fi rack shelves. Bamboo's composite cell structure results in its unique mechanical properties, vibration absorbing, and resonance dampening (suppressing), being the qualities of most interest to us as audiophiles. Bamboo also has natural antibacterial properties, making it naturally mold-resistant. It is a renewable and sustainable structural material.

However, commercial audiophile bamboo hi-fi racks and shelves are pretty pricey. For instance, UK-made Atacama Evoque Eco 60-40 SE base module (a single bamboo shelf on four legs) sells for AU$679.- in Australia, while the 4-shelf hi-fi rack costs a whopping $2,716.- (only £799.98 in the UK)!

As always, the DIY option will save you a small fortune. Although most bamboo chopping boards are smaller than 300mm in size (the longest dimension), there are larger ones, 400 x 300 and 500 x 400 mm, perfect for medium-sized amplifiers and other audio components. Buy a couple and try them out, and if you don't think they improve the sound of your system (I do!), you can always use them in your kitchen.

Fig. 14.9. The AU$9.99 IKEA Aptitlig bamboo chopping board measures 28 x 45 cm, perfect size for smaller audio components such as phono stages and preamplifiers.

No matter how fashionable it may be, bamboo isn't the only widely available timber. This acacia "Long Grain Cutting Board" (below right) measures 48 x 35 cm and sells for AU$55.- The hanging loop with metal insert may be used to thread a thinner cable or two and thus keep the wiring neater and tidier. You can easily add another one or two holes by drilling the board either next to it or on the other side of the board, as required.

Measuring 50 x 35 x 4 cm, the acacia "End Grain Cutting Board" (pictured below left) is slightly longer and thicker, but also more expensive, AU$69.95. It is made of 128 solid blocks, which were joined together, sanded and finished. Obviously, the two types of boards will resonate and absorb (dampen) vibrations in a different way. Assuming upward or perpendicular vibrations from the floor through the rack and across the thickness of the board, such disturbances will hit them either along or across the grain.

Fig. 14.10. "End grain" (ABOVE LEFT) and "long grain" (ABOVE RIGHT) cutting boards are affordable options for DIY audio equipment rack shelves. No cutting, sanding, coating or messing around in any way.

Carbon fiber shelves

Carbon fiber reinforced polymer (CFRP) is a composite material, a mix of carbon fibers and polymer resin. A 5 mm thick carbon fiber board, measuring 300 x 500 mm and weighing around 1.2 kg, sells for around AU$135.- including free shipping direct from China. A similar 5 mm thick but larger (500 x 500 mm) plate made from carbon fiber fabric, prepreg carbon fiber, and epoxy resin sells for AU$262.- in Australia. Plate thicknesses of up to 10 mm are available and can be CNC cut to any size or shape. Even the most expensive 10 mm thick boards are still cheaper than custom manufactured marble or stone shelves.

CFRPs are poor conductors of electricity, but they do conduct, so, in comparison to timber and hard stone rack shelves, such shelves provide some degree of RFI shielding.

Isolating equipment from external vibrations

There are two schools of thought resulting in two approaches to damping or reducing external vibrations and the resonances they may induce. We can tackle vibration at the global level, treating the whole equipment rack, for instance, or we can deal with various audio components separately, isolating each in the most optimal way. Of course, a combination of both approaches is also possible; in fact, it could be the most effective approach.

As discussed, with a single power conditioner or regenerator, one audio component can back feed interference through the common mains connection into other audio components connected to the same power conditioner output. Likewise, mounting the whole audio rack on spikes or cones can reduce external vibrations, such as those originating from speakers; yet, it cannot do much against vibrations generated by equipment on the rack itself!

Transformers in audio amps and preamps will still vibrate, and such vibrations will find their way into other components that share the same rack; turntables, MC step-up transformers, and phono stages are especially sensitive.

Unless, of course, each of the shelves on the rack rests on anti-vibrational mounts as well.

The obvious candidates for the anti-vibrational treatment are turntables and digital sources. Except those on high-cost models, the stock feet or supports are usually made of either rubber or plastic; these do nothing in terms of vibration reduction.

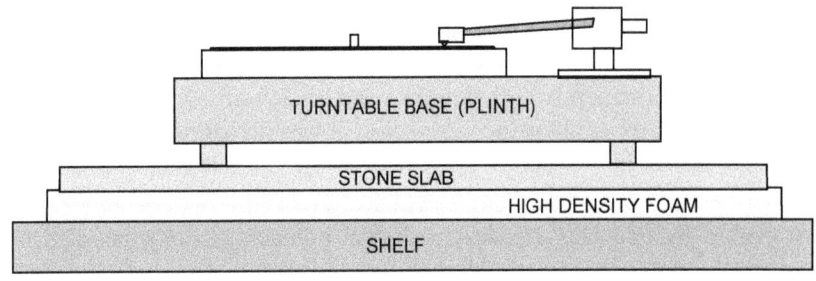

Fig. 14.11. The soft sandwich arrangement for turntable support features a stone slab on top of high-density foam, which acts as a cushion to dampen vibrations.

The simplest and easiest treatment is to replace the stock feet with Vibrapods, IsoPods, or similar cones or anti-vibration supports.

The illustrations on the right show a few variations on the same theme. A platform made from a suitable material (stone, timber, carbon fiber, bamboo, etc.) is interposed between the shelf of a hi-fi rack and a turntable.

The interface can vary, commercial anti-vibration cones (acoustic isolators) such as Vibrapods, a high-density foam panel, or coils (springs) acting as vibration absorbers.

For instance, Dunlop 29-200 is high-density packaging and seating foam available in various thickness levels, of which 25, 50, and 75 mm are the most suitable for this application. A 400x400 mm slab (25 mm thick) costs only around AU$6.-, or AU$12.- for a 50 mm thick piece.

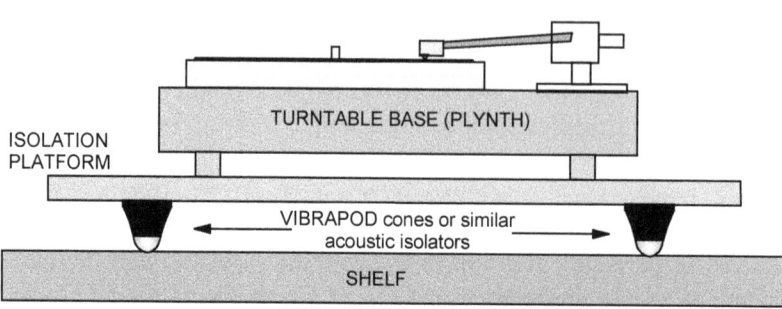

Fig. 14.12. The alternative equipment support arrangement requires an additional rigid platform (timber or stone) in-between the rack shelf and the audio component (in this case, a turntable) and anti-vibration cones or similar acoustic isolators.

Fig. 14.13. (BELOW) Springs (coils) are also quite effective as shock absorbers. You don't have to spend big. The illustration shows a floating platform for a turntable or a CD player. If the coils or springs you are using are stiff enough, you can even use such a system for heavier components such as power amps.

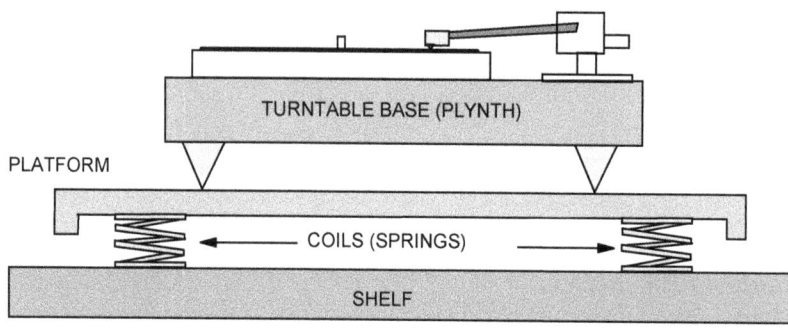

Fig. 14.14. (BELOW) The hard sandwich arrangement applied in practice as a turntable support: 1) The hi-fi rack (shelf-less type) 2) Polyurethane hi-fi rack mounts 3) Marble slab 4) Vibrapod cones 5) Solid timber platform

Fig. 14.15. Vibrapods are just one of many commercial anti-vibration cones (acoustic isolators) available.

Wall-mounted shelves

This solution (next page photo) works particularly well in music rooms with rigid (brick or concrete) walls and non-rigid floors prone to vibrations, such as the suspended timber floors on joists. Such flexing floors transmit vibrations from speakers and footsteps particularly well, so dispensing entirely with a floor-standing hi-fi rack may be the best choice.

IKEA sold the wall shelving system on the photo (next page) more than two decades ago. The vertical aluminium profiles are permanently bolted to the solid brick wall. Dozens and even hundreds of holes needed to be drilled and plugs inserted, most likely the reason why it was discontinued and sold on a clearance sale when we bought enough for three rooms (a bedroom, a large home office, and this music room).

Timber panels sit on top of those vertical aluminium runners, and each MDF shelf (1), covered in beech veneer, is supported by two cleverly designed steel brackets (cannot be seen from the top or front), which lock into the perforated slots (2). The spacing between the shelves can easily be changed and customized.

Placing a phono stage next to a turntable is a wise choice since it minimizes the length of the first interconnect in the signal chain, where the signal is the weakest.

However, even a physically small and well-designed & made power transformer in such a preamp (3) vibrates, and such vibrations may find their way through the shared shelf back to the cartridge. As always, an experiment is worth a thousand debates. Move the phono stage one shelf below (4), next to the power amp, and compare the noise levels and the sonics of the two arrangements.

Vibration damping of hi-fi racks and other audio components

The knock test (page 298) does not apply just to amplifiers, speakers, and turntables; it can also give you valuable insight into the vibrational character of hi-fi racks.

Especially troublesome are the racks with hollow or tubular legs or "uprights". When knocked on, they sound hollow, and the ringing takes a long time to die down.

Thus, damping such vibrations down is often more acoustically beneficial than throwing money at more expensive shelves or racks.

Some vertical supports can be opened at one end and filled with sand or any inert material. Even epoxy used for potting transformers can be used, which, after it hardens, effectively makes hollow tubular supports into solid ones.

Epoxy resins retain a certain degree of elasticity, in contrast to supports machined out of solid steel or aluminium, which are rigid and transmit vibrations well. Some hi-end equipment racks come with their tubes already filled with a damping material.

Fig. 14.16. In many listening rooms, walls vibrate less than the floor, so a wall-mounted audio rack or shelving system becomes a sonically superior and often cheaper alternative.

The easiest and cheapest sound improvement hack of all

At the start of my engineering career, I expected that engineers and project managers would make most (if not all) of the technical decisions. That proved to be a total delusion. Most decisions, even many technical ones, were made by upper management and accountants, people who controlled the money.

Fig. 14.17. Even some high-riced audio components come with cheap hard plastic (illustrated) or soft rubber feet. Upgrading those will result in one of the most cost-effective sonic improvements of all.

"You want gold-plated solid machined brass feet? What the heck for, these five cent plastic babies from our supplier in Shenzen will do just fine! Off you go, mate!" And that's the end of that "discussion."

Now, when you see a $5,000 CD player or a $3,000.- music streamer with four flimsy hard plastic or soft rubber feet, you know that your beloved possession was made using the cheapest parts possible.

And if that was done on the outside, where it can be seen, imagine the CCC (a highly technical term, an acronym for "cost-cutting crap") done on the inside!

So, your first duty as a self-respecting audiophile is to remove that rubbish like dentists excise rotten teeth and replace them with the best anti-vibrational feet or "equipment supports" you can afford. As a guide, look for a 20:1 price ratio; for a $2,000.- audio component, invest a minimum of $100.- on a set of three or four feet!

Find a friendly audio dealer who'd loan you two or three sets of different models and brands of equipment supports ("feet"). Try them under various components (CD players, streamers, preamps, lighter power amps) and see how much difference each makes compared to the stock feet.

Better supports will improve focus and transparency, make the background "darker" and quieter, improve sound staging, and pretty much everything else. Some sonic aspects will be significantly better, others only marginally, but that's OK; it will all still sound heaps better than before.

A vibration damping treatment, a magnetic soak, or both?

Early into my audiophile journey, I bought a used Denon CD player. It looked elegant, sounded quite smooth and musical (for a CD player, at least), and, above all, was very heavy. Solid-state amplifiers can have substantial heatsinks, tube amps have large and heavy output transformers, but what could be so heavy in a CD player? It certainly wasn't the fairly light folded steel top cover or the bottom chassis, so it had to be something on the inside. Intrigued, I removed the top cover, and sure enough, there was a large solid rectangular steel bar (about 20 cm long) bolted to its underside.

Can such primitive vibration damping measures result in sonic improvement? Again, it pays to carry out a few experiments. Place a brick, a heavy steel profile, or an unused power or output transformer on top of your CD player or amplifier. Some audiophiles claim that the magnetic coupling between the de-energized transformer (or choke) placed on top of a tube amplifier's power or output transformers improves the sound.

There are even commercial products (VPI's "Magic Brick," for instance) that use only the magnetic transformer core (the stack of laminations) without any windings. They enclosed the core in a timber box and claimed sonic effectiveness. Since there aren't any windings, it must be the magnetic coupling and its "redirection" of the stray magnetic flux (the magnetic field lines that "escape" the transformer windings) that makes such a difference.

DIY PROJECT: Make your own magnetic soak box

The VPI's website explains the operation of their "Magic Brick" this way: "By placing the Magic Brick as close as possible to the power supply you create a short circuit path for the stray magnetic fields. This keeps the field close to the power supply and away from the signal carrying circuit.

A further advantage of the high mass needed to accomplish this is the physical damping of the chassis."

So, the Magic Brick acts as both a magnetic soak (soaking up stray magnetic fields) and a mechanical damper (as a weight).

The Magic Brick is not available for purchase anymore and seems to be discontinued. Perhaps audiophiles could not muster enough belief (or rather set aside their disbelief) to buy them in high enough quantities.

Should you wish to make your own, get a nice wooden box, fill it with old EI-type transformers (depending on their size, one fits into the small and up to three in the large-sized box pictured), and place them on top of your audio components.

We used solid jarrah jewelry grade boxes, nicer looking than the VPI's, AU$30 for the small and AU$60 for the large one.

After placing them on top of a few different tube amps, did we notice any sonic improvements? No, but you may. Each amp or audio component is different.

The amps we tried the magnetic soak on had no other audio component above them and have been extremely well designed and made (by us); perhaps that's why no improvement was detected.

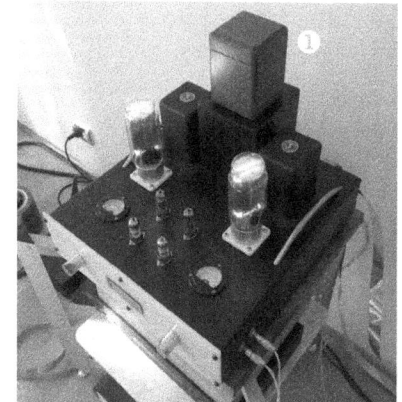

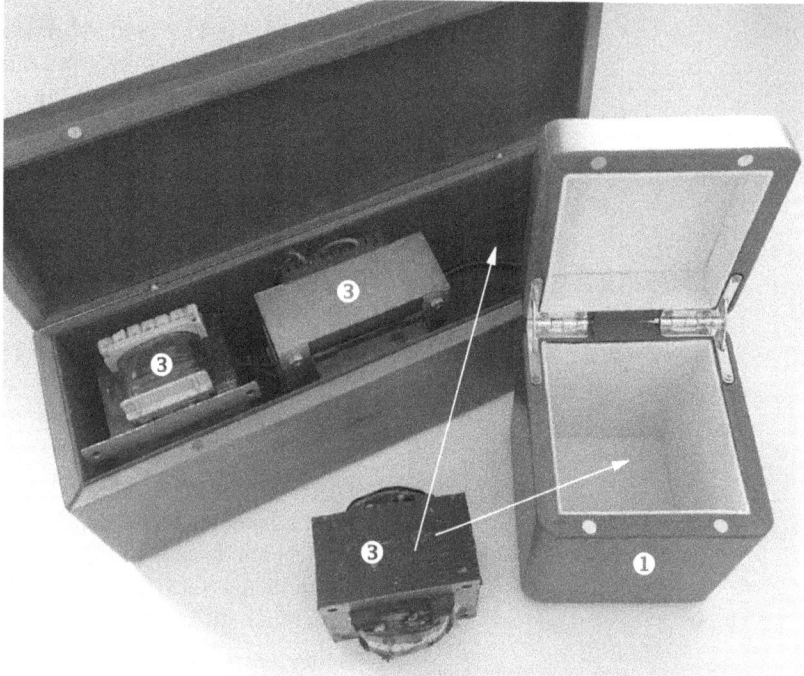

Fig. 14.18. "The Small Box" on top of a tube amp's power transformer (1), the original "Magic Brick" (2), and stuffing "The Large Box" and "The Small Box" with scrap power and output transformers of EI-construction (3).

CABLE LIFTERS (RAISERS)

Cable lifters improve the sound - fact or fiction?

Amongst all audiophile tweaks and "accessories," few cause so much controversy as cable lifters or raisers, those little critters made of wood or plastic that lift your cables an inch or so off the floor. Although usually used with long speaker cables, they can also support long interconnects in situations where monoblocks are placed right behind or next to the loudspeakers, and the equipment rack is far away.

The usual explanation of their effectiveness is based on two factors. Mechanically, lifting the audio cables off the floor reduces the vibrations they are exposed to (from the floor), which may have a small but noticeable impact on the cohesiveness, timing, and transparency of the sound. Electrically, lifting the cables up changes the dielectric properties around them, which in turn affects their capacitance or inductance, resulting in minute changes in their frequency-dependent behavior as filters and transmission lines.

From an engineering viewpoint, it all sounds plausible, albeit far-fetched. My initial reaction was dismissive. So, I performed two simple experiments.

The listening test was performed first since knowing the measurement test results could impact my perceptions. After completely lifting the two speaker cables (left and right channel) about two inches off the floor using the DIY cable lifters (described below), I could detect no sonic difference at all.

Disconnecting my speaker cables at both ends, I measured their capacitance and inductance using 1 kHz test frequency. Again, after lifting the cables off the floor as in the listening test, the measurements were repeated. Neither the capacitance nor the inductance changed one bit.

DIY PROJECT: Cable lifters for those who want to decide for themselves

Although some cable raisers aren't expensive, if you are DIY-inclined and enjoy tinkering around, Target and Amazon sell a game called Jenga. It's a bunch of nicely finished blocks, 74 x 24 x 14 mm pieces of solid hardwood, and for US$12.99, you get 54 of them. The game often goes on sale at 50% off (US$6.49).

Many retail stores and chains sell their version of this game, almost always cheaper than the original Jenga (sometimes even half price!), so shop around.

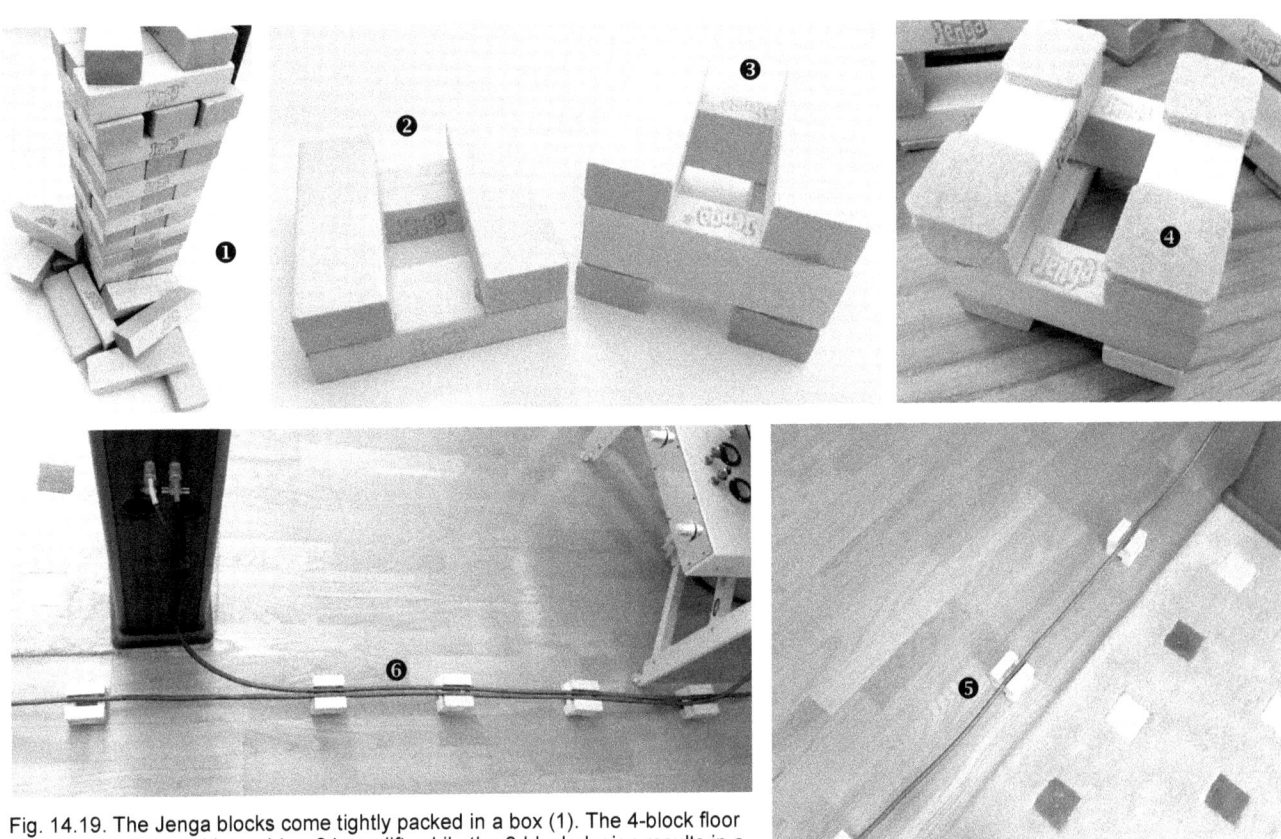

Fig. 14.19. The Jenga blocks come tightly packed in a box (1). The 4-block floor riser design (2) provides a 14 or 24 mm lift, while the 6-block design results in a 38 mm raise (3). 25 mm squares of self-adhesive felt (4) are cheap and add about 5mm to the 38mm of the block, for a total raise of 43mm. Since the equipment rack is to the left side, a single speaker cable stretches between the two speakers (5). Two speaker cables can fit into these DIY raisers (6).

One way to use such wooden blocks would be to simply place them under your cables, but since they have no V-shaped notch or groove of any kind, the cables could easily slip off if disturbed. A better option would be to cut such grooves or notches by hand or with a router, but that takes time and considerable effort. We decided to glue the blocks together.

Vibration-wise, rubber feet or felt pads (used for furniture and floor protection) can be attached to isolate the blocks from the floor. Square-shaped, beige color, 25 x 25mm self-adhesive felt pads (a pack of 80) sell for around US$5.-

Two possible arrangements are pictured, a 4-block design and a 6-block version. With a 4-block design, we have 54/4=13.5, so six raisers for one and six or seven for the other speaker cable. Or, with two game sets, you'd be able to make 28 raisers or 14 raisers per speaker cable. Depending on how the bottom pair of blocks are positioned, the cables would be either 14 or 24 mm above the floor.

With a 6-piece design, they'd be raised 14=14=28, 14+24=38, or even 24+24=48mm. After the initial improvement with low-level risers (14 or 24 mm, users report only slight additional benefits from higher supports (38 or 48 mm), but you may decide that such diminishing returns are still worth pursuing.

Since 54/6 = 9, two game sets will be sufficient for 18 raisers of the 6-block design, or nine risers per speaker cable. With a typical 2.5 m cable length, raisers would be about 28 cm apart, close enough to prevent the sag of most cables.

15 | TROUBLESHOOTING YOUR AUDIO SYSTEM

Every driver should be able to change a flat tire, a flat battery, or even perform an oil & filter change on their car. Likewise, every dedicated audiophile should be able to troubleshoot their audio system by using the basic faultfinding methods described in this chapter.

Due to their nature and inherent simplicity, tube amps and preamps are easier to troubleshoot than their solid-state (transistor) counterparts. However, even with tube amps and preamps, there aren't many faults that users can diagnose and fix themselves; most will require the amp to be serviced by a competent audio technician.

Since it requires technical knowledge, fault finding is an arduous, even impossible task for most audiophiles. As with other aspects of life, the more you know about audio in general, and electronics in particular, the easier it will be for you to fix things yourself and save heaps of money in the process, not to mention the inner satisfaction such self-reliance usually brings.

- TYPES OF FAULTS AND THE COMMON CAUSES
- THE BASIC EQUIPMENT - MULTIMETERS AND LCR METERS
- QUICK LOUDSPEAKER CHECKS
- QUICK AMPLIFIER AND PREAMPLIFIER CHECKS
- SIGNAL TRACING METHODS AND PRINCIPLES
- STRANGE AND HARD-TO-PINPOINT AUDIO SYMPTOMS
- TROUBLESHOOTING VALVE (TUBE) AMPLIFIERS

> "You can observe a lot just by watching."
> Lawrence (Yogi) Berra, baseball player, famous for his tautological one-liners such as this one

TYPES OF FAULTS AND THE COMMON CAUSES

The first step towards locating and fixing faults is the understanding of their causes and types. Some faults and failures have a singular cause; others are caused by two or more contributing factors. Often one such factor would not be sufficient to cause trouble, but two factors together certainly are. In no particular order of importance or commonality, here are the most common reasons audio components fail:

- POOR DESIGN: electronic components operating beyond their maximum rated voltage, currents or power levels, capacitors exploding or suffering breakdown in insulation, resistors overheating and burning out, printed circuit boards suffering heat damage and cracking, resulting in broken traces
- POOR WORKMANSHIP: contact problems, loose connections, cold solder joints
- COMPONENT FAILURE: aging, drying out of electrolytic capacitors, drifting values
- WEAR & TEAR + ROUGH HANDLING (vibration, rubbing, loose connections)
- IMPROPER PREVIOUS REPAIRS, MODIFICATIONS OR "IMPROVEMENTS": it is always faster, easier, and cheaper to fix amplifiers in their original condition than those butchered by ignorant "experts"!
- EXCESSIVE HEAT & HUMIDITY: Loudspeakers are especially susceptible to air humidity. Paper cones and rubber surrounds deteriorate, become brittle (lose flexibility) and even disintegrate ("crumble away").

Tubes and capacitors are the most troublesome electronic components, especially electrolytic capacitors, which age and dry out, become leaky and lose capacitance.

Due to their nature (moving parts and poor contacts due to dirt, grime, and oxidization), switches are also a common offender. Many current production tube sockets are of poor design and construction. Socket pins lose contact with tube pins, break off or even fall out.

Resistors and transformers are less likely to be the source of problems, although old carbon composition resistors (the molded type) drift significantly in value, causing distortion. Some circuits that require precise matching of resistors (for instance, phase splitters and phono stages) may not operate properly in such a case. However, the title of The Prime Suspect goes to wires, sockets, and contacts. Let's look at them in more detail.

Contact issues and intermittent troubles

Intermittent faults are the most frustrating of all. They suddenly appear for a while, a moment, a few seconds or minutes, and then all is well again. That unpredictability makes them difficult to trace and locate. The cause behind all intermittent faults is the loss of contact between two points in the circuit that should be connected or unwanted contact or a short-circuit between points that should not be connected together.

Quite a few factors can cause the loss of contact. "Cold" solder joints are probably the most common issue, especially in poorly constructed DIY equipment and vintage kits. Although a joint of two or more wires passes the visual inspection, it is uncertain what is happening inside the joint. Mechanical stress on the hookup wire or corrosion/oxidation of the conductor caused a break in a wire or the solder joint itself.

If in doubt, re-solder all the joints in the affected area, for instance, around a tube stage that does not work properly. This is often faster than signal troubleshooting. In amplifiers that use printed circuit boards, especially double-sided, there could be a poor solder joint or a break in the metal rivet that joins two sides via the plated-through hole. A quick ohmmeter check between two copper layers would confirm such an issue, illustrated below in a).

Illustration b) shows a situation that also passes a purely visual inspection. The hookup wire has a break under the insulation. While performing a visual inspection, use a small insulated screwdriver, tweezers, or a dental probe to jerk wires and component leads around.

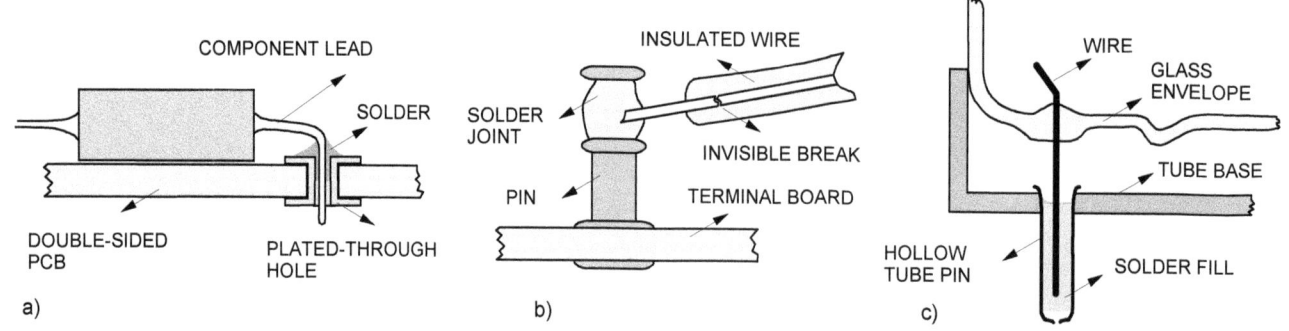

Fig. 15.1. Three common contact problems, with equipment that uses double-sided printed circuit boards (a), when point-to-point construction is used (b), and inside the hollow pins of vacuum tubes (c).

Three-piece dental tool kits are very cheap and include a probe, tweezers, and a mirror. No, you won't use them to perform the root canal surgery on yourself or your loved ones (best left to qualified dentists, don't you think?) You will use them for troubleshooting your precious audio gear! A mirror and a magnifying glass enable you to see what's happening in hard-to-reach places!

Some amplifiers use "Snap-On" and other types of connectors. These should always be on your list of suspects if a connection problem develops. A wire may seem properly plugged in, but you cannot see under the connector's insulation. Unplug all such connectors, if they seem loose, tighten them with small pliers and re-seat.

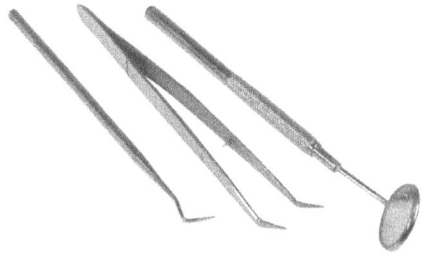

Fig. 15.2. These cheap basic dental kits are useful in identifying "cavities" in your solder joints and other contact problems.

A contact defect can also happen inside a tube.

If a loss of contact occurs inside a glass envelope, nothing can be done, and such a tube should be discarded. A common problem, illustrated in c), is poor solder joints inside the tube pins, especially in the currently produced tubes.

In octal and other larger tube sockets, the wires that bring tube elements out to the base must be soldered inside the hollow pins (illustration on the previous page). Due to mechanical stresses and thermal expansion and contraction, a poorly soldered contact will soon be lost. So, if a tube tests "bad" or "dead," don't throw it away in haste; always re-solder all the pins first and then test it again. It could be just a poor contact, and a tube could be perfectly healthy otherwise.

Many currently manufactured tube sockets suffer from contact problems. The metal contacts are often poorly designed. Instead of enveloping the tube pins around their entire perimeter, they make contact only in one or two points. The materials used are very brittle, so contacts deform, lose connection, or even break off after a dozen or so plugging and unplugging of tubes.

> HOW TO PROPERLY PULL TUBES OUT OF THEIR SOCKETS
> When pulling tubes out of their often very tight sockets, be extremely careful. Never jerk a tube sideways. You could bend or break delicate pins and loosen the socket contacts. Some tubes have a bottom glass pinch, which can be snapped off. Excessive force can easily crack the glass envelope. Both situations would render a tube useless. Always pull the tube straight up, with minimal lateral movement.

Another type of contact problem is where two points in a circuit should not be joined together but are short-circuited due to a mechanical, thermal, or construction fault. It is very easy to make a soldering mistake on smaller (7- and 9-pin) tube sockets or when soldering components onto a printed circuit board. Too much solder can make a short circuit between two PCB tracks or tube socket lugs.

Internal shorts can develop during the postage or transport of vacuum tubes due to shock and vibration. Filament-type triodes such as 300B and super-rare & expensive AD1 are especially vulnerable.

That is why you should always test tubes after transport before plugging them into an amplifier. Short of testing them in a tube tester, which most audiophiles do not possess, observe the newly installed power tubes for a few minutes after you power up a tube amp for the first time. If any red-plating (anodes becoming cherry-red), internal (inside a tube), or external (inside the amplifier) arcing, crackling noises, or strange smells or smoke happen, quickly switch it off and have it checked by an audio expert.

Checking order

1. VISUAL INSPECTION: Inspect the internals for any overheating, burned-out, missing, broken, loose or disconnected parts, links, and components. Identify non-original parts and previous repairs or modifications.

2. MECHANICAL CHECKS: Using a dental probe or a similar tool, jerk wires and component leads and check soldered connections for any loose or cold joints, broken or detached wires, and components, anything that could not be identified by a visual inspection only.

3. COLD CHECKS (equipment turned of & unplugged) & COMPONENT CHECKS: With an ohmmeter and LCR meter, check connections, continuity in various circuits (grid, cathode, anode) and components (inductors, capacitors, resistors, chokes, transformers), without powering the amplifier up.

4. HOT CHECKS (equipment turned ON): Check AC and DC voltages in critical points, such as the heater, cathode, screen, and anode voltages on tubes, emitter, base, and collector voltages of bipolar transistors.

5. SIGNAL TESTS: Measure hum, noise, distortion, output power. Study the waveforms on an oscilloscope.

THE BASIC EQUIPMENT - MULTIMETERS AND LCR METERS

Multimeters

A multimeter is the most-used piece of testing gear, so buy the best one you can afford. Most modern ones are digital and auto-ranging, and thus very easy and quick to use.

However, many amplifier builders and repair technicians prefer analog meters, not for their romantic and old-fashioned "feel," but for the smooth behavior of the analog indicator. For quick checks, it isn't even necessary to read the exact figures on the scale. One glance at the needle is enough to ascertain that the idle voltage is "in the ballpark," usually 1-5 V_{DC} on cathodes and 100-200 V_{DC} on the anodes of preamp tubes.

Also, when testing potentiometers for tracking and smooth resistance change (to detect any sudden jumps or breaks), analog ohmmeters are the only way to do it. Many digital meters take a long time to settle; the digits annoyingly bounce around for a few seconds. On some fast-changing signals, they never settle, rendering them useless! So, test drive your multimeter before you buy it.

Types of multimeters

- Passive analog non-RMS meters are very basic and thus cheap. The ranges are manually selectable. Their low input resistance affects the measured circuit resulting in a dismal lack of accuracy. Since they cannot measure frequencies above 2-3 kHz, they are not suitable for audio testing.
- Active analog non-RMS meters, usually of the VTVM (Vacuum Tube Volt Meters) or FET kind, have an internal tube or transistor differential amplifier, hence the word "active." Their high input impedance does not load the measured circuit, they can measure audio frequencies up to, and even way above 20 kHz and have manually selectable ranges.
- Active digital non-RMS meters are the most common kind, available in a wide range of quality and capability levels, from cheap and cheerful to so-so. However, I see no point in buying them when quality "True-RMS" meters are widely available and more affordable than ever!
- Active digital True-RMS meters: the best choice for the digital generation.

True RMS multimeters versus average-responding RMS-calibrated (ARRC) multimeters

The Root-Mean-Square (RMS) value of a varying or alternating signal such as an AC voltage or current is that value that produces the same heating effect (dissipates the same power) on a resistor as the equivalent DC voltage. For that reason, it is also called an effective value (same heating or work-producing effect). It is mathematically defined as

$$V_{RMS} = \sqrt{\frac{1}{T} \int_{t_0}^{t_0+T} v^2(t)\, dt}$$

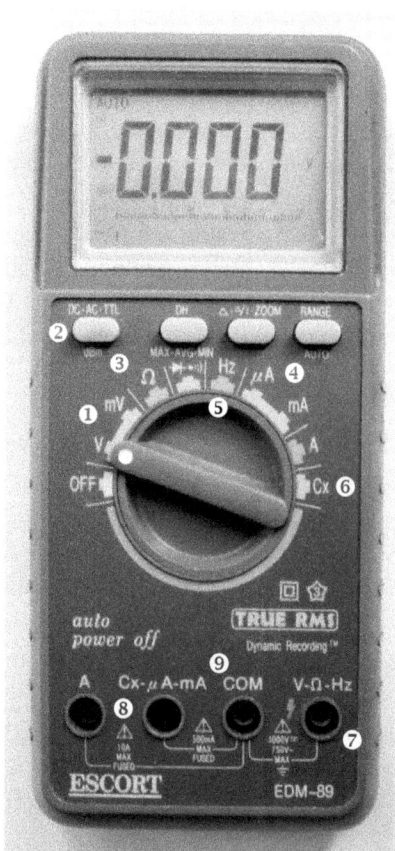

The waveform is first squared, then the average (or "mean") value is calculated through the integration over one period ("T") of the signal. Finally, the square root is taken of the mean value. For a voltage sine wave, whose mathematical formula is $v(t)=V_P \sin(\omega t)$, where $\omega=2\pi f$ and frequency $f=1/T$ (T=period), after integrating and taking the square root, the RMS voltage is $V_{RMS}= V_P/\sqrt{2}=0.71 V_P$, or 71% of the peak voltage.

Non-RMS multimeters measure *the average* value of the signal, but their scale is then calibrated in RMS value by multiplying it by 1.11 (since for the sine wave the RMS value is 11% higher than the average value). They are sometimes called average-responding-RMS-calibrated meters (ARRC). Therefore, they are *only* accurate for a pure (undistorted) sine wave! This is the reason they aren't suitable for audio work, where music signals have complex waveforms.

True-RMS digital multimeters

Escort's model EDM-89S is a typical example of the mid-to-late 1990s handheld digital multimeters. We have been using it continuously until the present day without any trouble. A quality instrument indeed, very reliable and dependable.

Fig. 15.3. A typical auto-ranging true-RMS digital multimeter

Apart from AC and DC voltage, with separate Volts and milliVolts ranges (1), selectable by a push-button (2), such instruments can also measure resistance (3), marked "Ω" for ohms, current (4), frequency (5), marked "Hz" for Hertz, and often even capacitance "C_X"(6).

The current and capacitance measurements require one of the test leads (the "hot" lead, usually red) to be unplugged from the usual socket (7) used for V- Ω - Hz measurements and moved to one of the two current test sockets, either "A" for high currents, Amperes, or "C_X - μA - mA" socket for low-level currents and capacitance (8). The black or "common" lead is always plugged into the "COM" socket (9).

How digital multimeters work

Conceptually, digital multimeters are easy to understand. The analog side includes the attenuators for DCV and ACV, the current-to-voltage converter on current measuring ranges, and the constant current source for the ohmmeter function.

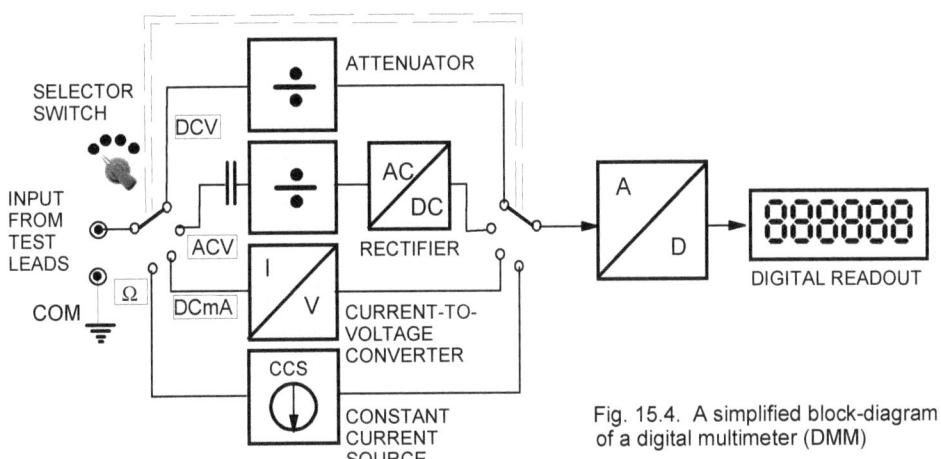

Fig. 15.4. A simplified block-diagram of a digital multimeter (DMM)

The ACV attenuator is capacitively coupled to remove the DC component of the voltage, followed by the rectifier (AC/DC converter).

The DC scaled voltage thus obtained is converted into a digital signal in the A/D (Analog-to-Digital) converter. Early chips needed a decade counter to interface them with a digital readout, but modern A/D converters drive display (digital readout) chips directly.

Digital LCR meters

In the most helpful handheld test instrument race, a digital LCR (inductance-capacitance-resistance) meter comes close second. Our Taiwan-made Escort LCR meter dates back to the mid-to-late 1990s, but things haven't progressed much since. This auto-ranging instrument measures L, C, and R parameters at two frequencies, 120 Hz and 1 kHz (1). The 120 Hz frequency was chosen since it's double the 60 Hz mains frequency in the USA. It is the fully rectified AC ripple frequency in a power supply of mains-powered audio components - amps, preamps, and power supplies.

Most LCR meters display two values simultaneously: C - capacitance on the main display and D - the dissipation factor on the secondary display, or L-inductance (2) and Q - the quality factor.

The "R" LCR meters measure isn't the DC resistance multimeters measure. Here the "resistance" is measured with an AC signal, so it's the impedance modulus of the measured component (capacitor, inductor, or resistor). If you perform the "R" measurement at both test frequencies, you will get very different results.

Some more modern LCR meters can also measure parameters at 10 kHz, which is a handy feature in some specialized tests, such as when measuring tube amp output transformer's leakage inductance.

Most LCR meters have slots (3) into which component leads can be inserted; alternatively, use plug-in test leads terminated with crocodile clips.

The instrument uses a 9-Volt battery which does not last long. For benchtop use, I'd recommend an external 12 VDC power supply (4); it will save you heaps on expensive 9-Volt batteries.

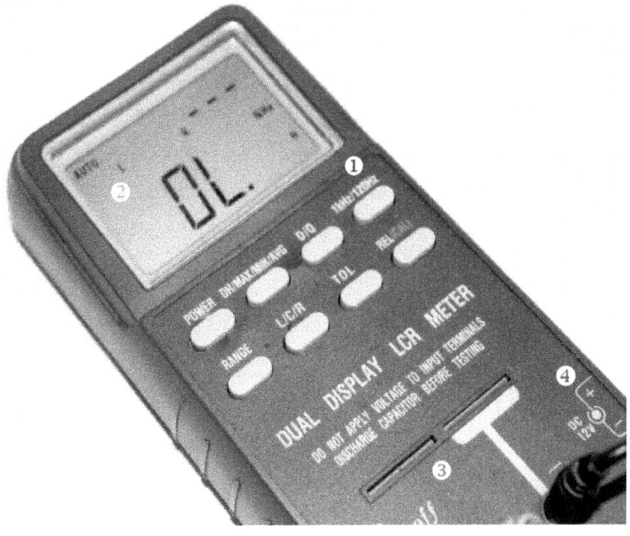

Fig. 15.5. A typical handheld digital auto-ranging LCR meter

QUICK LOUDSPEAKER CHECKS

Mechanical (manual) checks

There are four quick and simple tests to check loudspeaker drivers' mechanical/structural and electrical health. The first is a visual inspection. There should be no holes, tears, or rips in the cone. Paper cones are particularly susceptible to aging and damage; aluminium, hemp, Mylar, polypropylene, and composite cones are more resilient. If there's any noticeable damage, do not buy such a speaker. Sure, experts can re-cone it (if a speaker is rare or hard-to-replace), but it's a costly exercise.

The second test involves gently pushing the speaker cone inwards by hand. The cone should move freely and return to its original position once the force is removed.

If you feel any sudden resistance or hear a "scraping" sound, that means the voice coil is rubbing against its surroundings. Either the "spider" (speaker's suspension) has been damaged and misaligned, or the voice coil is deformed. Such a speaker may work to some extent but will sound distorted and will eventually fail.

Electrical speaker checks: battery, ohmmeter and LCR meter

Once a speaker passes these two mechanical tests, it is ready for electrical checks. If you have a multimeter, set it on a low resistance range and measure the resistance across the speaker's terminals. Since this is a DC resistance, it will be lower than the speaker's nominal (declared by the manufacturer) impedance.

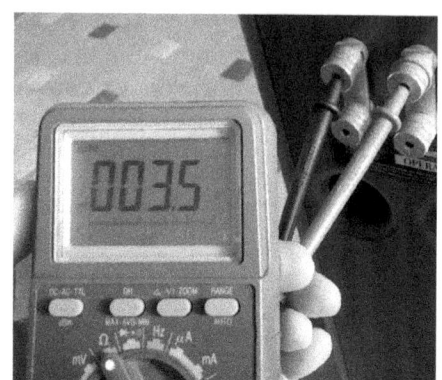

Opera Terza speakers were declared to be 6-ohm speakers. The photo (top right) shows their DC resistance of 3.5 ohms. This test confirms that the speaker's voice coil has continuity, in other words, that it hasn't burned out. There will be no sound from the speaker during this test.

You could use a 1.5 V or 9 V battery for another quick test. Take two leads terminated with crocodile clips and connect the battery across the speaker's terminals. With the battery's + pole on the speaker's + terminal, the speaker's cone should quickly move forward (outward) and stay there. With the battery's + pole on the speaker's - (negative) terminal, the speaker cone should quickly move back (inward) and remain there.

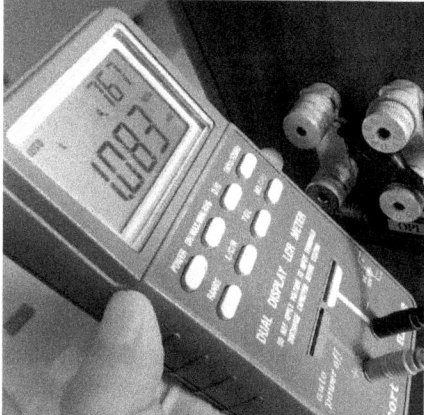

Finally, the best and most conclusive test is by connecting a digital LCR meter across speaker terminals. Make sure the LCR meter is on its "R" (AC resistance) range. The LCR meter will send pulses of current through the voice coil and show the AC impedance of the voice coil on its LCD display. As we have seen, a few selectable test frequencies are available, 60 Hz and 1 kHz, in almost all cases. These pulses should be audible; the speaker should go beep-beep-beep about once a second.

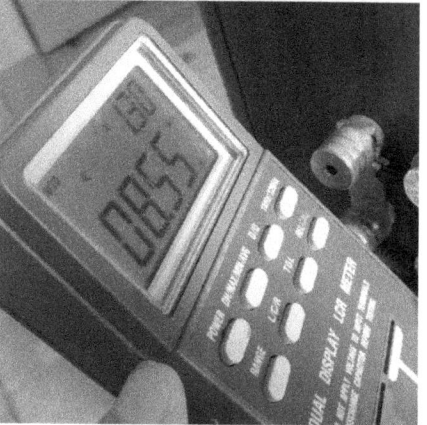

This LCR test can be done on individual drivers such as woofers, midrange drivers, and tweeters, not just the whole speaker boxes.

LCR meter tests of Opera Terza speakers at 1 kHz test frequency show 1.083 mH inductance, 8.55 µF capacitance, and 14.19 ohms AC impedance (photos on the right).

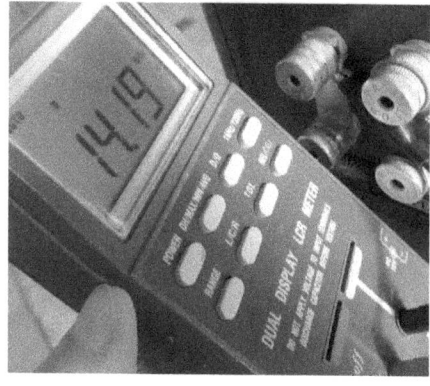

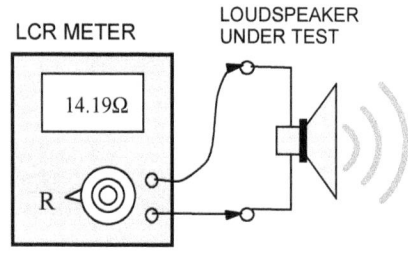

Fig. 15.6. (LEFT) With an LCR meter set on "R" or impedance range, a pulsating tone should be heard in the speaker. The higher the speaker's sensitivity or SPL level (dB/W), the louder that tone.

Fig. 15.7. RIGHT (TOP TO BOTTOM) Measuring DC resistance with a multimeter, inductance measurement using LCR meter at 1 kHz, LCR meter measuring capacitance at the same frequency, and LCR meter "R" measurement, showing the impedance modulus of 14.19 ohms.

QUICK AMPLIFIER AND PREAMPLIFIER CHECKS

Measuring amplifier noise & hum

Connect the amplifier to a loudspeaker or a dummy load. Since preamps almost always have an internal resistor connected across their outputs, there is no need to connect a load to a preamplifier.

Turn the amp on. If there is no volume control, short its input terminals. If there is volume control, turn it to zero (fully CCW). The AC millivoltmeter (multimeter on "mV" range) will show the residual hum and noise at the output. Single-ended tube amps with directly-heated tubes will measure the worst, 0.5-3 mV; tube preamps and solid-state amps should be much quieter, the hiss/hum should be much lower.

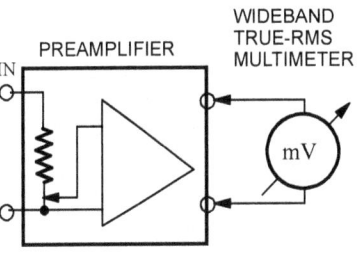

Fig. 15.8. Measuring the hum or noise at the amplifier's (BELOW) or preamplifier's (ABOVE) output.

After the voltage, measure the frequency (multimeter on "f" or "Hz" range). It will be around 50 Hz in Europe and Australia (60 Hz in the USA) or 100 Hz (120 Hz in the USA). 50 or 60 Hz hum most likely means that the mains transformer or the mains wiring is radiating the mains frequency AC signal, which is getting picked up by the tubes or the wiring. Better shielding or magnetic component positioning is needed.

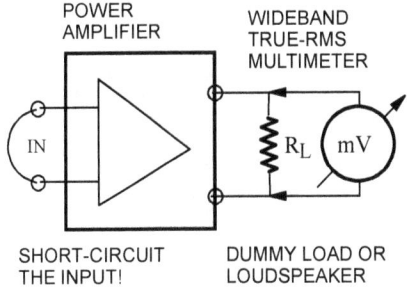

A 100 Hz hum (or 120 "cps" in the USA and other 60 Hz mains countries) means that the amp's DC power supply filtering & smoothing is inadequate. The AC ripple on the amplifier's internal DC power supply lines is too high and is being amplified by the audio circuit.

Measuring the input impedance of an amp or preamp

LCR meters are great for measuring the input impedance of various devices. Since these are usually of capacitive-resistive nature, you first measure the input capacitance and then change the LCR meter's range to resistance and measure the device's input impedance.

When LCR meters measure "resistance" on the "R" range, it is an AC test; the measured resistance is not a DC resistance but a total AC impedance at the test frequency. Measure DC resistance with a multimeter, and you will get a different result).

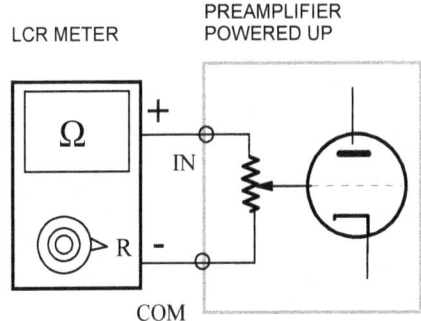

Fig. 15.9. Measuring the input impedance of an amplifier or preamplifier with an LCR meter

You can do the test while the device-under-test (amp or preamp) is powered up; the LCR meter injects a small audio-frequency signal into the input and measures the response, so you cannot damage anything.

DIY PROJECT: Dummy loads for amplifier testing

High power resistors used instead of loudspeakers for amp testing convert audio power into wasted heat, which is why they are called "dummy" loads. Speakers produce loud sounds during testing, and some tests could damage or destroy them, so a dummy load is always a prudent idea. One such made-in-China wirewound resistor (100 Watts, 8Ω) is pictured below. Alternatively, resistors of lower power rating can be connected in any combination (series, parallel, or series + parallel) to get the resistance and power rating you need. 7 Watt and 10 Watt wirewound resistors are cheap and commonly available. Two of many options are shown below.

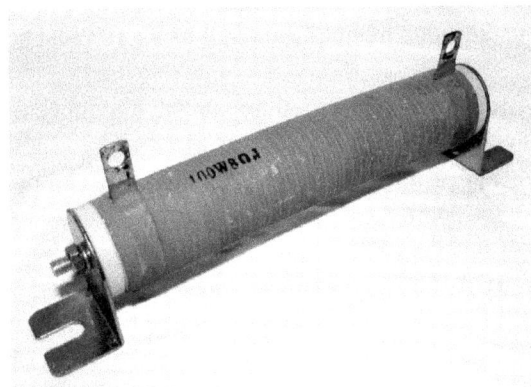

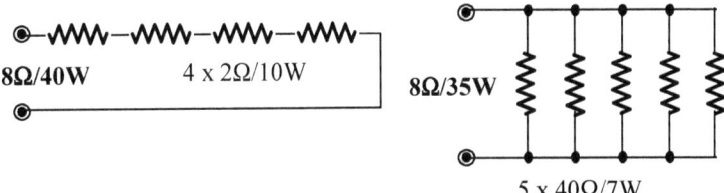

Fig. 15.10. Two examples of combining lower power resistors to make a higher power dummy load, a series and parallel arrangement. Of course, combinations of the two are also possible.

Fig. 15.11. 8 Ω wirewound resistor rated at 100 Watts (Made in China, available online)

Testing high power amplifiers

While only a few tube amplifiers exceed the 100 Watt per channel power rating, there are many solid state amps rated at 200, 400 or more Watts per channel. Sourcing 400 or 500 Watt power resistors for their dummy loads is hard, but again you can use a combination of identical resistors of a lower power rating, say four 100 Watt resistors as pictured. Two 8Ω resistors in series give you 16Ω resistance, and two such pairs in parallel give you half of 16Ω or 8Ω. How neat is that?

The output power is four times the measured output power or P = $4V^2/R$, where R is the resistance of one of the four identical resistors, 8Ω in this example, and "V" is the AC voltage measured by an AC voltmeter or multimeter on "AC Volts" range.

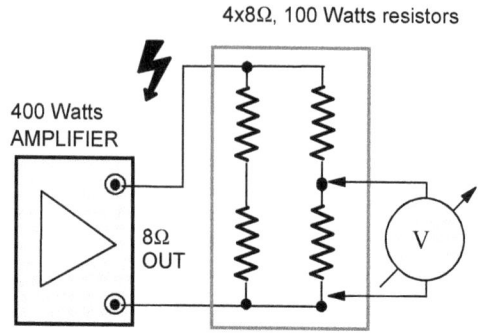

Fig. 15.12. Using four 100 Watts 8Ω resistors to construct an 8Ω 400 Watts dummy load.
WARNING: High AC voltages (50+ Volts) may be present!

Checking audio cables

For continuity checks and resistance measurements, a multimeter's selector switch must be in "Ω," "R," or "RESISTANCE" position (different brands and models use different markings). Check its calibration and zeroing by short-circuiting the red and black leads (pressing them together). The display should indicate 000.0 ohms or some very low value close to zero (0.1-0.7 ohm). If not zero, remember that residual value which is effectively the zero for your further measurements.

For instance, if shorted leads indicate 0.5 ohms and you measure a certain cable's resistance of 0.8 ohms, that means the actual cable resistance is 0.8 - 0.5 = 0.3 ohms.

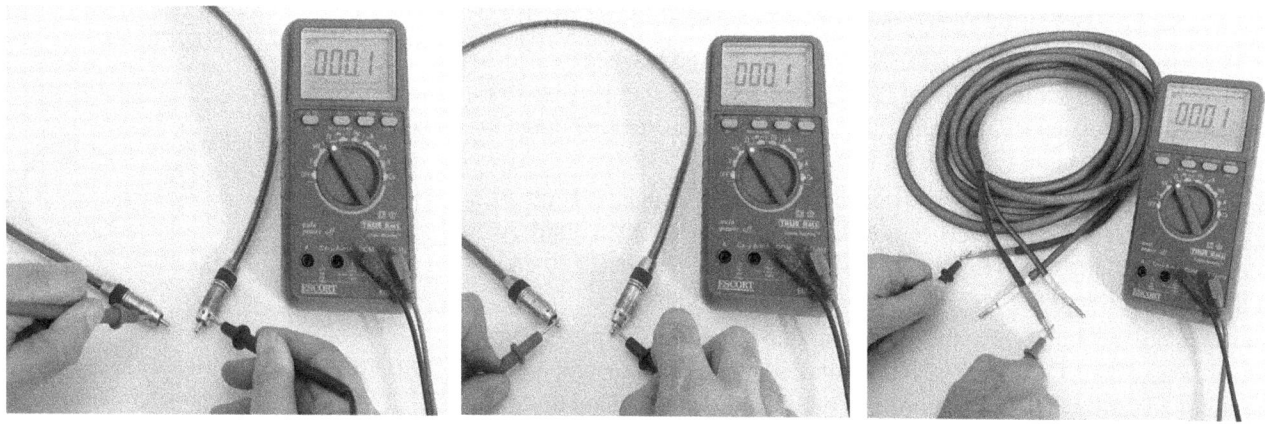

Fig. 15.13. (L-R) Checking the shield continuity and resistance of a RCA-RCA interconnect cable; checking the central conductor's continuity and resistance of a RCA-RCA interconnect cable; Checking the continuity and resistance of a speaker cable.

Speaker cables are checked by pressing test leads against their ends, which could be bare wires, banana plugs, or spades. There are only two conductors, plus (positive), usually with red color-coded en (heatshrink or sleeve) and minus (negative), usually with black or white color-coded end.

The resistance between two red ends and two white or black ends should be zero or close to it. As illustrated, in our case, it is 0.1 ohm, which is the lowest resistance this multimeter can measure and display.

The resistance between either red and either black or white end should be infinite (open circuit). In that case, most multimeters display "OL" and not the infinity sign. If there is any resistance, no matter how small or big (it must be an open circuit) between the + and - ends, the cable is faulty!

Interconnects are tested in the same manner. The resistance between the outer shell of an RCA connector ("negative") and the central pin ("positive") should be infinite (open circuit), "OL."

The cable is faulty if there is any resistance, no matter how small or big (it must be an open circuit) between the central pin and the outer metal sleeve! The resistance between the two pins at the opposite ends of a cable should be a fraction of an ohm, likewise for the resistance between the two outer metal sleeves.

You can touch and hold one side or one metal tip of a lead and the RCA, banana, or spade connector (for better contact), but not both. Holding both ends would effectively add your body's electrical resistance in parallel with the measured cable, thus lowering the measured resistance, resulting in erroneous readings and false results.

TROUBLESHOOTING YOUR AUDIO SYSTEM

Checking your power amplifier or the whole audio system for left-right channel balance

This check is an essential part of the process of obtaining the best possible sound stage. Positioning your speakers precisely with respect to the centerline of the room (the listening axis) is the main precondition for a stable and balanced soundstage. However, that assumes perfectly balanced audio signals between left and right channels. How can we check if that is indeed the case? What if it isn't?

This relatively simple test can be done on power or integrated amplifiers to ensure both channels are as balanced as possible or on the whole audio system (up to the speakers). The principle is the same. We feed an equal signal to both channels of a stereo amplifier (or to both monoblock amplifiers) and listen (with a third or auxiliary test speaker) for any sound "between" the two channels. The auxiliary speaker can be a smaller, less powerful unit.

Either the main speakers or the test (AUX) speaker must be positioned in another room so that you can only hear the AUX speaker, and the main speakers are not audible. With a perfectly balanced amplifier or an upstream system, there should be absolutely no sound of any kind from the test speaker.

The signal source can be a function generator (set on "sine wave"), whose signal amplitude and frequency are continuously variable, or a CD player with a test disc containing various fixed frequencies. That way, you can ascertain at what frequencies the imbalance happens, in other words, if the imbalance is present across the whole audio range or only in specific frequency bands.

In any case, whatever the signal source is, the louder the sound from the test speaker, the more significant the imbalance between the two amplifiers or amplifier channels.

An alternative setup is the "silent" way. The main speakers are replaced by "dummy" loads (high power resistors) which can be in the vicinity of the amplifier, so there is no need to run speaker cables into the adjacent room or to move your large and heavy speakers.

Again, the null indicator can be the third (test) speaker, or you can use an AC voltmeter or a multimeter set on the "AC Volts" range. The multimeter makes it possible to quantify the imbalance since you'd be able to precisely read the indication (AC volts) on its analog scale or digital display.

Remember that using dummy loads is not a conclusive test since they are almost purely resistive loads; an actual speaker (a complex resistive, inductive, and capacitive load) is a better option.

For subjective evaluation, the best test signal is music. If the difference signal (heard in the test speaker) is clear and undistorted, it is simply a matter of volume imbalance between the two channels. Any distortion or the absence of bass, midrange or treble, indicates that one or both amplifiers suffer from distortion and irregular frequency characteristics or nonlinear transfer function.

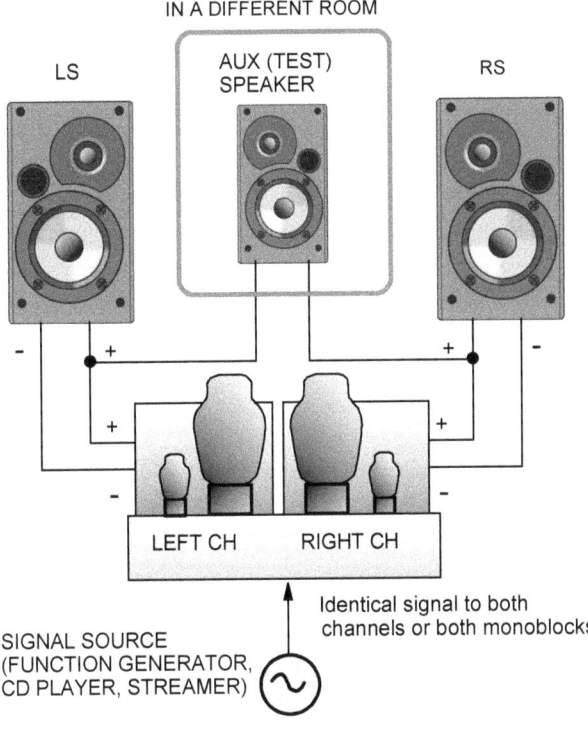

Fig. 15.14. The aural balance test of an amplifier or a whole audio system requires only a small auxiliary speaker.

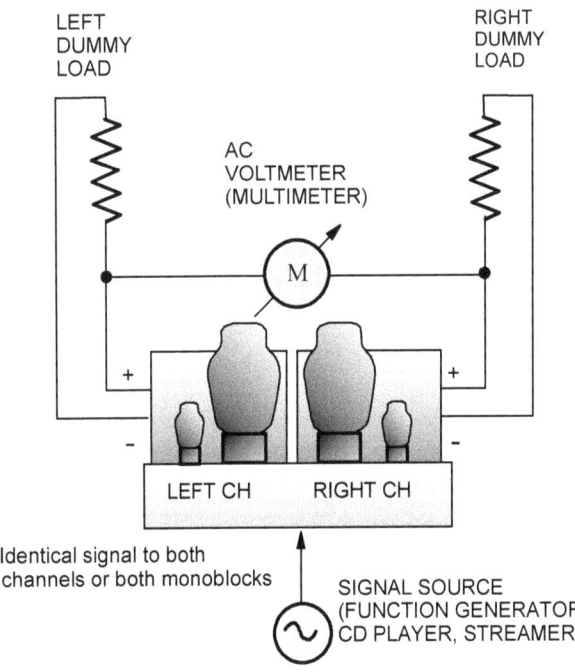

Fig. 15.15. The alternative or "silent" balance test. A combination of the two can also be used (dummy loads with an auxiliary speaker instead of the multimeter, for instance).

SIGNAL TRACING METHODS AND PRINCIPLES
No sound in one channel: Backward ("upstream") tracing and the halfway methods

This procedure assumes a turntable-based system, but the same sequence applies to systems with a CD player or digital streamer as a source. In that case, omit the phono preamplifier and the steps that relate to it. Likewise, if you don't have a line-level preamplifier but feed the signal from the source (whatever it may be) directly into power or integrated amplifier, you have a much simpler task; your system is the simplest possible, with only three blocks, the source, the amplifier, and the speakers.

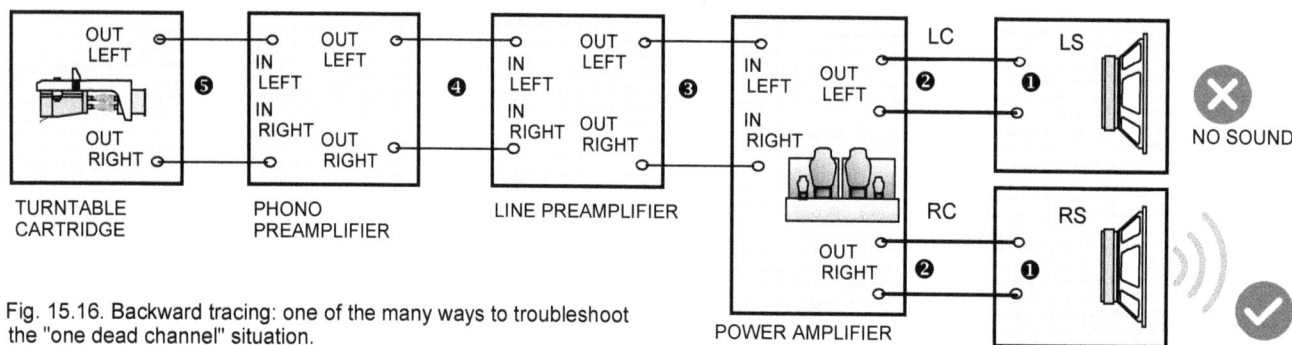

Fig. 15.16. Backward tracing: one of the many ways to troubleshoot the "one dead channel" situation.

1. Assuming that the left channel (LS) is "dead" (no sound), turn off the power amplifier (PA) and swap speaker wires at the back of the two speakers. Connect "OUT LEFT" to the right speaker RS and "OUT RIGHT" to the left speaker LS. Turn the amplifier on and play music. If there is still no sound from LS, the problem is in the left speaker. If there is now sound from LS but not from RS (the problem shifted to the other speaker), both speakers are OK, and the fault is somewhere upstream in the left channel. Turn the power amplifier off and return the connections as they originally were.

2. Disconnect the left speaker cable from the power amplifier's binding posts and, in its place, connect the right channel's speaker cable. Turn the amplifier on and play music. We know that the right speaker cable is OK, so if there is now sound from the right speaker, the problem is in the left speaker cable. If there is no sound from RS, the problem is in the power amplifier's left channel or somewhere upstream in that channel. Turn the power amplifier off and return the speaker connections as they originally were.

3. Disconnect both inputs to the power amplifier and swap them around (connect the left cable to the right channel input and vice versa. Turn the power amplifier on and play music. If there is still no sound from the LS, the problem is in the left channel of the power amplifier. If there is now no sound from the RS (the problem moved to the other channel), the power amplifier is OK (both channels), and the fault is upstream in the left channel.

4. Turn the power amp off. The procedure is now identical to step #3; swap around the input connections to the line preamplifier (LP). Turn the power amplifier on and play music. If there is still no sound from the LS, the problem is in the left channel of the line preamplifier. If there is now no sound from the RS (the problem moved to the other channel), the LP is OK (both channels), and the fault is upstream in the left channel.

5. Turn the power amp off. The procedure is now identical to steps #3 and #4; swap around the input connections to the phono stage (PP). Turn the power amplifier on and play music. If there is still no sound from the LS, the problem is in the left channel of the phono stage. If there is now no sound from the RS (the problem moved to the other channel), the fault is upstream, in the turntable cartridge or the internal tonearm/turntable wiring.

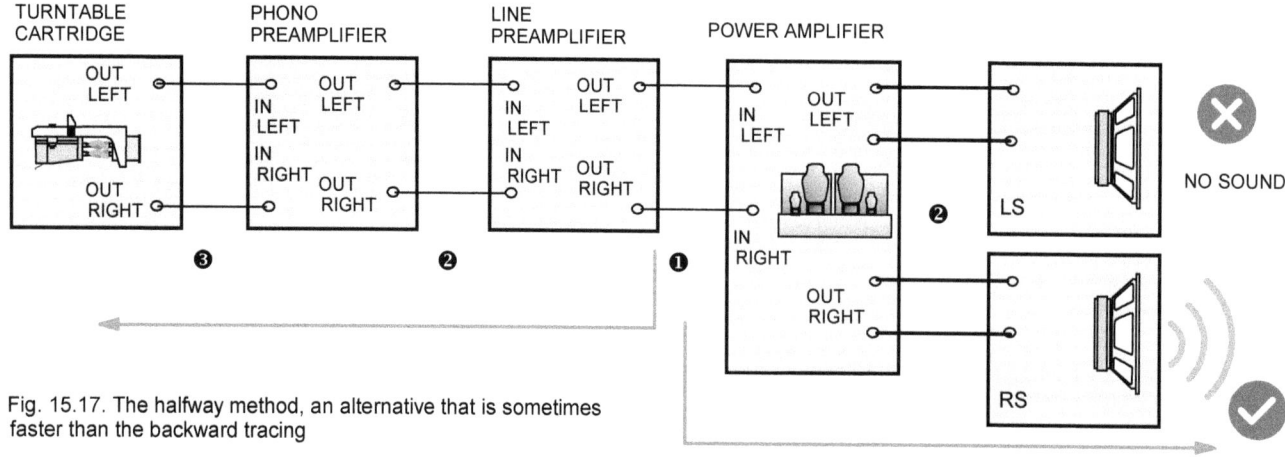

Fig. 15.17. The halfway method, an alternative that is sometimes faster than the backward tracing

TROUBLESHOOTING YOUR AUDIO SYSTEM

The halfway method is an alternative that is often faster than backward tracing the "one dead channel" problem. Instead of starting at the last link in the audio chain (amplifier-speaker), we start at one of the middle links, for instance, by swapping the interconnects at the input of the power amplifier.

If there is still no sound from the LS (the original problem), the cause is downstream in the left channel - in the power amplifier, the left speaker cable, or the left speaker. We can now use backward tracing for that section (2).

If there is now no sound from the RS (the problem moved to the other channel), the power amplifier/left speaker cable/left speaker (2) are OK (both channels), and the fault is upstream in the left channel (3).

Should you prefer, you can use forward tracing, starting by swapping the RCA cables between the turntable and the phono stage (or the CD player/streamer/DAC and the preamp in digital systems), and moving forward ("downstream"), following the signal flow.

Troubleshooting individual audio components

Once the faulty audio component is identified, the troubleshooting methodology will depend on what type of a component it is. With a burned tweeter, there would be sound from the midrange driver (if any) or the woofer.

If one speaker is completely dead (from my personal experience, a very rare occurrence), either all drivers have burned out (unlikely), or there is a loss of contact internally. Perhaps a wire (common to all drivers) came off due to speaker vibration, or there is a cold solder joint in the crossover. Some speakers are fused internally, so a burned-out fuse at the speaker's input is another possible cause of this problem.

STRANGE AND HARD-TO-PINPOINT AUDIO SYMPTOMS

When it just doesn't sound "right": the phasing issues

Some speakers' drivers are wired with inverted polarity, which is related to the design of their crossovers. For instance, a woofer is wired with positive polarity, the midrange is wired with inverted polarity, and the tweeter is again wired in phase with the woofer.

In one instance, when auditioning one of our amps with a speaker designed and made by a local Perth audio company, the three of us listeners could not agree when the system was in phase and when it was out of phase. Normally an amp/speaker combination sounds better one way and "strange" or unnatural the other way.

There are only two ways speakers can be hooked up, + of the speaker to + binding post of the amp, or "+ to -." Of course, both speakers must be connected the same way. However, in this case, the setup (with the guy's expensive DIY speakers using Focal drivers) sounded less than optimal both ways.

How do I know that the problem was in the speaker design? Because we had tried and auditioned that particular demo amplifier with dozens of speaker models and brands, and it always sounded faultless.

Generally speaking now, there is always another possibility, especially with DIY amplifiers or preamplifiers. Perhaps one channel had been wired incorrectly internally, out-of-phase with the other. Never just assume that a manufacturer has checked amplifier outputs for correct polarity (phasing) - double-check!

One channel sounds "different": high frequency oscillations as the possible culprit

In contrast with low-frequency oscillations (motorboating), HF (high frequency) oscillations are not always evident. The human ear cannot detect ultrasonic instability (above 20,000 Hz) directly; only instruments can.

When both channels of a stereo amplifier are oscillating at an ultrasonic frequency, this may not be obvious or even apparent. The amp may sound harsh or veiled or with a strange "metallic" edge to it. However, when only one channel or one of two monoblock amps is oscillating, it will sound different from the properly performing one; it simply won't sound right.

If, after all the obvious possibilities have been eliminated (poor input, output, and tube pins' contacts, aging and mismatched tubes, etc.) you suspect that HF instability and oscillation could be the culprit, have the amplifier checked by a competent technician. High-frequency oscillations can be seen on an oscilloscope as widening or smudging of the sine wave (signal voltage) at the output of an amplifier.

How warped records can deprive your amplifier of power and negatively affect its dynamics

The impedance of some power-hungry loudspeakers dips below a few ohms (even below 1 ohm in some cases) at specific frequencies, usually in the bass region. Such a low impedance is practically a short circuit across the amplifier's output terminals. The amplifier must have an ample current capability; otherwise, its power supply will not be able to sustain such high current levels. This voltage drop (sag) and reduced output power will result in dynamic compression and lifeless, "anemic" sound. An increased distortion is also likely.

Power-hungry speakers aren't the only possible culprit in such situations (amplifiers running out of power). Warped records produce high infrasonic (below the audible range) thumps of periodic nature.

Unless a phono stage or preamplifier has a subsonic filter, a high pass filter that attenuates very low frequencies, generally below 15 or 20 Hz (many don't), the power amplifier will try to amplify such infra-low frequency pulses with high energy content. This may overload its power supply, resulting in poor dynamics, increased distortion, and muddy or weak bass.

TROUBLESHOOTING VALVE (TUBE) AMPLIFIERS

Tube amplifiers are much easier to troubleshoot than their solid-state counterparts, both at the basic- or user-level, which will be covered here and at the technician level, which is obviously beyond the scope of this book. Such troubleshooting is covered in detail in Volume 2 of my book "Audiophile Vacuum Tube Amplifiers."

If your integrated or power amp is of the solid-state kind, apart from checking the fuse and replacing it if blown, there is nothing you can do as a user; it needs to go straight to a repair shop.

Some tube amps have a protective tube cage, which should be removed to enable basic troubleshooting. That may require turning the amp upside down and removing the bottom cover to access the fasteners (nuts and bolts) that hold the cage in place. The amp should be turned off and unplugged from the power outlet.

SYMPTOM: The mains fuse blows

When a fuse blows once, there is nothing to be concerned about. Fuses age and deteriorate; a slightly higher inrush current on power-up may be enough to melt their inner wire. However, if a replacement fuse blows, something is causing the higher than normal current draw. When replacing fuses, you must use a replacement fuse of the same voltage and current rating as the original.

Before replacing the fuse for the third time, unplug all output tubes. That effectively isolates the output stage (removes it from the circuit). If the fuse does not blow, the fault is either in one or more power tubes (internal short-circuit) or the related internal components (cathode resistors, bypass capacitors, or output transformers).

If the fuse still blows, the problem is most likely in the power supply. Unplug the rectifier tube - that will disconnect the filtering circuit in the power supply from the power transformer. If the fuse still blows, the fault is almost certainly in the power transformer, which would need to be checked by a technician and most likely replaced.

If the fuse does not blow after unplugging the rectifier tube, check the rectifier tube (if you have a tube tester) or simply replace it. If there's no rectifier tube, the solid-state rectifier (a bridge or discrete diodes) could be faulty, or there is a fault in the filtering circuit that follows it. There are no remedies at the user level; your amp needs to be fixed by a professional.

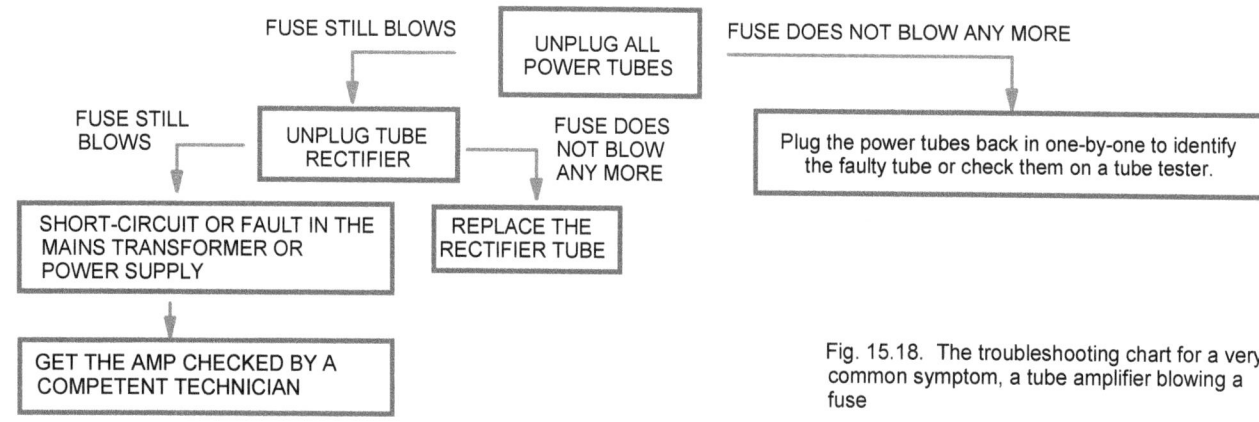

Fig. 15.18. The troubleshooting chart for a very common symptom, a tube amplifier blowing a fuse

SYMPTOM: No sound from both channels

If the amplifier powers up (does not blow a fuse) but there is no sound from either channel (assuming a stereo amp), check that all tubes are lighting up, that their heaters are glowing. The heater filament is always the innermost electrode, in the center of the vacuum tube.

If all tubes' heaters are lit up, without playing any music, put your ear to each loudspeaker in turn (tweeter or midrange driver's cone). You should hear a slight hiss or hum. If that is the case, the high voltage supply inside the amp is present and is most likely working correctly.

The heaters of some preamp tubes glow so faintly it's impossible to be sure, especially in a brightly lit room. Touch them briefly to ensure their heater's operation - they should be quite warm. Don't touch the power tubes!

TROUBLESHOOTING YOUR AUDIO SYSTEM

Check that there is an input signal into the amplifier (from a preamplifier, CD player, DAC, streamer, or whatever audio source you are using). Check both input and output connections and rotate the volume control potentiometer; it could have a break at the very point it is set at, although that is unlikely. With dicky potentiometers, there is usually some sound coming out of at least one channel.

If there is no hum or hiss at all from the speakers, there is a fault somewhere inside the amplifier's high voltage power supply. Some amplifiers have a dedicated HT (high tension or high voltage) fuse. Check that fuse first. You may try replacing the rectifier tube (if used); otherwise, the amp needs to be diagnosed by a technician. If the amp has a standby switch, check that it isn't activated. The standby mode keeps tube heaters on but kills the high voltage, so there is no hum or hiss of any type on the speakers!

SYMPTOM: *No sound from one channel*

When one channel is working while the other is producing no sound, check that all the tubes in the dead channel are lit up (heaters glowing). Replace the tube that is not glowing since it most likely means its heater has burned out.

If all tubes are lit up, put on a kitchen glove and gently check if the tubes are properly seated in their sockets. Carefully tilt them a few millimeters in all directions and listen for the sound reappearing or for a crackling noise. A loss of contact at one pin is enough for a complete loss of sound in that channel - the heater pins may be making contact, but one or more of the other pins may not.

If there's no apparent bad contact, power the amp down and swap the preamp tubes between the two channels. Power the amp up and play music. If the problem (no sound) moves to the other channel, that preamp tube (or one of them if two or more are used) is faulty. If there is still no sound from the same channel, the preamp tubes are fine.

Power the amp down and swap power tubes between channels. Power the amp up and play music. Again, by the same logic, if there is still no sound from the same channel, the power tubes are fine, and the problem is in the inside circuitry of the amp. A technician should be consulted.

If the problem (no sound) moves to the other channel, that power amp tube is faulty (assuming a single-ended amp with only one power tube). With push-pull amps with two-four-six power tubes per channel, it would seem that both (or all) tubes are faulty. The probability of both or all of the tubes (four, six, eight) failing simultaneously is extremely low. And here we get to the next possible problem ...

SYMPTOM: *Lower power and/or distortion from one channel*

Paradoxically, when an amplifier is dead, it usually isn't difficult to find the culprit and fix it. On the other hand, distortion is one of the more difficult faults to troubleshoot, mainly because it is the symptom of a problem, not the primary fault or problem itself. Many factors can cause it:

- Leaky coupling capacitors or gassy tubes (both conditions may cause a grid current to flow)
- Over- or under-biasing of the output stage
- Low anode or screen voltages
- DC or AC imbalance of push-pull driver and output stages
- Unmatched phase inverter or output tubes
- One faulty phase inverter or output tube, so a push-pull amp is working in a single-ended mode

When one of the two power tubes of a push-pull amplifier fails, that channel will continue working in a single-ended mode and will usually (but not always) start distorting. The extent of sonic degradation will depend on the quality of the amplifier. When it happened on one of our amps, since we listened to it at very low power levels at the time, we didn't notice anything for a while. The output transformers were large (oversized), so the unbalanced DC current (flowing through only one-half of the primary winding) did not saturate the one in the faulty channel

Higher power push-pull amps may have four, six, eight, and even more (always an even number) power tubes per channel. So, if one fails, the problem may go unnoticed for a while; There could be four tubes working fine in one channel and only three tubes working on the other, a relatively small imbalance.

SYMPTOM: *Tubes last a very short time, weeks and months instead of years)*

If an amp "eats" through tubes very quickly (meaning tubes' life is short), two possible causes come to mind. Either the heater voltage is way too high (or, less likely, too low), or tubes are working above their maximum voltage or power levels. The second case is more common with output or power tubes.

The first step is to check the heater voltage. Most commonly, it will be 6.3 V, usually AC, although some amps use DC heating. It should not be higher than 6.5 V or lower than 6.0 V. Some designers underheat preamp tubes (for instance, 5.9 - 6.1 V instead of 6.3 V) to reduce hum and noise. That is OK. However, if the voltage is too high, it must be reduced by modifying the heater supply circuit in the amp.

WARNING: Check the amplifier's nominal mains voltage

Many currently produced amplifiers sold in countries with 240 V nominal mains voltage (Australia, New Zealand, etc.) were designed for 220 Volts. For instance, Yaquin MS12B is a budget tube phono and line stage. It uses two 12AX7 duo-triodes in the phono stage, their heater voltages regulated at $12V_{DC}$. However, the heaters for the line stage's 12AU7 tubes are supplied from the unregulated supply whose output was 16.5 V_{DC}. Before we fixed this issue, the preamp's owner had to replace tubes every month or so.

The reason was evident from the 220 V stamping on the mains transformer. With our 248 V mains voltage, the heater voltage should be 248/220 = 1.127 or 12.7% higher - 14.2 V. This was still far away from the measured 16.5 V, meaning this design was faulty from the start, the heater voltage would be far too high even on 220 V mains.

Solving this preamp's design fault was easy. However, correcting such a problem in power amps isn't so simple. All voltages need to be reduced, not just the heater voltage but also the bias and anode voltages.

SYMPTOM: Tube's anode (plate) glowing red

Improper bias, say -5 V instead of -15 V, or a total loss of the negative bias voltage will cause power tubes to draw too much current and anodes (plates) to overheat and glow. This glow can be localized in one or two spots, or, in severe cases, the whole plate structure may glow bright red.

Fig. 15.19. The mains transformer cover removed from Yaquin MS12B. Made for 240V mains as claimed? Yeah, right!

When you power up a newly bought (after transportation), just constructed, or repaired amplifier for the first time, even if everything seems fine (no burning smell, no crackling sounds), switch the lights off or darken the room and observe the power tubes in the dark for a minute or two. If there is no red plate glow, proceed to play the music through it. If ordinarily black or gray anodes (plates) start glowing red, turn the amp off immediately and check the biasing circuit.

SYMPTOM: Milky white coating or blue glow inside a tube

A healthy tube has a shiny black coating, usually on top of the glass bulb (1), as in this case, but also on its sides (2), or even at the bottom (3). That is the getter, an active chemical substance whose job is to "get" or "bind" ions and impurities left over after the air was evacuated from the tube. The vacuum inside is never perfect; there are always some impurities.

The getter keeps working throughout a tube's life. However, if the vacuum is lost (due to porous glass but most often due to the broken or cracked glass bulb), the air gets inside the tube, and the getter turns milky white.

That is what happened to the tube in photo (1). Such a tube cannot possibly work (no vacuum inside) and cannot be repaired.

Fig. 15.20. (ABOVE) The ordinarily bright, shiny, and dark getter at the top of this power tube (1) turned white, indicating the loss of vacuum. The tube is beyond repair.

How tubes age

Just like old incandescent light bulbs (remember those?), tubes have filaments (heaters) which age with use, primarily due to the constant "cycling" between the cold (off) and hot (on) states. Burned-out heaters are the most common reason for tube failure.

As a tube clocks more operating hours in an amp, its getter will start shrinking, the edges will turn dull and gray, as illustrated (4).

These two vintage Philips 6550 power tubes are from the same amplifier. Although they worked at the same heater, bias, and anode voltages, the tube on the left (getter almost gone) worked harder and has less life left in it than the one on the right. A tube analyzer confirmed that its mutual conductance and cathode emission levels were lower.

Fig. 15.21. (RIGHT) Judging by the size and color of the getter, the tube on the left (5) worked harder and has less life left in it than its pair (4).

16 | THE END MATTER

If you've come here by quickly thumbing through this book's pages, shame on you - go back and read it properly and thoroughly! If, however, you've reached this point by doing just that, I congratulate you.

Thank you for persisting through my raving and ranting, my incessant "do this and don't do that" bad parenting attitude, and the occasional head-spinning, glassy-eyed mathematical exhibitionism.

No, seriously, reading a book of this type in its entirety is a serious and rare achievement! Apparently, 95% or so of people who buy or borrow a book do not finish reading it. Ever.

I hope you've had as much fun reading this book as I had writing it.

My warmest regards and best wishes!

- INDEX
- LITERATURE (FURTHER READING)
- ONLINE AUDIO MAGAZINES
- AUDIOPHILE TUBE AMPLIFIER BOOKS BY IGOR S. POPOVICH
- GUITAR AMPLIFIER BOOKS BY IGOR S. POPOVICH
- OTHER AUDIO-RELATED BOOKS BY IGOR S. POPOVICH

> "That is a good book which is opened with expectation and closed with profit."
> A. Bronson Alcott, writer and philosopher

INDEX

A

Absolute phase, 151
Absorption panels, 267
Acoustic insulation boards, 287
Acoustic suspension speakers, 209
Acoustic treatment rules, 278
AC polarity, 71
AC power supply, 70
Active bass traps, 285
Active crossovers, 239-240
Adaptation problem, 35
A/D converter, 44
Amplifier,
    bandwidth, 57-58, 163
    biasing, 162
    channel balance test, 315
    checks, 313
    damping factor, 233
    dual-mono, 162
    frequency characteristics, 57
    frequency range, 55
    headroom, 163
    input impedance, 164, 313
    input sensitivity, 164
    maximum output power, 163
    noise and hum, 313
    output impedance, 163, 233
    power supply, 165
    signal-to-noise ratio, 164
    square wave response, 60
    tone-burst tests, 57
AMT tweeters, 207, 227
Anti-skating, 132
Attenuators, 185
Audio bargains, 22
Audio consultants, 32
Audio dealers, 16
Audio magazine reviews, 20-21, 26
Audio marketing, 27-28
Audio synergy, 33
Audiophile, 9
Audio Research Corporation, 12
Audio system troubleshooting, 308-320
Autotransformers, 75-78
Azimuth alignment, 122, 133

B

Background noise levels, 274-275
Balanced outputs, 152
Balanced power supply, 75-78
Bamboo shelves, 301
Bass traps, 284-287
Bass-reflex speakers, 209-210
Battery power, 92-94
Beryllium-copper, 112
Bi-amping, 238-239
Bi-wiring, 241
Bookshelves as bass traps, 286
Break-in of audio equipment, 65, 108
Bridging, of amplifiers, 186-187
Broadband absorbers, 270

C

Cables
    checks, 314
    lifters, 305-306
Capacitor banks, 161
Capacitors, 64
    film & foil, 65
    PIO (Paper-In-Oil), 65
Carbon fibre shelves, 301
Carbon film resistors, 65
Carpets as sound absorbers, 278
Cary SLP-90 line preamplifier, 182
Cathode follower, 150-151
Cavity resonators, 283-284
Characteristic impedance of a cable, 98, 100, 107
"Cheater" plugs, 74, 89
Class A operation, 28, 159-160, 193-194
Class D amplifiers, 169-171
Clipping, 55
Closed - loop tape transport, 122
CMRR (Common Mode Rejection Ratio), 152
Complementary transistors, 27, 169
Contact
    cleaners, 114
    problems, 308-309
Convex & concave surfaces, 274
Copper foil loudspeaker cables, 105
Copper wire, 96
Critical (purposeful) listening, 34
Crossovers, 214-215, 222-226
Crossover distortion, 174-175
Crosstalk, 104, 133
Cryogenic treatment, 116

D

D'Appolito speaker design, 211
DAC, 149-154
DC-biased cables, 98
DC blockers, 79-81
DC coupling, 166
DC power supplies, 90-92
Delay distortion, 47
Dielectrics, 64, 98
Differential amplifier, 152
Diffusers, 288-291
Digital pressure gauge, 131
Digital room equalization, 291-292
Directly-heated tubes, 166
Disappearing Dealer Syndrome, 31
Dome tweeters, 204
Double-shielded cables, 87
Dual mono amplifiers, 162
Dummy loads, 313
Dynaco ST-70 amplifier, 12, 180-181
Dynamic compression, 61-63, 147
Dynamic loudspeakers, 202-205
Dynamic range, 55
Dynamic power test, 55

E

Early reflections, 261, 277-278
Earthing issues, 71
Electric shock, 74
Electrodynamic speakers, 203-204
Electromagnetic waves, 96
Electrostatic headphones, 191
Electrostatic speakers, 161, 206-207
EMI, 45, 70, 148
Equal loudness curves, 44
Equipment feet upgrade, 303
Equipment racks, 64, 266, 294, 299-303
Equipment reviews, 20-21

F

Fake cables, 108
Fast transients, 59
Faults and failures, 308-309
Fiberglass acoustic panels, 279, 287
Field coil speakers, 203-204
Field-Effect Transistors (FETs), 52
Fidelity magazine, 26
Film capacitors, 29, 165-166, 224
Fuses, 114-116, 319

G

Gain optimization, 230
Geopathic stress, 66-68
Gold-plated contacts, 111
Ground loops, 87-89

H

Hagerman Tuba headphone amplifier, 194-195
Half power (-3dB) frequency, 57, 64-65, 144, 161, 163
Harmonic distortion, 46, 49-50, 235
Harmonics, 56
Hartmann lines, 68
Headphones, 190-191
Headphone amplifiers, 29, 192-200
Headphone cables, 109-111
Headroom, 54-55
Helmholtz resonators, 283
Hi-fi racks, 301-303
High frequency oscillations, 317
Horn speakers, 211-213
Human ear & hearing, 44

I

Impulse buying, 22
Inductors, 224-225
Initial Time Delay Gap, 260-261
Input capacitance (phono stage), 141
Input impedance/resistance, 313
Interconnect cables, 99-101, 140
Intermittent faults, 308
Intermodulation distortion, 40, 46-47
Interstage transformers, 65
Isobarik speakers, 210
Isolation platforms, 301-302
Isolation transformers, 74-76

INDEX, continued

L

L-pad tweeter attenuator, 226-227
Lateral tracking angle, 130
LCR meter, 141, 311
LEDE approach, 261
Listening Evaluation & Auditioning Sheet, 37
Listening room,
    longitudinal arrangement, 266-267
    proportions, 261-262
    resonant modes, 263-270
    transversal arrangement, 267-268
Litz cables, 97
LP production, 119
Loudspeaker
    binding posts, 114
    cables, 105-109
    checks, 312
    dynamic compression, 62
    enclosures, 208-209
    floor interface, 297
    impedance curve, 205, 216
    modifications, 221-228
    placement methods, 256-258
    optimal spacing, 251
    radiation patterns, 251-252
    sensitivity, 220, 232
    shorting links, 114
    tests & measurements, 218-221
    time alignment, 252-253

M

Magnetic tape recording 121
Master tape production, 118
McIntosh tube amplifiers, 160
MC step-up transformer, 143-147
Measurement fallacy, 10, 24, 53
Mercury vapor tube rectifiers, 178
Microphony, 179, 250, 296
Modifying amps and preamps, 178-185
Moving coil cartridges, 134
Moving iron cartridges, 134
Moving magnet cartridges, 134
Multimeters, 310-311

N

Negative feedback, 51-52, 63, 138, 167
Nonlinearity (tubes and transistors), 61

O

Objectivists, 10, 204
Optical cartridges, 134-135
OTL tube amplifiers, 171-172
Output transformer for DAC, 154

P

Panel absorbers, 280-281
Parasitic resonators, 295
Passive preamplifiers, 168-169
Passive radiators, 210
Pentode, 48

P, cont.

Phase inverter, 158
Phono stage, 136-139
Piano test, 37
Planar magnetic drivers, 191
Planar magnetic speakers, 207
Point-to-point wiring, 184
Potentiometers, 185
Power boards, 78-79
Power cables, 101-104
Power factor, 85
Power regenerators, 70
Power supply,
    dynamic compression, 62
    of an amplifier, 156, 165
Printed circuit boards, 28, 179
Proximity effect in cables, 97
Push-pull amplifiers, 52, 158-159, 319

Q

QRD diffusers, 288-289
Quartz power filters, 83

R

Reactive loads, 160
Rectifier tubes, 165
Reel-to-reel tape decks, 120-125
Resistor color chart, 145
Radio Frequency Interference (RFI), 45
Retail pricing, 41
RFI filtering, 81, 85-86
Rhodium plating, 112
RIAA equalization, 136-137
Ribbon drivers, 191, 208
RMS value of a signal, 52

S

S-curve, 25
Safety precautions, 73
Schumann generators, 67
Sealed lead-acid batteries, 94
Shielded cables, 99, 105
Signal-to-noise ratio, 164
Signal tracing, 316-317
Single-ended amplification, 24, 26, 52, 158
Skin effect, 97
Solid-core cables, 97
Solid-state amplifiers, 49, 51
Sonic signature of materials, 64
Sound absorption coefficient, 272
Sound diffraction, reflection and absorption, 272-273
Sound envelope, 38
Soundproofing doors, 275
Square wave reproduction, 59-60
SPL meters, 40
Standing waves, 262-270
Stereophile magazine, 27, 171, 265
Stranded cables, 97
Subjectivists, 10

S, cont.

Subwoofer,
    integration, 242-244
    orientation, 244
    placement, 246
Switch-mode power supplies, 89-90
Synergy in audio, 245-246

T

Technophiles and technophobes, 12
Teflon®, 29
Test recordings, 38
THD, see harmonic distortion
Tone-burst test, 55
Transformer vibrations, 295
Transistor sound, 51
Transmission line speakers, 210
Triode, 48
Tube adapters, 175-176
Tube buffers, 50
Tube dampers, 296-297
Tube rolling, 173-178
Tube sound, 50
Tube test "certificates", 174
Turntables, 126-135
    anti-skating, 132
    azimuth, 133
    drive types, 128-129
    lateral tracking angle, 130
    styli, 135
    tonearm geometry, 127
    vertical tracking angle, 132
    vertical tracking force, 131

U

Ultralinear amplifiers, 159

V

Vacuum tube,
    aging, 320
    amplifier, 49, 156-167
    bias adjustment, 162
    mutual conductance, 166
    rectifiers, 9, 177-178
    red platting, 309, 320
Vertical tracking angle, 132
Vertical tracking force, 131
Vibrapods, 302
Vibration damping, 296-304
Voltage regulators, 81-82
Voltage regenerators, 82-84

W

Wall-mounted shelves, 302
Warped records, 317
Warranty, 22, 24

Z

Zenith alignment, 122
Zobel networks, 205

LITERATURE (FURTHER READING)

Acoustic Absorbers and Diffusers: Theory, Design and Application, 3rd ed., Cox and D'Antonio, 2020, 576 pages
Acoustic Design for the Home Studio, Gallagher, 2006, 272 pages
Acoustic Techniques for Home and Studio, Everest, 1974, 224 pages
Acoustics and Psychoacoustics, 5th Edition, Howard & Angus, 2017, 518 pages
Acoustics of Small Rooms, Kleiner & Tichy, 2014, 492 pages
Architectural Acoustics: Principles and Practice 2nd Edition, Cavanaugh, Tocci & Wilkes, 2009, 352 pages
Audio Engineering Handbook, Benson, 1988, 1040 pages
Get Better Sound, Smith, 2008, 292 pages
Good Sound, Laura Dearborn, 1987, 418 pages
Handbook for Sound Engineers, 5th Edition, G. Ballou (editor), 2015, 1784 pages
High Performance Loudspeakers (5th edition), Colloms, 1997, 494 pages
Home Recording Studio: Build It Like the Pros, Gervais, 2010, 368 pages
How to Build a Small Budget Recording Studio from Scratch, 4th edition, Shea and Everest, 2012, 464 pages
Introductory Guide to High-Performance Audio Systems, Harley, 2007, 240 pages
Loudspeaker and Headphone Handbook, Borwick, 3rd edition, 2001, 718 pages
Master Handbook of Acoustics, 6th Edition, Everest and Pohlmann, 2014, 640 pages
Principles and Applications of Room Acoustics, Vol. 1, Cremer & Muller, 2016, 670 pages
Principles and Applications of Room Acoustics, Vol. 2, Cremer & Muller, 2016, 458 pages
Room Acoustics, 4th edition, Kuttruff, 2016, 322 pages
Sound Reproduction: The Acoustics and Psychoacoustics of Loudspeakers and Rooms, Toole, 2017, 514 pages
Sound System Engineering, 4th ed., Davis at al., 2013, 644 pages
The Audio Expert: Everything You Need to Know about Audio, Winer, 2017, 808 pages
The Better Sound of the Phonograph: How come? How-to!, Miller, 2017, 142 pages
The Complete Guide to High-End Audio, 5th Edition, Harley, 2015, 578 pages
The New Stereo Soundbook, Streicher R. and Everest F., 1998, 272 pages
The Science and Applications of Acoustics, Raichel, 2000, 620 pages
The Sound Reinforcement Handbook, 2nd edition, Davis and Jones, 1988, 432 pages

ONLINE AUDIO MAGAZINES

Fidelity International: https://www.fidelity-magazine.com/
Fidelity Online: https://www.fidelity-online.de/
Haute Fidélité: https://www.hautefidelite-hifi.com/
Hi-Fi+: https://www.hifiplus.com/
Hi-Fi Choice: https://www.hifichoice.com/
Hi-Fi Critic: https://www.hificritic.com/
Hi-Fi News: https://www.hifinews.com/
HiFi Pig: https://hifipig.com/category/news/
Hifi-Stars:https://hifi-stars.de/
Hi-Fi World: https://www.hi-fiworld.co.uk/
Part-Time Audiophile: https://parttimeaudiophile.com/
Positive Feedback: https://positive-feedback.com/
Stereo: https://www.stereo.de/
StereoLife Magazine: https://www.stereolifemagazine.com/
Stereophile: https://www.stereophile.com
The Absolute Sound: http://www.theabsolutesound.com/
Tone Audio Magazine: https://www.tonepublications.com/

AUDIOPHILE TUBE AMPLIFIER BOOKS BY IGOR S. POPOVICH

Available from Amazon, Barnes & Noble, Book Depository and all other major online bookstores

Audiophile Vacuum Tube Amplifiers, Vol 1
ISBN: 978-0-9806223-2-4

- BASIC ELECTRONIC CIRCUIT THEORY
- ELECTRONIC COMPONENTS
- AUDIO FREQUENCY AMPLIFIERS
- PHYSICAL FUNDAMENTALS OF VACUUM TUBE OPERATION
- VOLTAGE AMPLIFICATION WITH TRIODES - THE COMMON CATHODE STAGE
- OTHER VOLTAGE AMPLIFICATION STAGES WITH TRIODES
- TETRODES AND PENTODES AS VOLTAGE AMPLIFIERS
- FREQUENCY RESPONSE OF VACUUM TUBE AMPLIFIERS
- IMPEDANCE-COUPLED STAGES AND INTERSTAGE TRANSFORMERS
- NEGATIVE FEEDBACK
- TONE CONTROLS, ACTIVE CROSSOVERS AND OTHER CIRCUITS
- PRACTICAL LINE-LEVEL PREAMPLIFIER DESIGNS
- PHONO PREAMPLIFIERS
- SINGLE-ENDED TRIODE OUTPUT STAGE
- PRACTICAL SINGLE-ENDED TRIODE AMPLIFIER DESIGNS
- SINGLE-ENDED PENTODE AND ULTRALINEAR OUTPUT STAGES

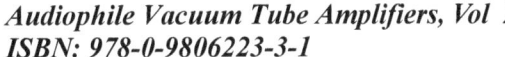

Audiophile Vacuum Tube Amplifiers, Vol 2
ISBN: 978-0-9806223-3-1

- PRACTICAL SINGLE-ENDED PENTODE AND ULTRALINEAR DESIGNS
- PUSH-PULL OUTPUT STAGES
- PRACTICAL PUSH-PULL AMPLIFIER DESIGNS
- BALANCED, BRIDGE AND OTL (OUTPUT TRANSFORMERLESS) AMPLIFIERS
- THE DESIGN PROCESS
- FUNDAMENTALS OF MAGNETIC CIRCUITS AND TRANSFORMERS
- MAINS TRANSFORMERS AND FILTERING CHOKES
- POWER SUPPLIES FOR TUBE AMPLIFIERS
- AUDIO TRANSFORMERS
- TROUBLESHOOTING AND REPAIRING TUBE AMPLIFIERS
- UPGRADING & IMPROVING TUBE AMPLIFIERS
- SOUND CONSTRUCTION PRACTICES
- AUDIO TESTS & MEASUREMENTS
- TESTING & MATCHING VACUUM TUBES

Audiophile Vacuum Tube Amplifiers, Vol 3
ISBN: 978-0-9806223-4-8

- THE FRONT-END: SUPERIOR INPUT & DRIVER STAGES
- FROM SHOCKING TO SUBLIME: LESSONS FROM COMMERCIAL LINE STAGES
- DIY LINE-LEVEL PREAMPLIFIERS: $10,000 SOUND ON $500-$1,000 BUDGET
- THE STARS OF THE AUDION ERA: ANCIENT TUBES IN MODERN AMPS
- CHEAP & CHEERFUL: PREAMP & DRIVER TUBES FOR AUDIO EXPLORERS
- SLEEPING GIANTS: OUTPUT TUBES FOR THOSE WHO WANT TO BE DIFFERENT
- THE QUEEN OF HEARTS: SINGLE-ENDED AMPLIFIERS WITH 300B TRIODES
- TRIODES, PENTODES AND BEAM TUBES: MORE SINGLE-ENDED DESIGNS
- BIG BOTTLES: SET AMPLIFIERS WITH HIGH VOLTAGE TRANSMITTING TUBES
- THE WAY IT USED TO BE: VINTAGE PUSH-PULL AMPLIFIERS
- NEW? IMPROVED? MODERN PUSH-PULL AMPLIFIER DESIGNS
- CUTE, CLEVER OR CONTROVERSIAL? INTERESTING IDEAS FROM TUBE AUDIO'S PAST AND PRESENT
- THRIFTY TIPS & TRICKS: TIME & MONEY SAVING IDEAS
- OUTPUT AND INTERSTAGE TRANSFORMERS: FROM COMMERCIAL BENCHMARKS TO YOUR OWN DESIGNS
- MEASUREMENTS VERSUS LISTENING AND OTHER AUDIO DESIGN DILEMMAS

GUITAR AMPLIFIER BOOKS BY IGOR S. POPOVICH

Available from Amazon, Barnes & Noble, Book Depository and all other major online bookstores

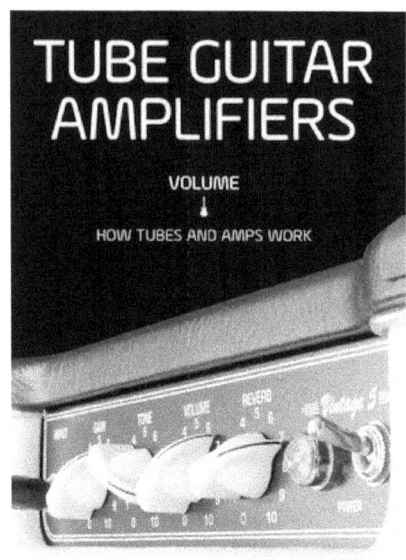

Tube Guitar Amplifiers Volume 1: How Tubes and Amps Work
ISBN: 978-0-9806223-5-5

- BASIC ELECTRONIC CIRCUIT THEORY
- AUDIO AMPLIFIERS
- ELECTRONIC COMPONENTS
- PHYSICAL FUNDAMENTALS OF VACUUM TUBE OPERATION
- TRIODES AS VOLTAGE AMPLIFIERS
- TETRODES, PENTODES AND BEAM-POWER TUBES
- INPUT CIRCUITS AND STAGES
- TONE CONTROLS
- ANALOG EFFECTS (TREMOLO, VIBRATO, REVERB) AND EFFECTS LOOPS
- POWER SUPPLIES FOR TUBE AMPLIFIERS
- SINGLE-ENDED TRIODE, PENTODE AND ULTRALINEAR OUTPUT STAGES
- PHASE SPLITTERS OR INVERTERS
- PUSH-PULL OUTPUT STAGES
- NEGATIVE FEEDBACK
- TRANSISTOR AND HYBRID GUITAR AMPLIFIERS

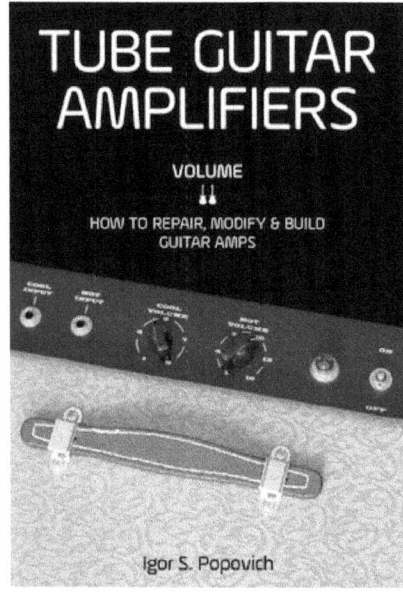

Tube Guitar Amplifiers Volume 2: How to Repair, Modify & Build Guitar Amps
ISBN: 978-0-9806223-6-2

- OUTPUT AND INTERSTAGE TRANSFORMERS FOR TUBE GUITAR AMPS
- LOUDSPEAKERS, OUTPUT ATTENUATORS & HEADPHONE CIRCUITS
- TROUBLESHOOTING AND REPAIRING TUBE GUITAR AMPLIFIERS
- WIRING, SOLDERING & MODIFICATION PRACTICES
- POWER SUPPLY MODIFICATIONS AND IMPROVEMENTS
- TONE TWEAKS
- MODERN PUSH-PULL AMPS
- DIY PROJECTS: CONVERTING SOLID STATE GUITAR AMPS TO TUBES
- DIY PROJECTS: ULTRA-SMALL AMPS
- REBUILDING COMMERCIAL AMPS IN A HANDWIRED (POINT-TO-POINT) FASHION
- DIY PROJECTS: QUIRKY & UNUSUAL DESIGNS
- CONVERTING VINTAGE TUBE GEAR INTO GUITAR AMPS

OTHER AUDIO-RELATED BOOKS BY IGOR S. POPOVICH

Available from Amazon, Barnes & Noble, Book Depository and all other major online bookstores

Audio Tests & Measurements: How to Test Electronic Components, Audiophile & Guitar Amplifiers and Loudspeakers Using Modern and Vintage Test Instruments
ISBN: 978-0-9806223-9-3

- TEST INSTRUMENTS, ERRORS, LIMITATIONS & SAFETY ISSUES
- SIGNAL SOURCES, TRACERS, POWER SUPPLIES AND FILTERS
- MULTIMETERS - TYPES, OPERATING PRINCIPLES AND FUNCTIONS
- OSCILLOSCOPES - HOW THEY WORK & HOW TO USE THEM
- TESTING PASSIVE ELECTRONIC COMPONENTS (RESISTORS, CAPACITORS & INDUCTORS)
- TESTING AUDIO AMPLIFIERS AND PREAMPLIFIERS
- DISTORTION MEASUREMENTS
- TRANSFORMER TESTS & MEASUREMENTS
- LOUDSPEAKER TESTS & MEASUREMENTS
- TRANSISTOR TESTERS AND CURVE TRACERS
- TESTING VACUUM TUBES (VALVES)

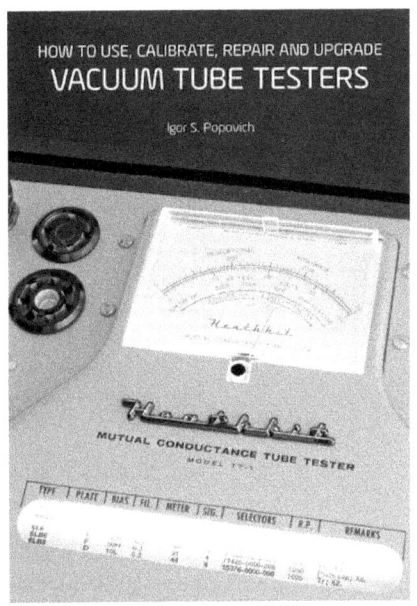

How to Use, Calibrate, Repair and Upgrade Vacuum Tube Testers
ISBN: 978-0-9806223-7-9

- HOW VACUUM TUBES WORK
- TESTING & MATCHING VACUUM TUBES
- EMISSION TESTERS
- GRID CIRCUIT TESTERS
- DYNAMIC CONDUCTANCE TESTERS
- PROPORTIONAL MUTUAL CONDUCTANCE TESTERS
- HICKOK-TYPE TESTERS
- TRUE MUTUAL CONDUCTANCE TESTERS
- REPAIRING & UPGRADING VINTAGE TUBE TESTERS
- TESTING & MATCHING TUBES WITHOUT A TUBE TESTER

Transformers For Tube Amplifiers: How to Design, Construct & Use Power, Output & Interstage Transformers and Chokes in Audiophile and Guitar Tube Amplifiers
ISBN: 978-0-9806223-8-6

- PHYSICAL FUNDAMENTALS OF MAGNETIC CIRCUITS AND TRANSFORMERS
- FILTERING CHOKES (INDUCTORS WITH DC CURRENT)
- TRANSFORMER MATERIALS, CONSTRUCTION METHODS AND ISSUES
- MAINS (POWER) TRANSFORMERS
- PHYSICAL FUNDAMENTALS OF AUDIO TRANSFORMERS
- SINGLE-ENDED OUTPUT TRANSFORMERS
- PUSH-PULL OUTPUT TRANSFORMERS
- SPECIAL MAGNETIC COMPONENTS: LOW POWER INPUT, PREAMP OUTPUT & DAC OUTPUT TRANSFORMERS, TRANSFORMER VOLUME CONTROL
- INTERSTAGE TRANSFORMERS, GRID & ANODE CHOKES
- OUTPUT AND INTERSTAGE TRANSFORMERS FOR TUBE GUITAR AMPS
- TRANSFORMER TESTS & MEASUREMENTS

www.ingramcontent.com/pod-product-compliance
Ingram Content Group UK Ltd.
Pitfield, Milton Keynes, MK11 3LW, UK
UKHW051525180426
11947UKWH00018B/1580